INCOMMENSURATE PHASES IN DIELECTRICS

2. Materials

MODERN PROBLEMS IN CONDENSED MATTER SCIENCES

Volume 14.2

NORTH-HOLLAND
AMSTERDAM·OXFORD·NEW YORK·TOKYO

INCOMMENSURATE PHASES IN DIELECTRICS
2. Materials

Volume editors

R. BLINC

Ljubljana, Yugoslavia

A.P. LEVANYUK

Moscow, USSR

1986

NORTH-HOLLAND
AMSTERDAM·OXFORD·NEW YORK·TOKYO

ISBN: 0 444 86970 0

Published by:

North-Holland Physics Publishing
a division of
Elsevier Science Publishers B.V.
P.O. Box 103
1000 AC Amsterdam
The Netherlands

Sole distributors for the U.S.A. and Canada:
Elsevier Science Publishing Company, Inc.
52 Vanderbilt Avenue
New York, N.Y. 10017
U.S.A.

Library of Congress Cataloging in Publication Data

Main entry under title:

Incommensurate phases in dielectrics.

(Modern problems in condensed matter sciences ; v. 14)
Includes bibliographies and indexes.
1. Dielectrics. 2. Lattice dynamics. 3. Crystal lattices. 4. Phase transformations (Statistical physics) I. Blinc, R. (Robert) II. Levanyuk, A. P., 1933- . III. Series.
QC585.I53 1985 537'.24 85-29693
ISBN 0-444-86970-0 (U.S. : v. 2)

Printed in The Netherlands

MODERN PROBLEMS IN CONDENSED MATTER SCIENCES

Vol. 1. SURFACE POLARITONS
V.M. Agranovich and D.L. Mills, *editors*

Vol. 2. EXCITONS
E.I. Rashba and M.D. Sturge, *editors*

Vol. 3. ELECTRONIC EXCITATION ENERGY TRANSFER IN CONDENSED MATTER
V.M. Agranovich and M.D. Galanin

Vol. 4. SPECTROSCOPY AND EXCITATION DYNAMICS OF CONDENSED MOLECULAR SYSTEMS
V.M. Agranovich and R.M. Hochstrasser, *editors*

Vol. 5. LIGHT SCATTERING NEAR PHASE TRANSITIONS
H.Z. Cummins and A.P. Levanyuk, *editors*

Vol. 6. ELECTRON–HOLE DROPLETS IN SEMICONDUCTORS
C.D. Jeffries and L.V. Keldysh, *editors*

Vol. 7. THE DYNAMICAL JAHN–TELLER EFFECT IN LOCALIZED SYSTEMS
Yu.E. Perlin and M. Wagner, *editors*

Vol. 8. OPTICAL ORIENTATION
F. Meier and B.P. Zakharchenya, *editors*

Vol. 9. SURFACE EXCITATIONS
V.M. Agranovich and R. Loudon, *editors*

Vol. 10. ELECTRON–ELECTRON INTERACTIONS IN DISORDERED SYSTEMS
A.L. Efros and M. Pollak, *editors*

Vol. 11. MEDIUM-ENERGY ION REFLECTION FROM SOLIDS
E.S. Mashkova and V.A. Molchanov

Vol. 12. NONEQUILIBRIUM SUPERCONDUCTIVITY
D.N. Langenberg and A.I. Larkin, *editors*

MODERN PROBLEMS IN CONDENSED MATTER SCIENCES

Vol. 13. PHYSICS OF RADIATION EFFECTS IN CRYSTALS
R.A. Johnson and A.N. Orlov, *editors*

Vol. 14. INCOMMENSURATE PHASES IN DIELECTRICS
(Two volumes)
R. Blinc and A.P. Levanyuk, *editors*

In preparation

Vol. 15. UNITARY TRANSFORMATIONS IN SOLID STATE PHYSICS
M. Wagner

TRANSPORT IN PHONON SYSTEMS
V.L. Gurevich

Oh, how many of them there
are in the fields!
But each flowers in its
own way —
In this is the highest achievement
of a flower!

Matsuo Bashó
1644–1694

PREFACE TO THE SERIES

Our understanding of condensed matter is developing rapidly at the present time, and the numerous new insights gained in this field define to a significant degree the face of contemporary science. Furthermore, discoveries made in this area are shaping present and future technology. This being so, it is clear that the most important results and directions for future developments can only be covered by an international group of authors working in cooperation.

"Modern Problems in Condensed Matter Sciences" is a series of contributed volumes and monographs on condensed matter science that is published by North-Holland Physics Publishing, a division of Elsevier Science Publishers. With the support of a distinguished Advisory Editorial Board, areas of current interest that have reached a maturity to be reviewed, are selected for the series. Both Soviet and Western scholars are contributing to the series, and each contributed volume has, accordingly, two editors. Monographs, written by either Western or Soviet authors, are also included. The complete series will provide the most comprehensive coverage available of condensed matter science.

Another important outcome of the foundation of this series is the emergence of a rather interesting and fruitful form of collaboration among scholars from different countries. We are deeply convinced that such international collaboration in the spheres of science and art, as well as other socially useful spheres of human activity, will assist in the establishment of a climate of confidence and peace.

The publishing house "Nauka" publishes the volumes in the Russian language. This way the broadest possible readership is ensured.

The General Editors of the Series,

V.M. Agranovich A.A. Maradudin

PREFACE

Translational lattice periodicity is usually considered one of the most basic characterizing features of a crystal. Depending on the arrangement of atoms inside the unit cell and the symmetry of the crystal lattice, each crystal has been traditionally assigned to one of the three-dimensional crystallographic space groups. The concepts of symmetry and translational periodicity of crystals have played a central role in the development of solid state physics. By and large these concepts have come to be accepted as universal properties of the crystalline state. This is in sharp contrast to the disordered and amorphous a-periodic states where there is no long-range order in the positions of the atoms.

In the past decade however, attention has focused on incommensurate materials which exhibit perfect three-dimensional long-range order but no translational periodicity in at least one direction, and which are thus intermediate between classical ideal crystals and a-periodic disordered or amorphous systems. In these materials a local property such as electric polarization, magnetization, charge density, mass density, atomic position or chemical composition is modulated with a periodicity q_{I} which is incommensurate to the periodicity q_{p} of the underlying lattice, i.e. the periodicity q_{I} of the order parameter $Q(x)$ and the periodicity of the basic lattice q_{p} cannot be expressed as a ratio of two integers:

$$\frac{q_{\mathrm{I}}}{q_{\mathrm{p}}} \neq \frac{M}{N}, \quad M, N = 1, 2, 3 \ldots$$

Here we are concerned with structurally incommensurate systems, where at least one atomic position is not exactly repeated from cell to cell and the translational symmetry of the crystal is lost in at least one direction, in spite of the persistence of long-range order. Translational lattice periodicity is generally restored at lower temperatures at a "lock-in" phase transition, where the lattice modulation changes from incommensurate to commensurate.

Incommensurate systems are interesting from the point of view of basic physics since they show a number of new phenomena—nonlinear multi-soliton lattice-type ground states, phason and amplitudon excitations, devil's staircase, solid-state chaos, etc.—which are not found in translationally periodic crystals. The study of incommensurate systems may thus lead to an improved understanding of the a-periodic state of matter.

Whereas the first volume of this monograph treats the macroscopic and microscopic theory of incommensurate systems and gives a review of the experimental techniques used to study these materials, this second volume presents the specific results obtained for some of the most representative examples. The various chapters on specific substances are written by the people who actually carried out the experiments and contributed most to the understanding of these systems.

The purpose of this book is thus to give a physical picture of incommensurate insulators, as well as to give the reader an idea of how the various experimental techniques can be used for a study of the dynamic and static properties of incommensurate systems on a microscopic level.

R. Blinc
Ljubljana, Yugoslavia

A.P. Levanyuk
Moscow, USSR

CONTENTS

Preface to the series vii
Preface ix
Contents xi

Part IV. Results for specific systems

10. Phase transformations in K_2SeO_4 and structurally related insulators 1
J.D. Axe, M. Iizumi and G. Shirane
11. Phenomenology of incommensurate phases in the A_2BX_4 family . 49
Y. Ishibashi
12. Incommensurate phases in an aromatic molecular crystal: biphenyl 71
H. Cailleau
13. Incommensurate phase in $NaNO_2$ 101
D. Durand, F. Dénoyer, R. Currat and M. Lambert
14. Modulated phases in thiourea 129
F. Dénoyer and R. Currat
15. Incommensurate phase in β-$ThBr_4$ 161
R. Currat, L. Bernard and P. Delamoye
16. The incommensurate phase of quartz 205
G. Dolino
17. Incommensurate phase of barium sodium niobate 233
J.C. Tolédano, J. Schneck and G. Errandonéa
18. The incommensurate phase of Na_2CO_3 253
P.M. de Wolff and F. Tuinstra
19. Statics and dynamics of incommensurate $BaMnF_4$ 283
J.F. Scott
20. Unique incommensurate–commensurate phase transitions in a layer-structure perovskite 301
R. Kind and P. Muralt

21. Incommensurate structures in liquid crystals 319
S.A. Pikin
Appendix. List of incommensurate crystals 367
A.U. Sheleg and V.V. Zaretskii
Author index 373
Subject index 391
Cumulative index 395

PART IV

Results for Specific Systems

CHAPTER 10

Phase Transformations in K_2SeO_4 and Structurally Related Insulators

J. D. AXE

Brookhaven National Laboratory

M. IIZUMI

Japan Atomic Energy Research Institute

and

G. SHIRANE

Brookhaven National Laboratory

Incommensurate Phases in Dielectrics 2
Edited by
R. Blinc and A.P. Levanyuk

Contents

1. Introduction . . . 3
2. Crystal structure of the normal (high-temperature) phase and symmetry analysis of the atomic displacements . . . 6
   2.1. Crystal structure of the normal phase of K_2SeO_4 . . . 6
   2.2. Group theoretical analysis of the symmetry of the collective atomic displacements in K_2SeO_4 . . . 7
   2.3. Use of the extended zone scheme . . . 10
3. Phonons and fluctuations in the normal (high temperature) phase . . . 12
   3.1. K_2SeO_4 . . . 12
   3.2. Other compounds . . . 13
   3.3. Origins of the incommensurate instability . . . 15
      3.3.1. Analysis of phonon dispersion . . . 16
      3.3.2. Analysis of atomic displacements . . . 18
4. Modulation in the incommensurate phases . . . 20
   4.1. Modulation in K_2SeO_4 and similar compounds . . . 21
   4.2. Modulation in $(NH_4)_2BeF_4$ . . . 23
   4.3. Modulation in tetramethyl ammonium compounds . . . 23
5. Fluctuations in the incommensurate phase . . . 33
   5.1. Static critical behavior . . . 33
   5.2. Dynamical fluctuations in the incommensurate phase in K_2SeO_4 . . . 37
6. Lock-in phase transformations . . . 42
   6.1. The soliton limit . . . 42
   6.2. Lock-in and macroscopic properties . . . 45
   6.3. Dynamics near the soliton limit . . . 45
References . . . 47

1. Introduction

In condensed matter, as a rule, structure dictates function, and we have learned no longer to be surprised when unusual, interesting (and possibly even useful) behavior is found in a particular compound and then, subsequently, in a broad class of structurally similar materials. Examples that come readily to mind are perovskite ferroelectrics or A-15 superconductors. The β-K_2SO_4 structure appears to hold such a position in the study of modulated incommensurate structures.

Within this group is found K_2SeO_4, which exhibits perhaps the best studied and prototypical example of an incommensurate soft mode instability in an insulator. Phonon instabilities have had a profound influence on our understanding of ferroelectric and antiferroelectric phase transformations in recent years. In these cases the wave vector of the unstable phonon is at a high symmetry point in the Brillouin zone. A necessary condition for a phonon instability is a local extremum in the phonon dispersion, $\nabla q(\omega(\boldsymbol{q}))=0$, which is automatically satisfied at such points. But there are no restrictions in principle on the number or position of additional minima which are possible in the phonon dispersion surface. They are determined entirely by the strength and range of the interatomic potentials.

Figure 1 shows the temperature dependent softening of the unstable phonon branch in K_2SeO_4 as the incommensurate transformation is approached from above. (Iizumi et al. 1977). The minimum in the phonon dispersion near the transformation temperature is $\boldsymbol{q}_s=(00\zeta)$, with $\zeta\approx 0.69$. It is a puzzling fact that although many other examples of similar phase transformations have been found in isostructural compounds, none of these other materials have shown clear evidence of well defined soft mode behavior.

Below the incommensurate transformation temperature T_i, static displacements associated with the condensed phonon mode give rise to satellite elastic Bragg reflections, which provide the most direct way to monitor subsequent development of the incommensurate state. As first pointed out by Moncton et al. (1975), in the incommensurate state there are still interactions at work which tend to restore periodicity. One of the ways these forces manifest themselves is well illustrated in fig. 2, which shows the variation of the satellite wavevector, $\boldsymbol{q}_1=(0,0,1-\frac{1}{3}(1-\delta))$, with temperature. Below $T_c=93\,\mathrm{K}$, $\delta=0$

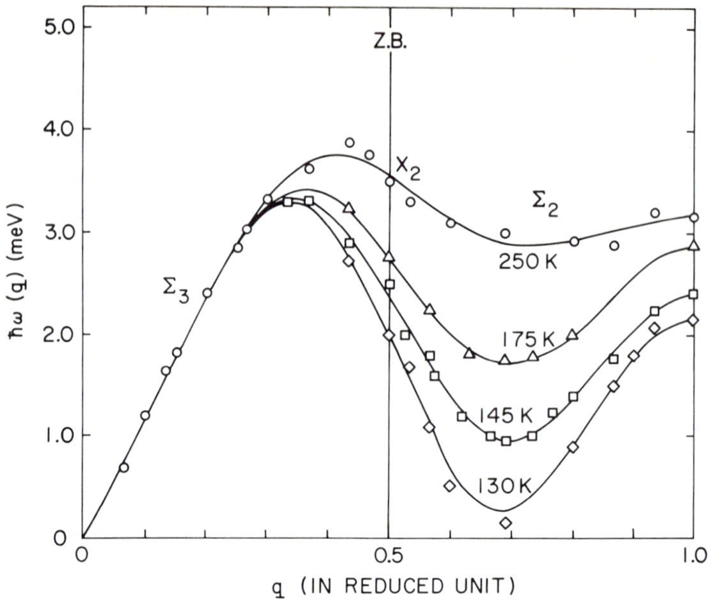

Fig. 1. Phonon dispersion in the soft branch of K_2SeO_4 versus temperature in the extended zone scheme.

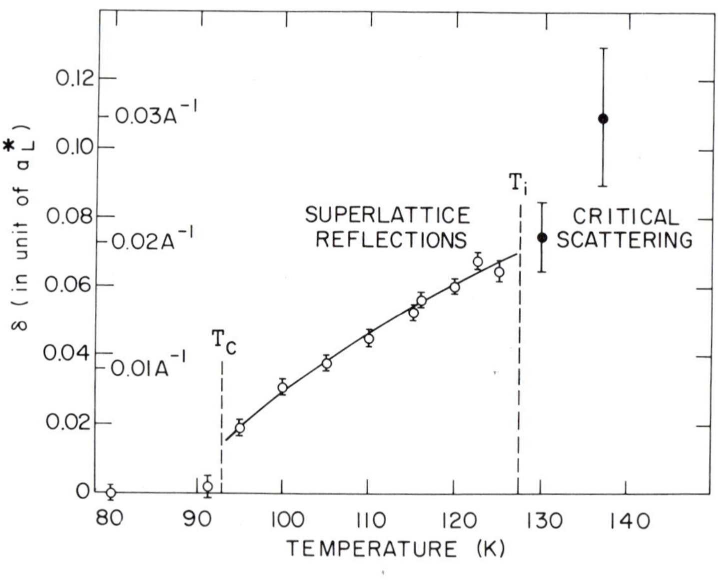

Fig. 2. Temperature dependence of the soft phonon (q_s) and incommensurate satellites (q_1) in K_2SeO_4.

Table 1
Typical crystals and their main characteristics.[a]

Crystal	T_i	T_c	ζ at T_i (δ)	ζ_c	e, a or anti-e
K_2SeO_4	93	130	0.310 (−0.023)	$\frac{1}{3}$	e
Rb_2ZnCl_4	303	192	0.305 (−0.028)	$\frac{1}{3}$	e
$(NH_4)_2BeF_4$	179.5	174	0.525 (0.025)	$\frac{1}{2}$	e
$[N(CH_3)_4]_2MnCl_4$	292.3	291.7	0.483 (−0.017)	$\frac{1}{2}$	a
$[N(CH_3)_4]_2ZnCl_4$	296.8	*280.2	0.420	*$\frac{2}{5}$	*e
		†275.5	(*0.020)	†$\frac{1}{3}$	†a
$[N(CD_3)_4]_2ZnCl_4$	298.7	275.0	0.424	$\frac{1}{3}$	a
$RbLiSO_4$	477	*475	0.403	*$\frac{1}{2}$	*anti-e
		†461	(+0.003)	†$\frac{2}{5}$	†e

[a] In crystals with more than one commensurate phase the different phases are distinguished by the symbols * and †. The abbreviations e, a and anti-e in the last column indicate that the crystal becomes ferroelectric, ferroelastic, or anti-ferroelectric in the commensurate phase.

and the structure is no longer incommensurate, the new unit cell being tripled along the c axis. This is the phenomenon of lock-in. Associated with this lock-in, and in fact intimately involved in the energetics of the process, is the appearance of a macroscopic dielectric polarization. This polarization is generated by secondary atomic displacements induced by anharmonic couplings to the primary lattice modulation. K_2SeO_4 is thus an improper ferroelectric. A peculiar aspect is that this polarization exists as well in the incommensurate state but with a small but non-zero wavevector, $q_p = 3\delta$.

As mentioned earlier K_2SeO_4 is but the best studied example of a large group of isostructural materials, growing in number and variety, that exhibit related behavior. They have a common chemical formula, A_2BX_n ($n = 3$ or 4), where A stands for an alkali metal ion or equivalent monovalent cation such as NH_4^+ or $N(CH_3)_4^+$ (tetramethyl ammonium or TMA). BX_n is a divalent tetrahedral ion such as SeO_4^{-2}, BeF_4^{-2}, $ZnCl_4^{-2}$ or $CuCl_4^{-2}$. Table 1 lists some typical examples, together with an indication of the type of modulation involved. They represent variations on a basic theme, in which the incommensurate wavevector lies along the pseudo hexagonal axis.

At sufficiently high temperatures, the crystals belonging to this group take a common high temperature crystal structure of the so-called beta-K_2SO_4 type. With decreasing temperature, at some temperature designated T_1 a phase transition (Aiki et al. 1969, 1970, Aiki and Hukuda 1969) to an incommensurately modulated structure (Iizumi et al. 1977) takes place.

In the following sections we will discuss these materials in greater detail. This is not intended as a comprehensive review of all areas of investigation of these materials. It is strongly biased toward scattering techniques (light, X-ray and neutron, both elastic and inelastic) since that is where our expertise lies. Many other techniques (heat capacity, dielectric response, quadrupole resonance, etc.) have contributed important insights but will be referred to only briefly. In some cases, these results are more fully discussed elsewhere in this volume.

2. *Crystal structure of the normal (high temperature) phase and symmetry analysis of the atomic displacements*

Knowledge of the crystal structure of the prototype phase is indispensable in investigating the microscopic basis of a displacive phase transition. The group-theoretical analysis of the atomic displacements affords a systematic way of characterizing and classifying the type of fluctuation involved in the phase transition. In this section a brief description of the orthorhombic crystal structure of K_2SeO_4 will be given together with the group-theoretical analysis of the normal modes of the crystal. The use of the extended zone scheme in describing the mode is also mentioned.

2.1. *Crystal structure of normal phase of K_2SeO_4*

As was mentioned in the Introduction, all the crystals under investigation in this chapter have the same crystal structure in their high temperature phase. The prototype of the crystal structures is β-K_2SO_4. Here the crystal structure of K_2SeO_4 in the normal phase is described as a representative instead of the β-K_2SO_4 structure.

The crystal structure of K_2SeO_4 at room temperature was first determined by Kalman et al. (1970), though von Gattow (1962) had pointed out that the structure is isomorphous with β-K_2SO_4.

The cell dimensions are as follows:

$$a = 6.003 \pm 0.003\,\text{Å}, \qquad b = 10.466 \pm 0.008\,\text{Å}, \qquad c = 7.661 \pm 0.004\,\text{Å}.$$

The space group is No. 62, D_{2h}^{16}-Pmcn, and there are 4 formula-units in the unit cell. There are two non-equivalent sites for the potassium ions which lie in special positions on the mirror planes at $x = \frac{1}{4}$ and $\frac{3}{4}$. The selenium ions sitting at the center of the tetragonal SeO_4 groups also occupy special positions on the mirror planes and therefore mirror planes of the tetragonal groups coincide with the mirror planes of the lattice. Thus two out of four oxygen atoms lie on

Table 2
Atomic parameters in N phase K_2SeO_4 with their e.s.d.'s (Kalman et al. 1970).

Atom	x	y	z
Se	$\frac{1}{4}$	0.4200(1)	0.2242(2)
K(1)	$\frac{1}{4}$	0.0843(3)	0.1705(5)
K(2)	$\frac{1}{4}$	−0.2905(3)	−0.0057(4)
O(1)	0.0271(10)	0.3471(6)	0.2931(10)
O(2)	$\frac{1}{4}$	−0.4356(8)	0.3024(16)
O(3)	$\frac{1}{4}$	0.4251(10)	0.0126(21)

the mirror planes. Atomic parameters according to Kalman et al. (1970) are given in table 2.

The crystal structure is most easily seen in projection on (001) as shown in fig. 3a. The projection on (100) shown in fig. 3b will be used later to indicate some of the modes relevant to the modulation we are concerned with. As is clearly seen in fig. 3a the c axis is a pseudo-hexagonal axis. This is also manifested by the ratio $b/a = 1.7437 \sim \sqrt{3}$. Therefore, the orthorhombic structure is a slightly distorted hexagonal structure. It is then more natural to take the pseudo-hexagonal axis as the c axis. This convention was already adopted above by interchanging the a and c directions originally used by Kalman et al. The modulation is in the c direction as was already mentioned in section 1. Throughout this chapter the c direction is always taken to be the pseudo-hexagonal axis.

The SeO_4 group is an almost regular tetrahedron within the estimated standard deviations. The two kinds of potassium ion have different environments. K(1) is surrounded by 11 oxygen atoms almost spherically with an average potassium–oxygen distance of 3.14 Å, while K(2) has only 9 oxygen atoms in its neighborhood with a significantly shorter average potassium–oxygen distance of 2.93 Å.

2.2. *Group theoretical analysis of the symmetry of the collective atomic displacements in* K_2SeO_4

As was mentioned in section 2.1, the normal phase of K_2SeO_4 takes a slightly distorted hexagonal structure. An actual phase transition to a true hexagonal phase has been observed in K_2SeO_4 at high temperature (Shiozaki et al. 1977b). There is thus a conceptual problem which of the two structures, hexagonal or orthorhombic, is to be taken as a prototype structure in considering the symmetry aspects of the displacive phase transition. Here we take the orthorhombic structure as the prototype because the hexagonal-to-ortho-

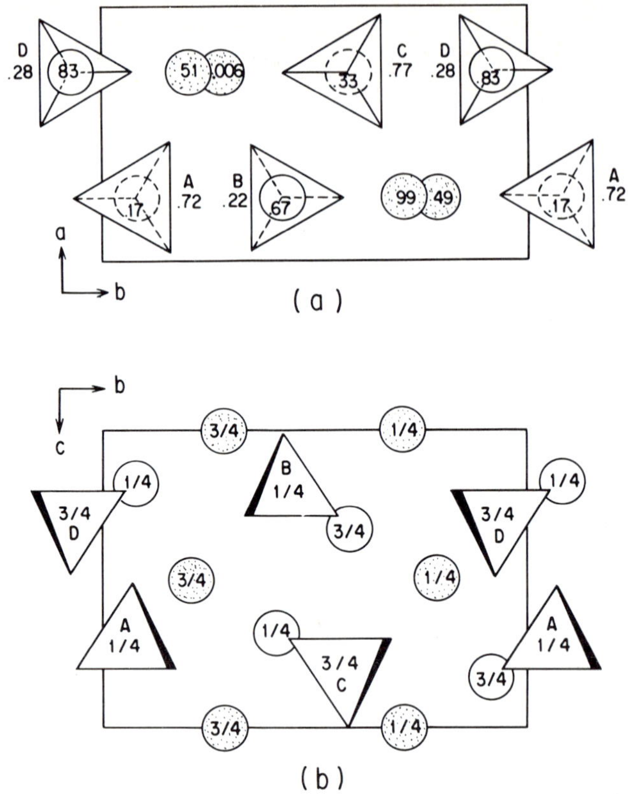

Fig. 3. Projections of the crystal structure of K_2SeO_4 in the normal phase. The open circles, hatched circles and triangles indicate K(1), K(2) ions and SeO_4 groups, respectively. The numbers indicate the level perpendicular to the projection axes.

rhombic phase transition is related to the ordering of the configuration of the SeO_4 groups and hence is not one of the series of successive phase transitions of displacive nature which occur at lower temperatures. It is the orthorhombic structure which gives the basic framework on which the atomic displacements takes place to give rise to low temperature phases.

The crystal structure analysis described in section 2.1 reveals that the distortion of the selenate group from the perfect tetrahedral coordination is very small. This suggests that the selenate groups are accommodated in the crystal as rigid units as they are in the free radical state. Since internal modes of the selenate group are well separated in energy from the external normal modes of lattice vibration, we take the selenate group as a rigid unit in considering the atomic displacements related to the structural phase transition. That is, we take into account only the translation and rotation of a selenate

Table 3
Irreducible representation of the group at $\Lambda:(0,0,\zeta)$

	$\{E|000\}$	$\{C_{2z}|\frac{1}{2}\frac{1}{2}\frac{1}{2}\}$	$\{\sigma_x|\frac{1}{2}00\}$	$\{\sigma_y|0\frac{1}{2}\frac{1}{2}\}$
Λ_1	1	1	1	1
Λ_2	1	1	−1	−1
Λ_3	1	−1	−1	1
Λ_4	1	−1	1	−1

group as a whole. Then the crystal has 48 degrees of freedom, 18 of which are rotational degrees of freedom. There are, therefore, 48 normal modes for each wave vector within the first Brillouin zone and any collective atomic displacements are described by a linear combination of them.

The group-theoretical analysis of the normal modes with vectors along the c^* axis, i.e. $q=(0,0,\zeta)c^*$ with ζ from 0 to $\frac{1}{2}$, has been carried out by Iizumi et al. (1977). Here we give a brief summary of their results. The a and c axes are interchanged in accordance with the choice of the crystal axes described in section 2.1.

The point group of q on the Λ line ($q=(0,0,\zeta)c^*, 0<\zeta<0.5$) is C_{2v} with elements E, C_{2z}, σ_x, and σ_y. There are four irreducible representations, designated Λ_1 to Λ_4 (table 3). The representations of the space group in the 48 normal modes are decomposed into

$$14\Lambda_1+10\Lambda_2+10\Lambda_3+14\Lambda_4. \tag{2.1}$$

Similar analysis gave the decomposition at the Λ point, $q=(0,0,0)$, as

$$7\Gamma_1+5\Gamma_2+5\Gamma_3+7\Gamma_4+5\Gamma_5+7\Gamma_6+7\Gamma_7+5\Gamma_8, \tag{2.2}$$

and at the Z point, $q=(0,0,\frac{1}{2})$, as

$$14Z_1+10Z_2, \tag{2.3}$$

where Γ_i and Z_i corresponds to t_i at k_{19} and k_{21} for D_{2h}^{16} in Kovalev's (1965) table. Z_1 and Z_2 are doubly degenerate modes. The compatibility relations at Γ and Z are illustrated in fig. 4.

The symmetry-adopted eigenvectors in the 48-dimensional atomic displacement space which transform according to the respective irreducible representations can be obtained by the projection operator technique. The most pertinent of these results are given in Iizumi et al. (1977).

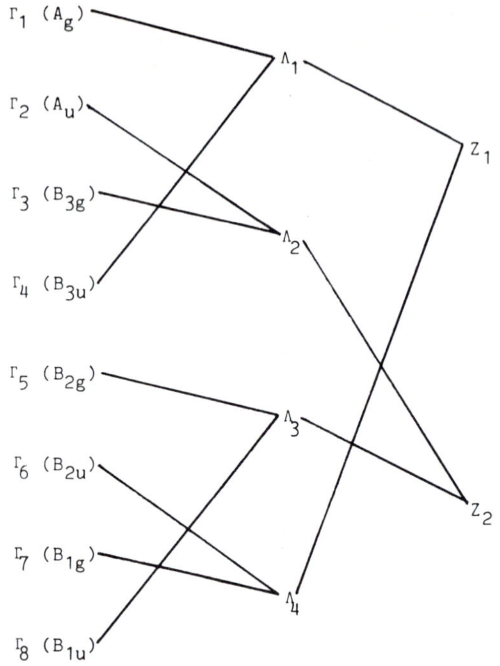

Fig. 4. Compatibility relations for irreducible representations at $\Gamma(0,0,0)$ and at Z(0,0,1). The irreducible representations Γ_1 to Γ_8 correspond to τ_1 to τ_8 for T32, and Z1, Z2 correspond to τ_1, τ_2 for T70, in Kovalev's (1965) compilation, A_g, etc. are the Milliken notations of the representations.

2.3. *Use of the extended zone scheme*

In a later section it will be pointed out that in the crystal structure under consideration it is more convenient to use a twofold-extended zone along the c^* direction. The appropriateness of this convention stems from the existence of a c-glide plane perpendicular to the b axis. Because of this glide plane, there is an atom at $(x, \frac{1}{2} - y, z + \frac{1}{2})$ in the reduced unit. Without loss of generality we can assume that there are only these two atoms in the unit cell. A displacive fluctuation propagating with wave vector $\boldsymbol{q}$ is represented by a plane wave:

$$\boldsymbol{u}(lk) = \boldsymbol{U}(k)\exp[\mathrm{i}(\boldsymbol{q}\cdot\boldsymbol{x}(lk) - \omega t]. \tag{2.4}$$

In the lattice we are considering we have two atoms in a unit cell and suffix k distinguishes those two, that is, $k = 1$ or 2. If we consider the fluctuation with wave vector mentioned above, the relevant phase factor in eq. (2.4) includes only the components of atomic coordinates projected on the (010) plane. Then

Table 4
Irreducible representation at Λ for the extended zone.

	E	σ_x
$\Lambda_1(\Lambda_4)$	1	1
$\Lambda_3(\Lambda_2)$	1	−1

for these particular wave vectors the fluctuation is effectively propagating on a two-dimensional lattice which has one half of the unit of translation of the original lattice in the c direction. (In case of wave vectors on the c^* axis the fluctuation is in effect propagating on a one-dimensional lattice with atoms separated by $c/2$.) Then the corresponding Brillouin zone is extended twofold in the c^* direction.

The point group of q at $q=(0,0,\zeta)$ in this extended zone scheme is C_s or C_{1h} which consists of E and σ_x, and the irreducible representation is that given in table 4.

In the language of lattice dynamics the normal zone scheme gives twofold degeneracy of a Λ_3 mode with a Λ_2 mode at the Z points, as shown in fig. 5a. In the Λ_3 mode $U(1)=U(2)$, while $U(1)=-U(2)$ in the Λ_2 mode. In the extended-zone scheme the Λ_3 branch is as shown in fig. 5b. Then the Λ_2 mode

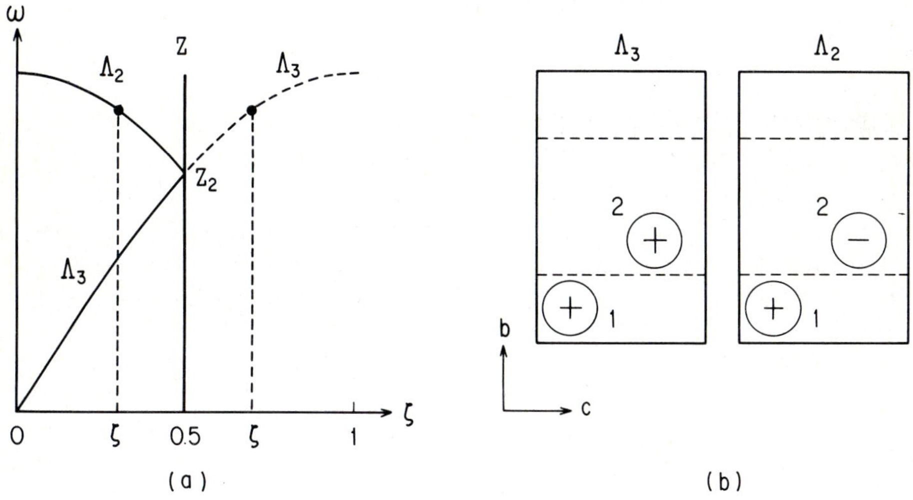

Fig. 5. Phonon dispersion in the normal and extended zone schemes. (a) The Λ_2 branch becomes the extension of the Λ_3 branch in the extended zone scheme. (b) Λ_3 and Λ_2 modes. Atom 1 is related to atom 2 by a glide plane. $U(1)=U(2)$ in the Λ_3 mode and $U(1)=-U(2)$ in the Λ_2 mode.

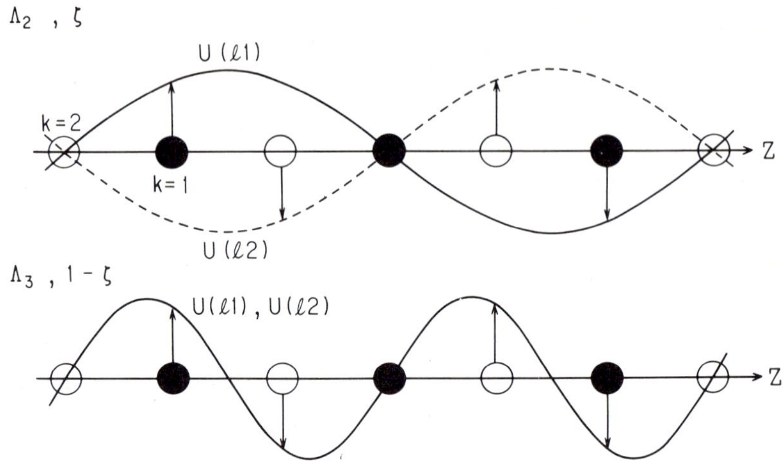

Fig. 6. Equivalence of the Λ_2 mode with wave number ζ to the Λ_3 mode with wave number $1-\zeta$.

with $U(1)=-U(2)$ at $q=(0,0,\zeta)$ in the normal-zone scheme becomes the Λ_3 mode with $U(1)=U(2)$ at $q=(0,0,1-\zeta)$ in the extended-zone scheme. The equivalence of atomic displacements in both schemes is illustrated in fig. 6. The continuity of the first derivatives of the dispersion relations at $q=(0,0,\frac{1}{2})$ results from the above mentioned fact that on the (010) projection the lattice has a halved translational periodicity. This continuity justifies the use of the extended-zone scheme beyond a mere convention.

3. *Phonons and fluctuations in the normal (high temperature) phase*

In this section we review the results of neutron scattering studies of the phonon dynamics and critical scattering in the disordered or high temperature phases ($T>T_i$). We ignore internal modes of the complex anions and cations and other modes not directly associated with the incommensurate transformation. In other words we concentrate on low lying modes with Λ_2–Λ_3 symmetry. Since these fluctuations are not near the center of the Brillouin zone, light scattering is of no help here. Discussion of fluctuations below T_i, where light scattering is in some respects the more suitable tool, is deferred to section 5.

3.1. *K_2SeO_4*

Our understanding of the dynamics of K_2SeO_4 is better than for related materials for the simple reason that there is a well defined soft phonon mode, first studied by Iizumi et al. (1977). A summary of these measurements as a function of $\boldsymbol{q} = (0, 0, \zeta)$ in a reduced zone scheme is shown in fig. 1. The lowest Λ_3 branch (designated Σ_3 in fig. 1) is one of the TA branches and shows little temperature dependence at small ζ. The Λ_2 (Σ_2 in fig. 1) branch, degenerate with Λ_3 at $\zeta = \frac{1}{2}$, is almost flat at high temperatures, but softens conspicuously as T approaches T_i. This softening is general and even extends to the upper Λ_3 modes. The wavevector, ζ_s, of the minimum of the dispersion relation decreases by about 10% on cooling from room temperature to 130 K. The low temperature value of ζ_s is nearly identical to (about 0.6% larger than) the wavevector of the incommensurate satellite just below T_i.

The dispersion of the soft mode was also investigated for a series of wavevectors $\boldsymbol{q} = (0, \xi, \zeta_s)$ off the pseudo-hexagonal axis. At 130 K the results could be described by

$$\omega(\boldsymbol{q})^2 = \omega_0^2 + v^2(q_z - q_s)^2 + v_\perp^2 q_y^2.$$

This anisotropy suggests that the softening primarily involves motion in which entire planes of atoms perpendicular to c take part as a whole, with motions which distort the planes costing higher energy.

The temperature dependence of the soft mode is approximately given by $\omega_0^2(T) \propto (T - T_i)$ for $145\,\mathrm{K} < T < 300\,\mathrm{K}$. At lower temperatures, ($145\,\mathrm{K} < T < T_i$), the phonon sidebands begin to disappear and the critical scattering peaks at $\omega = 0$. X-ray diffuse scattering intensity is reasonably represented by $I \propto (T - T_i)^{-\gamma}$ (Terauchi et al. (1975) in this region with $\gamma = 1$, consistent with classical overdamped phonons. (In the classical limit both under- and overdamped phonons obey the same X-ray scattering law.) More careful considerations of the critical behavior (Bruce and Cowley 1978) predict critical exponents identical to the $d = 3$ XY system, and thus $\gamma = 1.27$ as $T \to T_i$. The accuracy of existing measurements is such that no disagreement can be said to exist. As would be anticipated from the phonon dispersion, the distribution of the critical scattering, which lies precisely along the c axis, is also distributed more widely along this axis than along the a axis. (Terauchi et al. 1975, Iizumi et al. 1977). There appears to be no information about the scattering along the b axis.

3.2. *Other compounds*

Similar neutron scattering studies of the dynamics of the normal phase have been carried out for several isomorphous structures. They include AFB (Iizumi and Gesi 1977), Rb_2ZnBr_4 (de Pater and van Dijk 1978, de Pater et al. 1979, Iizumi and Gesi 1984), $TMA{\cdot}ZnCl_4$ (Iizumi and Gesi 1983a) and K_2ZnCl_4 (Gesi and Iizumi 1984). The results of these studies have indicated that (1) K_2SeO_4 is a unique example which shows the complete softening of the connected Λ_3 acoustic–Λ_2 optic branch, and that (2) other crystals present predominant quasi-elastic scattering which increases its intensity critically toward T_i from high temperature. [de Pater and van Dijk (1978) initially reported a double peak structure, suggesting underdamped soft phonon in Rb_2ZnBr_4 just above T_i, but later measurements (de Pater et al. 1979, Iizumi and Gesi 1983a, b) failed to reproduce this feature.]

The critical scattering is diffuse with a distribution elongated along the c^* axis around an incommensurate position which corresponds to q_1 where the incommensurate satellite appears below T_i. The energy spectrum of the scattering is elastic within the limit of the energy resolution of the neutron spectrometer, indicating that the dynamical nature of the displacive fluctuation is relaxational with time constant of order larger than 5×10^{-9} s. [A recent measurement on Rb_2ZnBr_4 (Gesi and Iizumi 1984) indicated finite energy width in the energy spectrum of the quasi-elastic scattering, suggesting that the relaxational time constant in this crystal is of the order of 10^{-11} s.]

The nature of the structure factor determining the scattering intensities among different zones in the reciprocal space is similar to that for one-phonon scattering in K_2SeO_4. This indicates that the critical fluctuation consists of atomic displacements similar to the soft phonon observed in K_2SeO_4 and that only the dynamical nature is different; the relevant atomic displacements are in an effectively double well potential in these crystals instead of the single harmonic potential well in the case of K_2SeO_4. In some crystals a small vestige of phonon softening was observed in the connected Λ_3 acoustic–Λ_2 optic branch at the position corresponding to the peak in the quasi-elastic scattering. In these cases the atomic displacements constituting the critical fluctuation have interaction with the phonon.

However the failure to observe well defined soft modes in the normal phases of any of the other materials shows that there are profound differences in the dynamics. One line of reasoning which may prove useful in understanding these differences is structural in nature. X-ray investigations of the tetrahalide compounds mentioned above reveal large and anisotropic mean-square displacements of the MX_4 tetrahedra. These data are more consistent with static orientational disorder than with thermal vibrations. Indeed, Rb_2ZnCl_4 (Itoh et al. 1984), $RbZnBr_4$ (de Pater et al. 1979) and $(NH_4)_2ZnCl_4$ (Matsunaga 1982) have been officially declared to have rotational disorder in the normal phase. Itoh et al. (1984) have made an interesting correlation of the mean square displacement amplitude of the anions with those of the cations in the same compound, as shown in fig. 7. It can be seen that as $\langle u^2(X)\rangle$ becomes large that $\langle u^2(A)\rangle$ does likewise, suggesting that the disordered motions of the tetrahedral anions are closely correlated with similar disorder in the neighboring cations. Itoh et al. point out that K_2SeO_4 has the smallest rms displacements and well defined soft modes. Extending this line of reasoning, they suggest that possibly $(NH_4)_2SO_4$ and AFB may prove to have a character intermediate between displacive and order–disorder, with the remaining

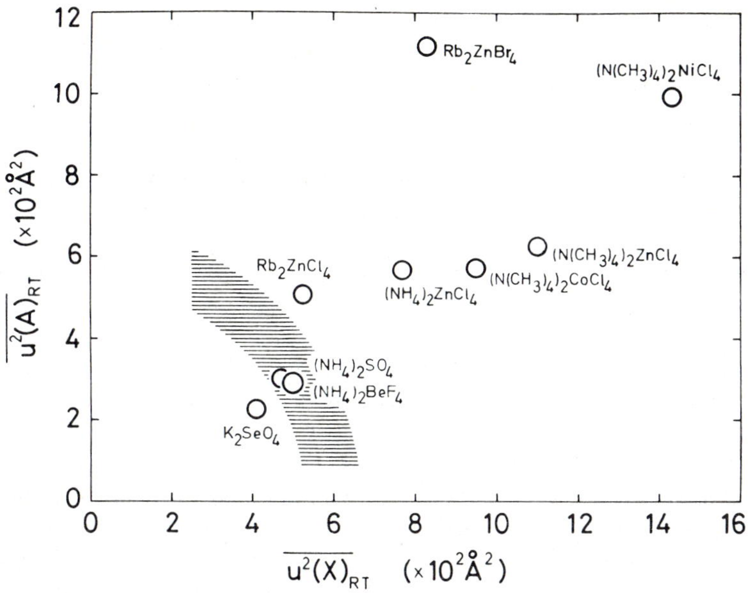

Fig. 7. Correlation of mean square vibrational amplitudes of anions and cations in A_2BX_n compounds.

materials with critical fluctuations of a strongly damped order–disorder type, rather than a propagating soft mode.

3.3. *Origins of the incommensurate instability*

We begin by admitting that we do not have a clear and compelling physical picture of the incommensurate instabilities in this class of compounds. Many investigators would not only agree to this assessment, but would broaden it to include incommensurate insulators generally. What is meant by this? We believe that the dissatisfaction results from an unfavorable comparison with the situation for incommensurate instabilities in metals. In metals, one needs simply invoke an additional characteristic dimension in reciprocal space supplied by a self-nesting Fermi surface, to provide an intuitive picture of charge density wave (CDW) instabilities. Sceptics may remind us that few of the many alleged CDW transformations have been satisfactorily described by microscopic calculations along the above lines, but this troubles us little. Of course, we say, the details are complicated, but we understand the "principle".

But what is the corresponding "principle" for understanding incommensurate insulators? Perhaps it is only this. Competing interactions of varying range (even relatively short range) can produce incommensurate instabilities. This was understood from the study of incommensurate magnetic structures (e.g. Kittel 1966), and includes the CDW case (short range versus RKKY-like interactions). But this strikes us as rather *too* general. At the other extreme, we have a lattice dynamic model calculation by Haque and Hardy (1980) for K_2SeO_4, which incorporates much of the specific detail of this structure to produce a "prediction" of the incommensurate instability. This seems too specialized to invite easy generalization.

There are two types of experimental information that might be hoped to provide clues to the origins of the instabilities. The first is dynamics; the nature of the soft mode dispersion can provide direct information on the interatomic forces. K_2SeO_4 is most important in this regard, since as we have seen the dynamics of the other compounds are unknown for reasons that are themselves obscure. Secondly we have structural information concerning the displacements. We will explore these two routes in turn.

3.3.1. *Analysis of phonon dispersion*

Iizumi et al. (1977) pointed out that when the phonon data for K_2SeO_4 were analyzed in the extended-zone scheme, the dispersion relation could be well represented by a Fourier decomposition

$$\omega_s(\zeta) = \sum_n F_n(1 - \cos n\pi\zeta), \tag{3.1}$$

with terms with $n < 3$ negligible. The coefficients F_n correspond to generalized effective *interlayer* force constants, coupling in this case layers in the crystal separated by distances $nc/2$. (We use the branch index, s, to designate both the lowest-lying Λ_2 and Λ_3 branches in the extended scheme). The behavior of the derived force constants as a function of temperature are shown in fig. 8. The fitted phonon dispersion curves are shown in fig. 1. This treatment demonstrates that the pertinent forces are in some sense short-ranged and that the softening near $\zeta = \frac{2}{3}$ results from the temperature dependent decrease of both F_1 and F_2 in the presence of a strong, persistent F_3. (Note that F_3 alone produces a commensurate instability at $\zeta = \frac{2}{3}$.) The importance of competing interaction among three layers is taken up again in section 5 in connection with stable commensurate phases.

Bussmann-Holder et al. (1981) considered a rather schematic one-dimensional lattice-dynamical model of K_2SeO_4 with highly anharmonic potentials which contains a structural instability, but the most extensive lattice-dynamical study of K_2SeO_4 was carried out by Haque and Hardy (1980). Using the known K_2SeO_4 structure, they combined general Coulomb interactions and phenomenological central short-range interactions between restricted sets of close neighbors into a Born–von Karman rigid ion model. The many adjustable constants were fixed by static equilibrium conditions and measured $\boldsymbol{q} = 0$

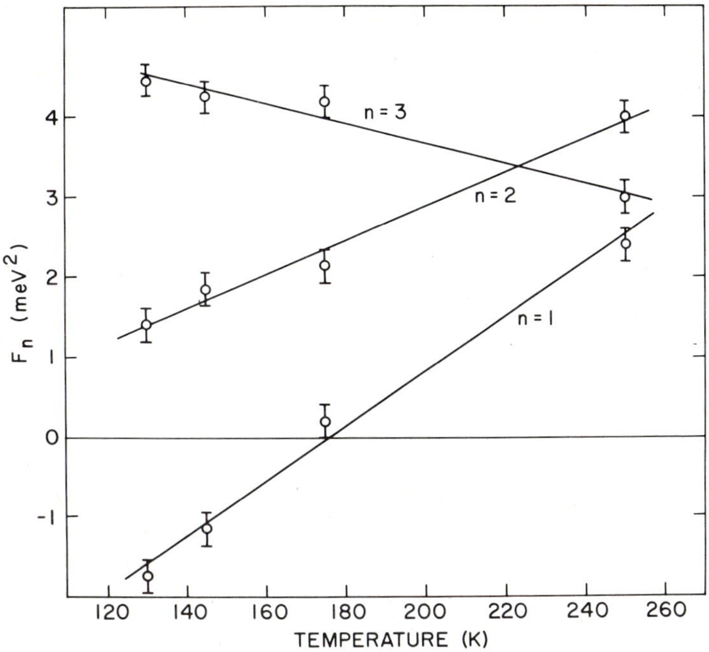

Fig. 8. Temperature dependence of interplanar force constants derived from phonon dispersion in K_2SeO_4.

phonon frequencies. When the frequencies of the lowest phonon branch $\omega_s^2(\zeta)$, with $\boldsymbol{q} = (00\zeta)$, are calculated using these parameters, the branch is found to be unstable ($\omega_s^2(\zeta) < 0$) near $\zeta \approx 0.7$. Haque and Hardy note that the instability is very sensitive to changes in a small number of potassium-oxygen and inter-anion oxygen-oxygen potentials. The potential labeled $\phi_{00}(7)$ by Haque and Hardy and shown in fig. 9, is particularly singled out in this regard and it is shown that the observed temperature dependence of the soft branch can be induced by temperature dependent changes in this force constant alone.

Haque and Hardy also furnish additional insight into the nature of the interplanar forces discussed by Iizumi et al. (1977). The dynamical matrix is first decomposed into Coulomb and short-ranged parts,

$$D_{ij} = D_{ij}^{c} + D_{ij}^{SR}.$$

The effective forces for the soft branch are then projected out using the soft-branch eigenvectors, $\boldsymbol{e}^s$,

$$(e^s)_i^* D_{ij}^c e_j^s = \omega_c^2, \qquad (e^s)_i^* D_{ij}^{SR} e_j^s = \omega_{SR}^2,$$

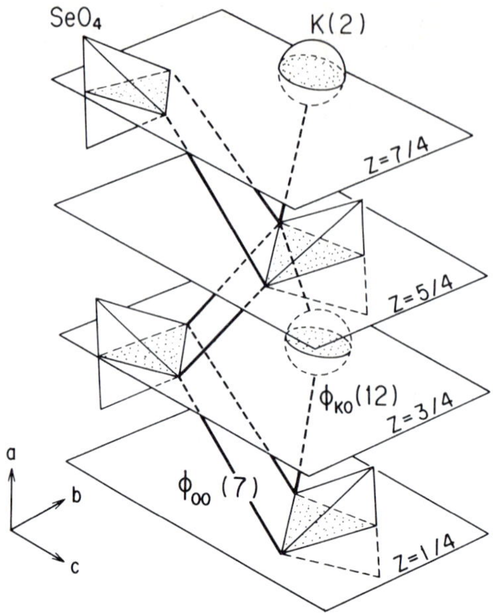

Fig. 9. K_2SeO_4 structure, showing critical short range force constants.

and

$$\omega_s^2(\zeta) = \omega_c^2(\zeta) + \omega_{SR}^2(\zeta).$$

The results for ω_c^2, ω_{SR}^2 and ω_s^2 for the extended soft branch are shown in fig. 10. Note in addition to the overall agreement of $\omega_s^2(\zeta)$ with the observed instability, the remarkable degree to which ω_c^2 and ω_{SR}^2 cancel. Over the whole soft branch $(\omega_c^2 - \omega_{SR}^2)/\omega_c^2 < 0.02$. Thus Haque and Hardy find that the instability results from a near cancellation of stabilizing Coulomb forces and destabilizing short range contributions.

3.3.2. *Analysis of atomic displacements*

The first attempt to discuss the displacements associated with the incommensurate instability was by Iizumi et al. (1977). They used group theoretical arguments to define a set of ten symmetry-adapted basis functions into which the soft-mode eigenvectors could be expanded, and estimated the coefficients of this expansion by a least square fit to a limited set of soft mode phonon intensities (dynamical structure analysis; see, e.g., Harada et al. 1970). Their analysis indicated that the major components of the displacements were translations of all constituents, but particularly K(1), along the *a* axis, in concert with rotations of the SeO_4 groups about the *b* axis.

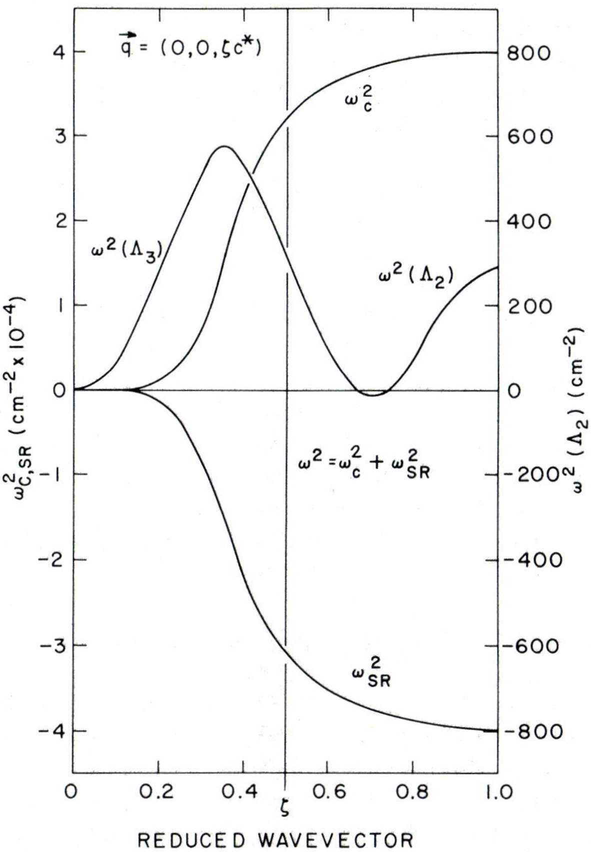

Fig. 10. Decomposition of soft mode dispersion in K_2SeO_4 into Coulomb and short range contributions.

These observations were extended and reinforced by the recent analysis of the structure of K_2SeO_4 by Yamada and Ikeda (1984). They measured the intensities of a large number of satellite as well as fundamental reflections at several temperatures in the incommensurate region using X-rays, and refined the structure based upon four-dimensional hyperspace methods of de Wolff (1974). The conclusions can be summarized as follows:

(1) The SeO_4 tetrahedra are almost regular and displace rigidly.

(2) Higher order satellites were not observed, indicating nearly sinusoidal modulation. (See, however, section 5).

(3) The displacements are of the expected Λ_2 character with prominent components as noted above. Quantitative differences exist when compared

with the determination of Iizumi et al. (1977). The new results have a significantly lower uncertainty and are probably to be preferred. Moreover, the new results are in excellent agreement with the theoretical eigenvectors of Haque and Hardy (1980), underscoring the impressive nature of these calculations.

(4) The bond lengths between oxygen atoms belonging to adjacent SeO_4 tetrahedra change remarkably from the normal phase and fluctuate widely in the incommensurate phase. These are precisely the bonds which Haque and Hardy found critical in their lattice dynamical model.

In an effort to provide a simple picture of the K_2SeO_4 instability Yamada and Ikeda (1984) took notice of a group of entities consisting of two K(2) ions and two SeO_4-groups. Consider, for example K2(1′), K2(3′), $(SeO_4)3$ and $(SeO_4)5^*$, shown in fig. 9, as a linked array. Within this array rotation of the SeO_4 groups in opposite sense about the *b*-axis is associated with the translation of the two K(2) ions in the same direction, while rotation in the same sense is associated with translation in the opposite direction. They point out that this is mediated by the short-range coupling through O(1)–K(2)–O(1).

4. *Modulation in the incommensurate phases*

The prototype structures considered up to this point transform into incommensurate structures below T_i. This is revealed by the appearance of satellite reflections in diffraction experiments. (Extra reflections which usually appear between fundamental reflections of the prototype structure are called superlattice reflections. They show up at some commensurate positions in the reciprocal space, say $\frac{1}{2}$ or $\frac{1}{3}$. On the other hand extra reflections which appear in the vicinity of the fundamental reflections are usually called satellite scattering and they usually locate at some incommensurate positions. The incommensurate reflections we are dealing with now do not fall in into either of above categories in the usual sense. However, in order to emphasize the incommensurate nature of the reflections and to distinguish them from the commensurate superlattice reflections we call them satellite reflections even though they appear, say, in the vicinity of $\frac{1}{2}$ or $\frac{1}{3}$.)

The wave vectors characterizing the modulated structure are deduced from the distances of the satellite positions in the reciprocal space from the reciprocal lattice points. The positions of the satellites in the reciprocal space are given by the scattering vectors:

$$Q = ha^* + kb^* + lc^* + nq_1 \tag{4.1}$$

where q_1 is the modulation wave vector and n $(=\pm 1, \pm 2, \ldots)$ indicates the order of the satellite reflections.

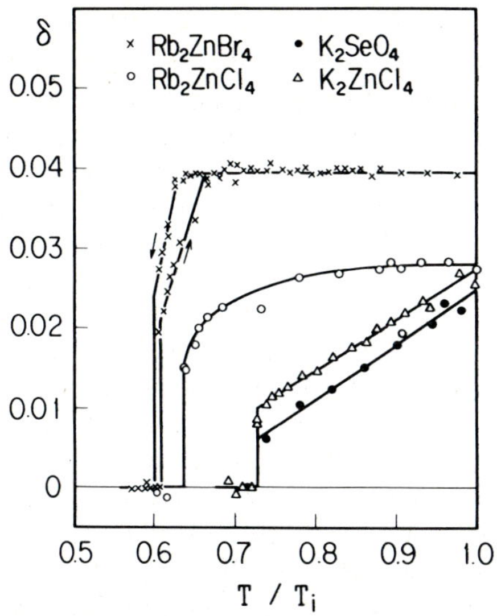

Fig. 11. Temperature dependence of the incommensurability parameters, δ, of K_2SeO_4 and related crystals. The temperatures are normalized to T_i for each crystal.

In the family of crystals belonging to the K_2SeO_4 type the modulation wave vectors are always along the c^* direction, $q_c = (0, 0, \zeta_c)$, where ζ_c indicates the nearby commensurate wave vector to which the lock-in transition actually takes place. (Quite recently $[N(CH_3)_4]_2ZnBr_4$ was found by Wada et al. (1981b) to develop modulation in the b direction. The detailed investigation of this exceptional behavior is a future problem.)

In the first group of crystals, which including K_2SeO_4, ζ_c equals $\frac{1}{3}$ and crystals show certain common features. We first compare the temperature dependence of their modulation wave vectors in fig. 11. As shown in fig. 12, ammonium fluoroberylate constitutes the second group in which the incommensurate wave vector is nearly $\zeta_c = \frac{1}{2}$. The tetramethyl compounds constitute a third distinct group in which more complicated lock-in occurs. In these compounds characteristic aspects of the incommensurate modulation have been disclosed by studying pressure as well as temperature dependence of modulation as mentioned in section 4.3.

4.1. *Modulation in K_2SeO_4 and similar compounds*

In K_2SeO_4 (Iizumi et al. 1977, Kudo 1982) and similar compounds, such as K_2ZnCl_4 (Gesi and Iizumi 1979), Rb_2ZnCl_4 (Gesi and Iizumi 1979, Mashiyama and Tanisaki 1982) and Rb_2ZnBr_4 (Gesi and Iizumi 1978,

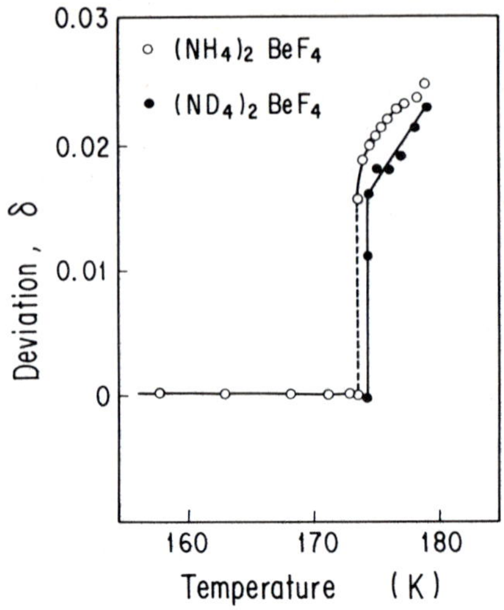

Fig. 12. Temperature dependence of the incommensurability parameters of AFB and d-AFB.

de Pater and van Dijk 1978, de Pater et al. 1979, Ueda et al. 1982), ζ_c is $\frac{1}{3}$:

$$q_1 = (\tfrac{1}{3} - \delta)c^*, \tag{4.2}$$

and the deviation of the incommensurate wave vector from the commensurate one is very small; in case of K_2SeO_4, $\delta = 0.02$ for temperature just below T_i. The temperature dependences of the incommensurate or misfit parameter, δ, are summarized in fig. 11 for these compounds. In this figure the temperature is scaled by T_i for each compound. Except for Rb_2ZnBr_4 the general tendency is very similar. Rb_2ZnBr_4 is peculiar; q_1 is temperature-independent for more than 140 K and then it changes very rapidly. This feature which was investigated later (Iizumi and Gesi 1983b) will be described in section 7.

Neutron scattering measurements on K_2SeO_4 (Iizumi et al. 1977) disclosed that the modulation is displacive and that the atomic displacements are given by the eigenvector of the soft mode belonging to the irreducible representation of Λ_2, and consist of (1) translation in the *a* direction of potassium, (2) translation of SeO_4 groups in the same direction, and (3) rotation of the groups around the *b* and *c* axes. Since the modulation is characterized by the Δ_2 irreducible representation, the modulation wave vector is

$$q_1 = (\tfrac{2}{3} + \delta)c^* \tag{4.3}$$

according to the extended-zone scheme mentioned in section 2.3. The effect of hydrostatic pressure on the incommensurate modulation was investigated by Press et al. (1980).

Tetramethyl-ammonium tetrachloro-cuprate shows a similar phase transition sequence as K_2SeO_4. The incommensurate structure (Gesi and Iizumi 1980b) is characterized by $q_1 = 0.326$ or $\frac{2}{3} + 0.007$ in the extended zone. The tripled commensurate phase appears below a lock-in transition. However this phase is not ferroelectric as in K_2SeO_4, but is ferroelastic (A. Sawada et al. 1980).

4.2. *Modulation in $(NH_4)_2BeF_4$ (AFB)*

The incommensurate phase of AFB was first observed (Iizumi and Gesi 1977) by neutron diffraction in the deuterated compounds, $(ND_4)_2BeF_4$ and is characterized by a wave vector which is very close to $\frac{1}{2}$ and changes with temperature as shown in fig. 12. The positions of the satellites indicate that the wave vector is slightly larger than $\frac{1}{2}$ in the extended-zone scheme. The incommensurate phase has a very narrow temperature range and locks in to a ferroelectric commensurate phase with $\zeta_c = \frac{1}{2}$. Measurements were subsequently carried out on the undeuterated compounds (Kudo 1982) and similar results were obtained. These are also indicated in fig. 12.

The crystal $[N(CH_3)_4]_2MnCl_4$ shows similar behavior (Mashiyama and Tanisaki 1981); the compound has an incommensurate phase in which ζ is very close to $\frac{1}{2}$ and locks in to a commensurate phase with $\frac{1}{2}$. However, in this compound the commensurate phase is ferroelastic. Moreover the compound shows more complicated phases under applied hydrostatic pressure and this may be better described in the section 4.3 together with other tetramethyl ammonium compounds.

4.3. *Modulation in tetramethyl ammonium compounds*

This is a group of compounds with chemical formula $[N(CH_3)_4]_2MX_4$ (hereafter TMA is used as an abbreviation of tetramethyl ammonium ions as follows, $TMA \cdot MX_4$) (S. Sawada et al. 1978a, b). The compounds with X = Cl and M = Mn, Fe, Co, Cu and Zn are known to have incommensurate phases. The deuterated compounds are also very interesting since they show conspicuous isotope effects in the appearance of incommensurate phases (deuterated tetramethyl ammonium ions are abbreviated as d-TMA). Some compounds with X = Br are beginning to be investigated, but in this section the systematic behavior of the mutually related tetrachloro compounds (X = Cl) with M = Mn, Fe, Co and Zn will be summarized. The compound with M = Cu shows a different behavior as has been described in section 4.1.

The incommensurate phases of these compounds are characterized by wave vectors with ζ distributing from about 0.4 to 0.5 and $\zeta_c = \frac{2}{5}$. The irreducible representation characterizing the atomic displacements in Λ_3 in contrast with

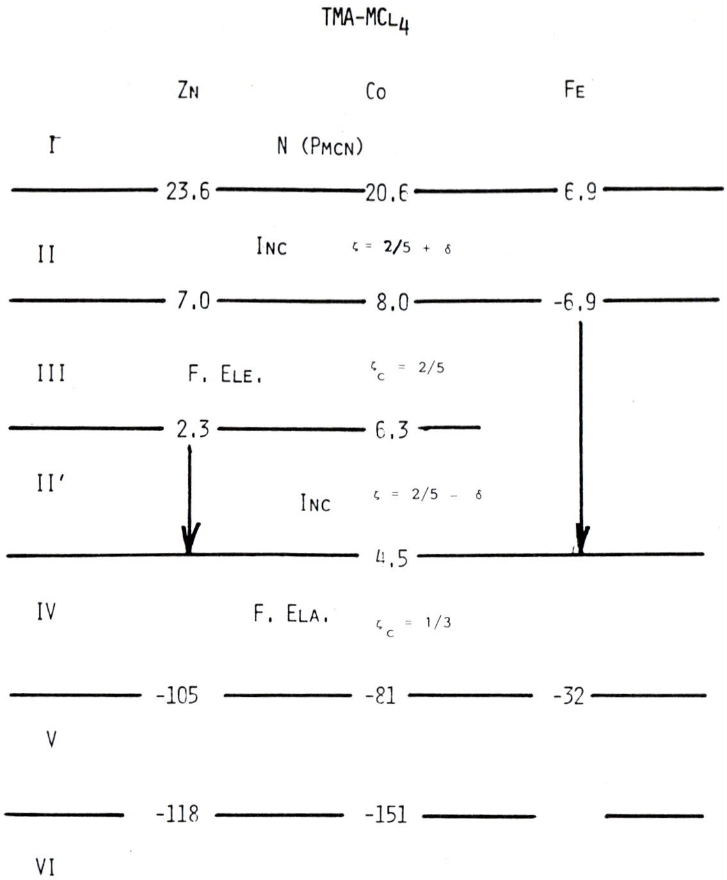

Fig. 13. Transition sequence of TMA compounds.

the K_2SeO_4 group in which the irreducible representation is Λ_2. This is known by how the satellite spots appear on the $(h01)$ scattering plane of the reciprocal space; the pattern is in accord with the extended zone scheme, not with the normal zone scheme. Therefore in K_2SeO_4 the satellite spots locate in the vicinity of $\boldsymbol{G}+\frac{2}{3}\boldsymbol{c}^*$, not in the vicinity of $\boldsymbol{G}+\frac{1}{3}\boldsymbol{c}^*$, and in TMA·$ZnCl_4$ they appear in the vicinity of $\boldsymbol{G}+\frac{2}{5}\boldsymbol{c}^*$, not in the vicinity of $\boldsymbol{G}+\frac{3}{5}c^*$. Here $\boldsymbol{G}$ is any one of the $(h01)$ reciprocal lattice vectors which satisfy the extinction rule $(h01:1=2n)$ originating from the glide plane.

A variety of successive phase transitions take place in these compounds as indicated in fig. 13. In the Zn salt (Tanisaki and Mashiyama 1980, Mashiyama and Tanisaki 1980, Mashiyama 1980, Mashiyama et al. 1980) phase II is

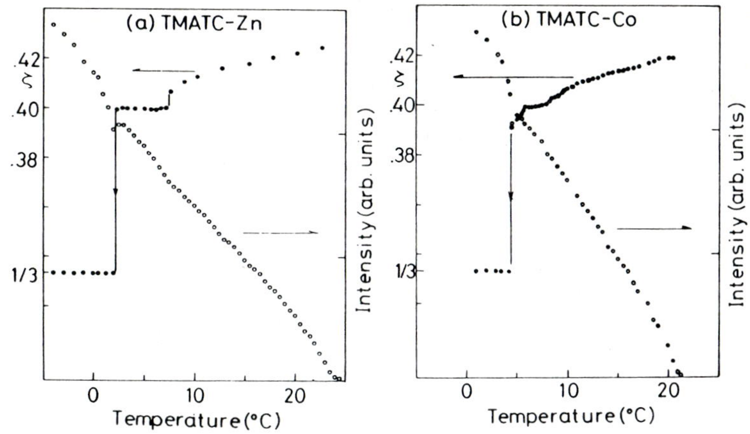

Fig. 14. Positions and intensities of satellites in $TMA \cdot ZnCl_4$ and $TMA \cdot CoCl_4$.

incommensurate with the wave vector about 0.42 and it locks in to a commensurate phase with $\zeta_c = \frac{2}{5}$ which is ferroelectric. This phase appears only in a narrow temperature range and it transforms into a ferroelastic monoclinic phase characterized by $\zeta_c = \frac{1}{3}$. This phase is the common low temperature phase of this kind of compounds. There are two more phases at low temperatures but they are outside the scope of this work.

The Co salt (Hasebe et al. 1980, 1982) exhibits quite a remarkable behavior; the incommensurate phase locks in to the ferroelectric commensurate phase with $\zeta_c = \frac{2}{5}$, but then transforms back to an incommensurate phase again. Except for this reentrant incommensurate phase the transition sequence is just the same as in the Zn salt. Figure 14 (Mashiyama et al. 1980) shows the temperature dependence of the modulation wave vectors of both compounds. In the Zn salt the lock-in transition indicates a small jump in ζ but for Co ζ does not show any discernible jump at either the lock-in or reentrant transitions. This is more clearly seen in fig. 15, which shows the data taken by Hasebe et al. (1980). They reported that the measurement was carried out simultaneously with the dielectric measurement and without applying an AC field they were unable to observe this change in the Co salt repeatedly. Therefore, the existence of the commensurate phase with $\zeta_c = \frac{2}{5}$ is subtle from the diffraction experiments but its existence has been established by the observation of hysteresis in the $D-E$ relation and observation of double peaks in the dielectric constant as a function of temperature. The peaks correspond to the lock-in and reentrant phase transitions, respectively.

In the Fe salt the ferroelectric commensurate phase is missing and the incommensurate phase transforms directly into a monoclinic phase with $\zeta_c = \frac{1}{3}$

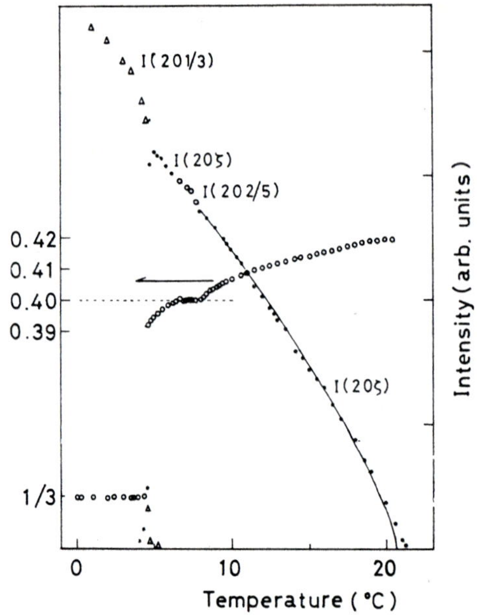

Fig. 15. Temperature dependence of the modulation wave vector of TMA·$CoCl_4$. The change of satellite intensity is also shown.

(Mashiyama and Tanisaki 1982). As was mentioned in the previous section the incommensurate phase in the Mn salt has a very narrow temperature range and locks in to a commensurate phase with $\zeta_c = \frac{1}{2}$, as was the case in AFB. The commensurate phase is not, however, ferroelectric but ferroelastic. The temperature dependence of the wave vector in the incommensurate phases of the Fe compound (Mashiyama and Tanisaki 1982) is shown in fig. 16. The corresponding data for the Mn salt is shown in fig. 20 (Itoh et al. 1984).

The wide variety of phases in these compounds stimulated a unified view by Shimizu et al. (1979a, b, 1980a, b) by extending the measurements to finite pressures. They determined phase boundaries on the pressure–temperature phase diagram by measuring anomalies in the dielectric and other properties under hydrostatic pressure. Figure 17 summarizes the diagram obtained for the four compounds whose behavior under normal pressure has been mentioned in this section.

Consider first the phase diagram of the Zn salt (Shimizu et al. 1979a, 1980b), fig. 17c. The phases I to IV are those identified at normal pressure; the normal, incommensurate, commensurate ($\frac{2}{5}$) and commensurate ($\frac{1}{3}$) phase, respectively. (The two commensurate phases characterized by $\zeta = \frac{2}{5}$ and $\frac{1}{3}$ are distinguished by the notation indicated.) The phase VII is a new phase which appears only

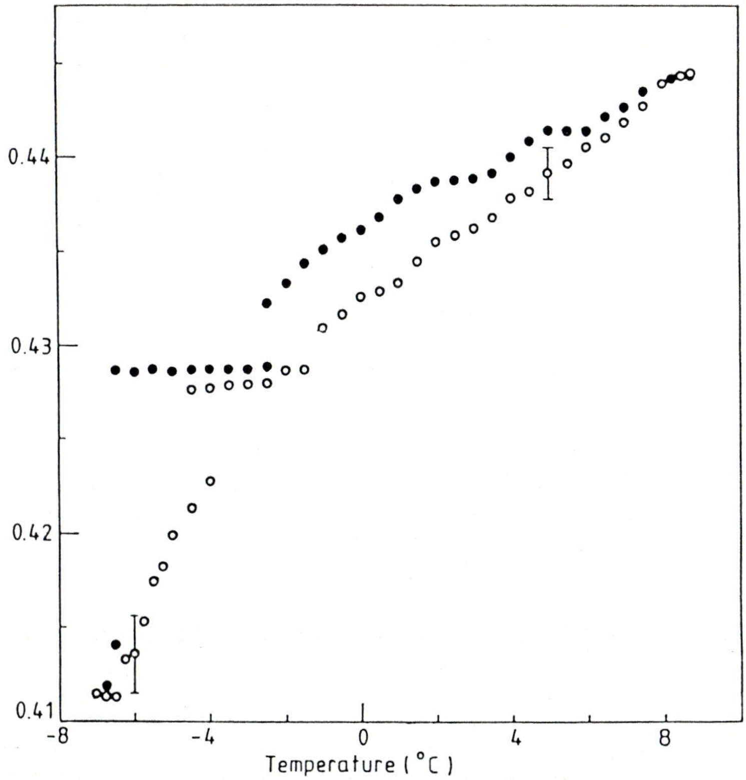

Fig. 16. Temperature dependence of the modulation wave vector of TMA·$FeCl_4$. The open and closed circles represent heating and cooling, respectively.

for pressures exceeding about 0.5 kbar. The temperature range of the commensurate ($\frac{2}{5}$) phase III becomes narrower with increasing pressure, and it finally disappears at about 1.2 kbar. The exact tip of the boundary of phase III is hard to determine experimentally; the boundaries below 1.2 kbar can be recognized by the double peak positions in the dielectric measurements, but above 1.2 kbar the measurements indicate single peaks which persist with some weakening up to fairly high pressure. The line extending from the tip of the phase III area indicates the trace of the peak in the dielectric measurements. As will be described later, diffraction measurements have disclosed that the line is just the contour for $\zeta = \frac{2}{5}$ on which the dielectric constants show anomaly. Thus the line should not be regarded as a phase boundary. Thus it may be said that the phase VII is a continuation of the phase II and that they constitute a single INC phase area on the two-dimensional phase diagram.

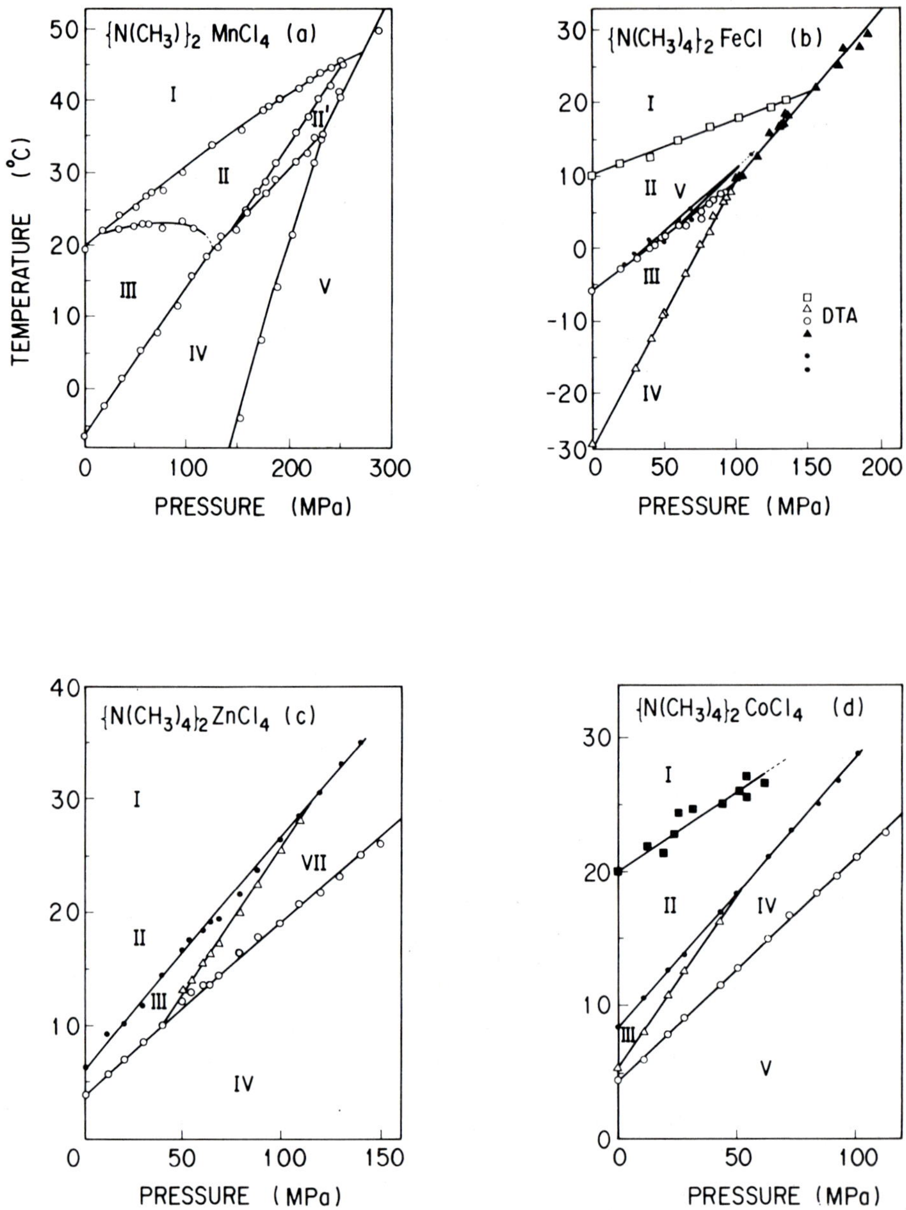

Fig. 17. Pressure–temperature phase diagram of TMA compounds.

The phase diagram for the Co salt (fig. 17d) closely resembles that of the Zn salt. The most important, as has been pointed out by Shimizu et al. (1980b), is that the phase diagram for the Co salt is obtained from that for the Zn salt simply by shifting the origin of the pressure axis to a finite pressure, about 0.6 kbar, in Zn salt. In other words, zero pressure in the Co salt corresponds to 0.6 kbar in the Zn salt. This gives some insight into the peculiar behavior of the reentrant incommensurate phase which is observed in the Co salt in normal pressure.

In case of the Fe salt the phase diagram (fig. 17b) (Shimizu et al. 1980a) can be interpreted as an extension to the negative direction of the pressure axis of the diagram for the Zn salt; a pressure of 0.5 kbar in the Fe salt corresponds to zero pressure in the Zn salt.

The phase diagram in Mn salt (fig. 17a), roughly determined by Shimizu et al. (1979b), was recently revised by Gesi (1982c) who determined the phase boundary of the commensurate phase, which has been found in the normal pressure by diffraction measurements mentioned earlier but was overlooked by Shimizu et al. This commensurate ($\frac{1}{2}$) phase seems to lie further to the negative side of the Zn and Fe salts in which the commensurate ($\frac{1}{2}$) phase does not appear. Then the origin of the pressure axis of the Mn salt is at some negative values on the pressure axes of the Zn and Fe salts.

Another striking point in the Mn salt is that the ferroelectric commensurate ($\frac{2}{5}$) phase disappears, leaving its trace only as single-peaked anomalies in the dielectric measurements. This was indicated by the line in the phase diagram (fig. 17a). The line can be identified with that extending from the tip of the ferroelectric phase in the Zn salts. Therefore the phases X and X′ separated by this line constitute a single incommensurate phase.

The interrelation of the $p-T$ phase diagram of the four compounds mentioned thus far is summarized in a single diagram shown in fig. 18. This schematic figure was obtained from the original one by Shimizu et al. (1980b) by adding the phase boundary of the commensurate ($\frac{1}{2}$) phase. The arrows marked by the element symbols indicate the zero pressure position of the compounds containing the respective elements. In order to attain the universal phase diagram the scales of the pressure as well as temperature axes should be adjusted as indicated by the segments attached to the abscissa as well as the ordinate.

This beautiful universality among various compounds cannot be pursued too far however. One cannot reproduce the details of the phase diagram of the respective compound simply by translating the origin of the pressure axis and changing the scales of both axes. The relative configuration of phase boundaries changes from one compound to another. For example, as was pointed out above, the ferroelectric commensurate ($\frac{2}{5}$) phase does not exist any more in the Mn salt. The boundary of the incommensurate/commensurate ($\frac{1}{3}$) phases shifts upward in this compound to such a position that the potential ferroelec-

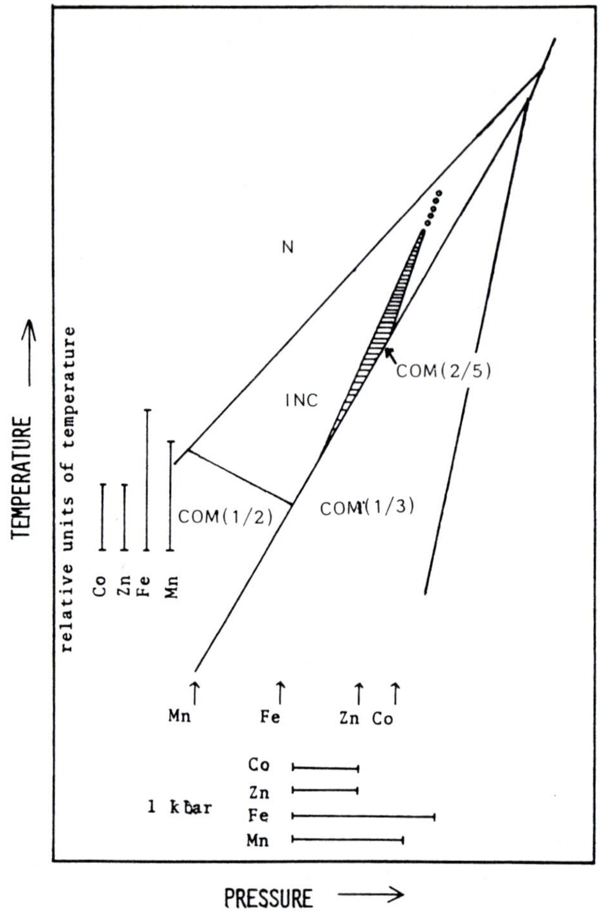

Fig. 18. Schematic phase diagram of TMA compounds. Arrows with the element names indicate the zero-pressure point for each compound. Relative units of the temperature and pressure axes are shown by bars along the axes.

tric phase is buried under the commensurate ($\frac{1}{3}$) phase, leaving only the trace of the ferroelectric phase.

The establishment of the universal phase diagram gives rise to an interesting question: how can the systematic difference among various compounds be explained? Although some speculation has been made, a complete answer has not been attained yet. In this regard the effect of deuteration of the tetramethyl ammonium groups gives another clue. The deuterated compound of Zn salt, $[N(CD_3)_4]_2ZnCl_4$, exhibits quite a different behavior than the undeuterated

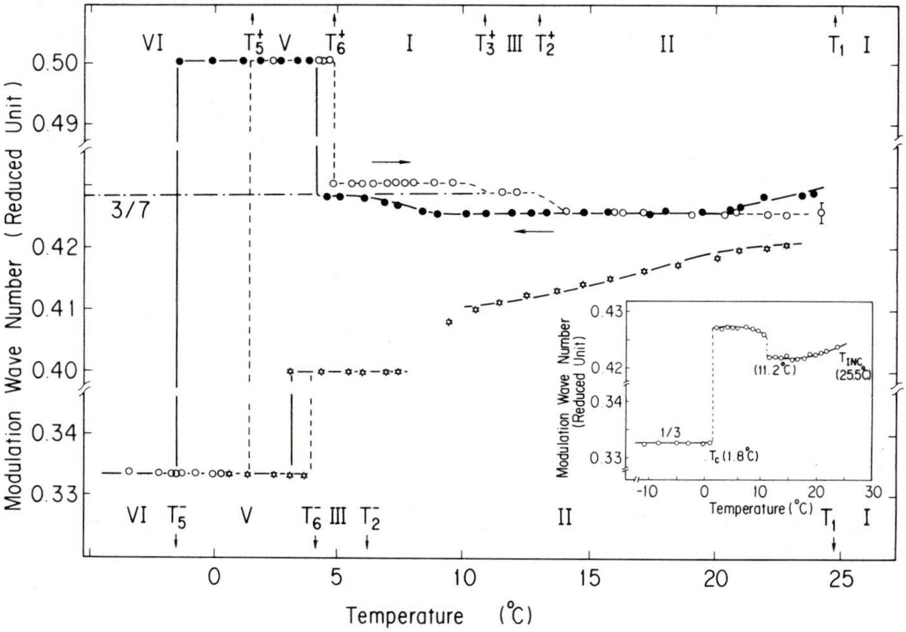

Fig. 19. Temperature dependence of the modulation wave vectors in TMA·$ZnCl_4$ (*) and d-TMA·$ZnCl_4$ (● on cooling, ○ on heating). The phase with $\zeta = \frac{1}{2}$ does not exist in a well prepared crystal.

compound mentioned earlier. Initially it was reported (Gesi and Iizumi 1980a, Marion et al. 1981) that a commensurate ($\frac{1}{2}$) phase like the one observed in the normal Mn salt coexists with the incommensurate phase. Later investigation (Mashiyama et al. 1981a) disclosed that the commensurate phase exists only under a small but finite value of applied uniaxial stress. The coexisting commensurate phase is then interpreted as being induced around imperfections in the crystal by residual stress. This suggests that the ordinate (zero pressure axis) of the phase diagram of the deuterated Zn salt lies very close to the pressure where the commensurate ($\frac{1}{2}$) phase disappears in the universal phase diagram. The effect of deuteration of other salts was investigated by Gesi (1982a, b, c), who determined the zero pressure positions of those compounds on the universal diagram and showed that deuteration causes a shift of the origin of the pressure axis in the negative direction.

The wave vector versus temperature relation of the deuterated Zn salt (Marion et al. 1981) is shown in fig. 19. The more precise measurements (Iizumi and Gesi 1983a) indicated in the inset of fig. 19 show that there is a jump in the wave vector within the incommensurate phase. The identification of the low temperature region with a "devil's staircase" is still controversial. This problem is discussed in section 6.

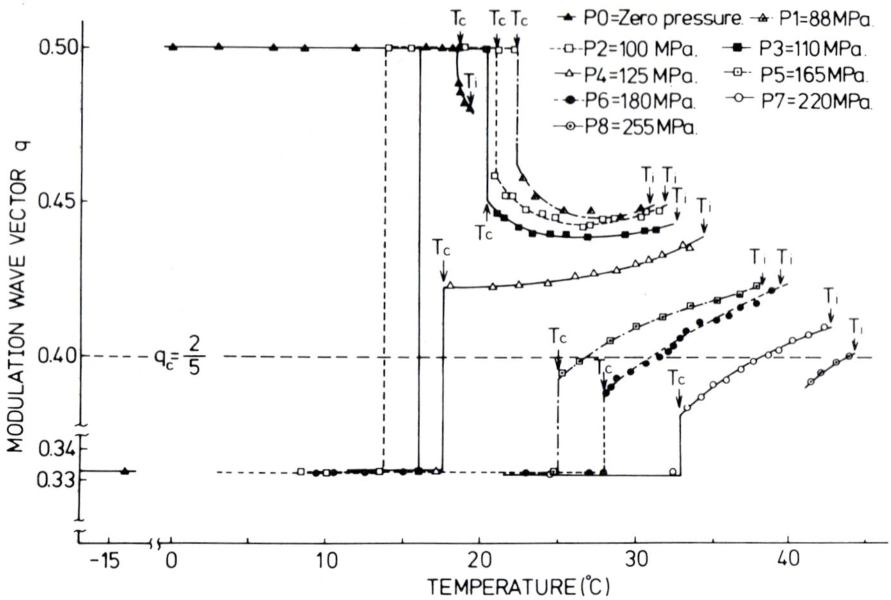

Fig. 20. Temperature dependence of modulation wave vector in $TMA \cdot MnCl_4$ under various applied hydrostatic pressures.

It is interesting to consider how the modulation wave vectors change within the incommensurate area on the $p-T$ phase diagram and to relate the changes with appearance of the ferroelectric commensurate ($\frac{2}{5}$) phase. A rough contour map can be sketched from the knowledge of the q_1 versus T relations on the respective zero pressure ordinate of each compound. An example of such a contour map has been published (Iizumi 1981). Quite recently the change of the modulation wave vectors as functions of pressure and temperature has been measured (Itoh 1984) for the Mn salt, which has the lowest pressure axis, and detailed knowledge of the contour is now available. Figure 20 shows the wave vector versus temperature relations for fixed pressures, and fig. 21 is the contour map obtained from them. In general the wave vector becomes smaller with increasing pressure or decreasing temperature. In the left and upper side of the incommensurate area the wave vector takes values larger than $\frac{2}{5}$ and in the right and lower side of the area it is smaller than $\frac{2}{5}$. The contour of $\frac{2}{5}$ coincides with the line where the dielectric constants show anomalies. The ferroelectric commensurate ($\frac{2}{5}$) phase, which is nonexistent in this salt, can be understood as the area which is generated when the $\zeta = \frac{2}{5}$ contour line acquires some finite width by some mechanism which stabilizes commensurate modulation. It also seems natural, from the general tendency of the contour map, that

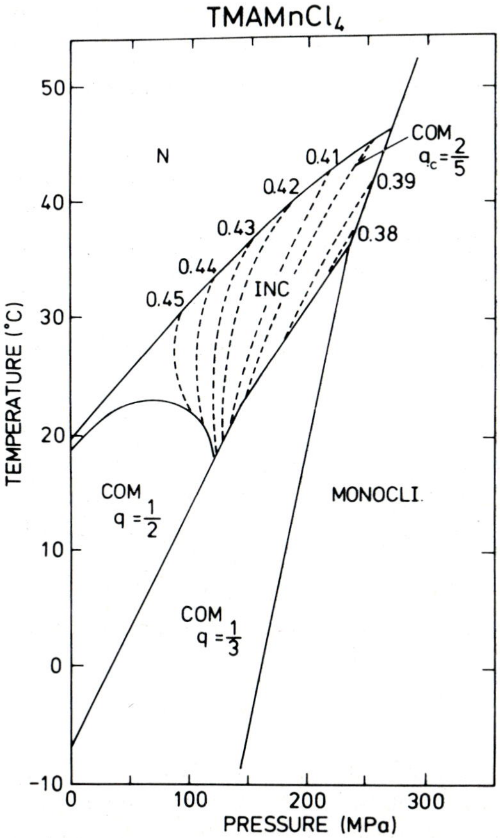

Fig. 21. Contour of equal wave vector within the incommensurate TMA·$MnCl_4$.

the lock-in transition to the commensurate ($\frac{1}{2}$) phase takes place in the left-most boundary of the incommensurate region and that to the commensurate ($\frac{1}{3}$) phase takes place near the low temperature boundary of the region.

5. *Fluctuations in the incommensurate phase*

5.1. *Static critical behavior*

We have seen elsewhere in this chapter and through the entire volume how powerful classical Landau order parameter expansions are in guiding our understanding of incommensurate phase transformations. However, as is now familiar, in the immediate vicinity of a second-order phase transformation a more careful treatment of fluctuation effects is necessary in order to explain the so-called critical behavior in a quantitative manner. Furthermore, the critical behavior of most interest is "universal" in character; i.e., it is independent of the microscopic details of the system under study and can thus be deduced theoretically by considering the simplest representative model system with the same requisite symmetry properties. We will see that for K_2SeO_4 and related materials in the vicinity of the normal–incommensurate transformation the expected critical behavior is that of a simple and much studied model, the *XY* model of ferromagnetism, for which relatively precise theoretical results are available. We first briefly review the theory (Bruce and Cowley 1978, Majkrzak et al. 1980) before turning to experiments bearing on critical behavior of these materials.

Incommensurate K_2SeO_4 can be viewed as resulting from a condensation of two degenerate phonon modes with wave vectors $\pm \boldsymbol{q}_\delta = \pm \frac{1}{3}(1-\delta)a^*$, and the transformation may be modeled by an $n = 2$-component Landau–Ginzburg–Wilson (LGW) Hamiltonian. As a simple model system consider a crystal with a single atom per unit cell. The elastic scattering at momentum transfer $\boldsymbol{Q}$ is then proportional to

$$S(\boldsymbol{Q}) = \sum_{\boldsymbol{x},\boldsymbol{x}'} \mathrm{e}^{-\mathrm{i}\boldsymbol{Q}\cdot(\boldsymbol{x}-\boldsymbol{x}')} \mathrm{d}_Q(\boldsymbol{x}) \mathrm{d}_Q^*(\boldsymbol{x}'), \tag{5.1}$$

where

$$\mathrm{d}_Q(\boldsymbol{x}) = \langle \mathrm{e}^{-\mathrm{i}\boldsymbol{Q}\cdot\boldsymbol{u}(\boldsymbol{x})} \rangle \tag{5.2}$$

and $\boldsymbol{u}(\boldsymbol{x})$ is the instantaneous displacement of an atom from its high temperature symmetry site, $\boldsymbol{x}$. If we choose to describe the displacements in terms of a local amplitude $A(\boldsymbol{x})$ and phase $\phi(\boldsymbol{x})$, the magnitude of $\boldsymbol{u}(\boldsymbol{x})$, $u(\boldsymbol{x}) = |\boldsymbol{u}(\boldsymbol{x})|$, is

$$u(\boldsymbol{x}) = A(\boldsymbol{x}) \cos[\boldsymbol{q}_1 \cdot \boldsymbol{x} + \phi(\boldsymbol{x})]; \tag{5.3}$$

then $\mathrm{d}_Q(\boldsymbol{x})$ can be rewritten as

$$\mathrm{d}_Q(\boldsymbol{x}) = 1 + \sum_{m=1}^{\infty} (-\mathrm{i})^m \langle \exp\{-\mathrm{i}m[\boldsymbol{q}_1 \cdot \boldsymbol{x} + \phi(\boldsymbol{x})]\} \times J(\boldsymbol{Q}\cdot\boldsymbol{A}(\boldsymbol{x})) + \mathrm{c.c.} \rangle \tag{5.4}$$

using a Bessel-function expansion; m labels the order of the Bragg satellite peaks, and the dominant contributions in the critical region where $\boldsymbol{Q}\cdot\boldsymbol{A}(\boldsymbol{x}) \ll 1$

are given by

$$J_m(\boldsymbol{Q}\cdot\boldsymbol{A}(\boldsymbol{x})) \approx \frac{1}{2^m m!}[\boldsymbol{Q}\cdot\boldsymbol{A}(\boldsymbol{x})]^m. \tag{5.5}$$

To go further it is convenient to introduce two new local coordinates $P_1(\boldsymbol{x})$ and $P_2(\boldsymbol{x})$, respectively local amplitude and phase fluctuations:

$$P_1(\boldsymbol{x}) = \int_{|\boldsymbol{k}|<\Lambda} [Q(\boldsymbol{q}_1+\boldsymbol{k})+Q(-\boldsymbol{q}_1+\boldsymbol{k})]\mathrm{e}^{-\mathrm{i}\boldsymbol{k}\cdot\boldsymbol{x}}\mathrm{d}\boldsymbol{k}, \tag{5.6a}$$

$$P_2(\boldsymbol{x}) = -\mathrm{i}\int_{|\boldsymbol{k}|<\Lambda} [Q(\boldsymbol{q}_1+\boldsymbol{k})-Q(-\boldsymbol{q}_1+\boldsymbol{k})]\mathrm{e}^{-\mathrm{i}\boldsymbol{k}\cdot\boldsymbol{x}}\mathrm{d}\boldsymbol{k}, \tag{5.6b}$$

defined in terms of the phonon modes $Q(\pm\boldsymbol{q}_1+\boldsymbol{k})$ in the neighborhood of $\pm\boldsymbol{q}_1$. (Λ is a small but otherwise arbitrary cut-off.) In this representation the displacement field is

$$u(\boldsymbol{x}) = P_1(\boldsymbol{x})\cos(\boldsymbol{q}_1\cdot\boldsymbol{x}) + P_2(\boldsymbol{x})\sin(\boldsymbol{q}_1\cdot\boldsymbol{x}). \tag{5.7}$$

Comparing eqs. (5.3) and (5.7) to rewrite eq. (5.4) in terms of (P_1, P_2), we find in particular that the primary $(m=1)$ and secondary $(m=2)$ satellites will have intensities proportional to

$$I_1(T) = \langle P_1(\boldsymbol{x})\rangle^2 + \langle P_2(\boldsymbol{x})\rangle^2 \tag{5.8}$$

and

$$I_2(T) = [\langle P_1^2(\boldsymbol{x})\rangle - \langle P_2^2(\boldsymbol{x})\rangle]^2 - \langle 2P_1(\boldsymbol{x})P_2(\boldsymbol{x})\rangle^2. \tag{5.9}$$

The significance of the (P_1, P_2) representation is that in terms of these coordinates the LGW Hamiltonian is precisely that of the ferromagnetic *XY* model, with $(P_1, P_2)\to(S_x, S_y)$. Thus one immediately concludes that the intensity of the primary satellite follows that of the square of the order parameter, and that asymptotically

$$I_1(T) \approx A_1(T_\mathrm{i}-T)^{2\beta} \tag{5.10}$$

where, for the *XY* model, making use of series expansion results and a scaling relation, β is given by

$$2\beta = (2-\alpha-\gamma) = 0.70\pm0.04. \tag{5.11}$$

The prediction for the behavior of the secondary satellite (eq. 5.9) is more subtle and interesting. A naive argument which disregards the effect of fluctuations would simply replace $\langle P_i P_j \rangle$ by $\langle P_i \rangle \langle P_j \rangle$ and lead to the prediction

$$I_2(T) \approx A_2(T_{\rm i} - T)^{4\beta}. \tag{5.12}$$

However, the operators $(P_1^2 - P_2^2, 2P_1P_2)$ represent a uniaxial perturbation of the XY model and a scaling argument establishes the correct result

$$I_2(T) \approx A_2(T_{\rm i} - T)^{2\tilde{\beta}}, \tag{5.13}$$

where

$$2\tilde{\beta} = 2(2 - \alpha - \phi) = 1.69 \pm 0.09, \tag{5.14}$$

and ϕ is the cross-over exponent associated with the uniaxial perturbation. The intensities of the higher order satellites can be related to other higher order perturbations whose critical exponents can, in principle, be calculated in a similar manner.

The most extensive investigation of satellite intensities in the critical region are the neutron scattering measurements of K_2SeO_4 by Majkrzak et al. (1980). In such measurements special care is necessary to insure that the much weaker secondary satellite intensities are not contaminated by two separate scattering events both involving primary satellites. (Such multiple scattering processes, which occur whenever two or more reciprocal-lattice points lie simultaneously on the Ewald sphere, are remarkably common for a material with large unit cell and a short wavelength scattering probe.) The principal remaining uncertainty was in the value of $T_{\rm i}(\pm 0.3\,{\rm K})$ due to the presence of unresolved inelastic critical scattering which masks the sharp onset of the Bragg satellites. The results of the study were

$$2\beta = 0.75(\pm .05) \quad \text{and} \quad 2\tilde{\beta} = 1.57(\pm .07).$$

The results for β are in satisfactory agreement with theory as well as with earlier less precise neutron measurements ($2\beta = 0.8(\pm 0.1)$) and an indirect measurement of anisotropic thermal expansion ($2\beta \simeq 0.7$) by Shiozaki et al. (1977). The measured value for $2\tilde{\beta}$ is also consistent with theory, and is inconsistent with the naive prediction $2\tilde{\beta} = 4\beta$, as long as one adopts the theoretical rather than the observed value of β.

Several observations have been made of the critical properties of other materials in the K_2SeO_4 family. Of these we mention first the result $2\beta = 0.69(\pm 0.01)$ obtained in an X-ray study of Rb_2ZnCl_4 by Mashiyama (1981).

He was able to improve the determination of T_i by simultaneous in-situ measurement of the dielectric constant, ε_a. The agreement with theory, and with K_2SeO_4, is striking, particularly in view of the pronounced difference in the dynamics above T_c. Regis et al. (1982) studied the optical birefringence, $\Delta n = n_a - n_c$ of TMA·$ZnCl_4$ and found $\Delta n(T) \propto (T_i - T)^{0.87}$. They note that the exponent agrees well with the theoretical value of β, but inasmuch as the coupling to Δn appears to be through $(P_1^2 + P_2^2)$, rather than $(P_1^2 - P_2^2; 2P_1P_2)$, the significance of this finding is unclear. Although it is plausible to conclude that these observations confirm non-classical XY behavior for these systems, Y. Levanyuk (private communication) has pointed out an alternative interpretation. In Landau theory the classical $2\beta = 1$ is not strictly universal but may have effective β-values that vary in the range $\frac{1}{2} \leqslant 2\beta \leqslant 1$ in the vicinity of a tricritical point. It would be coincidental but not impossible, for all K_2SeO_4-like systems (or at least all thus far studied) to be equally "near tricriticality" with an effective classical β value closely similar to the expected non-classical XY-value.

5.2. *Dynamical fluctuations in the incommensurate phase in* K_2SeO_4

Much of what is novel and unique about incommensurate structures centers around the nature of the time dependent fluctuations, and it is therefore hardly surprising that this subject has attracted considerable attention in K_2SeO_4 and related materials, since as we have seen, they represent examples of canonically simple incommensurate systems. Before we discuss the results of these observations, let us review in simple terms how the incommensurate distortions are expected to modify the soft phonon instability below T_i.

Above T_i, if damping effects are neglected, (quasi-) harmonic phonons with amplitudes $Q(\pm \boldsymbol{q}_1 + \boldsymbol{k})$ are adequate descriptions of the fluctuations around $\pm \boldsymbol{q}_1$. Below T_i the condensation of displacements with wavevector $\pm \boldsymbol{q}_1$ introduces new terms in the crystal potential which vary spatially as multiples of $\pm \boldsymbol{q}_1$. In particular fourth-order terms in the Landau expansion produce a new pseudo-harmonic term below T which couples $Q(\boldsymbol{q}_1 + \boldsymbol{k})$ with $Q(-\boldsymbol{q}_1 + \boldsymbol{k})$. If the static displacements are sinuosoidal (a good approximation close to T_i),

$$\langle \boldsymbol{u}(\boldsymbol{x}) \rangle = \boldsymbol{A} \cos(\boldsymbol{q}_1 \cdot \boldsymbol{x}), \tag{5.15}$$

then the new excitations are the linear combinations

$$P_1 = \frac{1}{\sqrt{2}} \left[Q(\boldsymbol{q}_1 + \boldsymbol{k}) + Q(-\boldsymbol{q}_1 + \boldsymbol{k}) \right], \tag{5.16a}$$

$$P_2 = \frac{\mathrm{i}}{\sqrt{2}} \left[Q(-\boldsymbol{q}_1 + \boldsymbol{k}) - Q(\boldsymbol{q}_1 + \boldsymbol{k}) \right], \tag{5.16b}$$

and associated with the two new modes are the displacements

$$\begin{aligned} \boldsymbol{u}_1(\boldsymbol{x}, t) &\approx [\mathrm{e}^{\mathrm{i}(-\boldsymbol{q}_1+\boldsymbol{k})\cdot \boldsymbol{x}} + \mathrm{e}^{\mathrm{i}(\boldsymbol{q}_1+\boldsymbol{k})\cdot \boldsymbol{x}}]\mathrm{e}^{\mathrm{i}\omega_1 t} \\ &\approx \cos(\boldsymbol{q}_1\cdot \boldsymbol{x})\mathrm{e}^{\mathrm{i}(\boldsymbol{k}\cdot \boldsymbol{x}+\omega_1 t)} \end{aligned}$$

and

$$\begin{aligned} \boldsymbol{u}_2(\boldsymbol{x}, t) &\approx [\mathrm{e}^{\mathrm{i}(-\boldsymbol{q}_1+\boldsymbol{k})\cdot \boldsymbol{x}} - \mathrm{e}^{\mathrm{i}(\boldsymbol{q}_1+\boldsymbol{k})\cdot \boldsymbol{x}}]\mathrm{e}^{\mathrm{i}\omega_2 t} \\ &\approx \sin(\boldsymbol{q}_1\cdot \boldsymbol{x})\mathrm{e}^{\mathrm{i}(\boldsymbol{k}\cdot \boldsymbol{x}+\omega_2 t)}. \end{aligned}$$

$\boldsymbol{u}_1(\boldsymbol{x}, t)$ clearly represents a modulation of the *amplitude* of the static distortion, eq. (6.15), and is found to have a dispersion relation with a temperature dependent gap, $\Delta(T)$:

$$\omega_1^2(T) = \Delta(T) + v^2k^2. \tag{5.17}$$

On the other hand, since $\cos(\boldsymbol{q}_1\cdot \boldsymbol{x} + \phi(t)) \approx \cos(\boldsymbol{q}_1\cdot \boldsymbol{x}) + \phi(t)\sin(q_1\cdot x)$, $u_2(\boldsymbol{x}, t)$ represents, *at least approximately*, a modulation of the *phase* of the static distortion, and what is most striking, the mode frequency has an acoustic-like dispersion relation,

$$\omega_2^2(T) = v^2k^2. \tag{5.18}$$

The gapless nature of the phase fluctuation mode is a direct result of the invariance of the free energy to a global shift in the phase of the order parameter in the incommensurate phase. If, as is the case for K_2SeO_4, a commensurate phase occurs at a lower temperature, T_c, a preferred phase is chosen and the phase branch develops a $k = 0$ gap, while at the same time it retains its phase-fluctuation character. We shall refer to such modes as pseudo-phase modes. (We have chosen not to introduce you to the chic and glamorous "phason". Not having to associate with her logical but ungainly sibling, the "amplitudon", should be compensation enough.) As T_c is approached from above, it may happen that the static displacements can no longer be well approximated by a single plane wave (eq. 5.15), and the above simple analysis of the dynamics may also be appreciably modified. We return to this topic in section 6.

There are no known symmetry selection rules that prevent observation of either amplitude or phase modes by light or neutron scattering in K_2SeO_4. In light scattering experiments the two modes may, however, be distinguished by polarization selection rules. The amplitude mode was first reported by Wada et al. (1977a) and this excitation has since been studied extensively by this technique. Inelastic neutron scattering measurments (Axe et al. 1980) extended

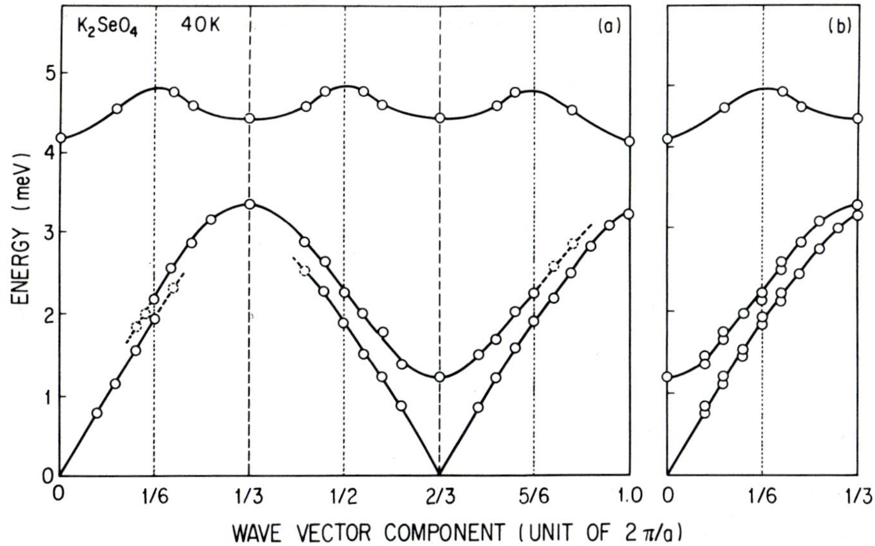

Fig. 22. Dispersion of (in order of increasing energy) transverse-acoustic, phase and amplitude fluctuation modes in K_2SeO_4 at 40 K: (a) extended zone; (b) folded zone.

the results to finite k (fig. 22). Agreement between neutron and all Raman frequencies is excellent in overlapping regions. Near T_i ($15\,K > T_i - T > 0\,K$) Fleury et al. (1979) report a steepening of the ω versus T curve over the smooth extrapolation shown in fig. 23. They find that a simple damped oscillator response with temperature independent damping adequately represents the lineshape. The mode is overdamped but still visible for $(T_i - T) < 6\,K$.

The above mentioned light and neutron scattering studies also observed the pseudo-phase modes ($T < T_c$) with little difficulty. (See fig. 23). However, in these cases the mode proved impossible to follow into the incommensurate phase. The phase mode is much weaker than the amplitude mode in Raman scattering (Wada et al. 1977a, b, Fleury et al. 1979) and is observed to overdamp $\sim 3\,K$ below T_c.

One complication which emerged from the neutron scattering studies was the necessity of including additional interaction between phase modes and acoustic phonons. In the absence of such interactions simple intensity arguments can be used to distinguish between acoustic phonon and phase mode scattering with neutrons, even though in principle at any particular wave vector both types of excitations appear simultaneously. The presence of such interactions nullifies these simple intensity arguments and leads to interesting observable interference effects. The situation in K_2SeO_4 is further complicated by

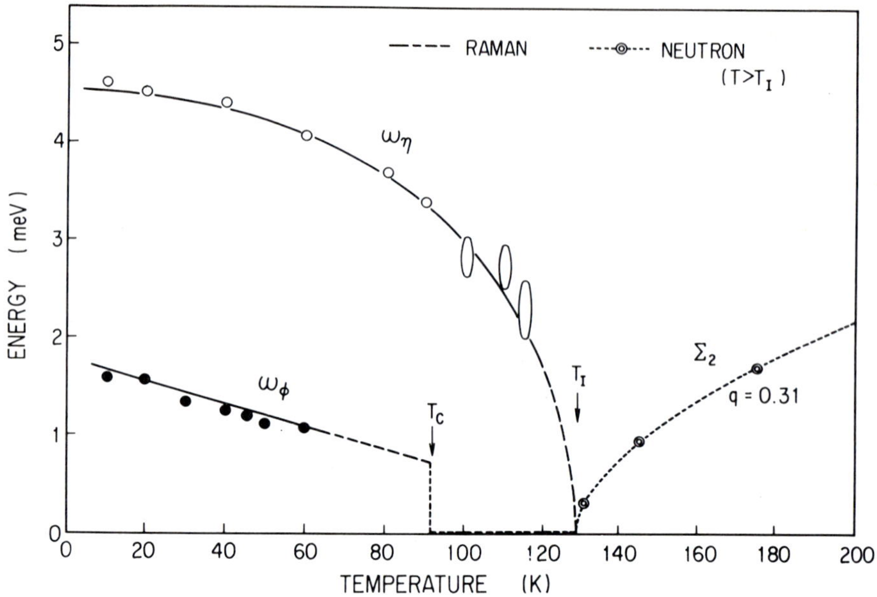

Fig. 23. Temperature dependence of $k = 0$ phase and amplitude mode frequencies. Solid lines represent Raman scattering results, data points are neutron results. The dashed portions of the ω_ϕ curve are schematic.

the fact that from high temperature measurements the phase mode velocity, v, (see eq. 6.18) along the *a* axis is expected to be very close to that of one of the transverse acoustic phonon branches.

None of these complications are sufficient to explain the failure to observe incommensurate phase modes. As is implied in our previous discussion of static critical phenomena, the free energy for K_2SeO_4 is that of an *XY* ferromagnet, which in its ordered state has gapless spinwaves that remain well defined (i.e. underdamped) as $k \to 0$. These are the so-called Goldstone modes associated with a broken continuous symmetry. The widespread belief was that since choosing a global phase for an incommensurate system also involved breaking a continuous symmetry the resulting phase fluctuations were also Goldstone modes and should be well behaved hydrodynamic modes of the system as $k \to 0$. Golovko and Levanyuk (1981) developed a theory which reached a contrary point of view, namely that dynamically the phase modes were no different than optic phonons which unlike (Goldstone) acoustic phonons could be expected to become overdamped as $k \to 0$. The two viewpoints were further clarified by Zehyer and Finger (1982), who showed that the global phase variable is a broken symmetry of the free energy but not of the Hamiltonian. Phase modes are therefore not true Gladstone modes and are always expected to be diffusive for sufficiently long wavelengths.

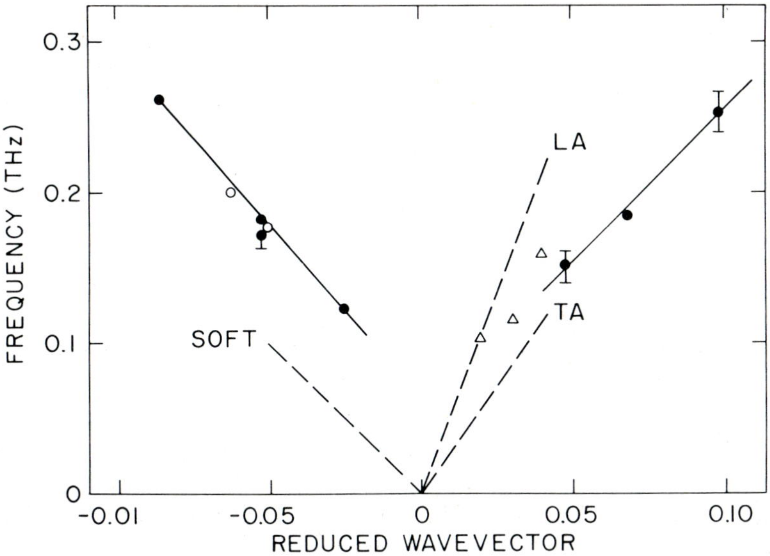

Fig. 24. Comparison of dispersion of the phase mode in incommensurate K_2SeO_4 from Raman scattering (open triangles) and neutron scattering (open and closed circles). The dashed lines indicate the dispersion of the acoustic phonons as well as that of the soft mode just above T_i. The wave vector is along the incommensurate (c) axis.

More recent Raman (Inoue and Ishibashi 1983) and neutron scattering (Quilichini and Currat 1983) experiments have provided direct evidence for overdamping of the phase mode in incommensurate K_2SeO_4 at small, but finite, wavevector, k. Both sets of observations are well described by a damped oscillator response,

$$S(k, \omega) = \frac{\Gamma}{\left[\omega^2 - \omega_2(k)^2\right]^2 + \omega^2\Gamma^2},$$

with k-independent damping, Γ. A summary of the quasiharmonic mode frequencies, $\omega_2(k)$, extracted from such fitting procedures for both the Raman and the neutron experiments, is shown in fig. 24. Note that contrary to naive expectation Raman scattering samples phase modes at a finite and temperature dependent k, $k(T) = 3(q_c - q(T)) = 3\delta(T)$, as was first pointed out by Dvořák and Petzelt (1978). Direct coupling of light to $k \approx 0$ phase fluctuations is also expected to give rise to a narrow central peak in the light scattering spectra, but it is uncertain whether or not these additional features have been observed. (Poulet and Pick 1981, Golovko and Levanyuk 1981).

6. Lock-in phase transformations

The single-plane-wave description of the incommensurate phase we have given in the previous section suffers from serious inadequacies which become increasingly important as the amplitudes of the incommensurate spontaneous displacements grow. Moncton et al. (1975) showed that anharmonic interactions conspire to produce secondary distortions that are harmonics of the primary distortion (i.e. the wave vectors of these distortions are given by $q_p = pq_1 - G_p$ where G_p is a reciprocal lattice vector). In the plane wave limit the higher harmonics ($p > 1$) are ignored. There is another limit, which although considerably more subtle, is relatively well understood. This is the so-called "soliton" limit, in which modulations of the amplitude of the order parameter are neglected. This limit is intended to apply near lock-in whereas the plane wave limit is intended to apply near T_i. The soliton limit was first described by McMillan (1976), clarified in some aspects by Bak and Emery (1976), and was discussed in some detail by Bruce et al. (1978) and Bruce and Cowley (1978). The cross-over between the two limits appears to be difficult to describe theoretically.

6.1. The soliton limit

The basic assumption in the soliton limit is that the displacements can be represented as $u(x) \sim A\cos(\boldsymbol{q}_c \cdot \boldsymbol{x} + \phi(x))$. The solutions which minimize the free energy near lock-in show $\phi(x)$ independent of x except at regularly spaced narrow intervals in which $\phi(x)$ changes by $2\pi/p$ (p an integer, such that $pq_c = G_p$). Thus the expectation is that near T_c the incipient commensurate phase is anticipated by the formation of commensurate domains. The order parameter for the transformation can be taken to be the incommensurability parameter $\delta = |\boldsymbol{q}_1 - \boldsymbol{q}_c|$, which is a measure of the number density of domain walls. In a continuum Landau theory $\delta(T) \to 0$ continuously as $T \to T_c$, and the lock-in transformation is of second order (McMillan 1976). However, coupling to other degrees of freedom (e.g. acoustic modes) renders the transformation first order (Bruce and Cowley 1978), as is in fact observed for all the lock-in transformations observed in β-K_2SeO_4 systems.

In general pth order commensurability is driven by pth order anharmonic terms in a Landau expansion. However in K_2SeO_4 this leading term is missing by symmetry (Iizumi et al. 1977). As a result the $q_c = \frac{1}{3}c^*$ lock-in derives from fourth-order coupling through long wavelength polarization and/or sixth-order coupling through adjacent Λ_3-modes. The former is, of course, also responsible for the (improper) ferroelectricity of the commensurate phase.

Even more complicated behaviors are predicted when the width of the domains become comparable with interatomic distances. Then the continuum approximation must be abandoned and domain walls can prefer to localize either at or between lattice sites, and phases of arbitrarily high order commensurability can be stablized. If an infinity of these commensurate phases completely exhaust the allowed wave vectors of the states, the sequence is known as a "devil's staircase." These, and less "devilish" possibilities, are reviewed by Bak (1982) in a general context.

As is shown in table 1 (section 1), a considerable variety occurs in the lock-in wave vectors exhibited by crystals with the β-K_2SO_4 structure. Often before the ultimate low temperature commensurate phase is realized, transient phases characterized by higher order commensurability (say $p < 3$) are observed, but persist only within a limited temperature region.

The commensurate ($\frac{2}{5}$) phase in TMA compounds mentioned in section 4 is a typical example of the transient lock-in phases. The wave vector in Rb_2ZnBr_4, which does not change with temperature at all for more than 140 K, was pointed out (Hogervorst and de Wolff 1982) and experimentally confirmed (Iizumi and Gesi 1983b) to be $\frac{5}{17}$. The latter authors searched for other transient lock-in phases and observed successive jumps in the wave vectors in the temperature region where the incommensurate–commensurate phase transition takes place. However they found that each "stair" is not necessarily specified by an exact fraction and moreover modulations characterized by those wave vectors seem to coexist in a certain temperature range (fig. 25).

Marion et al. (1981) pointed out that a part of the q_1 versus T curve of d-TMA·$ZnCl_4$ is commensurate and given by $\frac{3}{7}$, to which some objection was later raised (Iizumi and Gesi 1983b). Similar transient lock-in to the $\zeta = \frac{3}{7}$ phase was also noticed in TMA·$FeCl_4$ (Mashiyama and Tanisaki 1982).

A structurally related compound, $(NH_4)_2ZnCl_4$, presents an interesting further variation on the theme (Matsunaga and Nakamura 1981, Matsunaga 1982). The high temperature normal phase in this compound transforms directly to a commensurate ($\frac{1}{4}$) phase at 406 K. The transformation is continuous (2nd order) or nearly so. This phase eventually transforms at 270 K into the ferroelectric phase characterized by $\zeta = \frac{1}{3}$. The latter phase is similar to the commensurate phase seen in K_2SeO_4 or K_2ZnCl_4.

Yamada and Hamaya (1983) have attempted to provide a unified view of the various transient commensurate phase transformations which occur in the A_2BX_4 halides. They consider an Ising model with long range (i.e. greater than next nearest neighbor) interactions. The precise connection between the physical order parameter and the Ising variable is not elaborated in detail. The model couples spins ferromagnetically into sheets, with antiferromagnetic coupling between first and second neighbor sheets ($J_1 < 0$, $J_2 < 0$) and arbitrary third sheet interaction ($J_3 \lesseqgtr 0$). [By a trivial spin relabelling, the model with $J_3 = 0$ is equivalent to the much studied ANNNI model (see, for example, Bak 1982.)

Yamada and Hamaya studied the relative stability of various commensurate states as a function of temperature in the mean field approximation. Incommensurate states were not considered. The resulting phase diagrams are quite

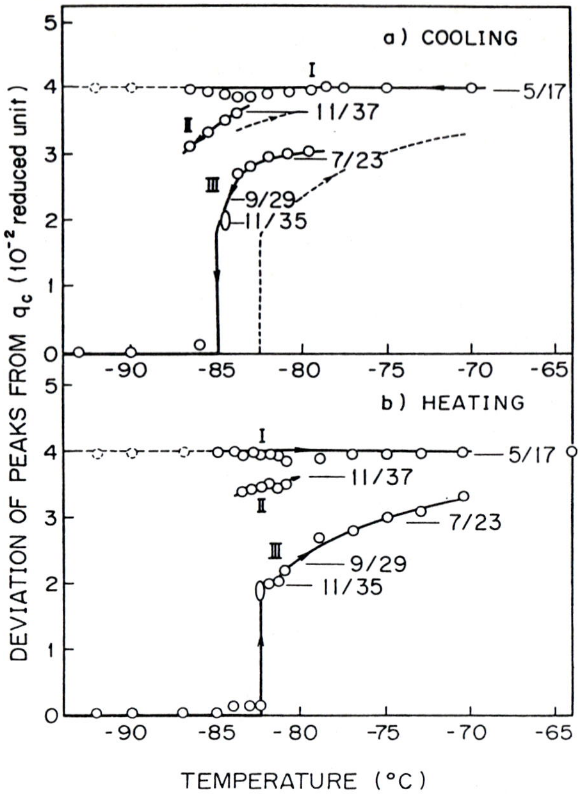

Fig. 25. Temperature dependence of satellite positions in Rb_2ZnBr_4. Horizontal bars indicate possible stairs on a devil's staircase.

complex, with many states of high order commensurability stable at high temperatures. But at $T = 0$ only states with reduced wave vector $\zeta = \frac{1}{2}$, $\frac{2}{3}$ (K_2SeO_4-like) or 1 (AFB-like) remain stable. By suitable choice of the coupling constants, J_n, phase diagrams with characteristics similar to those of the A_2BX_4 halides can be constructed. Different ranges of coupling constants are needed for the alkali-cation (A = K, Rb) and TMA-cation series. It appears that the detailed conclusions are compromised by the authors' failure to note that the TMA compounds (except for $TMA \cdot CuCl_4$) are characterized by wavevectors with $\zeta < \frac{1}{2}$ (Λ_2-mode) rather than $\zeta > \frac{1}{2}$ (Λ_3-mode) which is correct for the alkalis.

Nevertheless, the investigation of discrete site models unquestionably represents a productive step. For such models generated in a natural way temperature dependent lock-in potentials of arbitrarily high order which are necessary to explain the complicated sequences of transient commensurate phases observed in many systems. Villain and Gordon (1980), Axel and Aubrey (1981) and Mashiyama (1982) have studied long range Ising models in a more general context.

6.2. *Lock-in and macroscopic properties*

One of the secondary lattice distortions induced in the incommensurate phase by the interaction with the primary lattice distortion (order parameter) transforms into a macroscopic quantity. In the cases of K_2SeO_4 and similar crystals with $\zeta_c = \frac{1}{3}$, it is a spontaneous polarization along the *a* axis. In AFB with $\zeta_c = \frac{1}{2}$ it is also a spontaneous polarization but its direction is along the *b* axis. On the other hand, in case of TMA$\cdot CuCl_4$ the elastic strain S_{xz} which is compatible with the Λ_3 acoustic phonon is induced spontaneously below T_c and the crystal becomes ferroelastic (Sawada et al. 1980). In the incommensurate phase, both long wavelength polarization and strain waves can coexist; this is not so in the commensurate phase; the crystal must choose one or the other. (Iizumi et al. 1977).

Consider, for example, the ferroelectric case. Near T_c in the soliton limit, the domains carry, in their bulk, a uniform polarization characteristic of the commensurate phase. However the polarization is reversed in alternate domains, and since domains of both types are equal in extent, there is no net polarization. When an external field is applied, two responses are possible. The amplitude of the polarization can grow in the favored set of domains in comparison to the unfavored domains. Alternatively, the domain walls could move, with the volume of one domain set expanding at the expense of the other. Detailed consideration of the soliton limit (Dvořák and Petzelt 1978, de Pater et al. 1979, Levstik et al. 1982) show that the latter is the dominant effect. Observed dielectric anomalies can be explained nearly quantitatively in this manner, as can the modification of harmonic satellite intensities with applied field (de Pater et al. 1979).

6.3. *Dynamics near the soliton limit*

Just as the single-plane-wave limit is expected to provide an inadequate description of the static structure near lock-in, so the simple picture of dynamics based on perturbation of the plane wave state, as described in section 5.2, is also expected to fail. Qualitatively, one might expect the following. The fluctuations within the interior of the domains of commensurate structure should closely resemble the excitations in the fully commensurate phase below T_c. What is left to consider is the behavior of the material near or exactly within the domain walls, that is to say the dynamics of the walls themselves, which, because they interact weakly, dominate the fluctuations as $\omega \to 0$.

This qualitative picture is confirmed by detailed calculations. The situation has been reviewed by Bruce and Cowley (1978). In the continuum limit the concern is with small fluctuations about multisoliton solutions of the sine-Gordon equation. Amplitude fluctuations pass through T_c smoothly, as is observed in K_2SeO_4. (See fig. 23.) On the other hand a gap develops in the phase fluctuation branch at wave vector $q_{gap} = \frac{1}{2} q_p$. The upper of these frequencies, $\omega_+(q_{gap})$, passes smoothly to that of the (pinned) phase fluctuation mode of the commensurate phase. This isolates a lower phase branch with frequencies $0 \leqslant \omega \leqslant \omega_-(q_{gap})$, and $\omega_-(q_{gap}) \to 0$ as $T \to T_c$ from above. Simultaneously $q_{gap} \to 0$ as $T \to T_c$, so both the mean frequency and the number of modes in this lower (quasi Goldstone) branch vanish at T_c. (It is a remarkable property of the sine-Gordon equation that only a single gap appears in the phase fluctuation branch. More general potentials with the same periodicity will introduce in addition higher order gaps spaced by multiples of q_p.) An additional complication concerns the possible role of phase pinning of the incommensurate structure, either by impurities or intrinsically in the narrow domain wall limit. Very little has been done in this area.

It should be clear that experimental studies of the behavior of low lying phase fluctuations near lock-in present a challenging opportunity to open a window on some subtle and extraordinary physics. With this in mind, we again consider fig. 24, showing the measured phase fluctuation mode dispersion in K_2SeO_4 as measured by both Raman and neutron scattering. Data from both sources are in reasonable accord and suggest a weakly temperature dependent dispersion relation of the form

$$\omega_2^2(k) = \Delta' + v^2 k^2.$$

The observed value of v is in rough agreement with the simple model, which predicts a value identical to that describing the soft mode above T_i (Axe 1976). The finite gap ($\Delta' \neq 0$) even at relatively high temperature, if confirmed, is quite interesting. Such effects may be the result of stronger than expected harmonics in the static displacements (soliton or domain wall formation) or perhaps, alternatively, domain pinning effects.

Acknowledgment

This work was supported in part by the Division of Materials Sciences, U.S. Department of Energy, under contract DE-AC02-76CH00016.

References

Aiki, K., and K. Hukudu, 1969, J. Phys. Soc. Jpn **26**, 1066.
Aiki, K., K. Hukudu and O. Matsumura, 1969, J. Phys. Soc. Jpn **26**, 1064.
Aiki, K., K. Hukudu, H. Koga, and T. Kobayashi, 1970, J. Phys. Soc. Jpn **28**, 389.
Axe, J.D., 1976, Proc. of Gatlinburg Neutron Scattering Conf., CONF-760601-P1 (U.S. Dept. of Comm.) p. 353.
Axe, J.D., M. Iizumi and G. Shirane, 1980, Phys. Rev. **22**, 3408.
Axel, F., and S. Aubrey, 1981, J. Phys. **C14**, 5433.
Bak, P., 1982, Rep. Prog. Phys. **45**, 587.
Bak, P., and V.J. Emery, 1976, Phys. Rev. Lett. **36**, 978.
Bruce, A.D., and R.A. Cowley, 1978, J. Phys. **C11**, 3577.
Bruce, A.D., R.A. Cowley and A.F. Murray, 1978, J. Phys. **C11**, 3591.
Bussmann-Holder, A., H. Büttner and H. Bilz, 1981, Ferroelectrics **36**, 273.
de Pater, C.J., and C. van Dijk, 1978, Phys. Rev. **B16**, 1281.
de Pater, C.J., J.D. Axe and R. Currat, 1979, Phys. Rev. **B19**, 4684.
de Wolff, P.M., 1974, Acta. Cryst. **A30**, 777.
Dvořák, V., and J. Petzelt, 1978, J. Phys. **C11**, 4827.
Fleury, P.A., S. Chiang and K.B. Lyons, 1979, Solid State Commun. **31**, 279.
Gesi, K., 1982a, J. Phys. Soc. Jpn **51**, 1043.
Gesi, K., 1982b, J. Phys. Soc. Jpn **51**, 2051.
Gesi, K., 1982c, J. Phys. Soc. Jpn **51**, 2532.
Gesi, K., and M. Iizumi, 1978, J. Phys. Soc. Jpn **45**, 1777.
Gesi, K., and M. Iizumi, 1979, J. Phys. Soc. Jpn **46**, 697.
Gesi, K., and M. Iizumi, 1980a, unpublished, cited by Gesi (1982b).
Gesi, K., and M. Iizumi, 1980b, J. Phys. Soc. Jpn **48**, 1775.
Gesi, K., and M. Iizumi, 1984, J. Phys. Soc. Jpn **53**, 4271.
Gesi, K., and K. Ozawa, 1983, J. Phys. Soc. Jpn **52**, 2440.
Golovko, V.A., and A.P. Levanyuk, 1981, Sov. Phys. JETP **54**, 1217.
Haque, M.S., and J.R. Hardy, 1980, Phys. Rev. **B21**, 245.
Harada, J., J.D. Axe, and G. Shirane, 1970, Acta. Cryst. **A26**, 608.
Hasebe, K., H. Mashiyama and S. Tanisaki, 1980, J. Phys. Soc. Jpn **49**, 1633.
Hasebe, K., H. Mashiyama and S. Tanisaki, 1982, J. Phys. Soc. Jpn, **51**, 2049.
Hogervorst, A.C.R., and P.M. de Wolff, 1982, Solid State Commun. **43**, 179.
Iizumi, M., 1981, Jpn. Atomic Energy Res. Inst. **M**, 9843.
Iizumi, M., and K. Gesi, 1977, Solid State Commun. **22**, 37.
Iizumi, M., and K. Gesi, 1983a, Physica **B120**, 291.
Iizumi, M., and K. Gesi, 1983b, J. Phys. Soc. Jpn **52**, 2526.
Iizumi, M., J.D. Axe, G. Shirane and K. Shimaoka, 1977, Phys. Rev. **B15**, 4392.
Inoue, K., and Y. Ishibashi, 1983, J. Phys. Soc. Jpn **52**, 556.
Itoh, H., Y. Endoh, M. Iizumi and K. Gesi, 1984, unpublished.
Kalman, A., J.S. Stephens, and D.W.J. Cruickshank, 1970, Acta Cryst. **B26**, 1451.
Kittel, C., 1966, Introduction to Solid State Physics, 3rd Ed. (Wiley, New York) pp. 486–487.
Kovalev, O.V., 1965, Irreducible Representations of the Space Groups (Gordon Breach, New York).
Kudo, S., 1982, Jpn J. Appl. Phys. **21**, 255.
Levstik, A., P. Prelovsek, C. Filipic and B. Žekš, 1982, Phys. Rev. **25**, 3416.
Majkrzak, C.F., J.D. Axe, and A.D. Bruce, 1980, Phys. Rev. **B22**, 5278.
Marion, G., R. Almairac, J. Lefebvre and M. Ribet, 1981, J. Phys. **C14**, 3177.
Mashiyama, H., 1980, J. Phys. Soc. Jpn **49**, 2270.

Mashiyama, H., 1981, J. Phys. Soc. Jpn **50**, 2655.
Mashiyama, H., 1982, J. Phys. **C16**, 187.
Mashiyama, H., and S. Tanisaki, 1980, Phys. Lett. **76A**, 347.
Mashiyama, H., and S. Tanisaki, 1981, J. Phys. Soc. Jpn **50**, 1413.
Mashiyama, H., and S. Tanisaki, 1982, J. Phys. **C15**, L 455.
Mashiyama, H., K. Hasebe and S. Tanisaki, 1980, J. Phys. Soc. **49**, Suppl. B, 92.
Mashiyama, H., S. Tanisaki and K. Gesi, 1981a, J. Phys. Soc. Jpn **50**, 1415.
Mashiyama, H., S. Tanisaki and K. Hamano, 1981b, J. Phys. Soc. Jpn **50**, 2139.
Matsunaga, H., 1982, J. Phys. Soc. Jpn **51**, 873.
Matsunaga, H., and E. Nakamura, 1981, J. Phys. Soc. Jpn **50**, 2789.
McMillan, W.L., 1976, Phys. Rev. **B14**, 1496.
Moncton, D.E., J.D. Axe and F.J. DiSalvo, (1975), Phys. Rev. Lett. **34**, 734.
Poulet, H., and R. Pick, (1981), J. Phys. **C14**, 2675.
Press, W., C.F. Majkrzak, J.D. Axe, J.R. Hardy, N.E. Massa and F.G. Ullman, 1980, Phys. Rev. **B22**, 332.
Quilichini, M., and R. Currat, (1983), Solid State Commun., **48**, 1011.
Regis, M., J.L. Ribet and J.P. Jamet, 1982, J. Phys. Lett. **43**, L333.
Sawada, A., J. Sugiyama, M. Wada and Y. Ishibashi, 1980, J. Phys. Soc. Jpn **48**, 1773.
Sawada, S., Y. Shiroishi, A. Yamamato, M. Takashige and M. Matsuo, 1978a, J. Phys. Soc. Jpn **44**, 687.
Sawada, S., Y. Shiroishi, A. Yamamato, M. Takashige and M. Matsuo, 1978b, Phys. Lett. **67A**, 56.
Shimizu, H., A. Oguri, N. Abe, N. Yasuda, S. Fujimoto, S. Sawada, Y. Shiroichi and M. Takashige, 1979a, Solid State Commun. **29**, 125.
Shimizu, H., N. Abe, N. Yasuda, S. Fujimoto, S. Sawada and Y. Shiroichi, 1979b, Jpn J. Appl. Phys. **18**, 857.
Shimizu, H., N. Abe, N. Kokubo, N. Yasuda, S. Fujimoto, T. Yamaguchi and S. Sawada, 1980a, Solid State Commun. **34**, 363.
Shimizu, H., N. Kokubo, N. Yasuda and S. Fujimoto, 1980b, J. Phys. Soc. Jpn **49**, 223.
Shiozaki, S., A. Sawada, Y. Ishibashi and Y. Takagi, 1977a, J. Phys. Soc. Jpn **42**, 353.
Shiozaki, S., A. Sawada, Y. Ishibashi and Y. Takagi, 1977b, J. Phys. Soc. Jpn **43**, 1315.
Sugiyama, J., M. Wada, A. Sawada and Y. Ishibashi, 1980, J. Phys. Soc. Jpn **49**, 1405.
Tanisaki, S., and H. Mashiyama, 1980, J. Phys. Soc. Jpn **48**, 339.
Terauchi, H., H. Takenaka and K. Shimaoka, 1975, J. Phys. Soc. Jpn **39**, 435.
Ueda, T., S. Ida and H. Terauchi, 1982, J. Phys. Soc. Jpn **51**, 3953.
van Aalst, W., J. den Hollander, W.J.A.M. Peterse and P.M. de Wolff, 1976, Acta Cryst. **B32**, 47.
van den Berg, A.J., F. Tuinstra and J. Warczewski, 1973, Acta Cryst. **B29**, 586.
Villain, J., and M.B. Gordon, 1980, J. Phys. **C13**, 3117.
von Gattow, G., 1962, Acta Cryst. **15**, 419.
Wada, M., A. Sawada, Y. Ishibashi and Y. Takagi, 1977a, J. Phys. Soc. Jpn **42**, 1229.
Wada, M., H. Uwe, A. Sawada, Y. Ishibashi, Y. Takagi and T. Sakudo, 1977b, J. Phys. Soc. Jpn **43**, 554.
Wada, M., M. Suzuki, A. Sawada, Y. Ishibashi and K. Gesi, 1981, J. Phys. Soc. Jpn **50**, 1813.
Yamada, Y., and N. Hamaya, 1983, J. Phys. Soc. Jpn **52**, 3466.
Yamada, N., and T. Ikeda, 1984, J. Phys. Soc. Jpn **53**, 2555.
Zehyer, R., and W. Finger, 1982, Phys. Rev. Lett. **49**, 1883.

CHAPTER 11

Phenomenology of Incommensurate Phases in the A_2BX_4 Family

Yoshihiro ISHIBASHI

Synthetic Crystal Research Laboratory
Faculty of Engineering, Nagoya University
Chikusa-ku, Nagoya 464
Japan

Incommensurate Phases in Dielectrics 2
Edited by
R. Blinc and A.P. Levanyuk

Contents

1. Introduction . . . 51
2. The Landau theory applicable to the IC phase transitions in the A_2BX_4 family . . . 51
3. The temperature dependences of various physical quantities in the N–IC–C transition sequence . . . 56
3.1. Specific heat . . . 56
3.2. Strain and polarization . . . 58
3.3. Elastic constant and dielectric susceptibility . . . 60
3.4. Soft modes . . . 63
3.5. Dielectric dispersion . . . 65
4. Phase diagrams . . . 66
4.1. Pressure–temperature phase diagrams . . . 66
4.2. Electric field–temperature phase diagrams . . . 67
5. Concluding remarks . . . 68
References . . . 68

1. Introduction

The A_2BX_4 family belonging to the space group Pnam in its high temperature phase constitutes one of the largest groups of crystals which show incommensurate (IC) phases, and the physical properties related to the IC phase transitions in this family have been most thoroughly investigated. Table 1, in which the IC phases observed in the family are summarized, shows that crystals can be grouped on the basis of the wave vector $\boldsymbol{k}_C$ of an unstable mode which, by freezing-in, gives rise to the appearance of the commensurate (C) phase (not of the IC phase). Obviously the largest group is the one which is represented by K_2SeO_4, where the C phase induced by the instability of a mode with $\boldsymbol{k}_C = \boldsymbol{a}^*/3$ is (improper) ferroelectric (Iizumi et al. 1977). Rb_2ZnCl_4, Rb_2ZnBr_4 and K_2ZnCl_4 belong to this group (Gesi and Iizumi 1978, 1979). The C phase of $[N(CH_3)_4]_2CuCl_4$ (Sawada et al. 1980, Gesi and Iizumi 1980) is improper ferroelastic (but not ferroelectric), though the wavevector of the unstable mode is the same.

The physical properties show anomalies, more or less, associated with the transitions. Some show a drastic change at the normal (N)–IC transition temperature T_I, but only a minor change at the IC–C transition temperature T_C, while others show a remarkable anomaly at T_C but not at T_I. In addition, in the IC phase some physical quantities show a temperature dependence which reflects the physical nature of the C phase. We shall try in this chapter to obtain a phenomenological understanding of the IC phase transitions and the related physical properties in the crystals of the A_2BX_4 family.

2. The Landau theory applicable to the IC phase transitions in the A_2BX_4 family

To begin with, we consider the continuum theory of the IC phase transitions, which is valid unless the spatial change of the structure should be drastic. Let us adopt as the transition parameter the complex amplitude Q of the wave

Table 1
A_2BX_4-type crystals having incommensurate phases, classified according to the wavenumber of the frozen-in commensurate wave, $\boldsymbol{k}_C$. P and u in parentheses indicate the spontaneous polarization and strain, respectively, which characterize the commensurate phase.

$\boldsymbol{k}_C$	Crystals	
$\frac{1}{2}$	$(ND_4)_2BeF_4$	(P)
	$[N(CH_3)_4]_2CuBr_4$ [a]	(P)
$\frac{1}{3}$	K_2SeO_4	(P)
	Rb_2ZnCl_4	(P)
	Rb_2ZnBr_4	(P)
	K_2ZnCl_4	(P)
	$(NH_4)_2ZnCl_4$	(P)
	$[N(CH_3)_4]_2CuCl_4$	(u)
$\frac{2}{5}$	$[N(CH_3)_4]_2FeCl_4$	(P)
	$[N(CH_3)_4]_2ZnCl_4$	(P)
	$[N(CH_3)_4]_2CoCl_4$	(P)
	$RbLiSO_4$	(P)

[a] The direction of the modulation is along the b axis ($b \approx \sqrt{3}\, a$).

with a commensurate wavevector $\boldsymbol{k}_C$, which is frozen in the C phase. For $(NH_4)_2BeF_4$, K_2SeO_4 and $[N(CH_4)_3]_2ZnCl_4$, $\boldsymbol{k}_C$ is $\boldsymbol{a}^*/2$, $\boldsymbol{a}^*/3$ and $2\boldsymbol{a}^*/5$, respectively (Iizumi and Gesi 1977, Iizumi et al. 1977, Tanisaki and Mashiyama 1980). The IC structure can be described as one where Q is spatially modulated, i.e., $\mathrm{d}Q/\mathrm{d}x \neq 0$, ($x$ is the direction of modulation). Thus, in the thermodynamic potential function which describes the IC phase transition we have to take into account the contribution from such modulation.

The essential part of the thermodynamic potential functional f is written in terms of Q, Q^*, their x-derivatives, and some macroscopic quantities such as the spontaneous polarization P and the spontaneous strain u_j. We can divide f into two parts f_1 and f_2:

$$f(x) = f_1 + f_2, \tag{2.1}$$

f_1 being the part characteristic to each crystal, and f_2 the part common to all crystals, i.e.,

$$f_2 = \frac{\delta \mathrm{i}}{2}\left(Q^* \frac{\mathrm{d}Q}{\mathrm{d}x} - Q\frac{\mathrm{d}Q^*}{\mathrm{d}x}\right) + \frac{\kappa}{2}\frac{\mathrm{d}Q^*}{\mathrm{d}x}\frac{\mathrm{d}Q}{\mathrm{d}x} + \frac{P^2}{2\chi_0} + \frac{c_{jj}^0}{2}u_j^2 + \zeta_1 P^2 QQ^* + \zeta_2 u_j^2 QQ^*. \tag{2.2}$$

Here, the δ-term called the Lifshitz invariant favors the IC structure, while the κ-term suppresses the IC phase when $\kappa > 0$, the third and fourth terms represent the electric energy with the dielectric susceptibility χ_0 and the elastic energy with elastic stiffness constant c_{jj}^0, respectively. The last two terms are higher order coupling terms between P, u_j and Q, Q^*. (Here, to avoid complication we took into account only the shear component of the strain, u_j, ignoring the contribution from the tensile component Δu_i which is to be discussed separately in a following section.) The concrete form of f_1 differs, depending on the symmetry of the soft mode partly specified by $\boldsymbol{k}_{\mathrm{C}}$. For clarity let us write down the relevant thermodynamic potential functional for several cases: For $(NH_4)_2BeF_4$ where $\boldsymbol{k}_{\mathrm{C}} = \boldsymbol{a}^*/2$,

$$f(x) = \frac{\alpha}{2} QQ^* + \frac{\beta_1}{4}(QQ^*)^2 + \frac{\beta_2}{8}(Q^4 + Q^{*4}) + \xi_1 P(Q^2 + Q^{*2}) + \mathrm{i}\xi_2 u_j(Q^2 - Q^{*2}) + f_2, \tag{2.3}$$

for K_2SeO_4 where $\boldsymbol{k}_{\mathrm{C}} = \boldsymbol{a}^*/3$,

$$f(x) = \frac{\alpha}{2} QQ^* + \frac{\beta}{4}(QQ^*)^2 + \frac{\gamma_1}{6}(QQ^*)^3 + \frac{\gamma_2}{12}(Q^6 + Q^{*6}) + \xi_1 P(Q^3 + Q^{*3}) + \mathrm{i}\xi_2 u_j(Q^3 - Q^{*3}) + f_2, \tag{2.4}$$

and for $[N(CH_3)_4]_2ZnCl_4$ where $\boldsymbol{k}_{\mathrm{C}} = 2\boldsymbol{a}^*/5$,

$$f(x) = \frac{\alpha}{2} QQ^* + \frac{\beta}{4}(QQ^*)^2 + \frac{\gamma}{6}(QQ^*)^3 + \frac{\nu}{8}(QQ^*)^4 + \frac{\pi_1}{10}(QQ^*)^5 + \frac{\pi_2}{20}(Q^{10} + Q^{*10}) + \xi_1 P(Q^5 + Q^{*5}) + \mathrm{i}\xi_2 u_j(Q^5 - Q^{*5}) + f_2, \tag{2.5}$$

where only α is temperature-dependent,

$$\alpha = \alpha_0(T - T_0), \tag{2.6}$$

and $\beta, \ldots,$ are constants whose sign and magnitude should be assumed so that the stability condition is ensured even when $|Q|$ and $|\mathrm{d}Q/\mathrm{d}x|$ grow higher. Terms such as $Q^n \pm Q^{*n}$ (n: positive integer), including those obtained after eliminating P and u_j by the equilibrium conditions, are called anisotropic terms because they become θ-dependent terms as $\cos n\theta$ or $\sin n\theta$ if we put

$$Q(x) = \rho(x)\mathrm{e}^{\mathrm{i}\theta(x)}. \tag{2.7}$$

The anisotropy in f plays an important role in governing the temperature range of the IC phase, i.e., in general the stronger the anisotropy, the narrower the stability range of the IC phase. Here one should not overlook, however, the contribution from invariant terms such as

$$\eta_1 \mathrm{i} QQ^* \left(Q^* \frac{\mathrm{d}Q}{\mathrm{d}x} - Q \frac{\mathrm{d}Q^*}{\mathrm{d}x} \right) \quad \text{and} \quad \eta_2 QQ^* \frac{\mathrm{d}Q}{\mathrm{d}x} \frac{\mathrm{d}Q^*}{\mathrm{d}x},$$

which are not included in (2.2). Especially, the latter disfavors the IC phase if $\eta_2 > 0$ and will reduce the stability range of the IC phase even though it is isotropic (Ishibashi 1983).

Let us next consider the solution which minimizes the average thermodynamic potential

$$F = \frac{1}{L} \int_{-L/2}^{L/2} f(x)\,\mathrm{d}x, \tag{2.8}$$

where L is one period of the IC structure. For so doing, we have to solve the Lagrange–Euler equation

$$\delta F = 0.$$

The solutions corresponding to the N phase and the C phase are simple; the solution

$$Q = 0, \qquad P = 0, \qquad u_j = 0 \tag{2.9}$$

corresponds to the N phase;

$$Q = a_0 \neq 0, \qquad \frac{\mathrm{d}Q}{\mathrm{d}x} = 0, \qquad P \neq 0, \qquad u_j = 0 \tag{2.10}$$

corresponds to the C-ferroelectric phase; and

$$Q = a_0 \neq 0, \qquad \frac{\mathrm{d}Q}{\mathrm{d}x} = 0, \qquad P = 0, \qquad u_j \neq 0 \tag{2.11}$$

corresponds to the C-ferroelastic phase, where a_0 is some constant. Which of the two possible C phases is realized is governed by the values of the coefficients appearing in the thermodynamic potential. As to the case (2.4), for example, the C-ferroelastic is realized if $\gamma_2' > 0$, while the C-ferroelectric phase is stable if $\gamma_2' < 0$, where

$$\gamma_2' = \gamma_2 - 3\left(\xi_1^2 \chi_0 - \xi_2^2 / c_{jj}^0 \right). \tag{2.12}$$

On the other hand, the exact solution corresponding to the IC phase has not yet been found, and so various approximations or numerical calculations will be needed to discuss the structure and the physical properties of the IC phase.

The following can be said generally. When the temperature becomes lower, the N phase loses its stability at $T = T_I$ given by

$$\alpha_I = \alpha_0(T_I - T_0) = \delta^2/\kappa, \tag{2.13}$$

and the IC phase with the modulation wavenumber

$$q_I = \frac{|\delta|}{\kappa} \tag{2.14}$$

appears (Levanyuk and Sannikov 1976, Ishibashi et al. 1981). With further decrease of temperature, the modulation period becomes longer (the modulation wavenumber q becomes smaller). At the temperature T_C the thermodynamic potential of the IC phase becomes higher than that of the C phase, and the IC–C transition takes place.

Now, the temperature dependent features of Q, P and u_j in the IC phase are different according to whether the C phase is ferroelectric (as in K_2SeO_4) or ferroelastic (as in $[N(CH_3)_4]_2CuCl_4$). But the form of their Fourier expansions is determined by the symmetry of Q, irrespective of the structure of the C phase. For example, for the case of the thermodynamic potential (2.4),

$$Q = \sum_{n=0} a_{6n+1} e^{i(6n+1)qx} + \sum_{n=1} a_{6n-1} e^{-i(6n-1)qx}, \tag{2.15}$$

$$P = \sum_{n=0} b_{6n+3} \cos(6n+3)qx, \tag{2.16}$$

$$u_j = \sum_{n=0} d_{6n+3} \sin(6n+3)qx \tag{2.17}$$

(Ishibashi et al. 1981). The above mentioned difference results from the difference of the concrete numerical values of $a_{6n\pm1}$, b_{6n+3} and d_{6n+3}, reflecting the difference of the sign of γ_2', eq. (2.12). Just below T_I, we only have to consider a_1, b_3 and d_3. As the temperature becomes lower, higher harmonics in Q grow up, leading to a domain-like structure consisting of a sequence of almost-C regions separated by transient regions called discommensurations. If the C phase is ferroelectric, the polarization appears in the almost-C regions while the strain appears only in the region of discommensurations, and if the C phase is ferroelastic, the situation is reversed.

3. *Temperature dependence of various physical quantities in the N–IC–C transition sequence*

In this section we review the experimental data on various physical quantities from the viewpoint of the Landau theory. Though the concrete behaviors of such quantities derived from eqs. (2.3), (2.4) and (2.5) are different, one exemplification will be sufficient to understand the principles of the phenomena concerned. Therefore, we consider again the case of $\boldsymbol{k}_C = \boldsymbol{a}^*/3$, applicable to K_2SeO_4, Rb_2ZnCl_4, $[N(CH_3)_4]_2CuCl_4$ and so on.

3.1. Specific heat

The specific heat is one of the most fundamental properties to clarify phase transitions. Recently high quality data have been reported on K_2SeO_4 (Atake et al. 1983a), Rb_2ZnCl_4 (Atake et al. 1983b) and Rb_2ZnBr_4 (Nomoto et al. 1983). The excess specific heat due to the N–IC and IC–C transitions is shown in fig. 1. With decreasing temperature, it peaks at the N–IC transition of the second order, decreases first in the IC phase and then shows a slight increase just above the IC–C transition.

On the other hand, a theory based on the constant amplitude approximation (or phase-modulation-only approximation) has been presented by adopting a thermodynamic potential functional

$$f(x) = \frac{\alpha}{2}\rho^2 + \frac{\beta}{4}\rho^4 + \frac{\rho^6}{6}(\varepsilon_1 + \varepsilon_2\cos 6\theta) - \delta\rho^2\frac{\mathrm{d}\theta}{\mathrm{d}x} + \frac{\kappa}{2}\rho^2\left(\frac{\mathrm{d}\theta}{\mathrm{d}x}\right)^2, \tag{3.1}$$

which can be obtained from eq. (2.4) after eliminating P and u_j according to the equilibrium conditions $\partial F/\partial P = \partial F/\partial u_j = 0$, ignoring the η-term, and putting

$$Q(x) = \rho \mathrm{e}^{\mathrm{i}\theta(x)} \tag{3.2}$$

(Ishibashi 1982). After minimization of the average potential F first with respect to θ and then with respect to ρ for given temperatures, the temperature dependence of the specific heat can be calculated. Though there is a defect in the model intrinsic to the Landau theory that no anomaly can be described above the N–IC transition temperature, it is said that the model could simulate experimental data well for its extreme simplicity. Namely, it successfully reproduces the general shape of the excess or anomalous part of the specific heat, such as the first-order nature, the smallness of the anomaly at the IC–C transition (in comparison with that at the N–IC transition) and the existence of a considerable tail on the high temperature side of the IC–C transition (Nomoto et al. 1983).

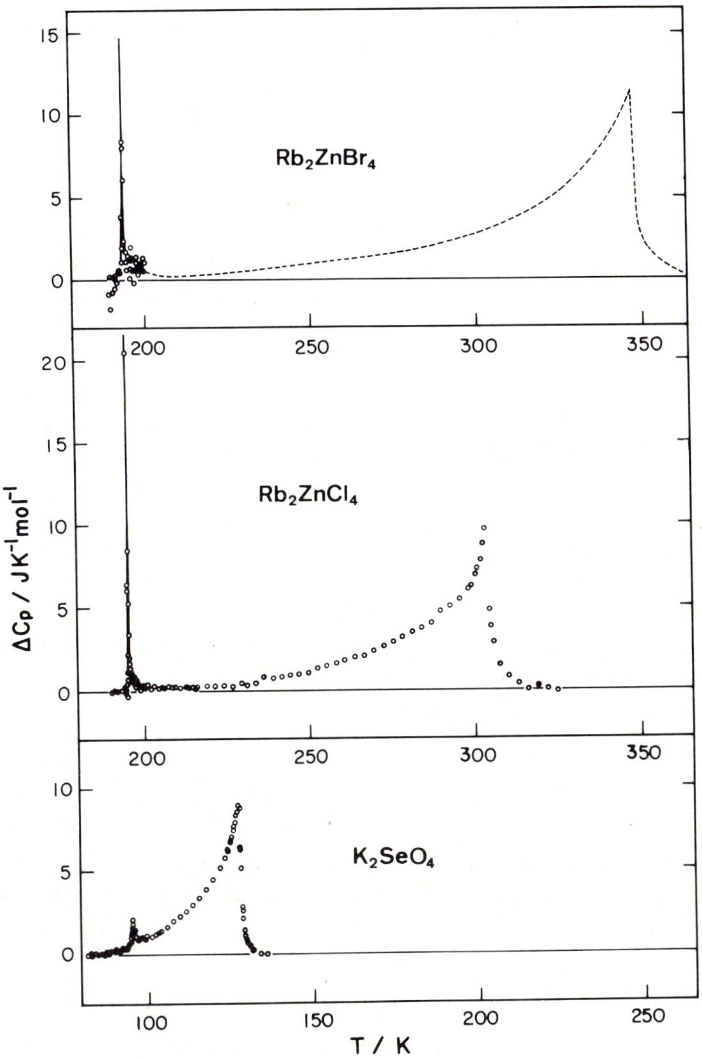

Fig. 1. The excess part of the specific heat in several crystals with the IC phase (Nomoto et al. 1983).

Needless to say, the present model can be applied even to crystals where the C phase is not ferroelectric but ferroelastic.

3.2. *Strain and polarization*

To make the arguments transparent, let us rewrite the thermodynamic potential (2.4) in a form valid around the N–IC transition temperature, adding to it the contribution of an induced tensile strain component Δu_i. Then we obtain

$$F=\frac{P^2}{2\chi_0}+\zeta_1 P^2 a_1^2+\frac{c_{jj}^0}{2}u_j^2+\zeta_2 u_j^2 a_1^2$$
$$+\frac{c_{ii}^0}{2}(\Delta u_i)^2+\sigma(\Delta u_i)a_1^2+\frac{\alpha-\alpha_{\mathrm{I}}}{2}a_1^2+\frac{\beta}{4}a_1^4+\cdots; \tag{3.3}$$

a_1 has been defined in (2.15). By the equilibrium condition $\partial F/\partial(\Delta u_i)=0$, we obtain that in the vicinity of T_{I}

$$\Delta u_i \approx -\frac{\sigma}{c_{ii}^0}a_1^2 \propto (T_{\mathrm{I}}-T)^{2\beta}, \tag{3.4}$$

($\propto$ implies "proportional to") where β is an index whose value in the Landau theory is $\frac{1}{2}$ [not to be confused with the coefficient to the $(QQ^*)^2$ terms in eqs. (2.4) and (3.3)]. Therefore, at the N–IC transition usually one should observe at least a kink-type anomaly in the tensile strain component. As the temperature becomes lower, the temperature dependence of Δu_i deviates from $T_{\mathrm{I}}-T$ gradually. The sign of Δu_i is governed by the sign of σ (Midorikawa 1981).

In order to discuss the phenomena to be observed in the C phase, let us write F valid in the C phase, using a_0 given in (2.10) and (2.11), as

$$F=\frac{\alpha}{2}a_0^2+\frac{\beta}{4}a_0^4+\cdots$$
$$+\frac{P^2}{2\chi_0}+2\xi_1 P a_0^3\cos 3\theta+\zeta_1 P^2 a_0^2$$
$$+\frac{c_{jj}^0}{2}u_j^2-2\xi_2 u_j a_0^3\sin 3\theta+\zeta_2 u_j^2 a_0^2+\frac{c_{ii}^0}{2}(\Delta u_i)^2$$
$$+\sigma(\Delta u_i)a_0^2+\cdots, \tag{3.5}$$

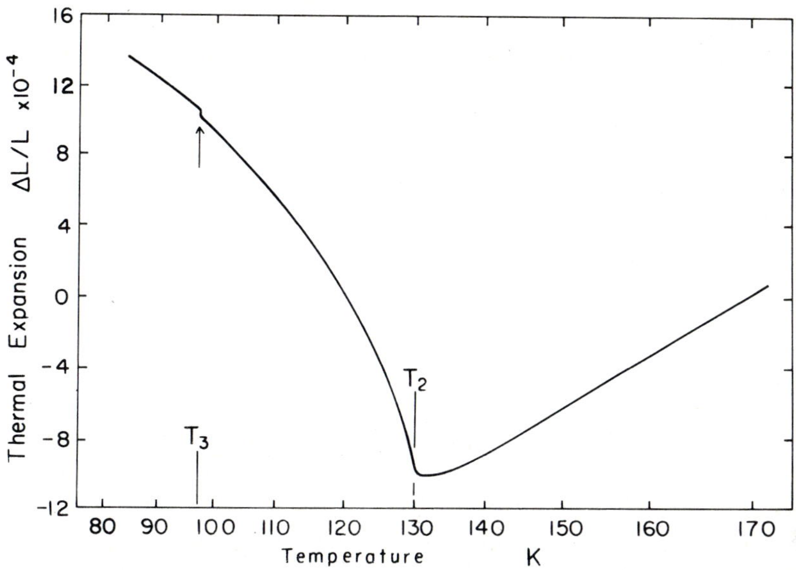

Fig. 2. Temperature dependence of the tensile strain component Δu_3 in K_2SeO_4 (Midorikawa 1981). T_2 and T_3 correspond to T_I and T_C, respectively.

where

$$\begin{aligned} \cos 3\theta &= 0 \quad \text{for the ferroelastic C phase,} \\ \sin 3\theta &= 0 \quad \text{for the ferroelectric C phase.} \end{aligned} \tag{3.6}$$

From this we obtain

$$\Delta u_i \approx -\frac{\sigma}{c_{ii}^0} a_0^2 + \cdots . \tag{3.7}$$

Near T_0, which is hidden in the IC phase range,

$$a_0 \propto (T_0 - T)^\beta , \tag{3.8}$$

and therefore an extrapolation of Δu_i should intersect the temperature axis at $T = T_0$ lower than T_I. At the IC–C transition temperature T_C a small jump should be observed if the transition is of the first order. The behavior of Δu_i in the whole temperature region is shown in fig. 2 (Midorikawa 1981).

The shear strain component u_j behaves differently. In the IC phase it is spatially modulated, and therefore its average must vanish. In the C phase,

$$u_j = \begin{cases} \pm \dfrac{2\xi_2}{c_{jj}^0} a_0^3 + \cdots & \text{for the ferroelastic phase,} \\ 0 & \text{for the ferroelectric phase.} \end{cases} \tag{3.9}$$

If the C phase is ferroelastic, as in $[N(CH_3)_4]_2CuCl_4$, u_j shows a jump at the IC–C transition temperature (even if the transition were of the second order), because a_0 does not vanish there. When such a shear strain appears, an orthorhombic crystal transforms to monoclinic, and domains appear. The angle between the extinction positions in these is approximately proportional to u_j, and therefore study of this angle will shed light on the temperature dependence of u_j (Sugiyama et al. 1980).

The polarization P behaves quite similarly as the shear strain u_j. In the IC phase it is spatially modulated, and its average must vanish. In the C phase,

$$P = \begin{cases} 0 & \text{for the ferroelastic C phase,} \\ \pm 2\chi_0 \xi_1 a_0^3 + \cdots & \text{for the ferroelectric C phase.} \end{cases} \tag{3.10}$$

The temperature dependence of the spontaneous polarization in K_2SeO_4 was measured a long time ago and a jump at $T = T_C$ was clearly observed (Aiki et al. 1970).

3.3. *Elastic constant and dielectric susceptibility*

The anomalies in the elastic constant and dielectric susceptibility can be derived from (2.4). To discuss the behavior of χ and c_{jj} at all temperatures of the IC phase, we have in principle to take into account all the higher harmonics in Q, P and u_j, and consider the effects caused by an electric field E and an external stress X_j conjugate to P and u_j, respectively. Note that both E and X_j are "anisotropic" in nature. On the basis of symmetry considerations one sees that when an electric field conjugate to P is applied, the variations in Q, P and u_j in the IC phase are (Ishibashi et al. 1981):

$$\begin{aligned} \Delta Q &= \sum_{n=0} g_{6n+2} e^{-i(6n+2)qx} + \sum_{n=0} g_{6n+4} e^{i(6n+4)qx}, \\ \Delta P &= \sum_{n=0} r_{6n} \cos 6ngx, \qquad \Delta u_j = \sum_{n=0} t_{6n} \sin 6ngx, \end{aligned} \tag{3.11}$$

while when a stress conjugate to u_j is applied, the variations in Q, P and u_j are:

$$\Delta Q = \mathrm{i}\left[\sum_{n=0} h_{6n+2}\mathrm{e}^{-\mathrm{i}(6n+2)qx} + \sum_{n=0} h_{6n+4}\mathrm{e}^{\mathrm{i}(6n+4)qx}\right],$$

$$\Delta P = \sum_{n=0} v_{6n}\sin 6nqx, \qquad \Delta u_j = \sum_{n=0} w_{6n}\cos 6nqx. \tag{3.12}$$

By determining the amplitude of these Fourier components so as to minimize the average thermodynamic potential of the crystal under applied field or stress, one can find χ and c_{jj} as

$$\chi = r_0/E, \qquad \frac{1}{c_{jj}} = \frac{w_0}{X_j}. \tag{3.13}$$

There are other analytical and more transparent methods based on some approximation by which one can avoid the procedure described above (Dvořák and Petzelt 1978, Sannikov 1980, Golovko 1980, Dvořák and Hudak 1982, Prelovsek 1983).

At the temperatures just below T_{I}, however, we can rely upon the expression (3.3) without resorting to (3.11) and (3.12)*. By making use of the golden rule utilizing the Hessian, we obtain

$$c_{ii} = c_{ii}^0 - \frac{2\sigma^2}{\beta}, \qquad c_{jj} = c_{jj}^0 + 2\zeta_2 a_{\mathrm{I}}^2, \qquad \frac{1}{\chi} = \frac{1}{\chi_0} + 2\zeta_1 a_{\mathrm{I}}^2, \tag{3.14}$$

where

$$a_{\mathrm{I}}^2 \sim (T_{\mathrm{I}} - T)^{2\beta}.$$

As is seen from eqs. (3.14), at the N–IC transition c_{ii} shows a downward jump, while χ and c_{jj} show at least kink-type anomalies (if $\beta \leqslant 1/2$). These features have been confirmed by experiments (Hoshizaki et al. 1980), as shown in fig. 3.

The behaviors of χ and c_{jj} just above the IC–C transition temperature T_{C} is especially noteworthy. They are very different according to whether the C phase is ferroelastic or ferroelectric. In general it can be said qualitatively that if the C phase is ferroelectric, χ shows a tendency towards divergence (Aiki

*Due care should be employed in this simplification. In the case of (2.3), the contribution to $c_{jj} - c_{jj}^0$ and $1/\chi - 1/\chi_0$ from the ξ_1 and the ξ_2 terms is of the order of a_{I}^2 just below T_{I} (Levanyuk and Sannikov 1976, Ishibashi and Dvořák 1978). In such a case, therefore, both terms must be retained, though it is allowed to omit them in (3.3).

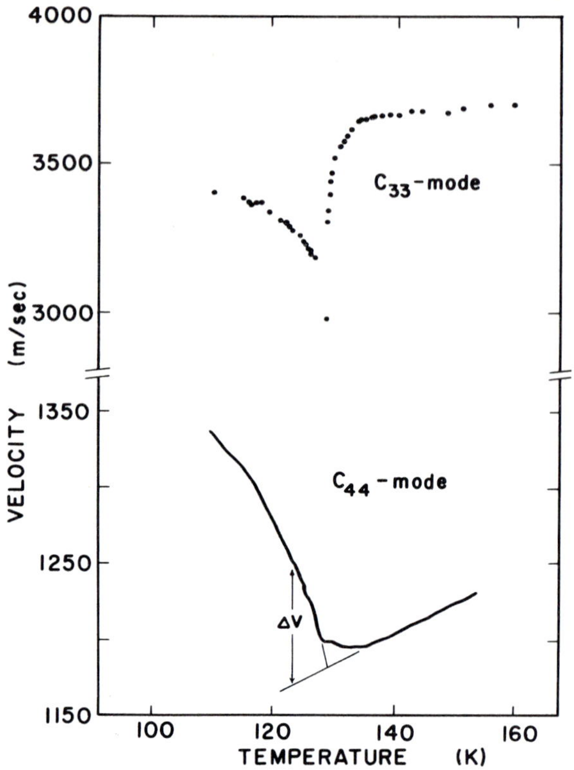

Fig. 3. Temperature dependences of elastic stiffness constants c_{33} and c_{44} in K_2SeO_4. (Hoshizaki et al. 1980.)

et al. 1970) and c_{jj} approaches its value extrapolated from the side of the low temperature C phase (Rehwald et al. 1980), while if the C phase is ferroelastic, χ approaches its value of the C phase and c_{jj} tends to vanish. It has been shown theoretically that the way of divergence of χ and $1/c_{jj}$ in approaching T_C from above is described by the Curie–Weiss law $(T-T_C)^{-1}$ (Sannikov 1980, Golovko 1980, Prelovsek 1983). In fact, as is shown in fig. 4, the elastic constant c_{55} in $[N(CH_3)_4]_2CuCl_4$ decreases to almost zero, obeying the elastic Curie–Weiss law (Sawada et al. 1980), and the dielectric susceptibility χ of, for example, Rb_2ZnCl_4 is experimentally shown to obey the Curie–Weiss law (Levstik et al. 1982).

The elastic constant and the dielectric susceptibility in the C phase are nothing but those for ordinary improper ferroelastics and ferroelectrics. They are easily derived from eq. (3.5).

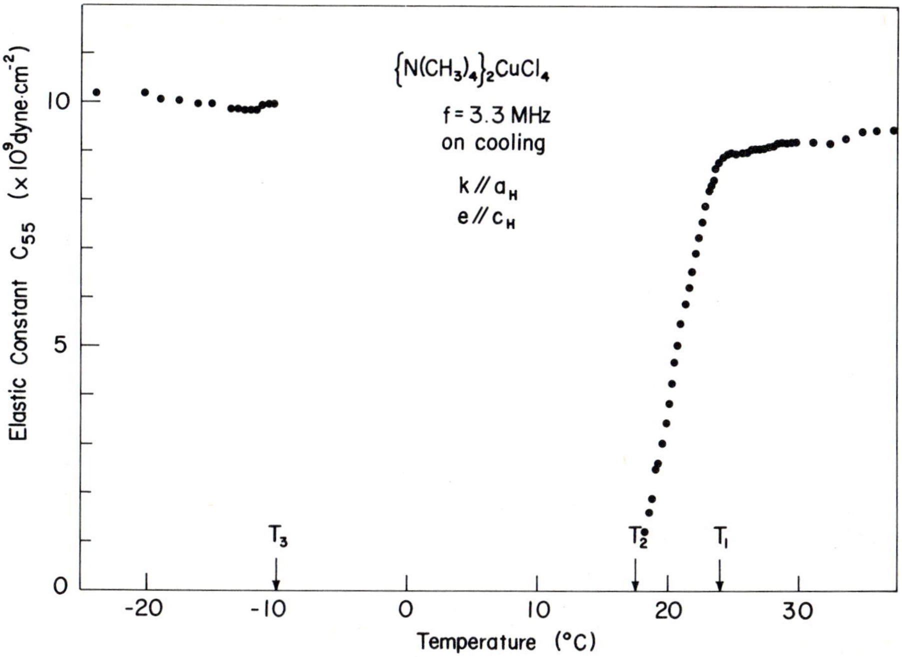

Fig. 4. Elastic softening in the IC phase of $[N(CH_3)_4]_2CuCl_4$. T_1 and T_2 correspond to T_I and T_C, respectively. (Sawada et al. 1980.)

3.4. *Soft modes*

It has been revealed by neutron and Raman scattering experiments that there exist soft modes inducing the N–IC phase transitions in K_2SeO_4 (Wada et al. 1977a, b, Iizumi et al. 1977), Rb_2ZnCl_4 (Wada et al. 1978) and Rb_2ZnBr_4 (Takashige et al. 1980). In some other A_2BX_4-type crystals listed in table 1 (section 1) the existence of soft vibrational modes is also quite likely. The soft mode associated with the N–IC phase transition is doubly degenerate $Q(\boldsymbol{k})$ and $Q(-\boldsymbol{k}) = Q^*(\boldsymbol{k})$ where $\boldsymbol{k}$ is close to a commensurate wavevector $\boldsymbol{k}_C$. It is needless to say that these modes are neither Raman-active nor infrared-active in the N phase since they have a nonzero wavenumber. In the IC phase below T_I this degeneracy is lifted, and two soft modes appear (see fig. 5). As is well known, they are written for small wavenumber q around the frozen-in wavenumber k_i as (Dvořák and Petzelt 1978)

$$A(q) = \frac{1}{\sqrt{2}}\left[Q(k_i + q) + Q(-k_i + q)\right],$$

$$\phi(q) = \frac{1}{\sqrt{2}}\left[Q(k_i + q) - Q(-k_i + q)\right]. \tag{3.15}$$

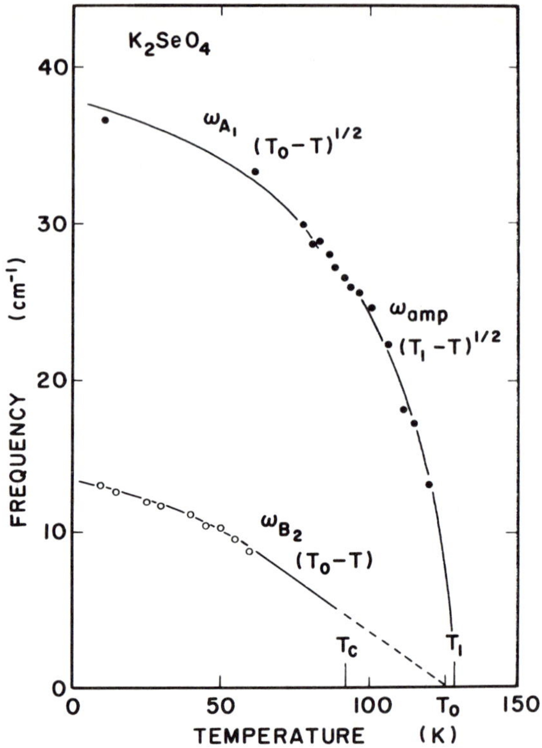

Fig. 5. Temperature dependences of the frequencies of the amplitude and the phase modes in K_2SeO_4 observed by means of Raman scattering. (Wada et al. 1977a, b.)

The $A(q)$ mode is called an amplitude mode, representing the fluctuation of the amplitude of the frozen-in wave while the $\phi(q)$ mode is called the phase mode, representing phase fluctuation. The $q=0$ amplitude mode has a non-zero frequency, and the frequency increases with decreasing temperature. This mode is totally symmetric in the sense of macroscopic symmetry and therefore it is Raman-active. The temperature dependence in the IC phase of the frequency of the soft amplitude with $q=0$ should be approximately expressed as

$$\omega_{A(0)} \sim (T_I - T)^{\beta}, \tag{3.16}$$

where β is the index for the order parameter.

On the other hand, the frequency of the $q=0$ phase mode is zero, i.e.,

$$\omega_{\phi(0)}=0, \tag{3.17}$$

reflecting the situation that the energy of the frozen-in IC wave does not depend on its position in an ideal lattice. It should be noted that the dispersion relation of the phase mode branch is expressed by a linear function of q as

$$\omega_{\phi(q)}=B_\phi|q| \quad (B_\phi;\ \text{constant}) \tag{3.18}$$

for small q.

There is nothing very special in the behavior of these modes in the C phase, since they are just the soft modes in ordinary improper ferroelectrics or ferroelastics. Both the amplitude and the phase modes have nonzero frequencies. The frequency of the amplitude mode may be expressed as

$$\omega_A \sim (T_0-T)^\beta, \tag{3.19}$$

while the one of the phase mode as

$$\omega_\phi \sim (T_0-T)^{2\beta'}. \tag{3.20}$$

According to the simple Landau-type model, the relation that $\beta=\beta'=1/2$ holds and some Raman experiments indeed support this relation in K_2SeO_4 (Wada et al. 1977a, b), though deviation of the β-value from the mean field result has been also reported (Unruh et al. 1979).

One comment should be added: in analyses of the temperature dependences of these frequencies in the C phase, any extrapolation of the ω versus T curve should intersect the temperature axis at T_0 (but not at T_I) as can be seen from eqs. (3.19) and (3.20).

3.5. *Dielectric dispersion*

Although the soft vibrational modes have been observed in many A_2BX_4-type crystals, Debye-type dispersion in the microwave frequency region has also been reported for K_2SeO_4 and Rb_2ZnCl_4 just above the IC–C transition temperature (Petzelt et al. 1979, Horioka et al. 1980). This very interesting phenomenon is interpreted as a manifestation of the formation of discommensurations in the temperature range concerned.

4. *Phase diagrams*

4.1. *Pressure – temperature phase diagrams*

Hydrostatic pressure, which is totally symmetric in nature, affects the temperatures of the N–IC and the IC–C transitions by giving rise to changes in the values of coefficients $\alpha, \beta, \ldots$, in (2.3)–(2.5), and sometimes it even suppresses appearance of the IC phase, inducing a direct transition from the N phase to the C phase, which must be of the first order according to the Lifshitz condition. As is shown in fig. 6, such features have in fact been revealed for

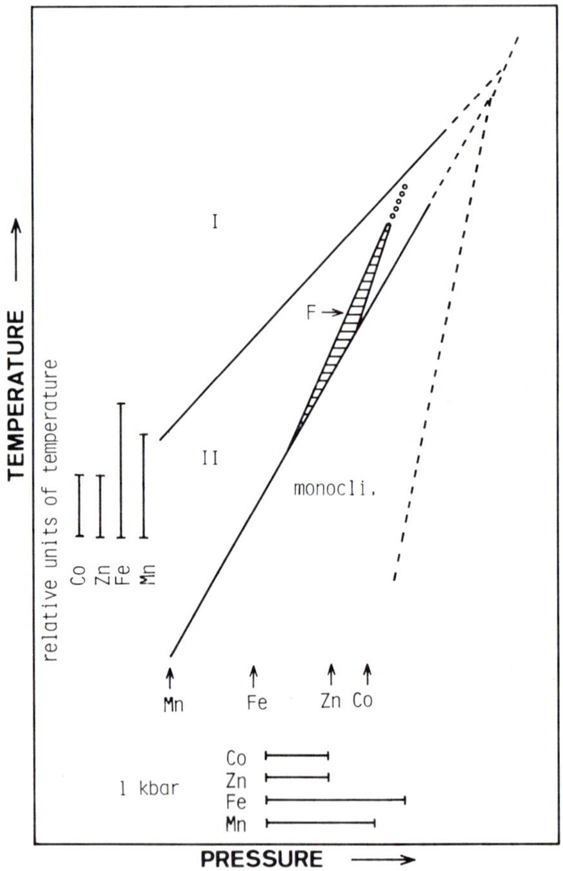

Fig. 6. Phase diagrams of $[N(CH_3)_4]_2XCl_4$ (X = Co, Zn, Fe and Mn). The phase denoted by II is the IC phase. Vertical arrows stand for zero pressure on the pressure axis for each salt. (Shimizu et al. 1980.)

crystals of the tetramethylammonium family such as $[N(CH_3)_4]_2XCl_4$ where X = Co, Fe, Mn and Zn (Shimizu et al. 1980).

In general, study of the pressure effect has a special significance, in addition to its intrinsic value, in that it provides a quantitative cross-check on other observed physical quantities through the Ehrenfest relation and the Clausius–Clapeyron relation for second- and first-order transitions, respectively. It has turned out, for example, that the shifts of T_I and T_C of K_2SeO_4 due to hydrostatic pressure are in good agreement with those estimated by relations using the values of the jump in the thermal expansion coefficient, the specific heat, the volume and the entropy (Press et al. 1980, Kudo and Ikeda 1981, Midorikawa 1981).

Sometimes uniaxial stress has been utilized to study the change of transitions. Certainly the application of uniaxial stress is a useful technique to study phase transitions because the shift of the transition temperature is usually more pronounced than that induced by hydrostatic pressure (Haque et al. 1978, Wada et al. 1982).

4.2. *Electric field – temperature phase diagrams*

Application of an electric field E is also a useful technique to study phase transitions.

Let us first consider the rate of change of the transition temperatures due to the applied field. The N–IC transition temperature T_I does not change in the first order in E:

$$\left.\frac{dT_I}{dE}\right|_{E=0} = 0, \tag{4.1}$$

because on both sides of T_I no spontaneous polarization exists and therefore the amount of change due to E in the thermodynamic potential of both phases is the same in the first order. On the other hand, the IC–C transition temperature T_C rises as a linear function of E (for small E) if the IC phase is ferroelectric (Aiki et al. 1970). The rate of the rise depends on the jump of the polarization P at the IC–C transition. (Note that there is a jump in P even when the IC–C transition is of the ideal second order.) By extrapolating the increase of T_C towards the higher field region it is reasonable to expect the disappearance of the IC phase. The phase diagram in the region of such high fields, however, has not yet been experimentally investigated in detail, probably due to a technical difficulty related to the dielectric breakdown of samples, though it must be a quite interesting problem.

The IC structure itself under the electric field provides an interesting problem. Unlike hydrostatic pressure, an electric field changes the overall symmetry of the system by inducing a macroscopic polarization. Therefore, the IC phase becomes a superposition of a polar structure and a modulated structure, which is a qualitatively different state from those encountered when only the temperature is controlled. In principle the structure is given by the Fourier components (2.15)–(2.17) and (3.11) which produce the lowest thermodynamic potential. To obtain a more transparent view on such a structure, the constant amplitude approximation has been adopted, where one has to treat a double sine-Gordon equation (Hudak 1983). We will not enter into detail here because the mathematics for this is involved.

5. *Concluding remarks*

In the present chapter incommensurate phase transitions in A_2BX_4-type crystals have been reviewed. It seems that the Landau-theoretical considerations could play the role of a guide, and the essential aspects of various phenomena related to the N–IC and the IC–C transitions have certainly been explained by simple considerations at least qualitatively on the phenomenological level. A close look at various experimental data, however, will soon disclose that there are many noticeable disagreements between reality and the result of such phenomenological considerations. For example, Landau theory says nothing about the anomaly in the specific heat observed above the N–IC transition temperature. This is obviously due to the fact that in Landau theory the entropy is considered only in a very rough approximation. We may therefore need much theoretical work to clarify these points in a deeper level.

In the Landau theory presented in section 2 the lock-in C wavenumber is the one to which Q belongs, and only one such special wavenumber is accommodated in the given thermodynamic potential. Therefore, the potential thus presented would be quite powerless in the case where many higher order lock-in wavenumbers appear. As for this point, we have to resort to models where the discreteness of lattices is properly taken into account (Bak 1982).

References

Aiki, K., K. Hukuda, H. Koga and T. Kobayashi, 1970, J. Phys. Soc. Jpn **28**, 389.
Atake, T., K. Nomoto, B.K. Chaudhuri and H. Chihara, 1983a, J. Chem. Thermodynamics **15**, 383.
Atake, T., K. Nomoto, B.K. Chaudhuri and H. Chihara, 1983b, J. Chem. Thermodynamics **15**, 339.
Bak, P., 1982, Rep. Prog. Phys. **45**, 587.
Dvořák V., and O. Hudak, 1982, Ferroelectrics **46**, 19.
Dvořák, V., and J. Petzelt, 1978, J. Phys. **C11**, 4827.
Gesi, K., and M. Iizumi, 1978, J. Phys. Soc. Jpn **45**, 1777.
Gesi, K., and M. Iizumi, 1979, J. Phys. Soc. Jpn **46**, 697.
Gesi, K., and M. Iizumi, 1980, J. Phys. Soc. Jpn **48**, 1775.

Golovko, V.A., 1980 Sov. Phys. Solid State **22**, 1729.
Haque, M.S., J.R. Hardy, Q. Kim and F.G. Ullman, 1978, Solid State Commun. **27**, 813.
Horioka, M., A. Sawada and R. Abe, 1980, Jpn J. Appl. Phys. **19**, L145.
Hoshizaki, H., A. Sawada, Y. Ishibashi, T. Matsuda and I. Hatta, 1980, Jpn J. Appl. Phys. **19**, L324.
Hudak, O., 1983, J. Phys. **C16**, 2641.
Iizumi, M., and K. Gesi, 1977, Solid State Commun. **22**, 37.
Iizumi, M., J.D. Axe, G. Shirane and K. Shimaoka, 1977, Phys. Rev. **B15**, 4392.
Ishibashi, Y., 1982, J. Phys. Soc. Jpn **51**, 1220.
Ishibashi, Y., 1983, J. Phys. Soc. Jpn **52**, 2634.
Ishibashi, Y., and V. Dvořák, 1978, J. Phys. Soc. Jpn **44**, 32.
Ishibashi, Y., J. Sugiyama and A. Sawada, 1981, J. Phys. Soc. Jpn **50**, 2500.
Kudo, S., and T. Ikeda, 1981, J. Phys. Soc. Jpn **50**, 733.
Levanyuk, A.P., and D.G. Sannikov, 1976, Sov. Phys. Solid State **18**, 245.
Levstik, A., P. Prelovsek, C. Filipic and B. Žekš, 1982, Phys. Rev. **B25**, 3416.
Midorikawa, M., 1981, J. Phys. Soc. Jpn **50**, 2141.
Nomoto, K., T. Atake, B.K. Chaudhuri and H. Chihara, 1983, J. Phys. Soc. Jpn **52**, 3475.
Petzelt, J., G.V. Kozlov, A.A. Volkov and Y. Ishibashi, 1979, Z. Phys. **B33**, 369.
Prelovsek, P., 1983, J. Phys. **C16**, 3257.
Press, W., C.F. Majkrzak, J.D. Axe, J.R. Hardy, N.E. Massa and F.G. Ullman, 1980, Phys. Rev. **B22**, 332.
Rehwald, W., A. Vonlanthen, J.K. Krüger, R. Wallerius and H.-G. Unruh, 1980, J. Phys. **C13**, 3823.
Sannikov, D.G., 1980, J. Phys. Soc. Jpn **49**, suppl., B75.
Sawada, A., J. Sugiyama, M. Wada and Y. Ishibashi, 1980, J. Phys. Soc. Jpn **48**, 1773.
Shimizu, H., N. Abe, N. Kokubo, N. Yasuda, S. Fujimoto, T. Yamaguchi and S. Sawada, 1980, Solid State Commun. **34**, 363.
Shiozaki, S., A. Sawada, Y. Ishibashi and Y. Takagi, 1977, J. Phys. Soc. Jpn **42**, 353.
Sugiyama, J., M. Wada, A. Sawada and Y. Ishibashi, 1980, J. Phys. Soc. Jpn **49**, 1405.
Takashige, M., T. Nakamura and S. Sawada, 1980, Ferroelectrics **24**, 143.
Tanisaki, S., and H. Mashiyama, 1980, J. Phys. Soc. Jpn **48**, 339.
Unruh, H.-G., W. Eller and G. Kirf, 1979, Phys. Stat. Sol. (a) **55**, 173.
Wada, M., A. Sawada, Y. Ishibashi and Y. Takagi, 1977a, J. Phys. Soc. Jpn **42**, 1229.
Wada, M., H. Ume, A. Sawada, Y. Ishibashi, Y. Takagi and T. Sakudo, 1977b, J. Phys. Soc. Jpn **43**, 544.
Wada, M., A. Sawada and Y. Ishibashi, 1978, J. Phys. Soc. Jpn **45**, 1429.
Wada, M., H. Shichi, A. Sawada and Y. Ishibashi, 1982, J. Phys. Soc. Jpn **51**, 3245.

CHAPTER 12

Incommensurate Phases in an Aromatic Molecular Crystal: Biphenyl

H. CAILLEAU

Groupe de Physique Cristalline, E.R.A. au C.N.R.S. 070015
Université de Rennes
Campus de Beaulieu, 35042 Rennes Cedex
France

Incommensurate Phases in Dielectrics 2
Edited by
R. Blinc and A.P. Levanyuk

Contents

1. Introduction and generalities . . . 73
2. Symmetry properties . . . 75
3. Molecular nature of the structural instability . . . 81
4. Satellite reflections . . . 85
5. Excitations in incommensurate phases . . . 89
6. Concluding remarks . . . 97
References . . . 97
Note added in proof . . . 99

1. *Introduction and generalities*

Among molecular dielectrics, only a few examples exhibiting incommensurate phase transitions are known. Biphenyl, of chemical formula $C_{12}H_{10}$, is one of them. It is an aromatic molecule belonging to the series of p-polyphenyls and consisting of two phenyl rings connected by a single C–C bond.

The ortho-hydrogen repulsion induces a non-planar conformation for the isolated molecule, as a torsional angle exists between the planes of the phenyl rings (fig. 1). The conformation strongly depends on the environment. In the crystalline state, around room temperature (phase I), the molecules are planar. As in other p-polyphenyls (Cailleau et al. 1980a) the competition between intramolecular and intermolecular forces gives rise to a molecular conformation instability. At low temperature, biphenyl crystals undergo structural phase transitions related to the appearance of a torsional angle.

A number of early experimental observations indicate that the low temperature structure must differ from that existing at room temperature. For example, the observation of line splittings in the electronic absorption and emission spectra of the pure biphenyl-d10 (deuterated) crystal, and also in that of biphenyl-h10 (hydrogenated) in a biphenyl-d10 crystal, suggested a structural modification upon cooling to 4.2 K (Hochstrasser et al. 1973). Raman scattering experiments (Friedman et al. 1974) revealed continuous spectral changes as the crystal was cooled down from 75 to 15 K. In an electron paramagnetic resonance study of triplet phenanthrene-d10 and naphthalene-d8 in biphenyl-h10 crystal, Cullick and Gerkin (1976) observed that each line splits into four lines around 39 K. They also reported the observation of a possible second phase transition around 15 K (Cullick and Gerkin 1977). Bree and Edelson (1977) extended the earlier Raman studies and inferred the occurrence of a soft mode mechanism for the first transition. By a birefringence experiment they confirmed the existence of the second transition (Bree and Edelson 1978). Later, using neutron diffraction, these transitions were found to lead to the existence of two low temperature incommensurate phases labelled II and III (Cailleau et al. 1979c).

In phase II, below the first transition temperature T_I, four satellite reflections appear in the vicinity of the zone boundary along $\boldsymbol{b}^*$ (fig. 2); the

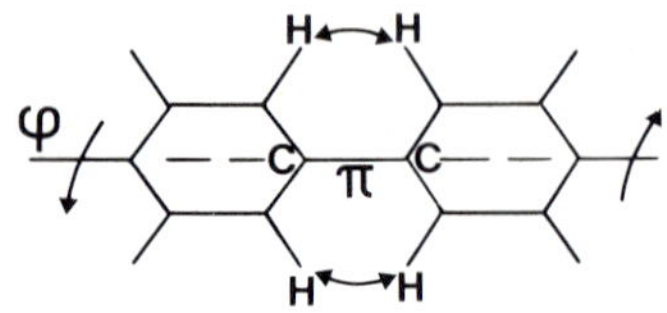

Fig. 1. Schematic drawing of the biphenyl molecule showing the existence of a torsional angle due to the ortho-hydrogen steric repulsion.

modulation wave vectors are

$$\boldsymbol{q}_s = \pm\,\delta_a \boldsymbol{a}^* \pm (1-\delta_b)\boldsymbol{b}^*/2.$$

In phase III, below the second transition temperature T_{II}, only two satellite reflections located along $\boldsymbol{b}^*$ remain (fig. 2); the modulation wave vectors become

$$\boldsymbol{q}_s = \pm(1-\delta_b)\boldsymbol{b}^*/2$$

The values of δ_a and δ_b are roughly equal to 0.05 and 0.085 at T_{I} and are weakly temperature dependent. No simple commensurate phase corresponding to

$$\boldsymbol{q}_s = \boldsymbol{b}^*/2$$

has been found so far in the lower temperature range. The sequences of phases in hydrogenated and deuterated biphenyl are given in table 1.

The incommensurate phases in biphenyl present a number of characteristic properties with respect to other incommensurate dielectric systems. On the one hand, the biphenyl crystal is a molecular crystal constituted of non-polar molecules and exhibiting an unusual type of structural instability. On the other hand, the successive appearance of two different incommensurate phases with the existence of a partial lock-in at one line of the Brillouin zone is, to our

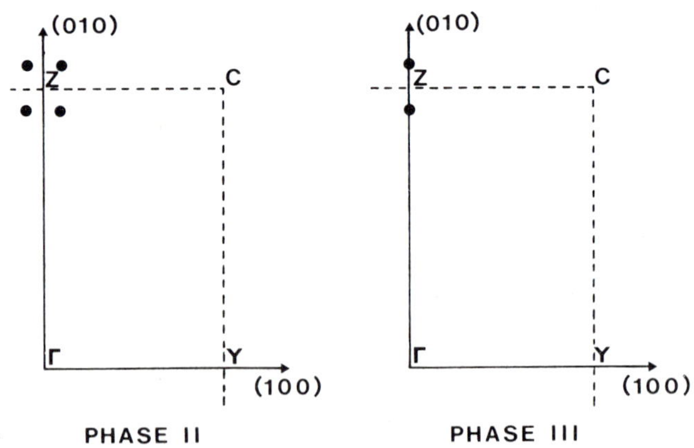

Fig. 2. Locations of satellite reflections for phase II and phase III. Dotted lines correspond to the limits of the first Brillouin zone.

Table 1
Sequence of phases in hydrogenated and deuterated biphenyl.

	Phase III		Phase II		Phase I
	→	T_{II}	↔	T_I	←
$C_{12}H_{10}$		17 K		40 K	
$C_{12}D_{10}$		20 K		37 K	

knowledge, unique in dielectric systems. In the same way, the persistence of an incommensurate modulation at very low temperature is not so common; this last point is particularly interesting for the studies of excitations in incommensurate phases.

All these aspects are discussed in the different parts of this chapter. Section 2 is devoted to symmetry considerations on these incommensurate phases. The molecular nature of the structural instability is discussed in section 3. Then, the study of satellite reflections is presented (section 4). Finally, in the last part (section 5), the observation of excitations in these incommensurate phases is considered.

2. *Symmetry properties*

In comparison with many other insulating incommensurate systems, the symmetry of biphenyl crystal is low. In phase I, the crystal structure is monoclinic with space group $P2_1/a(C_{2h}^5)$ and two molecules in the unit cell (fig. 3) (Trotter 1961, Robertson 1961, Hargreaves and Rizvi 1962). Symmetry elements of this non-symmorphic space group are:

$\{E\|0\}$	identity,
$\{I\|0\}$	inversion,
$\{C_2^b\|\boldsymbol{a}/2+\boldsymbol{b}/2\}$	two-fold screw axis,
$\{\sigma^{ac}\|\boldsymbol{a}/2+\boldsymbol{b}/2\}$	glide plane.

The point group is 2/m. The molecules occupy inversion sites, which implies a planar molecular conformation. In the unit cell, one molecule can be transformed into the other by the screw axis or the glide plane.

In the high temperature phase, the soft mode is located in the immediate vicinity of the $Z(0\,\frac{1}{2}\,0)$ zone boundary point. The $(0\,\xi\,0)$ direction is parallel to the two-fold screw axis. Hence for this direction both one symmetric and one antisymmetric representation with respect to the diad axis exist. At the zone boundary, the symmetric and the antisymmetric modes become degenerate and merge with opposite but finite slopes.

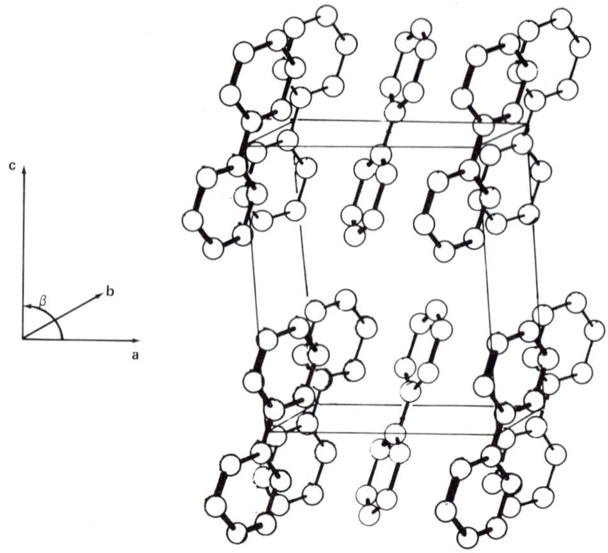

Fig. 3. Crystal structure of biphenyl in the high-temperature phase (phase I). At room temperature the cell parameters are: $a = 8.11$ Å; $b = 5.56$ Å; $c = 13.61$ Å; $\beta = 92°$.

Figure 4 shows the dispersion curves for the soft phonon branches in the $(0\,\xi\,0)$ direction as obtained from inelastic neutron scattering experiments (Cailleau et al. 1985). An extended zone scheme which is more convenient in this case has been used. The symmetric or antisymmetric character of the excitations has been determined from extinction rules for a momentum transfer Q parallel to the $(0\,\xi\,0)$ direction. The symmetric modes are extinct in $(0\,k\,0)$ zones for odd k and the antisymmetric ones for even k (Dorner 1982). The symmetric character of the soft mode can be proved in this way. It is worthwhile to mention that various lattice dynamics calculations have given this same result (Natkaniec et al. 1981, Takeuchi et al. 1981, Wasiutynski et al. 1981). Furthermore, it also agrees with that obtained by a determination of the modulated structure of phase III (Baudour and Sanquer 1983), the symmetry of this phase being discussed in terms of superspace groups.

The above-mentioned arguments can explain the existence of a single modulation along $\boldsymbol{b}$ as in phase III. But, an understanding of the appearance of four satellite reflections below T_I requires more sophisticated arguments. Along the $(\xi\,q_b\,0)$ direction, perpendicular to the ΓZ line, the slope of the phonon dispersion branches is zero for $\xi = 0$. The two related soft phonon branches, which belong to two different representations on the ΓZ line, belong to the same unique representation for a general point inside the Brillouin zone. Therefore, the corresponding modes can interact with an interaction energy

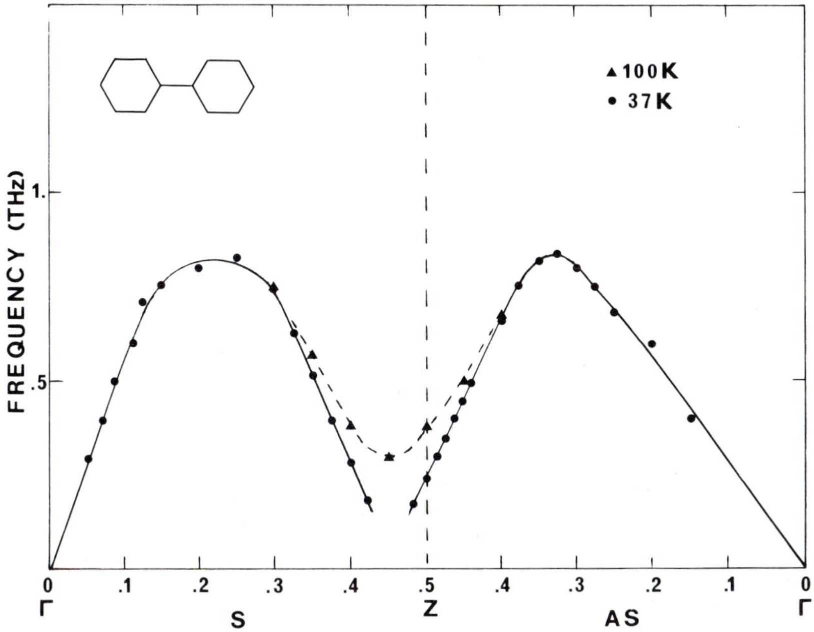

Fig. 4. Dispersion of the soft phonon branches in the $(0\,\xi\,0)$ direction in an extended zone scheme. S symmetric, AS antisymmetric.

which increases with the value of the ξ component. These simple symmetry considerations may explain the existence of minima in the soft phonon surface located away from the ΓZ line, in connection with the occurrence of four satellite reflections. Thus, biphenyl is an unusual incommensurate system which along the $\boldsymbol{b}^*$ direction presents a situation similar to that of $(NH_4)_2BeF_4$ (Levanyuk and Sannikov 1976) and along the $\boldsymbol{a}^*$ direction a situation similar to that of thiourea or $NaNO_2$ (Ishibashi and Shiba 1978).

Figure 5 presents some soft phonon dispersion curves around the satellite location at the transition temperature T_I. It shows the shape of the lower and upper soft phonon dispersion surfaces. Note that the two modes are continuously degenerate along the zone boundary line ZC.

It is well known that the situation for the soft phonon branch along the $\boldsymbol{b}^*$ direction corresponds to the existence of a Lifshitz invariant in the Landau free energy expansion (Levanyuk and Sannikov 1976, Ishibashi and Dvořák 1978). If we abstract the existence of the intermediate phase II, only a two-dimensional order parameter $\eta(\eta_1, \eta_2)$ must be considered; in phase III, the star of the satellite wave vector has two arms as the related points are located on the diad axis (fig. 6). The Landau free energy density can be written (Ecolivet et al.

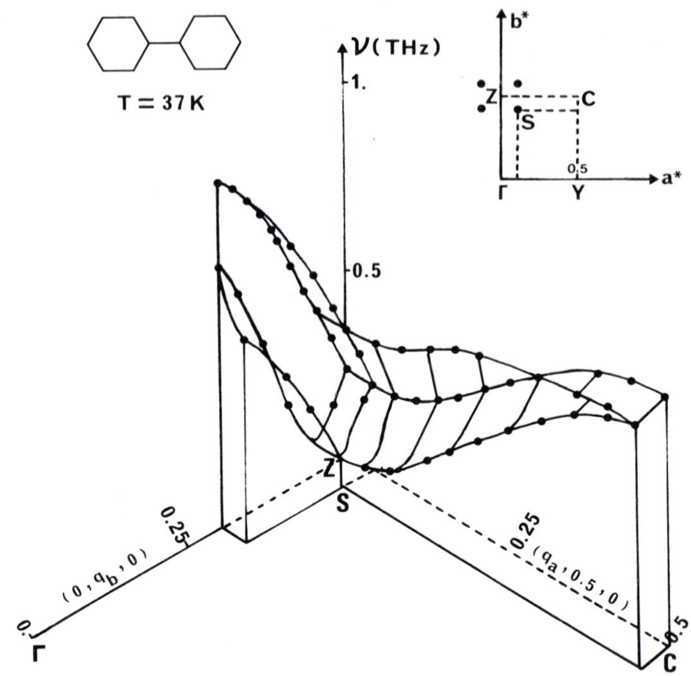

Fig. 5. Soft phonon dispersion surfaces around the satellite position at the transition temperature T_I.

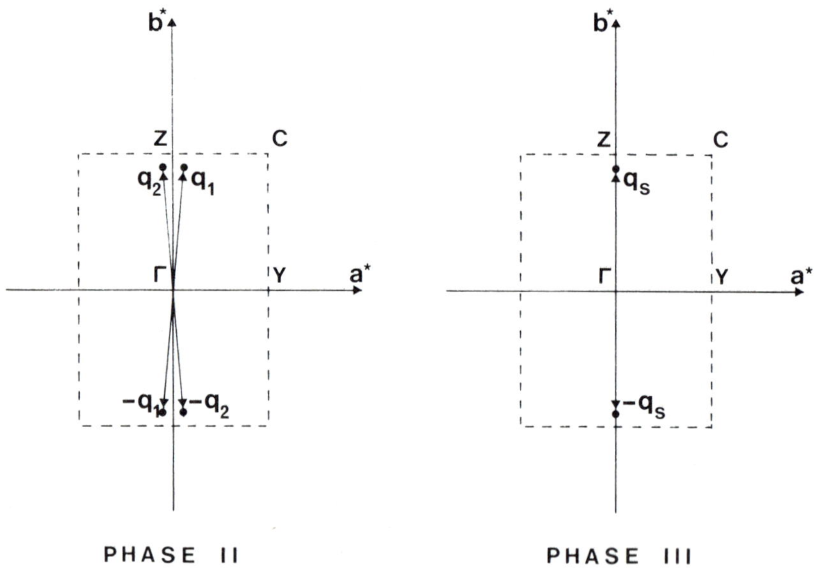

Fig. 6. Star of the satellite wave vector in the two incommensurate phases.

1983):

$$f = \frac{A}{2}\left(\eta_1^2 + \eta_2^2\right) + \frac{B_1}{4}\left(\eta_1^2 + \eta_2^2\right)^2 + \frac{B_2}{4}\eta_1^2\eta_2^2 + \delta\left(\eta_1\frac{\mathrm{d}\eta_2}{\mathrm{d}y} - \eta_2\frac{\mathrm{d}\eta_1}{\mathrm{d}y}\right)$$
$$+ \frac{K}{2}\left[\left(\frac{\mathrm{d}\eta_1}{\mathrm{d}y}\right)^2 + \left(\frac{\mathrm{d}\eta_2}{\mathrm{d}y}\right)^2\right] + \cdots,$$

where, as usual, $A = a(T - T_i)$. Ishibashi (1981) has tentatively interpreted the existence of phase II by adding supplementary terms which favour the incommensurate modulation along the $\boldsymbol{a}^*$ direction. In this way, he described the sequence of phase transitions in biphenyl. However, in this phenomenological theory a two-dimensional order parameter is conserved although, in phase II, the star of the satellite wave vector has four arms, corresponding to a general point inside the Brillouin zone (fig. 6). Toledano and Guilluy (1984) have recently discussed the application of Landau theory for systems with partial lock-in transitions and emphasized the importance of the choice of a four-dimensional order parameter for biphenyl.

As for other incommensurate systems with a four-component order parameter (Cowley 1980), it is interesting to discuss the symmetry of phase II. Indeed, there are two possibilities for the modulated structure:

(1) a combination of displacements associated with wave vectors $\pm\boldsymbol{q}_1$ *and* $\pm\boldsymbol{q}_2$.

(2) a domain structure corresponding to displacements associated with wave vectors $\pm\boldsymbol{q}_1$ *or* $\pm\boldsymbol{q}_2$.

These two structures have different types of second-order satellite reflections since, only in the first case, reflections for the wave vector $\boldsymbol{q}_1 + \boldsymbol{q}_2$ exist. As indicated in the third part of this chapter, a neutron scattering experiment performed on a triple-axis spectrometer (Cailleau et al. 1985) has shown the possibility for such a satellite reflection at the $(1|\boldsymbol{q}_1 + \boldsymbol{q}_2|0)$ position. However, the occurrence of multiple scattering effects cannot be ruled out and more extensive studies are necessary before a final conclusion can be reached. Furthermore, the symmetry of phase II has recently been discussed in terms of macroscopic properties (Dvořák et al. 1983) and of Bravais classes (Janner et al. 1983). The point group which describes the symmetry of macroscopic properties as well leaving invariant the set of all satellite reflections, is $2/\mathrm{m}$ in the case of a combination of displacements and $\bar{1}$ in the case of a domain structure.

In phase II, the possibility of a component for the modulation wave vector along $\boldsymbol{c}^*$ must be considered, since the locations of satellite reflections correspond to general points inside the Brillouin zone. Until now, such a possibility cannot be ruled out from the above-mentioned triple-axis experiments (Cailleau et al. 1979c, 1985). (See note added in proof, p. 99.)

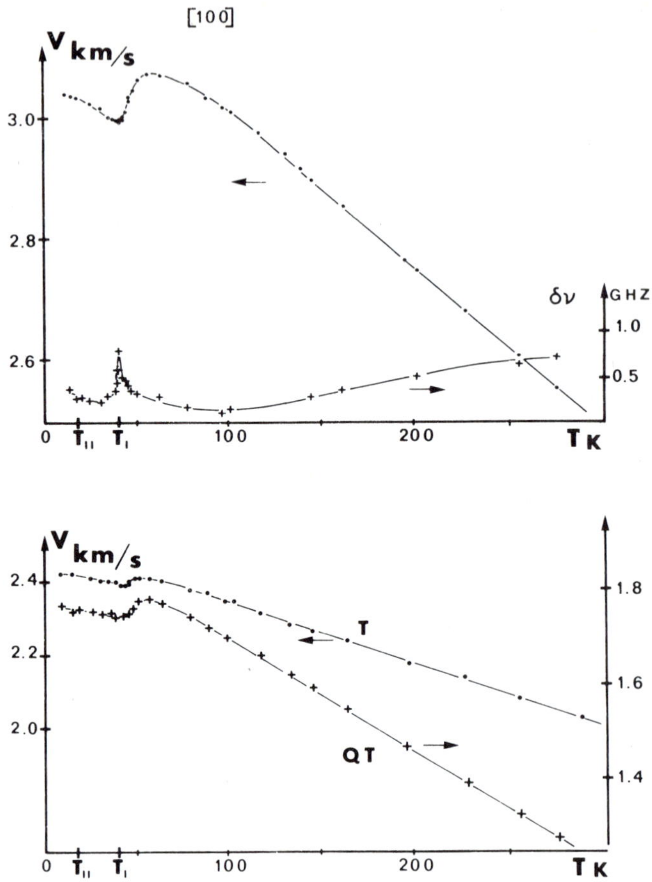

Fig. 7. Sound velocities of the three modes propagating along ***a*** from Brillouin scattering measurements. (Ecolivet et al. 1983.)

Ecolivet and co-workers (Ecolivet and Sanquer 1980, Ecolivet et al. 1983) have performed Brillouin scattering experiments from room temperature down to liquid helium temperature. The main features of the detected acoustic anomalies are, first, an almost perfect similarity between all the singularities recorded at T_I and, secondly, the absence of any apparent anomaly at T_{II}. A typical example of these results is shown in fig. 7. The elastic anomalies have been qualitatively interpreted within the framework of Landau theory using the two-dimensional Z representation. So, terms allowed by symmetry for the coupling of strains (in Voigt notation) with order parameter invariants are added to the free energy density considered above:

$$f_c = (\eta_1^2 + \eta_2^2) \sum_{i=1,2,3,5} g_i e_i + (\eta_1^2 - \eta_2^2) \sum_{i=4,6} g_i' e_i$$

$$+ (\eta_1^2 + \eta_2^2)\left(\sum_{i,j=1,2,3,5} h_{ij} e_i e_j + \sum_{i,j=4,6} h_{ij}' e_i e_j \right) + 2 \cdots .$$

The experimental data are consistent with this anharmonic quadratic coupling. It is important to bear in mind that this simple theory does not account for the exact form of the free energy expansion for phase II, since a four-dimensional order parameter should be used. However, since the soft mode occurs near the Z point, coupling terms and the resulting anomalies are not expected to be basically different.

3. *Molecular nature of the structural instability*

As underlined in the Introduction, the structural instability in biphenyl results from a molecular conformation change. It is interesting to discuss the specific aspects of this unusual soft mode mechanism.

The essential feature concerns the competition between intramolecular and intermolecular forces. The conjugation energy between phenyl rings is not sufficient to overcome the ortho-hydrogen steric repulsion, and this situation induces a non-planar conformation in the gaseous phase, with the existence of a phenyl–phenyl torsional angle. Electron diffraction studies (Karle and Brockway 1944, Bastiansen 1949, Bastiansen and Traetteberg 1962) yield a torsional angle of 42°. Furthermore, several computations of the corresponding intramolecular potential have been performed (Casalone et al. 1968, Franchini and Sbrana 1970, Dewar and Herget 1970, Almlöf 1974). The molecular conformation of biphenyl strongly depends on its environment. So, the torsional angle is reduced to 20° in the liquid phase (Suzuki 1959). In the crystalline state, early and recent X-ray diffraction studies (Trotter 1961, Robertson 1961, Hargreaves and Rizvi 1962, Charbonneau and Delugeard 1977) consistently show that, at room temperature, molecules are planar. Thus, the intermolecular forces are antagonistic to the intramolecular ones. The space group $P2_1/a$ is very common for planar aromatic molecules, and crystal packing calculations (Casalone et al. 1968, Messager 1976, Brock 1979, Busing 1983) have shown that intermolecular interactions may favour a planar conformation. The gain of energy packing by adopting a zero torsional angle geometry counterbalances the internal ortho-hydrogen repulsion. The structural phase transitions in biphenyl, as in other p-polyphenyls, are related to this delicate balance between competing intramolecular and intermolecular forces (Cailleau et al. 1980a). An illustration of this situation is given by the results of Busing (1983). Starting from a structure with non-planar biphenyl molecules, he has shown that increasing the importance of the conjugation energy with respect to the non-bonded interactions causes his model to undergo the reverse phase change to the $P2_1/a$ structure with planar molecules.

Contrary to other p-polyphenyls, which undergo phase transitions of order–disorder nature (Baudour et al. 1977, Cailleau et al. 1979a, Cailleau et al. 1980a, Toudic et al. 1983), in the case of biphenyl the nature of the phase transition at T_I is displacive. The experimental results on the torsional soft mode dynamics are given in section 5. However, careful structure refinements of thermal motion parameters have anticipated large librational motion about the long molecular axis (Charbonneau and Delugeard 1976, 1977). Much of this motion may be the result of low frequency torsional vibrations. The mean-square amplitudes were found equal to 45.7 and 109.2 $(\mathrm{deg})^2$ at 110 and 293 K respectively, thus showing near proportionality to temperature. It is interesting to mention that Bonadeo and Burgos (1982) have performed lattice dynamics calculations of these mean-square amplitudes which are in fair agreement with experimental values.

Of course, the rigid-molecule approximation cannot be used for phonon dynamics calculations in solid biphenyl (Burgos et al. 1976, Natkaniec et al. 1978). More elaborate models must be considered, taking into account the mixing between interphenyl and lattice vibrations. In this way, some lattice dynamics calculations have been performed at the Brillouin zone center (Burgos et al. 1976) and, more recently, in the whole Brillouin zone (Natkaniec et al. 1981, Takeuchi et al. 1981, Wasiutynski et al. 1981, Plakida et al. 1983). The results show that there is a strong coupling between phenyl–phenyl torsions and translations. In the $(0\xi 0)$ direction the torsional modes strongly depend on the q-value and couple to other branches of the same symmetry. The lowest symmetric branch exhibits a minimum located near the zone boundary point. At this minimum the corresponding mode has a predominant torsional motion character. Figure 8, from Plakida et al. (1983), shows the effects of mutual coupling between acoustic and torsional branches in the $(0\xi 0)$ direction. This scheme is consistent with the experimental results for the dispersion curves of the soft mode (see fig. 4). Furthermore, we can underline the good agreement between the calculated phonon density of states resulting from these lattice dynamics models and low temperature inelastic incoherent neutron scattering spectra (Natkaniec et al. 1981, Takeuchi et al. 1981).

Plakida et al. (1983) have presented a model for the internal soft mode phase transition in biphenyl. A self-consistent approximation is used, first by including the mixing of the anharmonic molecular torsion with all external modes, secondly by considering the mixing with the acoustic modes only. The two

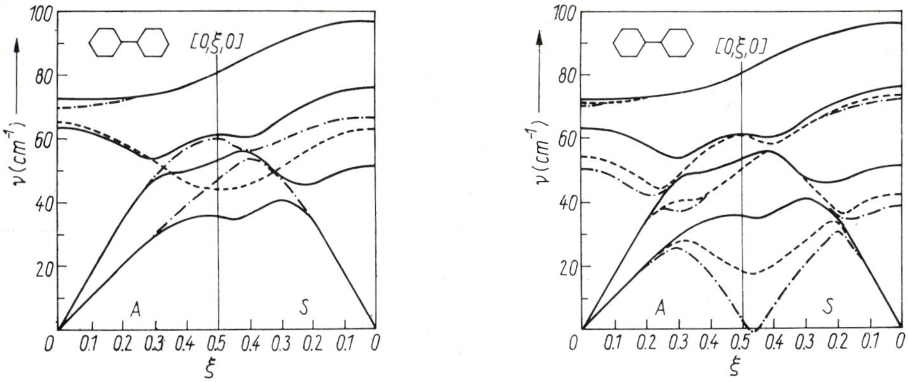

Fig. 8. Acoustic and torsional branches with (solid lines) and without (dashed lines: torsional, dash-dotted lines: acoustic) mutual coupling. (Plakida et al. 1983.)

results are similar and agree well with the experimental data for the temperature dependence of the soft mode (see section 5). Busing (1983) has proposed another way of modeling the phase change in crystalline biphenyl by using a temperature-dependent potential. In this potential the non-bonded terms are modified according to the thermal motions of the atoms involved. So, the motion perpendicular to the ortho-hydrogen interaction vector permits mean atomic locations to become closer to each other, the mean instantaneous distance being larger than the distance between mean atomic positions. Thus, at high temperatures, the molecules can be planar on the average. At low temperatures, the amplitude of thermal motions decreases and the molecules must have a torsional angle to increase the hydrogen–hydrogen distance.

In biphenyl, the study of the influence of pressure on its structural phase transitions yields experimental evidence for the competition between intramolecular and intermolecular forces. Indeed, pressure increases the relative strength of intermolecular interactions and, therefore, the first transition temperature T_I, corresponding to the lattice instability, is expected to decrease with increasing pressure (Cailleau et al. 1980a). The phase diagram (fig. 9) has been determined by neutron diffraction on a deuterated sample (Cailleau et al. 1984). As expected from the above considerations, the first transition temperature T_I decreases with pressure. At low pressure the behaviour of T_I is linear ($dT_I/dP \simeq -17\,K/kbar$); it becomes non-linear above, the transition vanishing at about 2 kbar. Above this pressure, the molecules of biphenyl remain planar at all temperatures. An interesting feature is the possible occurrence of quantum effects on approaching the so-called displacive limit (i.e. $|dT_I/dP| \to \infty$ as $T_I \to 0\,K$). This behaviour of biphenyl is very similar to that observed in K_2SeO_4 (Samara et al. 1981), but at much lower pressure. Furthermore, the

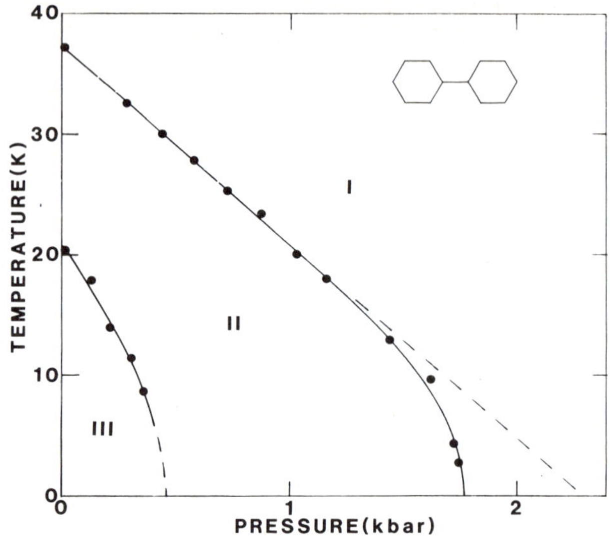

Fig. 9. Phase diagram of deuterated biphenyl.

partial lock-in transition temperature T_{II} also decreases with pressure and goes to 0 K at about 0.5 kbar. In contrast to the first transition which looks continuous at all pressures, the second one is always discontinuous. Besides, a study of the pressure dependence of the low frequency Raman spectra at room temperature (Kirin et al. 1983) shows the possibility of another structural change between 8 and 10 kbar.

Finally, we mention the results of neutron diffraction studies. In a first study (Cailleau et al. 1979b), the structural transition has been interpreted as resulting from the appearance of super-lattice reflections at the $Z(0\,\frac{1}{2}\,0)$ zone boundary point. This structural investigation showed the existence of a non-zero torsional angle in the low temperature phases. The modulated structure of phase III has later been determined by Baudour and Sanquer (1983). Starting from the basic structure, which is essentially the structure in the high temperature phase, with planar molecules, the modulated structure is described within a sinusoidal approximation. It results from the combination of three motions:

(a) A torsion around the long molecular axis with an amplitude of 5.5° for each phenyl ring, the maximum deformation angle being 11°. This torsion constitutes the essential aspect of the modulation.

(b) A rotation around the normal to the mean molecular plane with an amplitude of 1.0°.

(c) A translation along the long molecular axis with an amplitude of 0.035 Å.

The main features of the modulation are in good agreement with the motional character which results from lattice dynamics calculations described above.

4. *Satellite reflections*

A detailed study of the temperature dependence of the satellite reflections has been performed by neutron scattering measurements on a deuterated compound (Cailleau et al. 1979c, Cailleau 1981, Cailleau et al. 1985). Figure 10 shows in phase II typical profiles of the elastic scattering scanned along the $\boldsymbol{a}^*$ direction and along the $\boldsymbol{b}^*$ direction ($\boldsymbol{q}_s = \pm\,\delta_a \boldsymbol{a}^* \pm (1-\delta_b)\,\boldsymbol{b}^*/2$). The deviations δ_a and δ_b weakly decrease with decreasing temperature: δ_a lies between 0.043 and 0.051, while δ_b falls into the range 0.080–0.085 (fig. 11). In phase III, δ_a vanishes ($\boldsymbol{q}_s = \pm(1-\delta_b)\,\boldsymbol{b}^*/2$) and $\delta_b \simeq 0.075$ is practically temperature independent (fig. 11). Taking into account the precision of these measurements, the possibility to have a high-order commensurate phase can, as usual, not really be excluded. Here, the commensurate wave vector $\frac{6}{13}\boldsymbol{b}^*$ would correspond to a value of 0.077 for δ_b. The variation of the intensity of satellite reflections with temperature is given in fig. 12; in order to compare their behaviour in the two phases, the intensities for $\boldsymbol{q}_1 = +\,\delta_a \boldsymbol{a}^* + (1-\delta_b)\,\boldsymbol{b}^*/2$ and $\boldsymbol{q}_2 = -\,\delta_a \boldsymbol{a}^* + (1-\delta_b)\,\boldsymbol{b}^*/2$ have been added in phase II. The satellite intensity continuously evolves at the first transition temperature T_{I}, in consistence with its second-order character. No attempt has been made to determine a critical exponent, because measurements are probably performed outside the critical region. At T_{II}, a discontinuous jump in satellite intensities and locations is clearly observed, in agreement with the first-order nature of this transition. Around T_{II}, coexistence of the two phases II and III has also been observed using scans along $\boldsymbol{a}^*$ which present three peaks. No simple commensurate phase corresponding to $\boldsymbol{q}_s = \boldsymbol{b}^*/2$ has been found at least down to 1.4 K.

The main characteristics of the static behaviour of incommensurate phases in biphenyl are the remaining incommensurability along $\boldsymbol{b}^*$ and the existence of a partial lock-in transition along $\boldsymbol{a}^*$. They are in good agreement with the anisotropic shape of the soft mode dispersion surface observed in the high temperature phase (fig. 5). Indeed, the dispersion curve is practically flat along the $\boldsymbol{a}^*$ direction while the minimum is very well pronounced along the $\boldsymbol{b}^*$ direction, the incommensurate wave vector remaining close to that corresponding to the reciprocal location of the deep minimum.

Furthermore, low temperature heat capacity measurements (Atake and Chihara 1980, Atake et al. 1983) have shown the occurrence of a broad anomaly around T_{I} (fig. 13) which is related to the second-order nature of the soft mode phase transition as discussed above. The second anomaly due to the

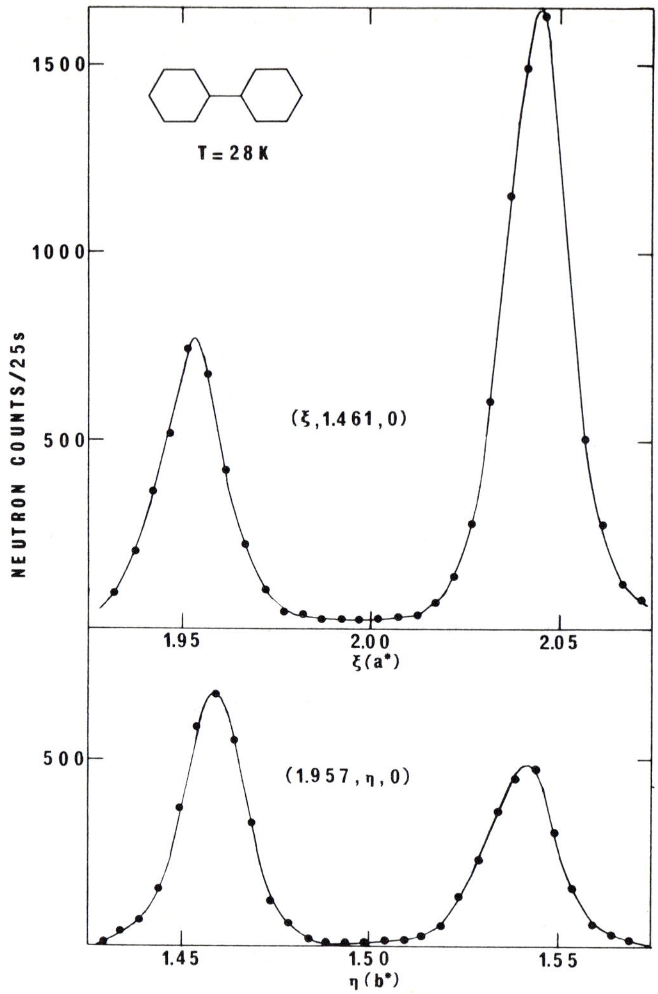

Fig. 10. Elastic scans performed along the $\boldsymbol{a}^*$ and $\boldsymbol{b}^*$ directions at 29 K. (Cailleau et al. 1979b.)

partial lock-in transition is also visible but much smaller, and it is therefore difficult to draw conclusions on the order of this transition on the basis of these data only.

Higher-order satellite reflections have also been studied using neutron scattering (Cailleau 1981, Cailleau et al. 1985). In phase III, the second-order satellites have a characteristic wave vector $2\boldsymbol{q}_{\mathrm{s}} = \pm\,\delta_b \boldsymbol{b}^*$ and are located

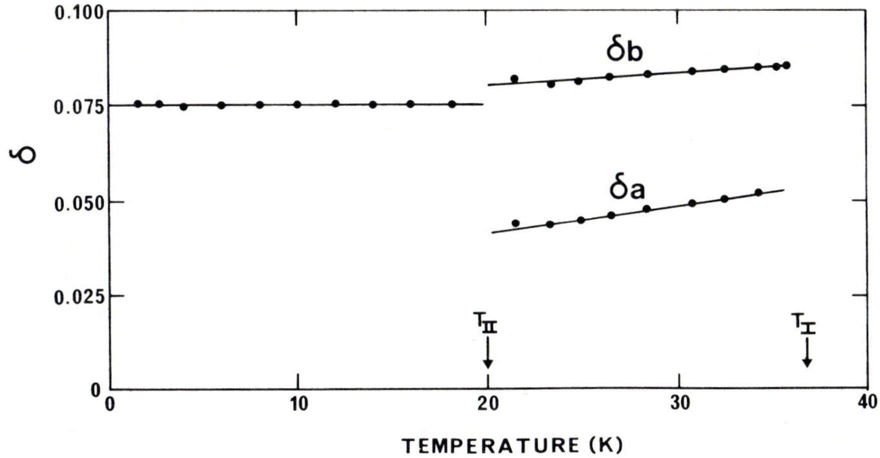

Fig. 11. Temperature dependence of the deviations δ_a and δ_b.

near the Brillouin zone center, the third-order ones correspond to $3\boldsymbol{q}_s = \pm(1-3\delta_b)\,\boldsymbol{b}^*/2$ and lie again near the zone boundary, and so on. In this phase, it has been possible to measure high-order satellites up to the third-order. Figure 14 shows typical elastic scans for different-order satellite reflections. It is not always easy to discriminate between the occurrence of diffraction harmonics and the possibility of higher-order components in the modulation

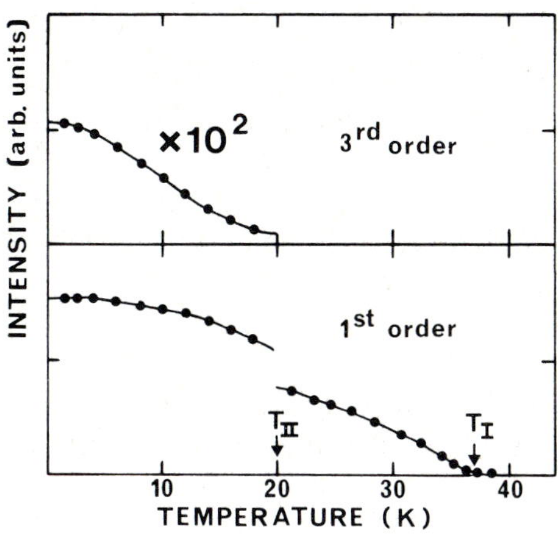

Fig. 12. Intensities of typical first- and third-order satellite reflections versus temperature. In phase II, the intensity of the two peaks found along $\boldsymbol{a}^*$ is added.

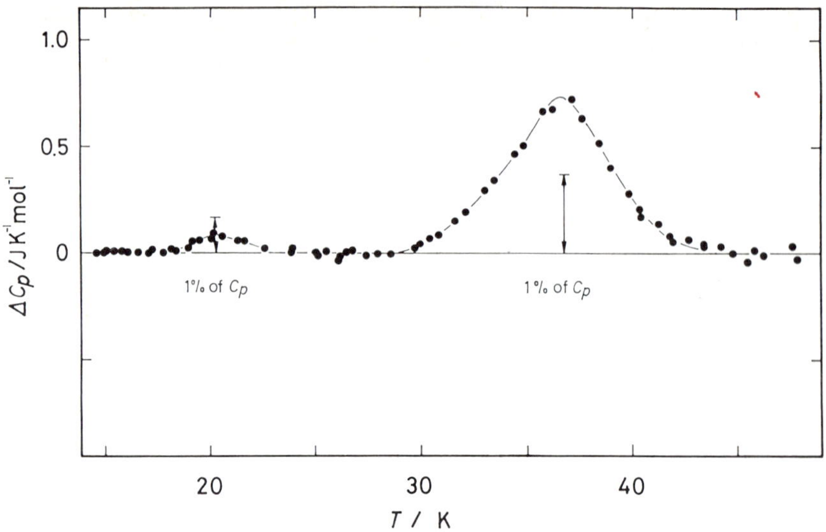

Fig. 13. Excess heat capacity of biphenyl-d 10. (Atake et al. 1983.)

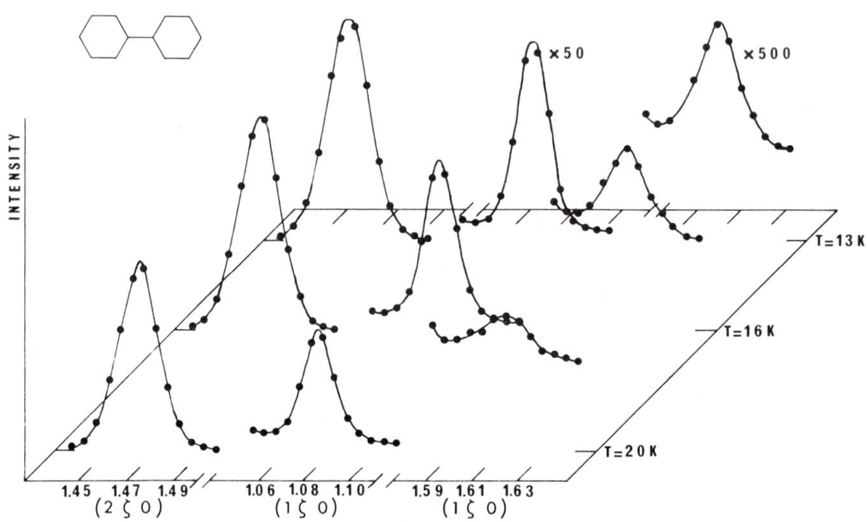

Fig. 14. Profiles of the elastic scattering scanned along $\boldsymbol{b}^*$ direction showing different order satellite reflections.

wave. However, the ratio of the intensities for neighbouring third- and first-order satellites is relatively high (between 10^{-2} and 10^{-3}), which is in favour of the second possibility. The temperature dependence of the intensity of these third-order satellites is also shown in fig. 12. As expected, this intensity strongly increases with decreasing temperature. Due to the occurrence of intensity saturation phenomena at very low temperature, it is difficult to determine the power law of the intensity behaviour of second- and third-order satellites and to compare it to the η^4 and η^6 behaviour. It is interesting to remember that the satellite reflections have essentially been measured near the $(2\,\frac{1}{2}\,0)$ and $(2\,\frac{3}{2}\,0)$ points and are found intense only on one side of these zone boundary points, in fact the side opposite to the one where the fundamental satellites are strong. This feature may correspond to the fact that the third-order component of the modulation wave is expected to be related to the upper soft mode branch (Dvořák and Ishibashi 1978). In phase II, the intensity of high-order satellites is probably very weak and only the satellite reflections at the $(1|\boldsymbol{q}_1+\boldsymbol{q}_2|0)$ position, already mentioned in section 2, have up to now been observed.

5. *Excitations in incommensurate phases*

In the Introduction we have indicated that, in incommensurate phases, the persistence of the modulation at very low temperature constitutes an interesting situation for the study of low frequency excitations. Indeed, if the phason branch has an acoustic-like dispersion relation, the damping constant of these excitations does not vanish when $\boldsymbol{q} \to 0$, but remains of the order of that of ordinary optical phonons (Golovko and Levanyuk 1981, 1983, Zeyher and Finger 1982). Therefore, long-wavelength phasons are necessarily overdamped. However, in biphenyl, the damping constant is small at very low temperature, and so, propagating phason branches have been clearly observed over a relatively large range of wave vectors by inelastic neutron scattering experiments. (Cailleau et al. 1980b, 1981, Cailleau 1981).

First, it is important to underline again the displacive character of the transition (Cailleau et al. 1979a). This is clearly confirmed by fig. 15 which shows the pronounced softening of the mode corresponding to the minimum of the appropriate dispersion surfaces presented in fig. 5. Only very close to the transition temperature does the soft mode become overdamped. Besides, heat capacity measurements (Dworkin and Cailleau 1980, Atake and Chihara 1980, Atake et al. 1983) have indicated that the thermal anomaly due to the soft mode transition is very small: $\Delta S_t \simeq 0.128\,\mathrm{J\,K^{-1}\,mol^{-1}}$ (cf. fig. 13).

In phase III, inelastic neutron scattering experiments have revealed the existence of low frequency excitations which present all known characteristics

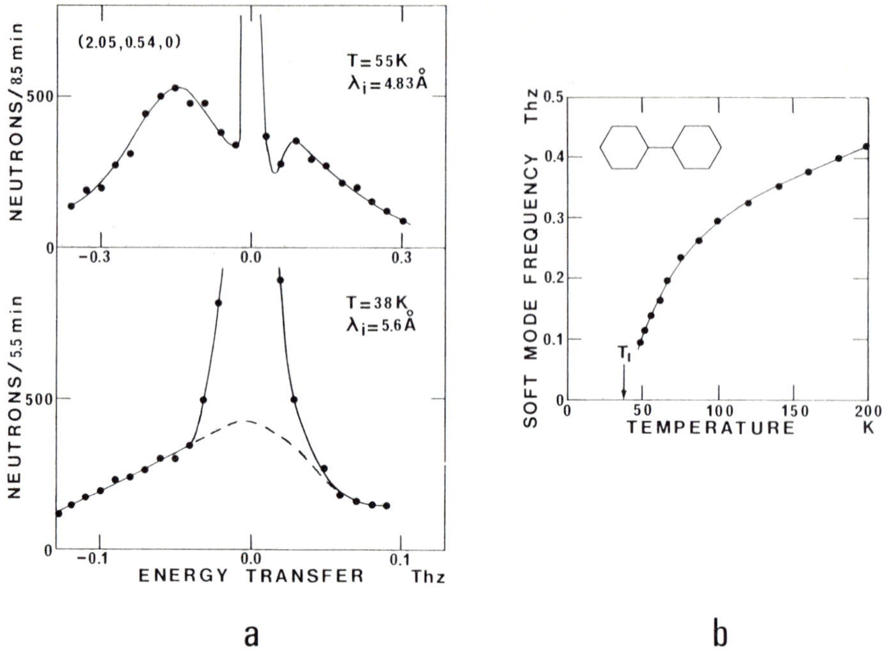

Fig. 15. (a) Energy scans for the soft mode at 55 K (underdamped) and at 38 K (overdamped). (b) Temperature dependence of the frequency of the soft mode.

of propagating phase modes (Cailleau et al. 1980b, 1981, Cailleau 1981). Figure 16 shows energy scans for different reduced wave vectors along the $\boldsymbol{a}^*$ direction, near a satellite reflection at 10 K. The inelastic peaks are well measured for both neutron energy-loss and energy-gain. The dispersion of these excitations is found to follow a linear law originating at the satellite reflection with a slope similar to that of the soft phonon branch in the high temperature phase. This slope is much lower than that of the lowest acoustic branch (fig. 17), the different sound velocities having been obtained from low temperature Brillouin scattering experiments (Ecolivet et al. 1983). In addition, fig. 17 shows energy scans performed under the same experimental conditions and corresponding to the same reduced wave vector; one is located near a satellite reflection, whereas the other one is obtained near an intense main Bragg reflection. The difference between this excitation and the transverse acoustic mode is clearly seen. So, it is possible to identify these observed excitations as inhomogeneous phasons. Along the $\boldsymbol{b}^*$ direction, an ambiguity may subsist because the phason dispersion slope is similar to that of an acoustic branch. However, the observed neutron intensity scattered by the phasons is much too strong to be related to acoustic phonon scattering, but is

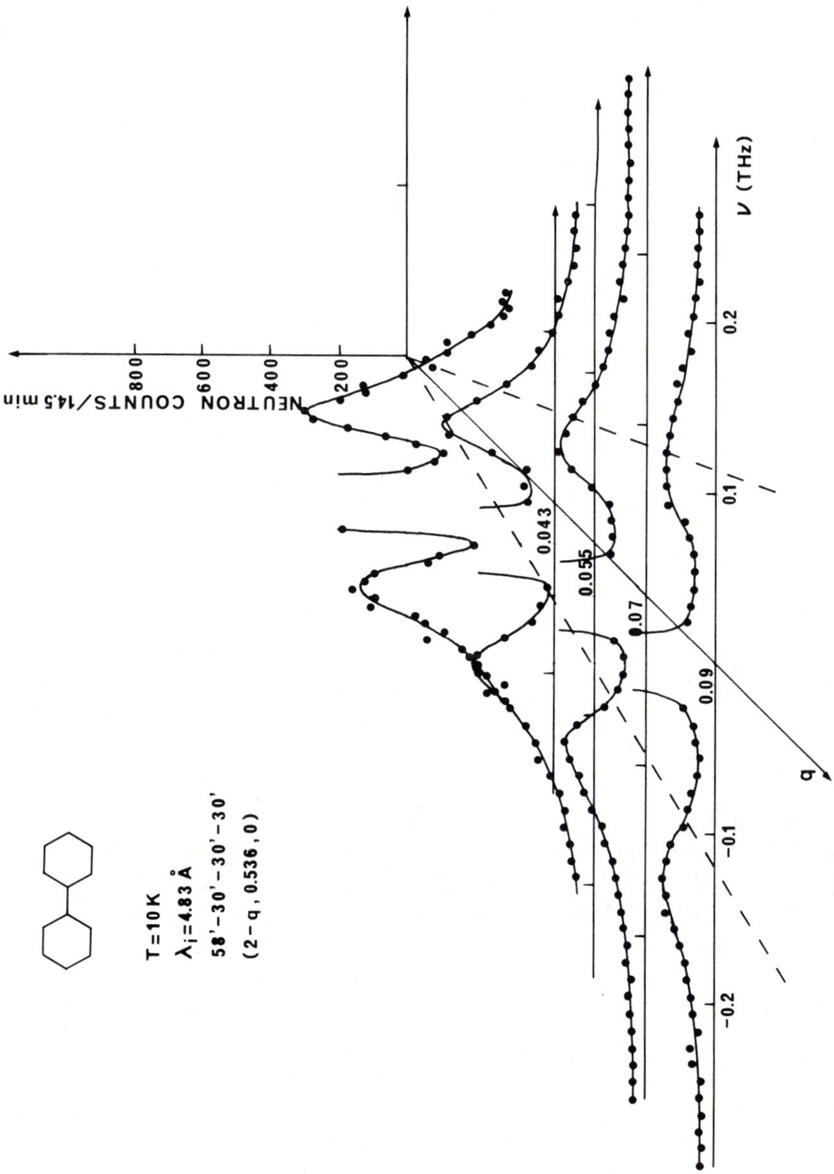

Fig. 16. Energy scans for different reduced wave vectors along $\boldsymbol{a}^*$ in phase III. (Cailleau et al. 1981.)

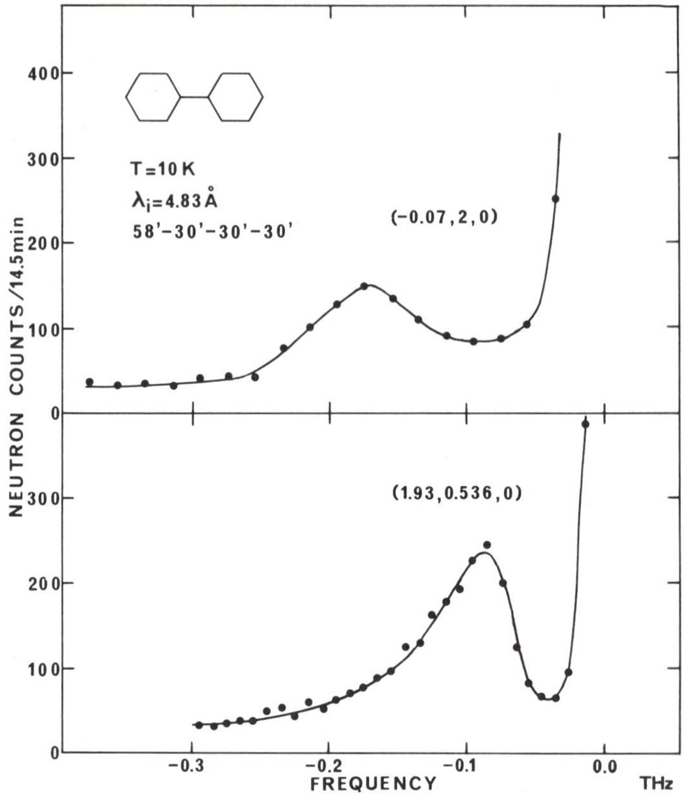

Fig. 17. Energy scans close to a satellite reflection (bottom) and to a main Bragg reflection (top). (facing page) Dispersion law for the phason branch (solid line) and the lowest acoustic branch (dotted line) in the $\boldsymbol{a}^*$ direction.

in agreement with that observed for phasons propagating along the $\boldsymbol{a}^*$ direction. Furthermore, it is important to notice that the existence of a small gap for the homogeneous phason at $\boldsymbol{q} = 0$ cannot be excluded on the basis of these neutron data.

Particular excitations at higher frequencies have also been observed (Cailleau et al. 1985). Due to the proximity above T_I between the upper and lower soft phonon branches, the excitation spectrum in the incommensurate phases of biphenyl is more complex than in usual cases. A sketch of phase and amplitude mode dispersion curves for the $\boldsymbol{b}^*$ direction, as expected if we abstract the existence of the intermediate phase II, is given in fig. 18 in an extended zone scheme. In phase III (fig. 19), the amplitudon branches are found at relatively high energy, and so, in inelastic neutron scattering scans for a fixed wave

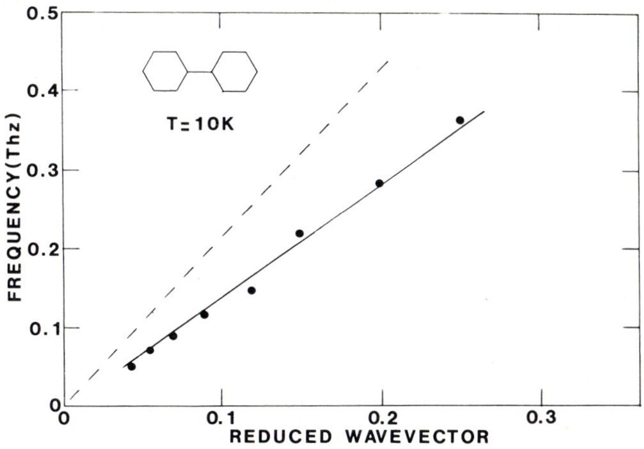

Fig. 17. (Continued)

vector $\boldsymbol{q}$ close to $\boldsymbol{q}_s$, we can clearly resolve the lower and upper phason branches and then the lower and upper amplitudon branches. Higher phason and amplitudon branches might exist, as well as acoustic branches originating at satellite positions, but their inelastic structure factors may be supposed smaller. A surprising feature is the occurrence of a frequency gap at the zone boundary point. This gap is unambiguously observed for the phason branches when energy scans are performed close to the $(2\,\frac{1}{2}\,0)$ zone boundary point. Indeed, as already indicated, near this zone boundary point the satellite

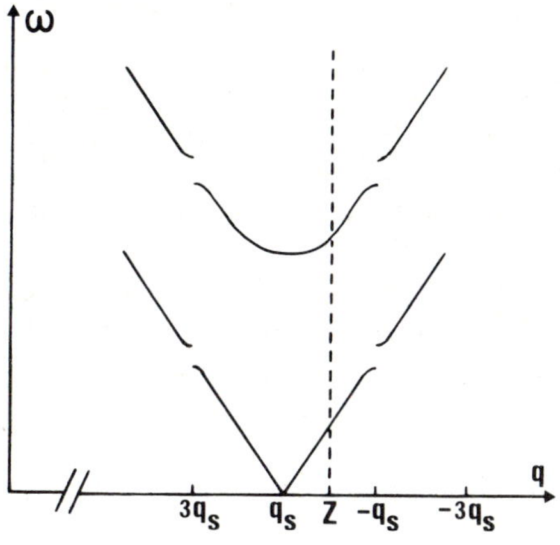

Fig. 18. Sketch of phason and amplitudon dispersions in the $\boldsymbol{b}^*$ direction (phase III).

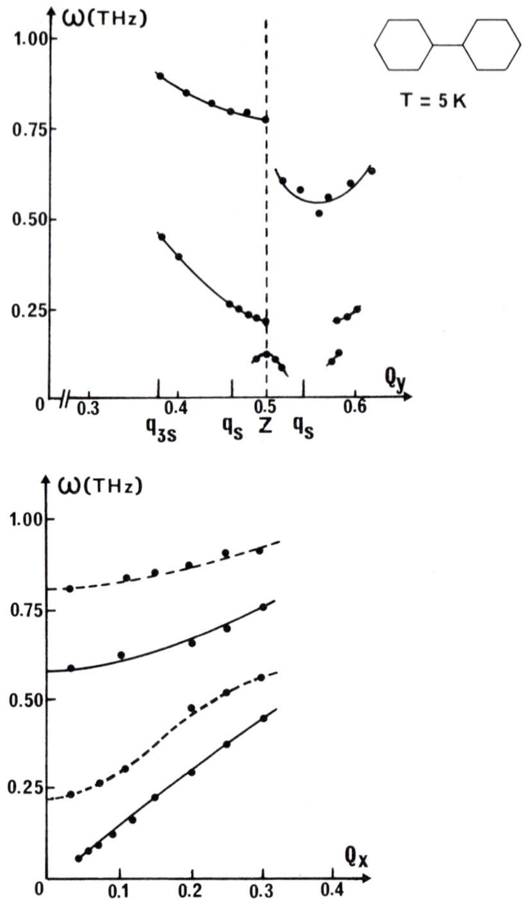

Fig. 19. Dispersion curves of phason and amplitudon branches as observed near the $(2\,\frac{1}{2}\,0)$ zone boundary point. For the dispersion curves in the $\boldsymbol{a}^*$ direction, the lower branches (solid lines) are measured close to the intense satellite reflection and the upper branches (dotted lines) close to the other satellite reflection (opposite side along $\boldsymbol{b}^*$).

reflections are intense on one side only and the lower phason branch is observed on this side, the upper branch on the opposite side. An analogous intensity behaviour for the amplitudon branches is not clearly seen. The origin of this frequency gap remains to be clarified, but the overall experimental dispersion curves of splitted phason and amplitudon branches are similar to that of fig. 18.

The situation of the intermediate phase II is less clear. In this phase, the excitation spectra are necessarily even more complex. Furthermore, the damping constants have increased with temperature and it becomes experimentally difficult to study these excitations. The evolution through the three phases is

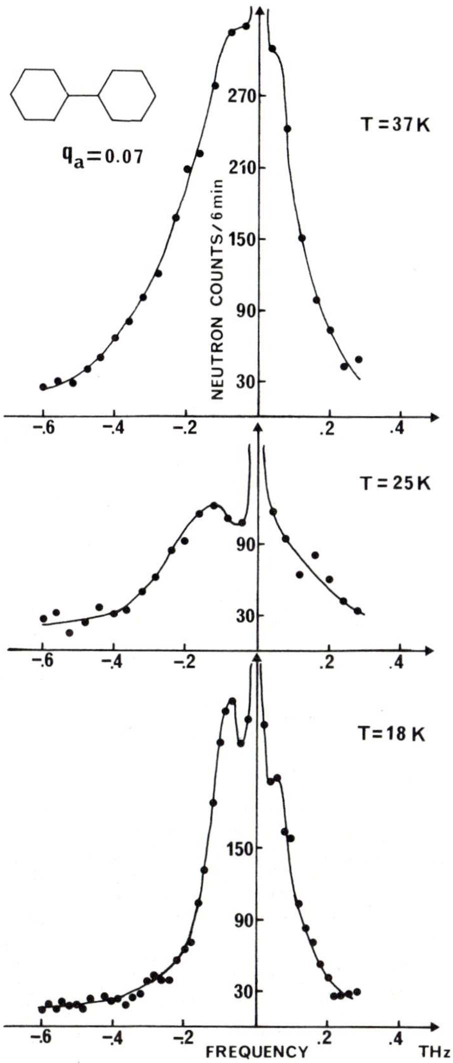

Fig. 20. Comparison of observed low frequency excitations for the same reduced wave vector at different temperatures.

given in fig. 20 for low frequency excitations measured at the same reduced wave vector.

The Raman scattering spectra of the incommensurate phases of biphenyl have been studied by different workers (Bree and Edelson 1977, Wada et al. 1981, Girard 1983). The results reported by Wada et al. (1981) are given in fig. 21. Many low frequency modes are observed in the range of neutron results

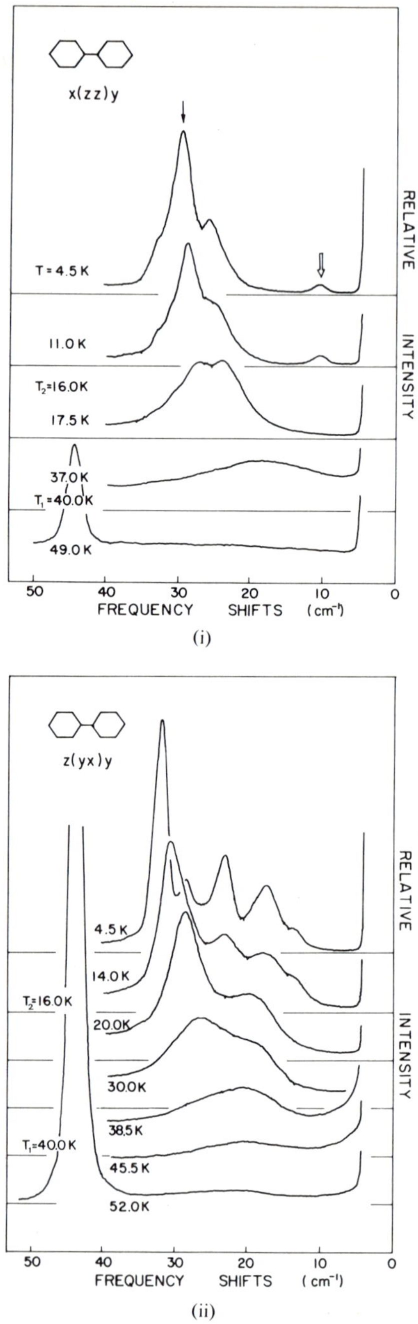

Fig. 21. Temperature dependence of low frequency Raman spectra of biphenyl. (Wada et al. 1981.)

discussed above. However, a detailed interpretation of these Raman spectra remains to be worked out. Moreover, the direct phason observation by the Brillouin scattering technique was unsuccessful (Ecolivet et al. 1983). This fact may be related to the necessarily overdamped nature of these excitations for the wave vectors used, which are at least one order of magnitude smaller than those used in neutron scattering experiments.

It is also useful to notice an interesting result obtained in recent investigations of low temperature heat capacities of biphenyl and p-terphenyl (Saito et al. 1984). Indeed, the heat capacity of biphenyl is found to decrease less steeply than that of p-terphenyl as temperature decreases. The heat capacity curves cross at about 12 K. The situation is similar for the deuterated compound. Such a behaviour is attributed to the contribution of the low frequency torsional mode in biphenyl. Therefore the enhanced heat capacity of biphenyl may be due to the existence of phason excitations.

6. Concluding remarks

The present understanding of the incommensurate phases of biphenyl has been reviewed by underlining the characteristic points of this molecular system composed of non-planar aromatic molecules, especially the unusual sequence of incommensurate phases. The direct neutron observation of propagating phason branches, related to the remaining incommensurability at low temperatures, has been discussed. However, many points remain to be clarified and further theoretical and experimental efforts are necessary. For example, the use of resonance methods would be useful, as well as a study of the excitations in the intermediate phase with an experimental investigation under pressure near the displacive limit ($T_I \rightarrow 0$ K).

Acknowledgments

The author is greatly indebted for invaluable assistance in the preparation of this manuscript to F. Moussa from L.L.B. at Saclay, C.M.E. Zeyen and J. Bouillot from I.L.L. at Grenoble and Y. Delugeard and C. Ecolivet of the Groupe de Physique Cristalline at Rennes.

References

Almlöf, J., 1974, Chem. Phys. **6**, 135.
Atake, T., and H. Chihara, 1980, Solid State Commun. **35**, 131.
Atake, T., K. Saito and H. Chihara, 1983, Chem. Lett. 1983, 493.

Baudour, J.L., H. Cailleau and W.B. Yelon, 1977, Acta Cryst. **B33**, 1773.
Baudour, J.L., and M. Sanquer, 1983, Acta Cryst. **B39**, 75.
Bastiansen, O., 1949, Acta Chem. Scand. **3**, 408.
Bastiansen, O., and M. Traetteberg, 1962, Tetrahedron **17**, 147.
Bonadeo, H., and E. Burgos, 1982, Acta Cryst. **A38**, 29.
Bree, A., and M. Edelson, 1977, Chem. Phys. Lett. **46**, 500.
Bree, A., and M. Edelson, 1978, Chem. Phys. Lett. **55**, 319.
Brock, C.P., 1979, Mol. Cryst. Liq. Cryst. **52**, 157.
Burgos, E., H. Bonadeo and E. D'Alessio, 1976, J. Chem. Phys. **65**, 2460.
Busing, W.R., 1983, Acta Cryst. **A39**, 340
Cailleau, H., 1981, Thesis: Transition structurale des p-polyphényles: Etude par diffusion des neutrons (Université de Rennes I) unpublished.
Cailleau, H., A. Girard, F. Moussa and C.M.E. Zeyen, 1979a, Solid State Commun. **29**, 259.
Cailleau, H., J.L. Baudour and C.M.E. Zeyen, 1979b, Acta Cryst. **B35**, 426.
Cailleau, H., F. Moussa and J. Mons, 1979c, Solid State Commun. **31**, 521.
Cailleau, H., J.L. Baudour, J. Meinnel, A. Dworkin, F. Moussa and C.M.E. Zeyen, 1980a, Farad. Discuss. Chem. Soc. **69**, 7.
Cailleau, H., F. Moussa, C.M.E. Zeyen and J. Bouillot, 1980b, Solid State Commun. **33**, 407.
Cailleau, H., F. Moussa, C.M.E. Zeyen and J. Bouillot, 1981, Proc. Int. Conf. on Phonon Physics, Bloomington, 1981, ed. W. Bron, J. Physique **42**, C6-704.
Cailleau, H., A. Girard, J.C. Messager, Y. Delugeard and C. Vettier, 1984a, Proc. 5th Eur. Meeting on Ferroelectricity, Benalmadena, 1983, ed. E. Maurer, Ferroelectrics **54**, 597.
Cailleau, H., F. Moussa, C.M.E. Zeyen and J. Bouillot, 1985, in preparation.
Casalone, G., C. Mariani, A. Mugnoli and M. Simonetta, 1968, Mol. Phys. **15**, 339.
Charbonneau, G.P., and Y. Delugeard, 1976, Acta Cryst. **B32**, 1420.
Charbonneau, G.P., and Y. Delugeard, 1977, Acta Cryst. **B33**, 1586.
Cowley, R.A., 1980, Adv. Phys. **29**, 1.
Cullick, A.S., and R.E. Gerkin, 1976, Chem. Phys. Lett. **42**, 589.
Cullick, A.S., and R.E. Gerkin, 1977, Chem. Phys. **23**, 217.
Dewar, M.J.S., and A.J. Herget, 1970, Proc. Roy. Soc. London **A315**, 443.
Dorner, B., 1982, Coherent Inelastic Neutron Scattering in Lattice Dynamics (Springer, Berlin) p. 21.
Dvořák, V., and Y. Ishibashi, 1978, J. Phys. Soc. Jpn **45**, 775.
Dvořák, V., V. Janovec and Y. Ishibashi, 1983, J. Phys. Soc. Jpn **52**, 2053.
Dworkin, A., and H. Cailleau, 1980, J. Physique Lett. **41**, L83.
Ecolivet, C., and M. Sanquer, 1980, J. Chem. Phys. **72**, 4145.
Ecolivet, C., M. Sanquer, J. Pellegrin and J. De Witte, 1983, J. Chem. Phys. **78**, 6317.
Franchini, P., and G. Sbrana, 1970, Trans. Faraday Soc. **66**, 572.
Friedman, P.S., R. Kopelman and P.N. Prasad, 1974, Chem. Phys. Lett. **24**, 15.
Girard, A., 1983, Thesis: Transitions structurales de cristaux moléculaires: Etude par diffusion Raman en fonction de la température et de la pression (Université de Rennes I) unpublished.
Golovko, V.A., and A.P. Levanyuk, 1981, Zh. Eksp. Teor. Fiz. **81**, 2296 [Sov. Phys. JETP **54**, 1217].
Golovko, V.A., and A.P. Levanyuk, 1983, in: Light Scattering near Phase Transitions, eds. H.Z. Cummins and A.P. Levanyuk (North-Holland, Amsterdam) ch. 3.
Hargreaves, A., and S.H. Rizvi, 1962, Acta Cryst. **15**, 365.
Hochstrasser, R.M., R.D. McAlpine and J.D. Whiteman, 1973, J. Chem. Phys. **58**, 5078.
Ishibashi, Y., 1981, J. Phys. Soc. Jpn **50**, 1255.
Ishibashi, Y., and V. Dvořák, 1978, J. Phys. Soc. Jpn **44**, 32.
Ishibashi, Y., and H. Shiba, 1978, J. Phys. Soc. Jpn **45**, 409.
Janner, A., T. Janssen and P.M. de Wolff, 1983, Acta Cryst. **A39**, 671.

Karle, I.L., and L.O. Brockway, 1944, J. Am. Chem. Soc. **66**, 1974.
Kirin, D., S.L. Chaplot, G.A. Mackenzie and G.S. Pawley, 1983, Chem. Phys. Lett. **102**, 105.
Levanyuk, A.P., and D.G. Sannikov, 1976, Fiz. Tverd. Tela **18**, 423 [Sov. Phys. Solid State **18**, 245].
Messager, J.C., 1976, Thesis: Contribution aux méthodes de résolution de structures moléculaires et cristallines par le calcul des phases invariantes et la minimisation de l'énergie potentielle d'interaction entre atomes non liés (Université de Rennes I) unpublished.
Natkaniec, I., M. Nartowski, A. Kulczycki, J. Mayer and M. Sudnik-Hrynkiewicz, 1978, J. Mol. Struct. **46**, 503.
Natkaniec, I., A.V. Bielushkin and T. Wasiutynski, 1981, Phys. Stat. Sol. (b) **105**, 413.
Plakida, N.M., A.V. Bielushkin, I. Natkaniec and T. Wasiutynski, 1983, Phys. Stat. Sol. (b) **118**, 129.
Robertson, G.B., 1961, Nature **191**, 593.
Saito, K., T. Atake and H. Chihara, 1984, Chem. Lett., in press.
Samara, G.A., N.E. Massa and F.G. Ullman, 1981, Ferroelectrics **36**, 335.
Suzuki, H., 1959, Bull. Chem. Soc. Jpn **32**, 1340.
Takeuchi, H., S. Suzuki, A.J. Dianoux and G. Allen, 1981, Chem. Phys. **55**, 153.
Toledano, P., and M. Guilluy, 1984, Proc. 5th Eur. Meeting on Ferroelectricity, Benalmadena, 1983, ed. E. Maurer, Ferroelectrics **54**, 311.
Toudic, B., J. Gallier, P. Rivet and H. Cailleau, 1983, Solid State Commun. **47**, 291.
Trotter, J., 1961, Acta Cryst. **14**, 1135.
Wada, M., A. Sawada and Y. Ishibashi, 1981, J. Phys. Soc. Jpn **50**, 737.
Wasiutynski, T., I. Natkaniec and A.V. Bielushkin, 1981, Proc. Int. Conf. on Phonon Physics, Bloomington, 1981, ed. W. Bron, J. Physique **42**, C6-599.
Zeyher, R., and W. Finger, 1982, Phys. Rev. Lett. **49**, 1833.

Note added in proof

The existence of a component for the modulation wave vector along $\boldsymbol{c}^*$ has in effect been observed in a recent neutron experiment associating triple-axis techniques with a full Eulerian cradle. See

Cailleau, H., J.C. Messager, F. Moussa, F. Bugaut, C.M.E. Zeyen and C. Vettier, 1985, Ferroelectrics, in press.

CHAPTER 13

Incommensurate Phase in $NaNO_2$

D. DURAND and F. DÉNOYER

*Laboratoire de Physique des Solides**
Université de Paris-Sud, Bâtiment 510
91405 Orsay Cedex
France

R. CURRAT

Institut Laue – Langevin, 156 X
38042 Grenoble Cedex
France

and

M. LAMBERT

*Laboratoire Léon Brillouin**, CEN Saclay*
91191 Gif-sur-Yvette Cedex
France

*Laboratoire Associé au CNRS.
**Laboratoire Commun CEA-CNRS.

Incommensurate Phases in Dielectrics 2
Edited by
R. Blinc and A.P. Levanyuk

Contents

1. Introduction 103
2. Static properties 104
   2.1. High- and low-temperature structures 104
   2.2. The intermediate modulated structure 105
      2.2.1. Satellite reflections 105
      2.2.2. Amplitude of the modulation 106
      2.2.3. Sinusoidal character of the modulation 107
      2.2.4. Structural analysis 108
   2.3. Static susceptibility 108
      2.3.1. Diffuse X-ray scattering 108
      2.3.2. Dielectric susceptibility 110
   2.4. Macroscopic properties 111
3. Dynamical properties 111
   3.1. Characteristic timescale 111
   3.2. Dielectric dispersion 112
   3.3. Ultrasonic dispersion 114
   3.4. Neutron Spin Echo 114
4. Theoretical studies 115
   4.1. Competing order-parameters 115
   4.2. The reduced free-energy 117
   4.3. Microscopic theories 118
5. Influence of external parameters 119
   5.1. Hydrostatic pressure 119
   5.2. Electric field ($\boldsymbol{E}\|\boldsymbol{b}$) 119
      5.2.1. The (E, T) phase diagram 119
      5.2.2. Field dependence of modulation wavevector and amplitude 121
      5.2.3. Phenomenological model 121
      5.2.4. Discussion 124
6. Concluding remarks 125
References 126

1. *Introduction*

Sodium nitrite, together with thiourea, was among the first insulating compounds where a modulated phase was discovered (Tanisaki 1961). Ever since, $NaNO_2$ has been extensively studied, both experimentally and theoretically. The reason for this continued attention lies in the remarkable structural simplicity of this compound, which enables realistic microscopic models to be constructed and tested against experiment.

The ordering process which begins in the modulated phase ($T_f < T < T_i$) and continues in the ferroelectric phase ($T < T_f$), involves large amplitude reorientations of the NO_2^- groups. The dynamics of these reorientations is clearly of the order–disorder type in so far as it involves long relaxation times, on a lattice-vibrational timescale. Thus, the ordering NO_2^- groups can be treated as a set of interacting pseudo-spin variables, to which well-known theoretical methods are applicable (Suzuki and Kubo 1968).

The phenomenological approach has also contributed to our understanding of the structural instabilities in $NaNO_2$. Levanyuk and Sannikov (1976b), and later Ishibashi and Shiba (1978) and Sannikov (1981) have developed phenomenological free energies which account for the transitions at T_i and T_f in a unified way. The original argument of Levanyuk and Sannikov is based on a free-energy functional written in terms of two competing order parameters of different symmetry and coupled through a gradient-interaction term. This term can be viewed as an extension to the case of one-dimensional representations of the familiar Lifshitz-invariant term.

The nature of the two competing variables and the consequences of the proposed mechanism for the description of the modulated structure have been investigated by Heine and McConnell (1981, 1984), Lynden-Bell et al. (1984) and Heine et al. (1984). In particular, these authors have shown how the predictions of the phenomenological model can serve as a guide for structural studies in the modulated regime, thereby establishing a connection with previous experimental work by Kucharczyk et al. (1978).

Finally, a similar type of free-energy functional was used by Durand et al. (1983) to describe the behaviour of $NaNO_2$ in an applied electric field.

The rest of this chapter is organized as follows. In sections 2.1 and 2.2 the various structural aspects of $NaNO_2$ are reviewed on the basis of available X-ray and neutron diffraction results. The behaviour of the static susceptibility, as obtained from diffuse X-ray scattering and low-frequency dielectric measurements, is discussed in section 2.3. Thermodynamic anomalies near T_i and T_f are summarized in section 2.4. Dynamical aspects are discussed in section 3, based on dielectric and ultrasonic dispersion data as well as on recent neutron spin-echo results. Section 4 deals with current theoretical developments, while the influence of external parameters (pressure and electric field) on the modulated phase of $NaNO_2$ are considered in section 5.

2. *Static properties*

2.1. *High- and low-temperature structures*

Sodium nitrite crystallizes in the body-centered orthorhombic lattice. The V-shaped NO_2^- groups lie in (b, c) planes with 2 in-plane and 4 out-of-plane Na^+ neighbors.

The low-temperature phase ($T < T_f = 162.5°C$) is ferroelectric with space group Im2m (C_{2v}^{20}). The spontaneous polarization results from the parallel alignment of the NO_2^- dipoles, together with a relative shift of the NO_2^- and Na^+ sublattices (cf. fig. 1).

In the high-temperature phase ($T > T_i = 164°C$), an additional mirror plane appears (space group Immm-D_{2h}^{15}), corresponding to the orientational disordering of the NO_2^- dipoles. The nature of the disorder was first established by neutron diffraction (Kay et al. 1962): each dipole possesses two alternate and

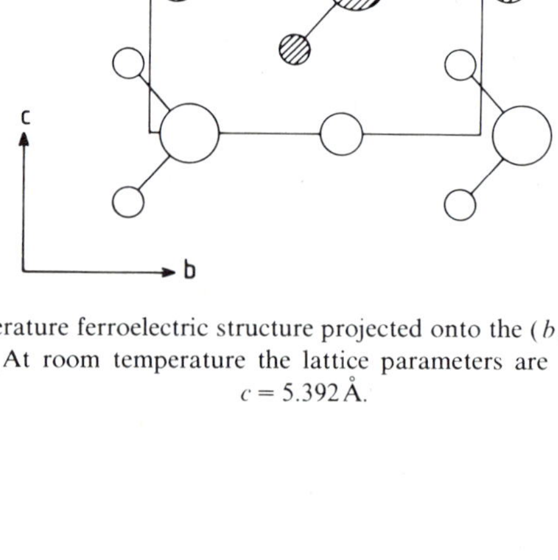

Fig. 1. The low temperature ferroelectric structure projected onto the (b, c) plane. The dark atoms are placed at $x = \frac{1}{2}$. At room temperature the lattice parameters are $a = 3.568$ Å; $b = 5.576$ Å; $c = 5.392$ Å.

equally probable orientations (parallel or antiparallel to $\boldsymbol{b}$). As a result of this statistical disorder the average dipole moment per cell vanishes and the average structure becomes centrosymmetric.

The order parameter of the transition may be defined microscopically in terms of the NO_2-group orientational probability, or macroscopically in terms of the spontaneous dielectric polarization.

2.2. *The intermediate modulated structure*

The existence of an intermediate phase, stable over a narrow temperature interval between the paraelectric and ferroelectric phases, was originally revealed by X-ray diffraction measurements (Tanisaki 1961, 1963, Yamada et al. 1963).

2.2.1. *Satellite reflections*

Between $T_i = 164°C$ and $T_f = 162.5°C$, the diffraction pattern of $NaNO_2$ (cf. fig. 2) exhibits new Bragg spots (satellite spots) with indices $(h \pm \delta, k, l)$, where h, k, and l are integers $(k \neq 0)$ and the modulation wavevector δ is found to vary with temperature (Hoshino and Motegi 1967).

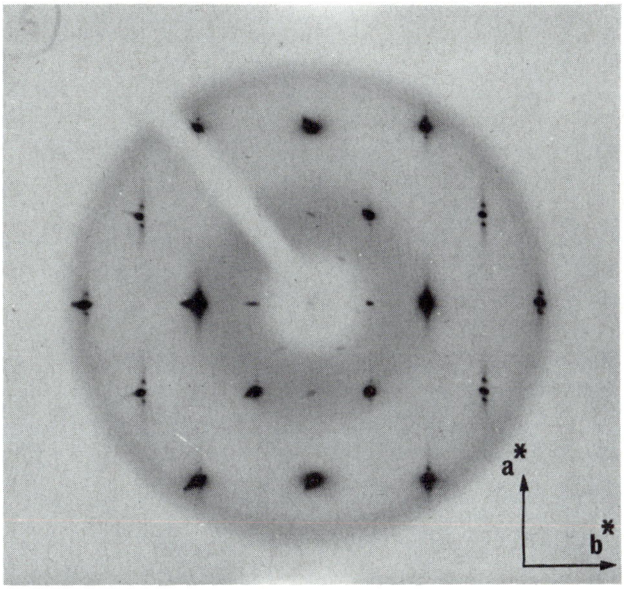

Fig. 2. X-ray diagram of the $(hk0)$ reciprocal plane in the incommensurate phase. Weak diffuse streaks coexist with the satellite reflections $(h \pm \delta, k, l)$ $(k \neq 0)$.

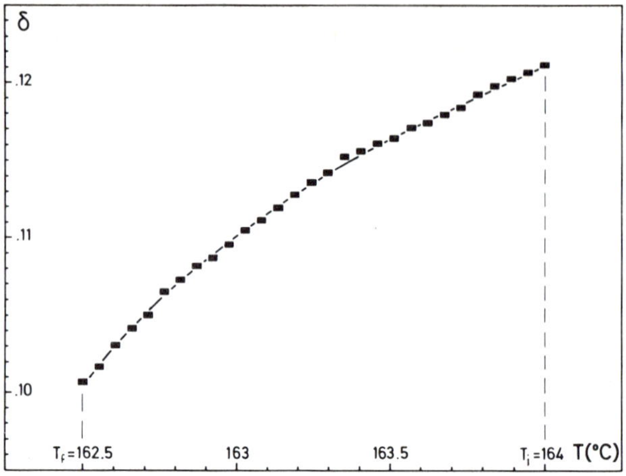

Fig. 3. Modulation wavevector δ as a function of temperature (Durand et al. 1982).

This variation has been remeasured recently by high-resolution neutron diffraction (Durand et al. 1982); the result is shown in fig. 3: a smooth decrease is observed from a value close to $\frac{1}{8}$ at T_i down to about $\frac{1}{10}$ at T_f. No evidence is found for commensurate plateaux at simple rational values such as $\frac{2}{17}$, $\frac{1}{9}$ or $\frac{2}{19}$. If such plateaux exist their range of stability must be of the order of 0.1 K at most. We shall comment again on this point in section 5.

At T_f, δ vanishes discontinuously. A small hysteresis (~ 0.1 K) is observed in the value of T_f. No wavevector hysteresis ("global" hysteresis) could be detected in the modulated state, in contrast with observations in thiourea and quartz (cf. relevant chapters, this volume).

2.2.2. *Amplitude of the modulation*

Below T_i the orientational ordering of the NO_2^- groups proceeds rapidly. The integrated intensity of the satellite reflections grows as $(T_i - T)^{2\beta}$, where the order-parameter critical exponent β is found to be 0.35 ± 0.05 (Durand 1982). More accurate intensity measurements by Noda and Yamada (1982) have yielded $\beta = 0.337 \pm 0.012$. These values are consistent with the critical behaviour of the three-dimensional *XY*-model ($\beta = 0.345$), which is the relevant model here, since the star of the modulation wavevector has two arms ($\pm \delta \boldsymbol{a}^*$) (cf. Cowley and Bruce 1978).

The possibility of the entire modulated phase lying outside of the classical regime is not, in itself, surprising. Indeed, its extent, when expressed in reduced temperature

$$\Delta t = \frac{T_i - T_f}{T_i},$$

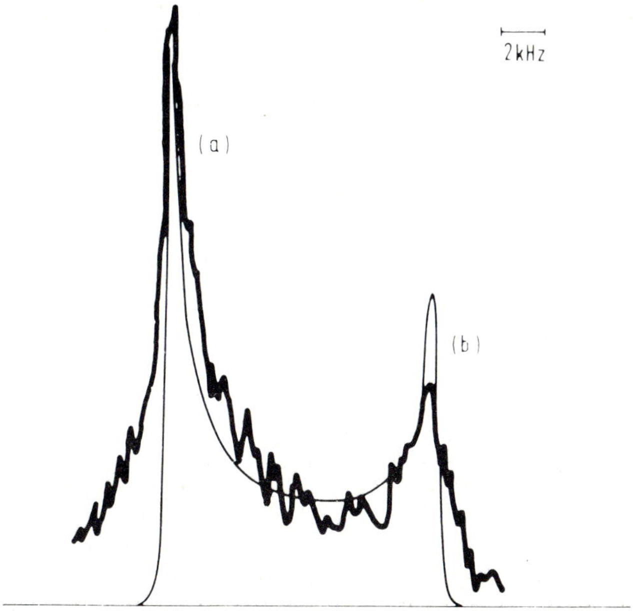

Fig. 4. The thick line represents the experimentally observed N^{14} NQR spectrum in the incommensurate phase at $T = T_i - 0.7°C$ (Ambrosetti et al. 1977) and the thin line is the theoretically predicted lineshape in the "plane wave" limit (Aleksandrova et al. 1980).

is only $\sim 3 \times 10^{-3}$. In other incommensurate systems such as thiourea (Moudden and Als-Nielsen 1981), K_2SeO_4 (Iizumi et al. 1977) and Rb_2ZnCl_4 (Mashiyama 1981) apparent deviations from classical behaviour ($\beta = \frac{1}{2}$) have been observed over much wider Δt ranges.

These deviations, however, are not necessarily to be ascribed to the effect of fluctuations. Cross-over effects, as encountered in systems close to a tricritical point may also give rise to low effective values for the order-parameter critical exponent (see also Ishibashi et al. 1981). Further discussion on the critical behaviour of $NaNO_2$ near T_i is given in section 2.4.

2.2.3. *Sinusoidal character of the modulation*

Even close to T_f, where the modulation amplitude is fairly large (cf. section 2.2.4), no higher-order satellite reflections have been detected (Yamada et al. 1963). This result, which again contrasts with the case of thiourea or quartz, implies that the NO_2^- orientational probability wave can be described as a simple sine wave.

Similar conclusions have been derived from NQR and NMR lineshape analysis (Aleksandrova et al. 1980, Buchheit et al. 1981, Blinc 1981). Figure 4 shows a typical NQR spectrum, as observed in the modulated state: a broad continuous frequency distribution bounded by two edge-singularities.

This type of profile is characteristic of inhomogeneous line broadening due to a sinusoidal distribution of local environments. The development as $T \to T_f$ of odd-order harmonic distorsions, leading to a squaring-up of the modulation profile, would result in the growth of one singularity at the expense of the other, an effect which is not observed here.

2.2.4. *Structural analysis*

A detailed analysis of the satellite intensities has been attempted by several authors (Böhm 1978, Kucharczyk et al. 1978, Harada et al. 1979). The main contribution to the satellite intensities arises from the polarization wave or, equivalently, from the NO_2^- orientational probability wave $p(x)$:

$$p(x) = \tfrac{1}{2} + P_\delta \cos 2\pi\delta x, \tag{1}$$

where the probability amplitude P_δ is strongly temperature dependent. Its maximum value is found to be ~ 0.25 just above T_f. Physically this means that in the planes of maximum polarization the degree of ordering, just above T_f, is considerable (~ 3 aligned dipoles out of 4) and comparable in magnitude to the degree of uniform ordering just below T_f.

Kucharczyk and co-workers (1978) have pointed out that the modulated structure can be further refined by introducing an additional displacive modulation of transverse acoustic character:

$$\boldsymbol{Y}(x) = \boldsymbol{Y}_\delta \sin 2\pi\delta x \quad (\boldsymbol{Y}_\delta \| \boldsymbol{b}). \tag{2}$$

The amplitude of the acoustic modulation is quite small ($\simeq 10^{-2}$ b at T_f) but its addition gives rise to a significant improvement in the agreement factor.

The existence of two modulated order parameters varying in *quadrature* is indeed a basic feature in Levanyuk and Sannikov's original work (1976b). However, as pointed out by Heine and McConnell (1981, 1984, see also McConnell 1983), symmetry arguments show that the waves to be considered here are the polarization wave and the shear wave $\mathrm{d}\boldsymbol{Y}(x)/\mathrm{d}x$. The displacement wave $\boldsymbol{Y}(x)$ should therefore be taken *in phase* with the polarization wave and hence the numerical values for $\boldsymbol{Y}_\delta$ deduced from Kucharczyk et al.'s (1978) coupled-wave analysis should be considered with some care.

2.3. *Static susceptibility*

2.3.1. *Diffuse X-ray scattering*

The static order-parameter susceptibility $\chi(\boldsymbol{q})$ can be obtained from diffuse X-ray scattering measurements (Yamada and Yamada 1966). Diffuse scattering is observable over a large temperature interval above T_i ($T_i < T < T_i + 60\,\mathrm{K}$).

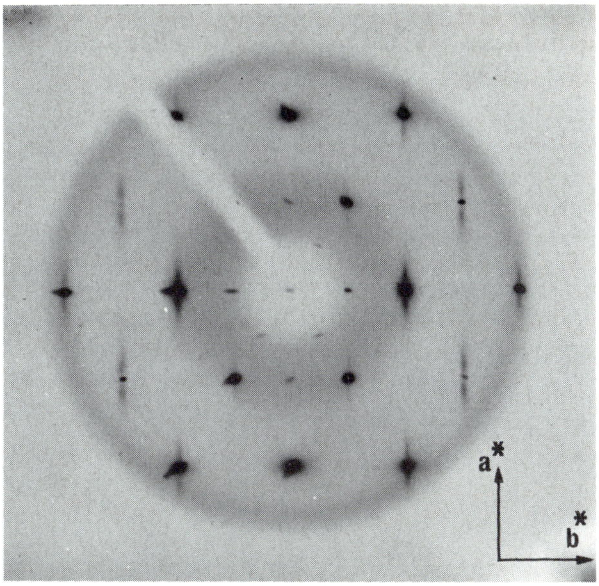

Fig. 5. X-ray diagram of the $(hk\,0)$ reciprocal plane in the paraelectric phase ($T = T_i + 4°C$).

Its temperature dependence is typical of critical scattering above a second-order phase transition. Its wavevector dependence is characterized by a marked anisotropy along the $\boldsymbol{a}^*$ direction, corresponding to strong correlations in the $(\boldsymbol{b}, \boldsymbol{c})$ planes (cf. fig. 5). As usual, when dipolar interactions are present, the largest correlations occur along the direction of the dipoles, here the $\boldsymbol{b}$ direction. For example, at $T_i + 3\,K$ the correlation lengths ξ_b and ξ_c in the $(\boldsymbol{b}, \boldsymbol{c})$ planes are of the order of 80 and 20 cells, respectively (Yamada and Yamada 1966). The maxima of the diffuse streaks shown in fig. 5 are located at $\boldsymbol{G}_{hkl} \pm \delta \boldsymbol{a}^*$, where $\boldsymbol{G}_{hkl}$ is a reciprocal lattice vector. As in the case of the satellite position below T_i, the peak position of the diffuse intensity above T_i varies with temperature: δ decreases on cooling from a high-temperature maximum value of $\simeq 0.2$ to a minimum of 0.12 at T_i. Most of the variation takes place during the last few degrees above T_i, which is suggestive of a renormalization process due to critical fluctuations. Recent work by Michel (1984) aims at accounting for the renormalization of δ just above T_i, within the framework of the self-consistent phonon approximation (see also Ueno et al. 1979). Below T_i, remnants of the diffuse streaks are observed in addition to the satellite peaks. The streaks may be due to the phase and amplitude fluctuations of the modulation. The separation between the two contributions has not been attempted so far, due to practical difficulties.

2.3.2. Dielectric susceptibility

The temperature behaviour of the order parameter susceptibility $\chi(\boldsymbol{q})$ can be probed at $\boldsymbol{q}=0$ by dielectric measurements (Hatta 1970, 1975). The dielectric susceptibility $\chi_0=\chi(\boldsymbol{q}=0)$ does not diverge at T_i since the ordering occurs at finite wavevector,

$$\boldsymbol{q}_\mathrm{i}=\delta(T_\mathrm{i})\boldsymbol{a}^*, \tag{3}$$

but nevertheless it increases by more than one order of magnitude as $T\to T_\mathrm{i}^+$. Clear deviation from a classical Curie–Weiss law is observed below 230°C. Early measurements by Hatta (1970) were fitted with a modified Curie–Weiss law of the type

$$\chi_0=\frac{C}{(T-T_0)^{\gamma}}, \tag{4}$$

with $T_0=163.1°\mathrm{C}$ ($T_\mathrm{i}=165.5°\mathrm{C}$, $T_\mathrm{f}=164.3°\mathrm{C}$) and $\gamma=1.11\pm0.05$.

In a later study, based on a more extensive set of measurements, Hatta (1975) proposed the alternate expression

$$\chi_0=\frac{C'}{(T-T_\mathrm{i})^{\gamma'}+\kappa^2}\qquad (T>T_\mathrm{i}+0.2\,\mathrm{K}), \tag{5}$$

with $\gamma'\simeq1.24$.

Both expressions are of interest since they illustrate two different viewpoints:

(i) in eq. (4) χ_0 is written as the order-parameter susceptibility associated with the (virtual) *direct* transition between the paraelectric and ferroelectric states. Obviously if the intermediate modulated state could be inhibited, this direct transition would occur at a temperature T_f' such that:

$$T_\mathrm{f}<T_\mathrm{f}'<T_\mathrm{i},$$

and since (cf. eq. 4)

$$T_0<T_\mathrm{f}<T_\mathrm{f}'$$

we may conclude that the direct transition would be of first-order character.

(ii) Equation (5) expresses the uniform dielectric susceptibility χ_0 in terms of the order-parameter susceptibility $\chi(\boldsymbol{q})$, associated with the transition at T_i, via an Ornstein–Zernicke-type expression:

$$\chi(\boldsymbol{q}) = \frac{C'}{(T-T_i)^{\gamma'} + \lambda|\boldsymbol{q}-\boldsymbol{q}_\delta|^2}. \tag{6}$$

From this expression κ in eq. (5) must be understood as proportional to q_δ.

Note in passing that, within scaling theory, the expected value for γ' is that of the XY-antiferromagnet, i.e. $\gamma' = 1.316$, and not, as erroneously stated by Hatta (1975), that of the Ising-antiferromagnet ($\gamma' = 1.241$).

2.4. *Macroscopic properties*

The behaviour of a number of thermodynamic quantities (specific heat, elastic compliances, expansion coefficients) has been studied extensively in the vicinity of the transitions at T_i and T_f (Hatta and Ikushima 1973, Hamano and Ema 1978, Ema et al. 1975).

In particular, a strong divergence of the specific heat C_p is observed at T_i with critical exponents $\alpha = 0.38$ ($T > T_i$) and $\alpha' = 0.18$ ($T < T_i$). Anomalously large exponent values are also seen on the isothermal elastic compliances and the thermal expansion coefficients, as expected from Pippard's relations.

On the other hand, the specific heat at constant volume C_v (which can be calculated from C_p and the ratio of the isothermal versus adiabatic compressibilities) only shows a cusp at T_i. The same is true for the adiabatic elastic compliances (Hamano and Ema 1978). This difference in behaviour between C_p and C_v as well as the large value of the exponent α deviate from the predictions of scaling theory and probably reflect the proximity of a tricritical point.

3. *Dynamical properties*

3.1. *Characteristic timescale*

The order-parameter fluctuations involve 180° reorientations of NO_2^- groups which imply large amplitude motions for the N and O atoms. The dynamics of the fluctuations is therefore expected to be characterized by slow relaxational processes rather than soft-phonon modes. No indication for a soft phonon branch was obtained either by inelastic neutron scattering (Sakurai et al. 1970), Raman scattering (Gorelik et al. 1967, Hartwig et al. 1972) or infrared spectroscopy (Barnoski and Ballantyne, 1968, Holah 1971).

As stated before, the NO_2^- dipoles can be treated as Ising variables and their collective dynamical behaviour near T_i is expected to display the main features of a system of Ising spins close to its ordering temperature. Suzuki and Kubo (1968) have shown that for such a system the order parameter susceptibility could be described in terms of a single Debye relaxation, viz.:

$$\chi(\boldsymbol{q},\omega)=\frac{\chi(\boldsymbol{q},0)}{1+\mathrm{i}\omega\tau(\boldsymbol{q})},\tag{7}$$

where $\tau(\boldsymbol{q})$ diverges at the ordering temperature T_{i} for the critical wavevector $\boldsymbol{q}_\delta$:

$$\tau^{-1}(\boldsymbol{q})=\alpha(T-T_{\mathrm{i}})^{\Delta'}+\lambda|\boldsymbol{q}-\boldsymbol{q}_\delta|^2.\tag{8}$$

Equation (7) may also be derived from a general phenomenological approach in which the thermodynamically irreversible processes associated with the polarization fluctuations are explicitly included (see Yamada and Fujii 1966).

The dynamical exponent Δ' which controls the rate of slowing-down of the order-parameter fluctuations as $T\to T_{\mathrm{i}}$, is, within mean-field theory, the same as the static susceptibility exponent γ'.

Below T_{i}, one expects to observe two relaxation mechanisms, with different characteristic relaxation times τ_{A} and τ_ϕ corresponding to fluctuations in the amplitude and phase of the modulation, respectively:

$$\tau_{\mathrm{A}}^{-1}(\boldsymbol{q})=2\alpha(T_{\mathrm{i}}-T)^{\Delta'}+\lambda|\boldsymbol{q}-\boldsymbol{q}_\delta|^2,\tag{9a}$$

$$\tau_\phi^{-1}(\boldsymbol{q})=\lambda|\boldsymbol{q}-\boldsymbol{q}_\delta|^2.\tag{9b}$$

Below T_{f}, the concept of phase fluctuations becomes unphysical and only one relaxation time, $\tau_{\mathrm{A}}(\boldsymbol{q})_{q_\delta=0}$, should be observable.

The available experimental results on the dynamics of the fluctuations near T_{i} are summarized below.

3.2. *Dielectric dispersion*

The $\boldsymbol{q}=0$ order-parameter susceptibility can be probed by dielectric measurements:

$$\chi(\boldsymbol{q}=0,\omega)=\frac{\chi_0}{1+\mathrm{i}\omega\tau_{\mathrm{P}}},\tag{10}$$

where χ_0 stands for $\chi(\vec{q}=0,\ \omega=0)$ and τ_{P} ($=\tau(\boldsymbol{q}=0)$) is the relaxation time for homogeneous polarization fluctuations.

Hatta (1970) was able to verify the validity of the single Debye-relaxation model above T_{i} and established the temperature dependence of τ_{P}:

$$\tau_{\mathrm{P}}=2.4\times10^{-8}(T-T_0)^{-1,3}\ \mathrm{s}.\tag{11}$$

The temperature T_0 appearing above, as in eq. (4), is the Curie–Weiss temperature associated with the direct para–ferro (virtual) transition. No attempt was made by Hatta to fit the observed values for τ_P with a law of the type shown in eq. (8). The value for the dynamical exponent shown in eq. (11) ($\Delta' \simeq 1.3$) differs from that of the static susceptibility ($\gamma = 1.1$), indicating a deviation from mean-field behaviour in both the static and the dynamical properties. The value of the relaxation time at T_i is 0.7×10^{-8} s corresponding to a relaxation rate $(2\pi\tau)^{-1} \simeq 20$ MHz.

Below T_i, the dielectric data should in principle be analyzed in terms of two relaxation mechanisms (cf. eq. 9). If, following Hatta (1970), one uses a single effective relaxation time, the resulting temperature behaviour is shown on the left-hand side of fig. 6, with the slowest dynamics occurring at T_f(~15 MHz). This is the behaviour expected for $\tau_\phi(0)$ alone (cf. eq. 9b) due to the decrease of q_δ from T_i to T_f. We however see no reason why phase relaxations should dominate the dielectric response at $\boldsymbol{q} = 0$.

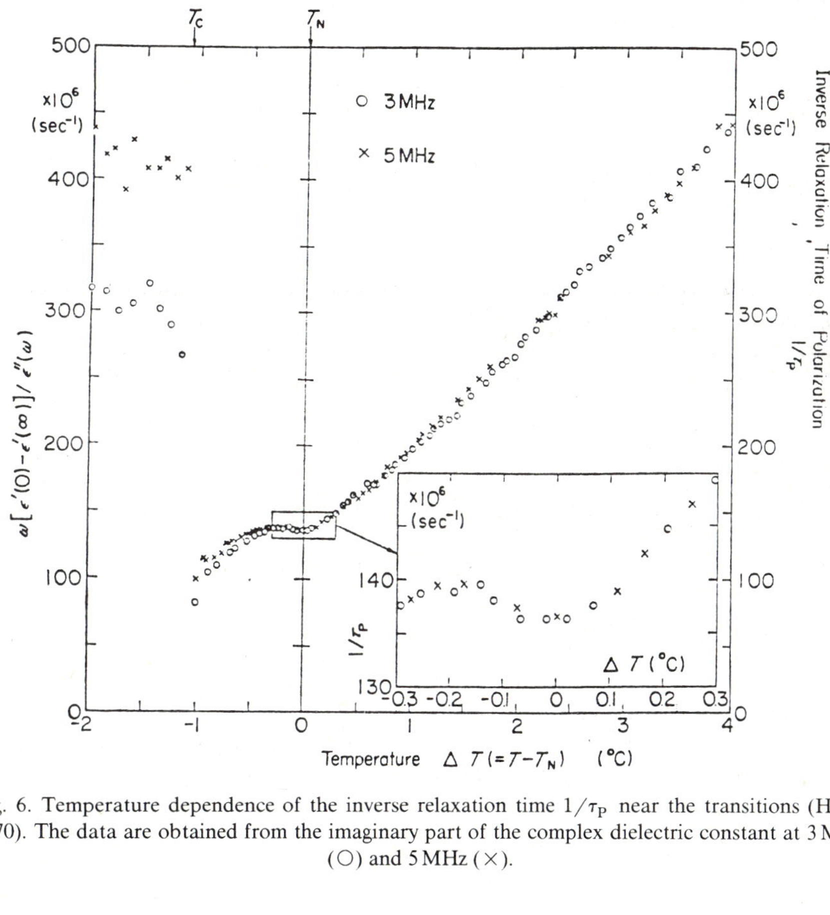

Fig. 6. Temperature dependence of the inverse relaxation time $1/\tau_P$ near the transitions (Hatta 1970). The data are obtained from the imaginary part of the complex dielectric constant at 3 MHz (○) and 5 MHz (×).

3.3. *Ultrasonic dispersion*

The frequency dispersion and attenuation of long-wavelength acoustic waves give access to the relaxation time associated with amplitude fluctuations $\tau_A(\boldsymbol{q}_\delta)$. The coupling mechanism between long-wavelength acoustic coordinates $Q_{ac}(\boldsymbol{q})$ and "soft-mode" normal coordinates $Q_s(\pm\boldsymbol{q}_\delta+\boldsymbol{q})$, has been discussed by several authors (Bruce and Cowley 1978, Hatta et al. 1980). Symmetry arguments indicate that the essential acoustic attenuation mechanism below T_i arises from the interaction of longitudinal acoustic waves with amplitude fluctuations of the frozen polarization wave.

From the peak position of the longitudinal sound attenuation just below T_i, measured at different frequencies, Hatta et al. (1980) deduced the relaxation time of the amplitude fluctuations at $\boldsymbol{q}=\boldsymbol{q}_\delta$. In the narrow region $T_i-0.4<T<T_i$, the data can be fitted by the law

$$\tau_A^{-1}=5\times10^8(T_i-T)\,\mathrm{s}^{-1}. \tag{12}$$

Similar results with a more detailed theoretical analysis have recently been obtained by Esayan et al. (1981). They obtain

$$\tau_A^{-1}=8\times10^8(T_i-T)^{0.8}\,\mathrm{s}^{-1}. \tag{13}$$

The propagation and attenuation of longitudinal sound waves have also been investigated by Brillouin scattering (Shimizu et al. 1974, Yagi et al. 1980, 1982). Weak velocity anomalies are observed at T_i and T_f, but no anomalous damping, in contrast to the ultrasonic results. This is expected since the characteristic frequencies associated with the order-parameter fluctuations lie much below the hypersonic frequencies ($\sim 10\,\mathrm{GHz}$) relevant to right angle Brillouin scattering.

3.4. *Neutron Spin Echo (NSE)*

In principle the Neutron Spin Echo (NSE) technique can be used to probe the order-parameter relaxation time $\tau(\boldsymbol{q})$, at all $\boldsymbol{q}$. The NSE technique measures directly the intermediate scattering function $S(\boldsymbol{q},t)$ defined by

$$S(\boldsymbol{q},t)\propto\int \mathrm{e}^{\mathrm{i}qx}\langle\eta(x,t)\,\eta(0,0)\rangle\,\mathrm{d}x, \tag{14}$$

where $\eta(x,t)$ is the local value of the order parameter at time t.

For a system of localized electric dipoles with Ising-like dynamics (Suzuki and Kubo 1968), $S(\boldsymbol{q},t)$ reduces to

$$S(\boldsymbol{q},t)\propto\mathrm{e}^{-t/\tau(\boldsymbol{q})}, \tag{15}$$

where the relaxation time $\tau(\boldsymbol{q})$, above the ordering temperature T_i, is given by eq. (8).

Under realistic experimental conditions, as presently available at the IN 11 facility of the Institut Laue–Langevin, with a 5 Å-wavelength neutron beam, relaxation processes faster than ~10 ns can be detected and analyzed. In a recent study (Durand et al. 1985), relaxation times from 0.5 to 1 ns have been measured at $T_i + 1.6$ K and for wavevectors in the range $0 < q < q_\delta$. As expected, $\tau(q_\delta)$ was found to be larger than $\tau(0) = \tau_P$. The neutron-determined values for τ_P are considerably lower than the values derived from Hatta's dielectric measurements. A few spectra obtained in the incommensurate phase are currently being analyzed but their statistical accuracy will probably not be sufficient to enable the amplitude and phase relaxation times to be determined simultaneously. In addition, the NSE technique detects a very low frequency (<1 MHz) resonance ("central peak") which is observed in all three phases and whose physical origin has not been clarified so far.

4. *Theoretical studies*

4.1. *Competing order-parameters*

A number of physical systems undergo a structural instability corresponding to the onset of a long-wavelength modulation followed by a lock-in transition at $q = 0$. A few of the better known examples are discussed in this volume (thiourea, quartz, $NaNO_2$).

Although these compounds appear to follow the general sequence of normal–incommensurate–commensurate phases, Levanyuk and Sannikov (1976b) and Bruce et al. (1978) have pointed out that they in fact belong to a separate class of modulated structures.

The reason for this distinction becomes clear when considering the direct (virtual) phase transition from the normal high-temperature state to the ferrodistorted or commensurate low-temperature state. For $NaNO_2$, thiourea and quartz this transition would be Ising-like ($n = 1$) corresponding to the condensation of a one-dimensional zone-center irreducible representation, whereas for all other systems, e.g. K_2SeO_4, the same transition would correspond to $n \geqslant 2$.

In the language of continuum free-energy functionals the class $n \geqslant 2$ is characterized by the presence of Lifshitz gradient terms of the type

$$\int \left[\eta_1 \frac{\partial \eta_2}{\partial x} - \eta_2 \frac{\partial \eta_1}{\partial x} \right] \mathrm{d}x, \tag{16}$$

which couple bilinearly the degenerate order-parameter components $(\eta_1, \eta_2, \ldots)$ and their spatial derivatives. As shown by Levanyuk and Sannikov (1976a) the presence of Lifshitz invariants, as in eq. (16) above, provides a phenomenological justification for the occurrence of an intermediate modulated structure between the normal and commensurate states.

Since in the case $n=1$ no such terms can be constructed, Levanyuk and Sannikov (1976b) have postulated the existence of two competing one-dimensional order-parameters, say η and ξ, with *different* symmetry properties, but which are allowed to interact through a Lifshitz-like gradient term of the type

$$\int \left[\eta \frac{\delta \xi}{\delta x} - \xi \frac{\partial \eta}{\partial x} \right] \mathrm{d}x. \tag{17}$$

Similar arguments have been developed by Axe et al. (1970) in the context of ferroelectric phase transitions in cubic perovskites.

The existence of a free-energy invariant as shown above imposes some restrictions on the symmetry properties of the variables η and ξ. In the case of $NaNO_2$, Heine and McConnell (1981, 1984) have identified the two interacting modes as the y-component of the local polarization vector and the xy component of the elastic strain tensor:

$$\eta = P_y(x), \qquad \xi = \frac{\mathrm{d}Y}{\mathrm{d}x}. \tag{18}$$

The free-energy functional proposed by Levanyuk and Sannikov (1976b) has the following form:

$$\Phi = \int \phi(x)\, \mathrm{d}x,$$

with

$$\phi(x) = \frac{1}{2} A \eta^2 + \frac{1}{2} A' \xi^2 + \frac{B}{4} \eta^4 + \frac{B'}{4} \xi^4 + \frac{B''}{2} \eta^2 \xi^2 + \sigma \left(\frac{\partial \eta}{\partial x} \xi - \eta \frac{\partial \xi}{\partial x} \right) + \frac{\delta}{2} \left(\frac{\partial \eta}{\partial x} \right)^2 + \frac{\delta'}{2} \left(\frac{\partial \xi}{\partial x} \right)^2. \tag{19}$$

All coefficients in the expression above are assumed to be positive and temperature-independent, except for A which vanishes at a temperature T_0:

$$A = a(T - T_0) \quad (a > 0). \tag{20}$$

Under the condition

$$A' < A'_L = 4\sigma^2/\delta \tag{21}$$

eq. (19) predicts a continuous phase transition at a temperature $T_i > T_0$ given by

$$a(T_i - T_0) = \frac{\delta}{\delta'}\left(\sqrt{A'_L} - \sqrt{A'}\right)^2 \tag{22}$$

towards a modulated structure characterized by a modulation wavevector

$$q_i = \left[\frac{aA'(T_i - T_0)}{\delta\delta'}\right]^{1/4}. \tag{23}$$

Also, a first-order transition to the ferroelectric state is predicted to occur at a temperature $T_f < T_0$.

The free energy in eq. (19) suffers from a number of shortcomings which we shall attempt to amend in section 5.3. Its essential merit, however, is to provide the correct description of the modulated structure, with a polarization wave and a shear wave oscillating 90° out-of-phase. As pointed out in section 2.2.4. the two component waves have indeed been detected experimentally (Kucharczyk et al. 1978). The specific prediction concerning the relative phase of the two waves, which is a direct consequence of the gradient interaction term appearing in eq. (19), should serve as a useful guide for future refined structural studies (Heine et al. 1984).

4.2. *The reduced free-energy*

When the coefficient A' in eq. (19) approaches the limiting value A'_L, both transition temperatures T_i and T_f converge towards T_0, i.e. the stability range of the modulated phase decreases and eventually vanishes for $A' = A'_L$. For $A' > A'_L$ a direct second-order transition occurs at T_0 between the paraelectric and ferroelectric states. Thus the point ($A = 0$; $A' = A'_L$) is a Lifshitz point in the (A, A') phase diagram (Hornreich et al. 1975).

In $NaNO_2$ at ambient pressure the range of stability of the modulated phase is quite narrow, which in the present context implies

$$A'_L - A' \ll A'_L. \tag{24}$$

With the help of the above condition, the variable ξ can easily be eliminated from eq. (19), which then reads (Sannikov 1981)

$$\phi(x)=\frac{1}{2}A\eta^2+\frac{1}{4}B\eta^4-\delta\left(\frac{\sqrt{A'_{\mathrm{L}}}-\sqrt{A'}}{\sqrt{A'}}\right)\left(\frac{\partial\eta}{\partial x}\right)^2+\frac{\delta\delta'}{2A'_{\mathrm{L}}}\left(\frac{\partial^2\eta}{\partial x^2}\right)^2$$
$$+\frac{B''}{2}\frac{\delta}{A'}\left(\eta\frac{\partial\eta}{\partial x}\right)^2. \tag{25}$$

Reduced free-energies of the above type have been used by a number of authors in the context of $NaNO_2$ (Ishibashi and Shiba 1978) and thiourea (Lederer and Chaves 1981). The condition (cf. eq. 21) which guarantees the stability of the modulated phase, in the original coupled mode picture, is seen to control the sign of the $(\partial\eta/\partial x)^2$ term in eq. (25) above. Thus, provided one is not interested in the detailed structural description of the modulated phase, the two types of free-energies are completely equivalent. We shall make use of eq. (25) in discussing the phase diagram of $NaNO_2$ in an applied electric field.

4.3. Microscopic theories

Ehrhardt and Michel (1981) and Michel (1981) have carried out a thorough microscopic theory of $NaNO_2$ in both the paraelectric and modulated phases.

Starting from a sterical hindrance potential between the NO_2^- groups and the six neighbouring Na^+ ions, they show that the NO_2^- reorientations take place essentially through 180° rotations around the crystallographic ***c***-axis.

In addition to the large amplitude reorientational displacements of the NO_2^- groups, the microscopic model of Ehrhardt and Michel takes into account the small amplitude translational displacements of the Na^+ and NO_2^- ions. This gives rise to a bilinear cross-term in the sterical hindrance potential which couples the translational together with the rotational displacements and mediates an effective interaction between neighbouring NO_2^- dipoles. It is the combined effect of the lattice-mediated dipole–dipole interaction with the true electrostatic dipolar interactions which leads to the appearance of the modulated structure.

Furthermore, the Fourier analysis of the bilinear coupling term shows that at finite wavevector the reorientational motions around ***c*** are mostly coupled to acoustic-like displacements along ***b***, with a coupling term of the form

$$q^2 Y_q P_q$$

where Y_q and P_q are the Fourier components of the acoustic and reorientational displacement fields. Such a term represents the interaction between the gradient of an acoustic displacement (iqY_q), i.e. a shear wave, and the gradient of a polarization wave (iqP_q). The connection with the phenomenological approach of Levanuyk and Sannikov (1976b), follows therefrom.

Fivez and Michel (1983) have investigated the dynamical consequences of the translation–rotation coupling mechanism. They show in particular that, as $T \to T_i$ in the paraelectric phase, the critical reorientational relaxations near $q \simeq q_i$ are much too slow to induce a softening of the transverse acoustic phonon branch at $q = q_i$, in agreement with inelastic neutron scattering observations (Durand et al. 1982). One only expects a small additional contribution to the phonon damping, not observable by conventional techniques.

The dynamical behaviour of $NaNO_2$ has also been studied by molecular dynamics simulations (Klein et al. 1982, Lynden-Bell et al. 1984, Heine et al. 1984). The results are consistent with the other approaches. The growth of the reorientational fluctuations, as T decreases, is found to be accompanied by a correlative growth in acoustic-like fluctuations of the same wavevector and with displacements along $\boldsymbol{b}$, and by the progressive development of a phase correlation between the two types of fluctuations. At the ordering temperature the two fluctuations, P_q and Y_q, are phase-locked and diverge simultaneously.

5. *Influence of external parameters*

5.1. *Hydrostatic pressure*

The (P, T) phase diagram of $NaNO_2$ has been determined by dielectric (Gesi et al. 1965) and calorimetric (Rapoport 1966) measurements. The width of the incommensurate phase increases with pressure ($T_i - T_f \simeq 8\,\mathrm{K}$ at 10 kbar). Up to the highest applied pressures (~ 40 kbar) no additional anomalies were detected within the modulated region, corresponding to possible commensurate lock-in phases.

A surprising and still unexplained result has been reported by Qiu et al. (1983) in the low pressure region, where $T_i - T_f$ is observed to decrease from 1.55 K at 1 bar to a minimum of 1.25 K at 70 bar and to increase for higher pressures.

5.2. *Electric field ($\boldsymbol{E} \| \boldsymbol{b}$)*

5.2.1. *The (E, T) phase diagram*

The effect of an applied electric field E ($\| \boldsymbol{b}$) on the stability of the modulated phase has been investigated independently by means of various techniques: dielectric constant measurements (Hamano 1964, Gesi 1965, Qiu et al. 1983), X-ray diffraction (Yamada 1981) and neutron diffraction (Durand et al. 1983).

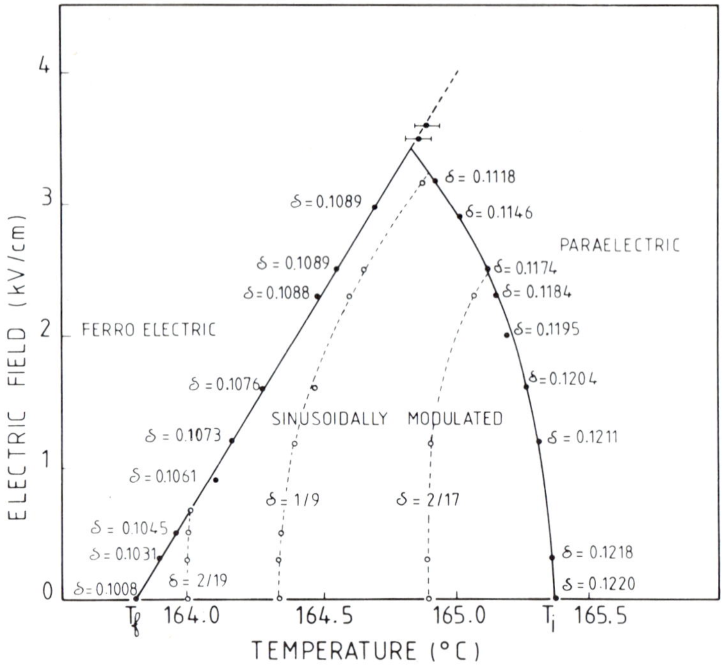

Fig. 7. Experimental (E, T) phase diagram of $NaNO_2$ investigated by means of neutron diffraction (Durand et al. 1983). The commensurate iso-δ lines are shown as dotted lines.

These various determinations are not in complete quantitative agreement, partly because of the finite electrical conductivity of the specimen (giving rise to Joule heating under an applied electric field) and partly due to uncertainties on the actual value of the applied field.

The main features of the experimental phase diagram are shown in fig. 7.

(i) as the applied field is turned up the domain of stability of the modulated phase decreases and eventually disappears for $E > E_T$. Experimental values for E_T range from 2.6 to 3.6 kV/cm. For $E > E_T$, the presence of a resistivity anomaly and a sharp peak in the dielectric constant indicate the existence of a first-order transition line between the paraelectric and ferroelectric regions. This line should end at a critical point (E_c, T_c), whose position has not been determined yet.

(ii) the paraelectric-to-modulated transition appears to remain continuous all along the $T_i(E)$ line up to the critical end-point (E_T, T_T).

(iii) the modulated-to-ferroelectric transition remains clearly first-order under applied field. The $T = T_f(E)$ boundary is essentially linear, with slope

$$\left[\frac{dT_f}{dE}\right]^{-1} = 3.5\,\mathrm{kV/cmK}.$$

5.2.2. *Field dependence of modulation wavevector and amplitude*

The modulation wavevector,

$$\boldsymbol{q}_\delta = \delta \boldsymbol{a}^*,$$

is a decreasing function of the applied field. More precisely, the neutron diffraction results of Durand et al. (1983) indicate that along the $T = T_i(E)$ line $\delta_i^2(E)$ varies quadratically with E:

$$\delta_i^2(0) - \delta_i^2(E) = \lambda E^2 \quad (E < 2\,\text{kV/cm}),$$

with $\lambda = 1.6 \times 10^{-4}\,\text{kV}^{-2}\text{cm}^2\,(\text{r.u.})^2$ (r.u. = reduced wavevector units). At the intersection point (E_T, T_T), $\delta_T = 0.109$.

Within the modulated region, $\delta(E,T)$ appears to vary smoothly as a function of E and T. No evidence is found for a lock-in at a commensurate value such as $\frac{2}{17}, \frac{1}{9}, \frac{2}{19}, \ldots$. This is in contrast to the case of thiourea where a field-induced commensurate phase at $\delta = \frac{1}{8}$ is observed. The situation here is very different however, since the shortest Umklapp vector along $\boldsymbol{a}^*$ is $2\boldsymbol{a}^*$, due to the body-centered nature of the parent lattice.

Hence the lowest order lock-in term for $\delta = \frac{1}{9}$,

$$\mathrm{V}^{(n)}\mathrm{P}^{(n)}_{1/9},$$

is of order $n = 18$, the term of order 9 vanishing due to both rotational and translational selection rules. In the case of the $\delta = \frac{1}{8}$ phase of thiourea the term of order 8 is forbidden due to rotational selection rules only. The applied electric field lifts the rotational selection rule and the order of the lowest-order lock-in term is thus reduced from 16 to 8. No such reduction can occur here since the translational selection rule is unaffected by the applied field.

As far as the modulation amplitude is concerned, neutron diffraction studies of the $(\pm\delta, 2, 0)$ first-order satellite intensities indicate a monotonic variation with temperature and field, the iso-intensity curves being roughly parallel to the second-order line $T = T_i(E)$.

No indication is found for the development of higher-order harmonics under field, implying that the sinusoidal approximation holds throughout the entire modulated region.

5.2.3. *Phenomenological model*

We start from a reduced free-energy of the type shown in eq. (25):

$$\phi(x) = \frac{1}{2}AP_y^2 + \frac{1}{4}BP_y^4 + \frac{1}{6}CP_y^6 + \frac{1}{2}\alpha\left(\frac{\partial P_y}{\partial x}\right)^2 + \frac{1}{4}\beta\left(\frac{\partial^2 P_y}{\partial x^2}\right)^2 + \gamma\left(P_y\frac{\partial P_y}{\partial x}\right)^2 - EP_y, \tag{26}$$

with:

$$A = a(T - T_0), \qquad B < 0, \qquad C > 0, \qquad \alpha < 0, \qquad \beta > 0, \qquad \gamma > 0.$$

The choice $B < 0$ (Ishibashi et al. 1981) is a consequence of the experimental observation $T_0 < T_f$ (cf. section 2.3.2). The behaviour of the spontaneous polarization $P_0(T)$, for $T < T_f$, also suggests a first-order direct para–ferro transition (Buchheit and Petersson 1980).

The adjunction of a term $\propto P_y^6$ follows from the negative sign of B. The negative sign of α has been discussed in section 4.2.

On the basis of experimental observations (cf. preceding section) one may assume a sinusoidal modulation:

$$P_y(x) = P_0 + 2P_\delta \cos(2\pi\delta x). \tag{27}$$

Inserting eq. (27) into eq. (26) yields:

$$\phi = \int \phi(x)\,\mathrm{d}x = \tfrac{1}{2}AP_0^2 + \tfrac{1}{4}BP_0^4 + \tfrac{1}{6}CP_0^6 + A_\delta P_\delta^2 + \tfrac{3}{2}B_\delta P_\delta^4 + \tfrac{3}{2}(B + B_\delta)P_0^2P_\delta^2 - P_0E, \tag{28}$$

with

$$A_\delta = A + \alpha\delta^2 + \tfrac{1}{2}\beta\delta^4, \tag{29}$$

$$B_\delta = B + \tfrac{4}{3}\gamma\delta^2. \tag{30}$$

Equation (28) can be minimized with respect to P_0, P_δ and δ in order to yield:

(i) The value of the transition temperature $T_i(E = 0)$,

$$a(T_i - T_0) = \alpha^2/2\beta. \tag{31}$$

(ii) The value of δ at $T_i(E = 0)$,

$$\delta^2(T_i) = \delta_i^2 = -\alpha/\beta. \tag{32}$$

(iii) The temperature dependence of δ below $T_i(E = 0)$,

$$\delta^2 - \delta_i^2 = \frac{1}{2}\left\{\left[\frac{B_i^2}{\gamma^2} + \frac{8a}{3\beta}(T - T_i)\right]^{1/2} - \frac{B_i}{\gamma}\right\} \simeq \mu(T - T_i) \tag{33}$$

with

$$\mu = \frac{2\gamma a}{3\beta B_i} \quad (B_i = B_{\delta=\delta_i} > 0). \tag{34}$$

(iv) the value $\delta_i(E)$ of δ along the $T_i(E)$ line:

$$\delta_i^2(E) - \delta_i^2 = -\frac{2\gamma}{\beta} P_0^2 = -\lambda E^2 + O(E^4) \tag{35}$$

with

$$\lambda = \frac{2\gamma}{\beta a^2 (T_i - T_0)^2}. \tag{36}$$

(v) the equation for the $T_i(E)$ line:

$$a(T_i(E) - T_i) = -\tfrac{3}{2}(B + B_i)\frac{E^2}{a^2(T_i - T_0)^2} + O(E^4). \tag{37}$$

A number of observables can be used in order to determine the values of the coefficients appearing in eq. (26). Durand et al. (1983) have chosen to use:

(i) the paraelectric Curie constant ($\mathscr{C} = 5130\,\mathrm{K}$) which fixes a,

$$a = 2.2 \times 10^7 \text{ MKS}.$$

(ii) the values of $T_i - T_0(2.45\,\mathrm{K})$ and $T_i - T_f(1.60\,\mathrm{K})$.

(iii) the value of δ_i (0.122 in reduced units).

(iv) the observed values for the coefficients λ and μ defined in eqs. (36) and (34), respectively [$\lambda = 1.6 \times 10^{-4}\,(\mathrm{r.u.})^2\,\mathrm{kV}^{-2}\mathrm{cm}^2$; $\mu = 0.22 \times 10^{-2}\,(\mathrm{r.u.})^2\,\mathrm{K}^{-1}$].

The corresponding values for the coefficients $\alpha, \beta, \gamma, B, C$ are:

$$\begin{aligned}
&\alpha = -7.3 \times 10^9\,\mathrm{MKS}\,(\mathrm{r.u.})^{-2},\\
&\beta = 4.9 \times 10^{11}\,\mathrm{MKS}\,(\mathrm{r.u.})^{-4},\\
&\gamma = 1.15 \times 10^{13}\,\mathrm{MKS}\,(\mathrm{r.u.})^{-2},\\
&B = B_i - \tfrac{4}{3}\gamma\delta_i^2 = -0.70 \times 10^{11}\,\mathrm{MKS}, \quad B_i = 1.55 \times 10^{11}\,\mathrm{MKS},\\
&C = 4.3 \times 10^{13}\,\mathrm{MKS}.
\end{aligned} \tag{38}$$

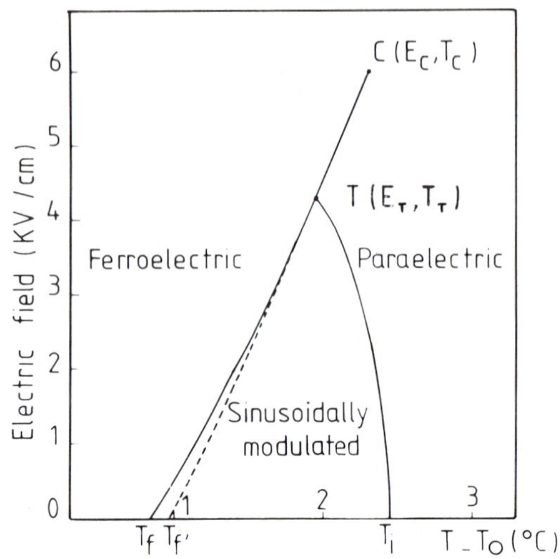

Fig. 8. (E, T) phase diagram calculated on the basis of a phenomenological Landau model (Durand et al. 1983). The model is constrained to yield the correct value of $T_f - T_i$. The dotted line corresponds to the virtual transition line $T_f'(E)$.

5.2.4. Discussion

Figure 8 shows the phase diagram calculated on the basis of the above numerical values. The agreement with the observed phase diagram is only semi-quantitative. In particular the value of E_T is overestimated in the model.

The free energy in eq. (28) can also be used to calculate zero-field observables such as the spontaneous homogeneous polarization $P_0(T)$ $(T < T_f)$ or the modulated polarization $P_\delta(T)(T_f < T < T_i)$. The comparison with available data from Hamano (1973) and Kucharczyk et al. (1978) is not fully satisfactory either. A closer agreement can only be achieved by modifying the numerical values of the model coefficients, which, in turn, creates discrepancies for some of the other observables.

We believe that the inability of the above model to account for the whole set of experimental observations in a quantitative manner, lies less in the approximate character of a finite free-energy expansion [as in eqs. (26) and (28) above], than in the *mean-field* approximation in which it is implicitly grounded. In the present case, fluctuations are likely to reduce the value of the ordering temperature T_i (and probably not that of T_f) and to influence the temperature dependence of $q_\delta(T)$ and $P_\delta(T)$ while $P_0(T)$ is probably less affected. These deviations from mean-field behaviour are bound to introduce inconsistencies in a mean-field model adjusted on the basis of the corresponding observables.

Recently, Selke and Duxbury (1984) have attempted to apply the formalism of the discrete (mean-field) ANNNI model to modulated ferroelectrics such as

$NaNO_2$. The only adjustable parameter in the model is the ratio $\kappa = -J_2/J_1$, between nearest- and next-nearest-neighbour exchange interactions. With a suitable choice for the value of κ, the gross features of the (E, T) phase diagram of $NaNO_2$ can be reproduced. However, the model predicts the existence of several high-order commensurate phases ($\delta = \frac{1}{9}, \frac{5}{44}, \frac{3}{26}$) which are not observed experimentally (cf. fig. 3). As pointed out by the authors the origin of this discrepancy is probably related to the use of the mean-field approximation to describe a modulated system in the vicinity of its ordering temperature.

Going back to the phenomenological model developed in section 5.2.3 above, we may conclude by pointing out that its essential merit is to establish a connection between two apparently unrelated observations, namely:

(i) the decrease of $\boldsymbol{q}_\delta$ from T_i to T_f and with field.

(ii) the continuous character of the ordering transition at T_i whereas the direct para–ferro transition would have been discontinuous.

Both features are accounted for *simultaneously* with the help of the amplitude-gradient coupling term $\propto (P_y\, \partial P_y / \partial_x)^2$ in eq. (26).

One may also point out that the same type of mechanism should be relevant to the case of quartz and thiourea, where similar or related situations exist. In addition, the amplitude-gradient coupling term has been found useful in describing the interaction of the modulation with relaxing defects, in the case of thiourea (Lederer et al. 1984).

6. *Concluding remarks*

In this review, we have attempted to summarize the most significant aspects of the structural instabilities and modulated phase of $NaNO_2$. As a result of the considerable amount of effort devoted to the study of its physical properties, sodium nitrite is currently one of the best understood modulated systems.

A number of features, however, remain unclear and will require further investigation. As an example, the structural analysis of the modulated phase should be reconsidered using the correct phase relationship between the two component waves (cf. section 2.2.4). More complete and accurate diffuse X-ray measurements may also be useful as they would help to characterize the nature of the modulated short-range order between $(\boldsymbol{b}, \boldsymbol{c})$-planes, as well as the nature of the *in-plane* correlations. In the same context one should mention the existence of weak diffuse X-ray streaks (Durand et al. 1982) connecting satellite pairs at $\boldsymbol{G}_{hkl} + \boldsymbol{q}_\delta$ and $\boldsymbol{G}_{hkl} + (\boldsymbol{a}^* \pm \boldsymbol{c}^*) - \boldsymbol{q}_\delta$, which have not been satisfactorily interpreted so far.

Some discrepancies remain between results obtained with different techniques as far as relaxation times are concerned (section 3) as well as for the (E, T) phase diagram (section 5.2.1).

On the theoretical side one clearly needs to develop microscopic models to the extent where connections can be established with the phenomenological approach. This should allow useful cross-checks to be made on the respective assumptions inherent to either approach. In both cases the need to go beyond the usual mean-field approximation should be recognized. Possibly the most puzzling experimental feature which theory should try and clarify is the apparent absence of higher-order satellite reflections close to T_f, where the large amplitude of the modulation (cf. section 2.2.4) should lead to a seemingly measurable "squaring" of the wave profile.

Acknowledgments

We wish to acknowledge Professor K. Michel from Antwerpen for several useful discussions and L. Bernard from Grenoble (I.L.L.) for good cooperation during our neutron experiments and for helpful comments and suggestions.

References

Aleksandrova, I.P., R. Blinc., B. Topic, S. Zumer and A. Rigamonti, 1980, Phys. Stat. Sol. (a) **61**, 85.
Ambrosetti, R., R. Angelone, A. Colligiani and A. Rigamonti, 1977, Phys. Rev. **B15**, 4318.
Axe, J.D., M. Harada and G. Shirane, 1970, Phys. Rev. **B1**, 1227.
Barnoski, M.K., and J.M. Ballantyne, 1968, Phys. Rev. **174**, 946.
Blinc, R., 1981, Phys. Rep. **79**, 331.
Böhm, H., 1978, Z. Kristallogr. **148**, 207.
Bruce, A.D., and R.A. Cowley, 1978, J. Phys. **C11**, 3609.
Bruce, A.D., R.A. Cowley and A.F. Murray, 1978, J. Phys. **C11**, 3591.
Buchheit, W., and J. Petersson, 1980, Solid State Commun. **34**, 649.
Buchheit, W., G. Herth and J. Petersson, 1981, Solid State Commun. **40**, 411.
Cowley, R.A., and A.D. Bruce, 1978, J. Phys. **C11**, 3577.
Durand, D., 1982, Thesis (Orsay University) unpublished.
Durand, D., F. Dénoyer, M. Lambert, L. Bernard and R. Currat, 1982, J. Physique **43**, 149.
Durand, D., F. Dénoyer, D. Lefur, R. Currat and L. Bernard, 1983, J. Physique Lett. **44**, L207.
Durand, D., F. Mezei, L. Bernard, M. Lambert, R. Currat and F. Dénoyer, 1985, submitted to Physica B and C.
Ehrhardt, K.D., and K.H. Michel, 1981, Z. Phys. **B41**, 329.
Ema, K., K. Hamano and I. Hatta, 1975, J. Phys. Soc. Jpn **39**, 726.
Esayan, S. Kh., V.V. Lemanov and N. Mamatkulov, 1981, Sov. Phys. Solid State **23**, 1195.
Fivez, J., and K.H. Michel, 1983, Z. Phys. **B51**, 127.
Gesi, K., 1965, Jpn J. Appl. Phys. **4**, 818.
Gesi, K., K. Ozawa and Y. Takagi, 1965, J. Phys. Soc. Jpn **20**, 1773.
Gorelik, V.S., I.S. Zheludev and M.M. Srishchinskii, 1967, Sov. Phys. Crystallogr. **11**, 527.
Hamano, K., 1964, J. Phys. Soc. Jpn **19**, 945.
Hamano, K., 1973, J. Phys. Soc. Jpn **35**, 157.
Hamano, K., and K. Ema, 1978, J. Phys. Soc. Jpn **45**, 923.
Harada, M., N. Koyano, S. Mitani, Y. Iwata and I. Shibuya, 1979, Ann. Rep. Res. Reactor Inst., Kyoto Univ. **12**, 1.

Hartwig, C.M., E. Wiener-Avnear and S.P.S. Porto, 1972, Phys. Rev. **B5**, 79.
Hatta, I., 1970, J. Phys. Soc. Jpn **28**, 1266.
Hatta, I., 1975, J. Phys. Soc. Jpn **38**, 1430.
Hatta, I., and A. Ikushima, 1973, J. Phys. Chem. Solids **34**, 57.
Hatta, I., M. Hanami and K. Hamano, 1980, J. Phys. Soc. Jpn **48**, 160.
Heine, V., and J.D.C. McConnell, 1981, Phys. Rev. Lett. **46**, 1092.
Heine, V., and J.D.C. McConnell, 1984, J. Phys. **C17**, 1199.
Heine, V., R.M. Lynden-Bell, J.D.C. McConnell and I.R. McDonald, 1984, Z. Phys. **B56**, 229.
Holah, G.D., 1971, J. Phys. **C4**, 2191.
Hornreich, R.M., M. Luban and S. Shtrikman, 1975, Phys. Rev. Lett. **35**, 1678.
Hoshino, S., and H. Motegi, 1967, Jpn J. Appl. Phys. **6**, 708.
Iizumi, M., J.D. Axe, G. Shirane and K. Shimaoka, 1977, Phys. Rev. **B15**, 4392.
Ishibashi, Y., and H. Shiba, 1978, J. Phys. Soc. Jpn **45**, 409.
Ishibashi, Y., W. Buchheit and J. Petersson, 1981, Solid State Commun. **38**, 1277.
Kay, M.I., B.C. Frazer and R. Ueda, 1962, Acta Cryst. **15**, 506.
Klein, M.L., I.R. McDonald and Y. Ozaki, 1982, Phys. Rev. Lett. **48**, 1197.
Kucharczyk, D., A. Pietrasko and K. Lukaszewicz, 1978, Ferroelectrics **21**, 445.
Lederer, P., and C.M. Chaves, 1981, J. Physique Lett. **42**, L127.
Lederer, P., G. Montambaux, J.P. Jamet and M. Chauvin, 1984, J. Phys. Lett. **45**, L-627.
Levanyuk, A.P., and D.G. Sannikov, 1976a, Sov. Phys. Solid State **18**, 245.
Levanyuk, A.P., and D.G. Sannikov, 1976b, Sov. Phys. Solid State **18**, 1122.
Lynden-Bell, R., M.L. Klein and I.R. McDonald, 1984, Z. Phys. **B54**, 325.
Mashiyama, H., 1981, **50**, 2655.
McConnell, J.D.C., 1983, Am. Miner **68**, 1.
Michel, K.H., 1981, Phys. Rev. **B24**, 3998.
Michel, K.H., 1984, private communication.
Moudden, A.H., and J. Als-Nielsen, 1981, unpublished.
Noda, Y., and Y. Yamada, 1982, Annual Meeting of the Physical Society of Japan.
Qiu, S.L., H.Z. Cummins, S.M. Shapiro and J.C. Steiner, 1983, Ferroelectrics **52**, 181.
Rapoport, E., 1966, J. Chem. Phys. **45**, 2721.
Sakurai, J., J.A. Cowley and G. Dolling, 1970, J. Phys. Soc. Jpn **28**, 1426.
Sannikov, D.G., 1981, Sov. Phys. Solid State **23**, 1827.
Selke, W., and P.M. Duxbury, 1984, Z. Phys. **B57**, 49.
Shimizu, H., M. Tsukamoto, Y. Ishibashi and M. Umeno, 1974, J. Phys. Soc. Jpn **36**, 498.
Suzuki, M., and R. Kubo, 1968, J. Phys. Soc. Jpn **24**, 51.
Tanisaki, S., 1961, J. Phys. Soc. Jpn **16**, 579.
Tanisaki, S., 1963, J. Phys. Soc. Jpn **18**, 1181.
Ueno, Y., S. Muto and T. Oguchi, 1979, **62**, 573.
Yagi, T., Y. Hidaka and K. Miura, 1980, J. Phys. Soc. Jpn **48**, 2165.
Yagi, T., Y. Hidaka and K. Miura, 1982, J. Phys. Soc. Jpn **51**, 3562.
Yamada, Y., 1981, Ferroelectrics **35**, 51.
Yamada, Y., and Y. Fujii, 1966, J. Phys. Soc. Jpn **21**, 1613.
Yamada, Y., and T. Yamada, 1966, J. Phys. Soc. Jpn **21**, 2167.
Yamada, Y., I. Shibuya and S. Hoshino, 1963, J. Phys. Soc. Jpn **18**, 1594.

CHAPTER 14

Modulated Phases in Thiourea

F. DÉNOYER

Laboratoire de Physique des Solides
Bâtiment 510, Université Paris Sud
91405 Orsay
France

and

R. CURRAT

Institut Laue – Langevin
BP 156 X, 38042 Grenoble Cedex
France

Incommensurate Phases in Dielectrics 2
Edited by
R. Blinc and A.P. Levanyuk

Contents

1. Introduction . . . 131
2. Static properties of the intermediate phases of thiourea . . . 135
   2.1. Temperature dependence of modulation wave vector and phase diagram . . . 135
   2.2. Amplitude and character of the modulation . . . 137
   2.3. Discussion . . . 139
      2.3.1. A model thermodynamic potential for the incommensurate phase . . . 139
      2.3.2. Temperature dependence of the modulation wave vector $\delta(T)$ . . . 141
      2.3.3. Commensurate phases and lock-in transitions . . . 142
      2.3.4. Umklapp lock-in terms . . . 143
      2.3.5. Microscopic models . . . 144
3. Effect of hydrostatic pressure on the modulated phases . . . 145
   3.1. Early results on the pressure–temperature phase diagram . . . 145
   3.2. Neutron diffraction results . . . 146
   3.3. Summary . . . 148
4. Effect of applied electric field on the modulated phases . . . 150
   4.1. General features of the (E, T) phase diagram . . . 150
   4.2. Field-induced commensurate phases . . . 151
5. Dynamical properties . . . 154
   5.1. General . . . 154
   5.2. Soft mode above T_i . . . 156
   5.3. The ferroelectric phase $(T < T_c)$ . . . 157
   5.4. The modulated phases $(T_c < T < T_i)$ . . . 157
6. Concluding remarks . . . 157
References . . . 158

1. *Introduction*

Thiourea, $SC(NH_2)_2$, and deuterated thiourea (d-thiourea), $SC(ND_2)_2$, are well known ferroelectric molecular crystals (Solomon 1956). The molecule (cf. fig. 1) is planar with C_{2v} symmetry and possesses a large ($\simeq 4.9$ Debye) permanent dipole moment directed along its 2-fold axis.

The temperature dependence of the dielectric constant ε'_a in h-thiourea (or d-thiourea) is characterized by three important maxima (cf. fig. 2) occurring at $T_i = 202$ K (218.1 K), 177 K (198.8 K) and $T_c = 169$ K (190.7 K) (Goldsmith and White 1959). The first anomaly is associated with a second-order phase transition from the paraelectric state (phase labeled V) to an intermediate phase (IV) while the last one corresponds to a first-order phase transformation from an intermediate phase (II) to a ferroelectric phase (I) with spontaneous polarization along $\boldsymbol{a}$ ($P_s \simeq 3 \times 10^{-6}$ C/cm^2). The intermediate phase III, sandwiched between phases IV and II, extends over about 1 K on either side of 177 K (198.8 K) and is also ferroelectric. The spontaneous polarization (also along $\boldsymbol{a}$) is however of the order of 10^{-3} that of phase I. In addition to these dielectric anomalies, an inflexion in the $\varepsilon'_a(T)$ curve appears at $T_1 \simeq 170.5$ K and is more pronounced in the deuterated compound ($T_1 \simeq 193.5$ K). A careful re-examination of the dielectric properties by McKenzie and Dryden (1973) has revealed an additional new small peak in $\varepsilon'_a(T)$ around 172.3 K, which seems to correspond to a new narrow intermediate phase with an even weaker spontaneous polarization ($\simeq 3.5 \times 10^{-10}$ C/cm^2).

Thus between the high temperature paraelectric phase (V) and the low temperature ferroelectric phase (I), the phase diagram of thiourea appears a priori as rather complicated with several intermediate phases corresponding to either non-ferroelectric or weakly ferroelectric states. However, a common feature characterizes all the intermediate phases, namely the presence in the diffraction patterns of satellite reflections $(h, k \pm n\delta, l)$ (h, k, l: Miller indices, n: integer), suggesting the existence of an *incommensurate or long period commensurate structural modulation of wavelength* $(1/\delta)b$, where δ is an increasing function of temperature (Futama and Chiba 1963, Mitsui et al. 1966, Futama et al. 1967).

At room temperature (phase V) thiourea crystallizes in the orthorhombic centrosymmetric space group Pnma (D_{2h}^{16}) ($a = 7.655$ Å, $b = 8.537$ Å, $c =$ 5.520 Å) with four molecules per unit cell. The structure of phase V is now well

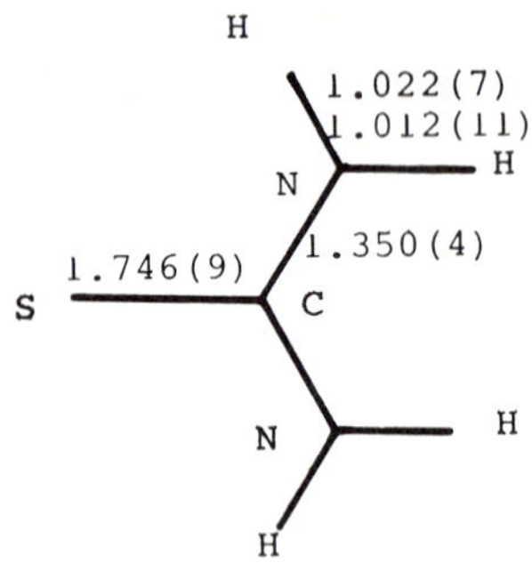

Fig. 1. Thiourea molecule. (Elcombe and Taylor 1968.)

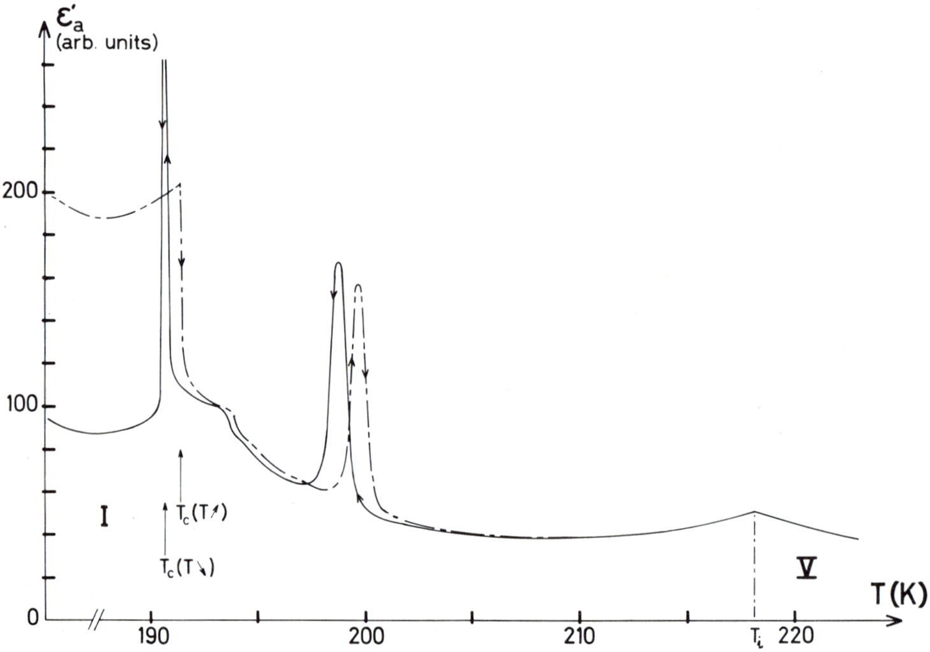

Fig. 2. d-thiourea: dielectric constant for increasing and decreasing temperature. (André et al. 1985a.)

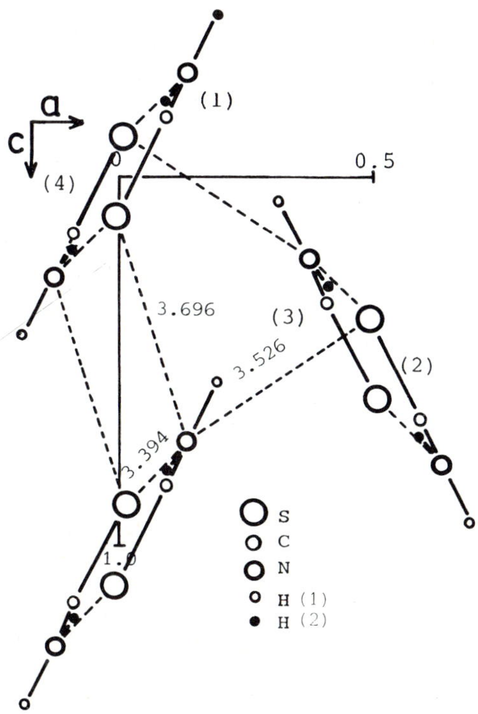

Fig. 3. The room temperature paraelectric structure projected on the (a, c) plane. (Elcombe and Taylor 1968.)

known, following successive studies by Kunchur and Truter (1958), Goldsmith and White (1959) and Truter (1967) using X-ray diffraction, Dvoryankin and Vainshtein (1960) using electron diffraction, and Elcombe and Taylor (1968) using neutron diffraction. The projection in the (a, c) plane of the room temperature structure is shown in fig. 3, the dipole moments being parallel to the (a, c) plane. The resulting moment per cell cancels due to the n- and a-glide planes.

In the lower ferroelectric region the cell is still orthorhombic ($a = 7.516$ Å, $b = 8.519$ Å, $c = 5.494$ Å), but the n-glide is no longer present and the space group becomes $P2_1ma$ (C_{2v}^2). The four molecules per unit cell are arranged in two non equivalent pairs, with different inclinations to the ferroelectric axis, a configuration which results in a net dipole moment along the $\boldsymbol{a}$ direction. Figure 4 gives the low temperature structure obtained by neutron diffraction by Elcombe and Taylor (1968).

Because of its ferroelectric properties, thiourea has been the object of a number of lattice dynamical studies. The phase transition at T_i was initially believed to be of the order–disorder type (Calvo 1960, Futama 1962,

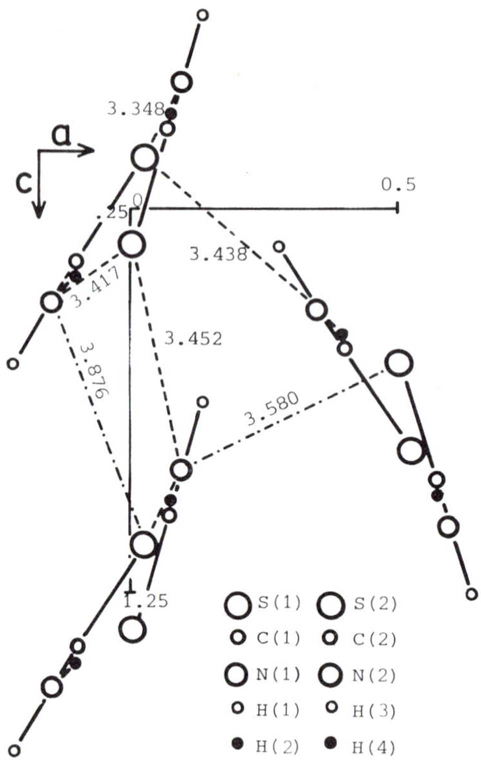

Fig. 4. The low temperature ferroelectric structure projected on the (a, c) plane. (Elcombe and Taylor 1968.)

McKenzie et al. 1970, Bréhat and Hadni 1971), until the Raman experiments of Benoit et al. (1972) and Benoit (1973), in which an A_1-type soft mode was observed below T_i, and in the ferroelectric phase I. This soft mode, which is Raman-inactive in phase V, was later observed by infrared reflectivity and shown to be of B_{3u}-type (Bréhat et al. 1976). Note however that the infrared results in phase V refer to the Brillouin zone center while the structural instability at T_i is related to a general wave vector along the $\boldsymbol{b}^*$ direction of the Brillouin zone.

The soft-mode eigenvector consists mainly (Takahashi et al. 1967) in a molecular libration around the $\boldsymbol{b}$ axis [molecules (1) and (3) rotating in antiphase to molecules (2) and (4)] combined with a translation along the $\boldsymbol{c}$ axis [molecules (1) and (4) moving in antiphase to molecules (2) and (3)] and a translation along the $\boldsymbol{a}$ axis (all molecules moving together). X-ray diffraction measurements in the intermediate phases have shown that in first approximation the modulated structure could be described as deriving from phase V by condensation of molecular displacements of τ_4 symmetry in Kovalev's notation (Kovalev 1965), varying sinusoidally in space with wavelength $(1/\delta)b$ (Futama 1970, Shiozaki and Hosoya 1970, Tanisaki and Nakamura 1970, Shiozaki 1971, McKenzie 1975a). Although these results were consistent with a soft-mode mechanism, the inelastic neutron scattering measurements of McKenzie (1975b) failed to provide a direct confirmation for a phonon branch softening.

Such was the situation up until 1975, with many questions left unanswered. The remainder of this chapter is devoted to recent developments on h- and d-thiourea and attempts to present an up-to-date account of the current state of understanding of this modulated insulator. In section 2 the nature of the intermediate phases is examined; in particular we discuss how a frozen electrical polarization wave develops in the intermediate phases and we describe its characteristic features: symmetry analysis of the atomic displacements, amplitude and wave vector evolution, lock-in phase transitions, etc. All these features can be accounted for, at least on a phenomenological level, on the basis of a Landau free energy expansion, which allows explicitly for the space group symmetry of the parent phase (V). In sections 3 and 4 we discuss the effect of pressure and electric field on the intermediate phases. Finally, in section 5, the dynamical properties of the paraelectric, ferroelectric and intermediate phases will be briefly reviewed.

2. *Static properties of the intermediate phases of thiourea*

2.1. *Temperature dependence of modulation wave vector and phase diagram*

The temperature dependence of the modulation wave vector $q_\delta = (2\pi/b)\delta$, was obtained from a recent series of neutron diffraction studies on deuterated thiourea (Moudden et al. 1979, Dénoyer et al. 1980). δ shows a smooth decrease from $\delta(T_i = 218.1\,\text{K}) \simeq 0.141$ to $\delta(T_1 = 193.3\,\text{K}) \simeq 0.115$. At T_1, δ locks-in discontinuously onto the rational value $\delta = \frac{1}{9}$, down to $T_c = 190.7\,\text{K}$. At T_c, δ drops to zero, as the proper ferroelectric state (phase I) sets in. Similar results are obtained on the hydrogeneous compound, apart from the quasi-continuous character of the lock-in transition at T_1 (cf. fig. 5).

Except for the intermediate $\delta = \frac{1}{9}$ phase ($T_c < T < T_1$) no structural evidence was found for additional commensurate phases which one might expect to be stable over narrow temperature intervals around $T = 198.5\,\text{K}$ ($\delta = \frac{1}{8}$), 206 K ($\frac{2}{15}$) and 194.5 K ($\frac{2}{17}$).

Both lock-in transitions at T_1 and T_c are affected by a temperature hysteresis of approximately 1 K. In addition, hysteresis on the value of $\delta(T)$ is observed throughout the incommensurate phase. This type of behaviour, sometimes

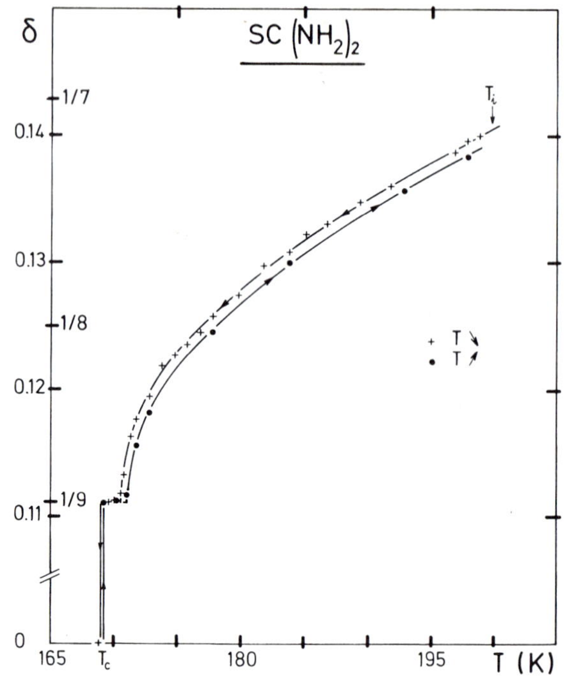

Fig. 5. h-thiourea: the temperature dependence of the modulation wave vector δ. (Durand et al. 1985.)

referred to as "global hysteresis" (Aubry, 1978, 1980) is a characteristic feature of many modulated systems (see for example Comès et al. 1976, de Pater et al. 1979, and Schneck and Dénoyer 1981).

The transitions at T_i, T_1 and T_c can be well correlated to specific heat (Chang and Westrum 1963) and birefringence anomalies (Jamet et al. 1980). As mentioned earlier, the sequence of phase transitions paraelectric → incommensurate → commensurate ($\delta = \frac{1}{9}$) → proper ferroelectric ($\delta = 0$) manifests itself also as anomalies in the static permittivity ε'_a (cf. fig. 2). One puzzling feature of the dielectric results is the suggestion of an additional weakly ferroelectric phase stable over 1 or 2 K around 198.5 K ("phase III"). If this additional dielectric anomaly is to be associated with an additional phase, why cannot one find any hint for it either in the diffraction results or in the calorimetric and birefringence data? As discussed further in section 4, the new phase, which is only observed in dielectric measurements (i.e. in the presence of a small applied electric field) turns out to be an electric-field induced commensurate lock-in phase with $\delta = \frac{1}{8}$.

2.2. *Amplitude and character of the modulation*

The study of first-order satellite intensities in d-thiourea, by neutron diffraction (Dénoyer et al. 1980, Moudden and Als-Nielsen 1980), shows that in the incommensurate phase the order parameter varies monotonically as a function of temperature, with only a small jump at T_1. When the temperature decreases, weak secondary peaks progressively appear at $\boldsymbol{q}_{2\delta} = 2\delta \boldsymbol{b}^*$ and $\boldsymbol{q}_{3\delta} = 3\delta \boldsymbol{b}^*$. These peaks can be interpreted either as diffraction harmonics arising from the non-linear relationship between the scattered amplitude and the molecular displacements or as distorsion harmonics arising from the non-sinusoidal character of these displacements, or as a combination of both effects. Critical exponents for the first- and second-order satellite intensities are found (Moudden and Als-Nielsen 1980) to be in good agreement with the critical behavior of the $d = 3$ XY model, including a uniaxial perturbation (Cowley and Bruce 1978). Alternately, the observed deviation from classical behaviour ($\beta \neq \frac{1}{2}$) may be ascribed to the proximity of a tricritical point.

Near T_i, the incommensurate phase can be viewed as a perturbation on the "basic" Pnma structure by means of a frozen-in soft-mode normal coordinate $Q(\boldsymbol{q}_\delta, \tau_4)$ whose symmetry properties are characterized by its wave vector $\boldsymbol{q}_\delta$ and by its irreducible representation τ_4. The symmetry elements of the space group Pnma are listed in table 1. The character of the irreducible representations of the group of the wave vector $\boldsymbol{q}_\delta = (2\pi/b)(0\,\delta\,0)$ in Pnma symmetry are given in table 2. The proper choice between the four possible irreducible representations is dictated by the observed systematic extinction rules for the satellite diffraction peaks.

Table 1
Symmetry elements of the space group Pnma.

Symmetry elements	Equivalent positions			Atoms labelled (m) and (m') belong to the same molecule (see also fig. 3)
$\{E\|0\}$	x	y	z	(1)
$\{C_{2x}\|\frac{1}{2}(\boldsymbol{a}+\boldsymbol{b}+\boldsymbol{c})\}$	$\frac{1}{2}+x$	$\frac{1}{2}-y$	$\frac{1}{2}-z$	(2)
$\{C_{2y}\|\frac{1}{2}\boldsymbol{b}\}$	$\bar{x}$	$\frac{1}{2}+y$	$\bar{z}$	(4)
$\{C_{2z}\|\frac{1}{2}(\boldsymbol{a}+\boldsymbol{c})\}$	$\frac{1}{2}-x$	$\bar{y}$	$\frac{1}{2}+z$	(3)
$\{I\|0\}$	$\bar{x}$	$\bar{y}$	$\bar{z}$	(4′)
$\{\sigma_x\|\frac{1}{2}\boldsymbol{a}+\frac{1}{2}\boldsymbol{b}+\frac{1}{2}\boldsymbol{c}\}$	$\frac{1}{2}-x$	$\frac{1}{2}+y$	$\frac{1}{2}+z$	(3′)
$\{\sigma_y\|\frac{1}{2}\boldsymbol{b}\}$	x	$\frac{1}{2}-y$	z	(1′)
$\{\sigma_z\|\frac{1}{2}\boldsymbol{a}+\frac{1}{2}\boldsymbol{c}\}$	$\frac{1}{2}+x$	y	$\frac{1}{2}-z$	(2′)

Table 2
Character of irreducible representations of the group of $\boldsymbol{q}_\delta = (2\pi/b)(0\ \delta\ 0)$ in Pnma symmetry.

$G_\delta(0\ \delta\ 0)$	$\{E\|0\}$	$\{C_{2y}\|\frac{1}{2}\boldsymbol{b}\}$	$\{\sigma_x\|\frac{1}{2}\boldsymbol{a}+\frac{1}{2}\boldsymbol{b}+\frac{1}{2}\boldsymbol{c}\}$	$\{\sigma_z\|\frac{1}{2}\boldsymbol{a}+\frac{1}{2}\boldsymbol{c}\}$
τ_1	1	1	1	1
τ_2	1	1	−1	−1
τ_3	1	−1	1	−1
τ_4	1	−1	−1	1

Assuming the molecules to behave as rigid units, the molecular displacement $\boldsymbol{u}_{lj}$ of molecule j ($j=1,4$) in cell l can be cast into the general form

$$\boldsymbol{u}_{lj} = \eta \sum_{\lambda=1,3,5} \alpha_\lambda \boldsymbol{s}_\lambda(j)\cos\left(\boldsymbol{q}_\delta\cdot\boldsymbol{r}_{lj}^0+\varphi\right) + \eta \sum_{\lambda=2,4,6} \alpha_\lambda \boldsymbol{s}_\lambda(j)\sin\left(\boldsymbol{q}_\delta\cdot\boldsymbol{r}_{lj}^0+\varphi\right), \quad (1)$$

where η and φ are the amplitude and phase of the frozen-in soft-mode normal coordinate,

$$Q(\vec{q}_\delta, \tau_4) = \eta e^{i\varphi},$$

and the $\boldsymbol{s}_\lambda(j)$ are symmetry-adapted eigenvectors whose expressions in terms of molecular translations (T_a, T_b, T_c) and rotations (R_a, R_b, R_c) are given in table 3.

Equation (1) shows that the modulation consists of two waves in quadrature, a result which has been derived by Heine and McConnell (1981) in a more general context, and which here follows as a direct consequence of time-reversal symmetry. Note also that the $P2_1ma$ structure of the commensurate phase, i.e. phase I, can be deduced from eq. (1) using only one of the two component

Table 3
Symmetry-adapted eigenvectors $\boldsymbol{s}_\lambda(j)$. The molecules are labelled as in fig. 3.

λ \ j	(1)	(2)	(3)	(4)
1	T_a	T_a	T_a	T_a
2	T_b	T_b	$-T_b$	$-T_b$
3	T_c	$-T_c$	$-T_c$	T_c
4	R_a	$-R_a$	$-R_a$	R_a
5	R_b	$-R_b$	R_b	$-R_b$
6	R_c	R_c	R_c	R_c

Table 4
Numerical values for the quantities $\eta\alpha_\lambda$ from satellite intensity measurements.

Displacement in reduced units or radians	Value	
	$SC(NH_2)_2$, 187 K (Shiozaki 1971)	$SC(ND_2)_2$, 205 K (McKenzie 1975a)
$\eta\alpha_3$ (T_c)	0.025	0.020
$\eta\alpha_5$ (R_b)	0.119	0.110
$\eta\alpha_1$ (T_a)	0.025	0.013

waves, here the cosine wave; [the sine-wave component would give rise to a ferroelastic commensurate phase ($P2_1/a$), in contradiction with experiment].

Numerical values for the α_λ's (or rather for the quantities $\eta\alpha_\lambda$) as obtained from satellite intensity measurements, are shown in table 4. Both authors listed therein have assumed that the sine-wave component was negligible ($\alpha_2 \simeq \alpha_4 \simeq \alpha_6 \simeq 0$), which leads to purely transverse displacements.

Subsequent diffraction measurements (Moudden et al. 1978a) have revealed the existence of longitudinal satellites associated with a small $\eta\alpha_2$.

2.3. *Discussion*

2.3.1. *A model thermodynamic potential for the incommensurate phase*

The least ambitious approach for discussing the stability of modulated structures consists in writing down a phenomenological expression for the free energy of the system as a functional of the order parameter [here the x-component of the dielectric polarization density, $P_x(\boldsymbol{r})$] and of its spatial derivatives. This Landau–Ginzburg (continuum) approach relies on the assumption that the spatial variations of the order parameter are slow on an atomic scale. Thus it is well suited for discussing the relative stability of the paraelectric, ferroelectric and (long-period) incommensurate phases. On the other hand, commensurate lock-in phases, such as $\delta = \frac{1}{9}$, which occur as a direct consequence of the microscopic lattice periodicity, cannot be treated within this same formalism (cf. section 2.3.4).

Following Michelson (1977), Ishibashi and Shiba (1978), Lederer and Chaves (1981) and Currat (1984), we consider a free-energy functional of the type

$$F = \int \left\{ \tfrac{1}{2}A_0 P_x^2 + \tfrac{1}{4}BP_x^4 + \tfrac{1}{2}\alpha\left(\frac{\mathrm{d}P_x}{\mathrm{d}y}\right)^2 + \tfrac{1}{4}\beta\left(\frac{\mathrm{d}^2P_x}{\mathrm{d}y^2}\right) + \gamma\left(P_x\frac{\mathrm{d}P_x}{\mathrm{d}y}\right)^2 \right\} \mathrm{d}y, \tag{2}$$

with

$$A_0 = a(T - T_0), \qquad B > 0, \qquad \alpha < 0, \qquad \beta > 0, \qquad \gamma > 0.$$

Gradient terms in directions other than that of the modulation wave vector (y) are not relevant to the present discussion and have been omitted from eq. (2). The essential feature of eq. (2) lies in the negative sign of the coefficient α, which gives rise to a soft-mode minimum at finite wavevector. The underlying physical picture has been discussed by Levanyuk and Sannikov (1976b) in the case of thiourea and $NaNO_2$ and by Aslanyan and Levanyuk (1978, 1979) in the case of quartz. These authors have argued that a symmetry-allowed gradient-interaction term between the primary distorsion wave and an ε_{xy} acoustic shear-wave leads to a Lifshitz-like free-energy invariant of the form

$$\frac{\mathrm{d}P_x}{\mathrm{d}y}\varepsilon_{xy},$$

which after minimization with respect to ε_{xy} reduces to

$$\alpha(\mathrm{d}P_x/\mathrm{d}y)^2 \quad (\alpha<0).$$

The physical relevance of this type of mechanism is reasonably well-established in the case of $NaNO_2$ and quartz [cf. the chapter by Dolino in the present volume (1985)]. The case of thiourea is not as clear. The available inelastic neutron scattering results (cf. section 5.2) on the soft-optic and acoustic mode dispersions do not seem to support the above picture. Alternatively, the soft-mode minimum at small but finite wave vector could be due to the long-range nature of the dipolar interactions (Parlinski and Michel 1984). The implications of eq. (2) are easiest to visualize assuming a purely sinusoidal modulation with wave vector $q_\delta=(2\pi/b)\delta$:

$$P_x(y)=P_0+2P_\delta\cos 2\pi\delta y. \tag{3}$$

Equation (2) then becomes:

$$F=\tfrac{1}{2}A_0P_0^2+\tfrac{1}{4}BP_0^4+A_\delta P_\delta^4+\tfrac{3}{2}B_\delta P_\delta^4+\tfrac{3}{2}(B+B_\delta)P_0^2P_\delta^2+\cdots, \tag{4}$$

with

$$A_\delta=A_0+\alpha\delta^2+\tfrac{1}{2}\beta\delta^4 \quad \text{and} \quad B_\delta=B+\tfrac{4}{3}\gamma\delta^2.$$

Analysis of eq. (4) yields

(i) a second-order phase transition at $T_\mathrm{i}=T_0+\alpha^2/2\beta a$ from the para-electric state (characterized by $P_0=P_\delta=0$) to the modulated state ($P_0=0$, $P_\delta\neq 0$);

(ii) a first-order transition at $T_\mathrm{c}<T_0$ from the modulated state to the ferroelectric state ($P_0\neq 0,\ P_\delta=0$).

The simple model outlined above is directly applicable to cases such as sodium nitrite (Ishibashi and Shiba 1978, Ishibashi et al. 1981, Durand et al. 1983) for which the single-harmonic approximation (cf. eq. 3) appears to be adequate. For a *quantitative* discussion of the phase diagram of thiourea, a somewhat more general version of the model should be used, taking into account second- and third-order harmonic distorsions (section 2.3.3).

2.3.2. Temperature dependence of the modulation wave vector δ(T)

The last term in eq. (2) couples the wave vector to the amplitude of the modulation. One thus expects the former to decrease continuously as the latter grows. More specifically, minimization of eq. (4) with respect to P_δ and δ yields for the T dependence of the modulation wave vector $\delta(T)$:

$$\delta^2(T)-\delta_i^2=\frac{1}{2}\left\{\left[\frac{B_i^2}{\gamma^2}+\frac{8a}{3\beta}(T-T_i)\right]^{1/2}-\frac{B_i}{\gamma}\right\}, \tag{5}$$

where

$$\delta_i=\delta(T_i)=(|\alpha|/\beta)^{1/2}, \qquad B_i=B_\delta(\delta=\delta_i)=B+\tfrac{4}{3}|\alpha|\frac{\gamma}{\beta}.$$

The above result is in good agreement with experimental observations (cf. fig. 5). In particular, the downward curvature of the experimental $\delta(T)$ curve is correctly predicted.

Close to T_1, $\delta(T)$ varies quite rapidly, in a way which probably cannot be described by means of eq. (5) with realistic values for the coefficients α, β, γ, etc. This rapid decrease of $\delta(T)$ towards the commensurate value $\frac{1}{9}$, is not unexpected and is commonly observed near lock-in transitions. In fact, for a pure system treated in the mean-field continuum limit one would expect a second-order lock-in transition with a very steep logarithmic law for $\delta(T)$ (McMillan 1976, Bruce 1980):

$$\delta(T)-\tfrac{1}{9}\sim-\left[\ln(T-T_1)\right]^{-1}.$$

One puzzling feature which has stirred a considerable amount of controversy is the fact that in $SC(ND_2)_2$ (Moudden and Als-Nielsen 1980) as well as in $SC(NH_2)_2$ (Durand et al. 1985), the modulation wave vector appears to follow a square-root law across the *entire* incommensurate regime:

$$\delta(T)-\tfrac{1}{9}\sim(T-T_1)^{1/2} \quad (T_1<T<T_i).$$

A number of authors have attempted to account for this unorthodox result in terms of wandering 9b-domain walls, by analogy with the two-dimensional case (Pokrovsky and Talapov 1979). The situation was eventually clarified by Fisher and Fisher (1982) who showed that in three dimensions no corrections to the classical logarithmic law could arise from domain wall fluctuations.

The present authors are of the opinion that the experimental "square-root" law results from the interaction of the modulation with defects (Feigel'man and Lyuksyutov 1983) or arises from a cross-over between several regimes. At least close to T_i, where the modulation is known to be quasi-sinusoidal, we believe that the T dependence of δ is more likely to be governed by eq. (5) than by the fluctuations of hypothetical 9b-domain walls.

2.3.3. Commensurate phases and lock-in transitions

It is important to distinguish between lock-in transitions at a general commensurate wave vector, such as $\delta = \frac{1}{3}$ or $\frac{1}{9}$, and lock-in transitions at a high-symmetry point in the Brillouin zone, such as, say, the Γ-point ($\delta = 0$).

The first case (so-called type I) corresponds to the case where a Lifshitz invariant is allowed at the commensurate wave vector (Levanyuk and Sannikov 1976a). The discommensuration picture is in principle applicable (Dzyaloshinskii 1964, McMillan 1976) and one expects a continuous or quasi-continuous lock-in transition.

In the second case (type II), no Lifshitz invariant is allowed, at least when the Γ-point soft-mode representation is non-degenerate. From the numerical analysis of Bruce et al. (1978), one expects an abrupt first-order transition from a quasi-sinusoidal incommensurate state to the ferrodistorted state. Although these predictions are all based on continuum-type arguments, recent results on discrete models (Hoegh-Jensen and Bak 1983) appear to confirm this general trend.

In the case of thiourea both types of transitions are present and indeed the lock-in at $\delta = 0$ is more strongly discontinuous than that at $\delta = \frac{1}{9}$. Deviations from the purely sinusoidal state are apparent at low temperatures, but the observed third-order satellite intensities are much too weak to be consistent with the square-wave type of modulation which would prevail in the "narrow soliton" limit. A quantitative analysis of the structure of the commensurate 9b-phase is currently underway (Simonson et al. 1985) in order to establish this point more firmly. Whereas the "ferroelectric-domain" model for the 9b-phase of thiourea (i.e. 9 planes polarized up followed by 9 planes polarized down) appears to be unfounded, both theoretically and experimentally, the existence just above T_1 of quasi-commensurate 9b-phase domains separated by domain walls is not to be dismissed a priori. Order of magnitude estimates indicate that the corresponding domain walls would be as narrow as a few lattice spacings (Barreto et al. 1983). Unfortunately, direct experimental confirmation, as attempted by Andrews and Mashiyama (1983) for the case of Rb_2ZnCl_4, seems to be beyond present experimental capabilities.

2.3.4. Umklapp lock-in terms

Lock-in transitions at $\delta \neq 0$ occur as a consequence of lattice translational periodicity. In fact, both the translational and the rotational invariance of the parent structure play a role in determining the most favorable rational values for δ. This point is well illustrated in the case of thiourea, where the presence of a lock-in phase at $\delta = \frac{1}{9}$ and none at $\delta = \frac{1}{8}$ may, a priori, seem surprising, given the fact that $\delta(T)$ passes through the value $\frac{1}{8}$.

In order to understand why odd superstructures appear to be more stable than even ones, we consider the corresponding Umklapp terms in the free-energy expansion (Dénoyer et al. 1982):

$$V^{(n)}(\boldsymbol{q}_\delta, \tau_4) \quad Q^n(\boldsymbol{q}_\delta, \tau_4) + \text{c.c.} \tag{6}$$

where $V^{(n)}$ is a nth-order anharmonic coefficient which vanishes by translational symmetry unless

$$n\boldsymbol{q}_\delta = \boldsymbol{G} \quad (\boldsymbol{G} = \text{reciprocal lattice vector}). \tag{7}$$

$Q(\boldsymbol{q}_\delta, \tau_4)$ is a microscopic order parameter corresponding to the complex amplitude of the primary Fourier component of the modulation (see for example Moncton et al. 1975 and Iizumi et al. 1977). Its transformation properties under space group operations $\{\mathrm{S}|\boldsymbol{V}(\mathrm{S})\}$ which leave $\boldsymbol{q}_\delta$ invariant, are expressed by:

$$\{\mathrm{S}|\boldsymbol{V}(\mathrm{S})\}\, Q(\boldsymbol{q}_\delta, \tau_4) = \chi_\mathrm{S}(\tau_4)\, \mathrm{e}^{\mathrm{i}\boldsymbol{q}_\delta \cdot \boldsymbol{V}(\mathrm{S})} Q(\boldsymbol{q}_\delta, \tau_4), \tag{8}$$

where the point-group characters $\chi_\mathrm{S}(\tau_4)$ are given in table 2.

Taking

$$\boldsymbol{q}_\delta = \delta \boldsymbol{b}^* \quad \text{and} \quad \boldsymbol{G} = m\boldsymbol{b}^*,$$

equation (7) can be rewritten as

$$\delta = m/n$$

A necessary condition for expression (6) to be rotationally invariant is

$$\{\mathrm{S}|\boldsymbol{V}(\mathrm{S})\}\, Q^n(\boldsymbol{q}_\delta, \tau_4) = Q^n(\boldsymbol{q}_\delta, \tau_4) \tag{9}$$

for all space-group operations $\{\mathrm{S}|\boldsymbol{V}(\mathrm{S})\}$ such that

$$\mathrm{S}\boldsymbol{q}_\delta = \boldsymbol{q}_\delta.$$

With the help of eqs. (7) and (8), equation (9) becomes:

$$[\chi_\mathrm{S}(\tau_4)]^n\, \mathrm{e}^{\mathrm{i}\boldsymbol{G} \cdot \boldsymbol{V}(\mathrm{S})} = 1. \tag{10}$$

The above condition is obviously satisfied for $\{S|\boldsymbol{V}(S)\} = \{E|0\}$ and $\{\sigma_z|\frac{1}{2}\boldsymbol{a} + \frac{1}{2}\boldsymbol{c}\}$, since $\chi_S(\tau_4) = 1$ and $\boldsymbol{G}\cdot\boldsymbol{V}(S) = 0$ (cf. table 2). However for the remaining two operations, namely $\{\sigma_x|\frac{1}{2}\boldsymbol{a} + \frac{1}{2}\boldsymbol{b} + \frac{1}{2}\boldsymbol{c}\}$ and $\{C_{2y}|\frac{1}{2}\boldsymbol{b}\}$, condition (10) reduces to:

$$(-1)^{n+m} = 1,$$

that is, n and m should have the same parity.

Hence the $\delta = \frac{1}{9}$ commensurate phase corresponds to a lowest-order lock-in term of order 9 while the $\delta = \frac{1}{8}$ phase would involve a lock-in term of order 16, too weak to have a measurable effect on the $\delta(T)$ curve. Further remarks on this point will be given in section 4.2.1.

2.3.5. *Microscopic models*

Recently Parlinski and Michel (1984), and Parlinski (1984) have developed a microscopic model for a one-dimensional analog of thiourea. The model includes a short-range interaction parameter between adjacent molecules, the only temperature-dependent parameter in the model, and long-range dipolar-type interactions between first-, and second- up to ninth-nearest neighbor molecules, along the chain direction ($\boldsymbol{b}$-axis). With the addition of two local third- and fourth-order anharmonic parameters the (mean field) model provides a good qualitative description of the modulated phases of thiourea. Commensurate phases corresponding to $\delta = \frac{1}{7}$, $\frac{1}{8}$ and $\frac{1}{9}$ are predicted, consistent with experimental observations under pressure and electric field (cf. sections 3 and 4). In the model the main lock-in mechanism is provided by the third-order local anharmonic coefficient, which gives rise to a series of Umklapp terms, involving appropriate combinations of distorsion harmonics. For example, the $\frac{1}{9}$-phase is stabilized by lock-in terms of the type

$$V^{(3)}Q^3(3\boldsymbol{q}_\delta, \tau_4) + \text{c.c.}$$

and

$$V^{(3)}Q(5\boldsymbol{q}_\delta, \tau_4)Q(3\boldsymbol{q}_\delta, \tau_4)Q(\boldsymbol{q}_\delta, \tau_4) + \text{c.c.},$$

with

$$\boldsymbol{q}_\delta = \tfrac{1}{9}\boldsymbol{b}^*.$$

In general the model tends to overestimate the relative amplitude of the higher-order odd harmonics, such as $Q(3\boldsymbol{q}_\delta, \tau_4)$ and $Q(5\boldsymbol{q}_\delta, \tau_4)$, and consequently it also overestimates the stability range of the lock-in phases with respect to the incommensurate region.

Nevertheless, the model provides some valuable insight into the physical mechanisms involved. In particular, it illustrates the relationship between the harmonic content of the modulation and the commensurability potential which it experiences. Undoubtedly, the model could be made more quantitative if needed, but only at some cost in term of added complexity.

3. *Effect of hydrostatic pressure on the modulated phases*

3.1. Early results on the pressure – temperature phase diagram

The first pressure investigation of thiourea dates back to the early volumetric measurements of Bridgman (1938, 1948); these revealed the presence of a new structure, which was stable above 2 kbar, and whose onset was marked by a strongly discontinuous phase transition. The position of the transition line was later determined by dielectric measurements (Leonidova 1964, Gesi 1969) and Raman spectroscopy (Figuière et al. 1975). These techniques did not allow any conclusion to be drawn concerning the nature of the high-pressure phase (sometimes labeled phase VI) and its possible ferroelectric character.

In order to study the behavior of the modulated intermediate phases under pressure, dielectric measurements were undertaken by Gesi (1969) and Klimowski et al. (1976). Figure 6a shows the overall phase diagram obtained. More details on the modulated region are shown in fig. 6b: the essential results

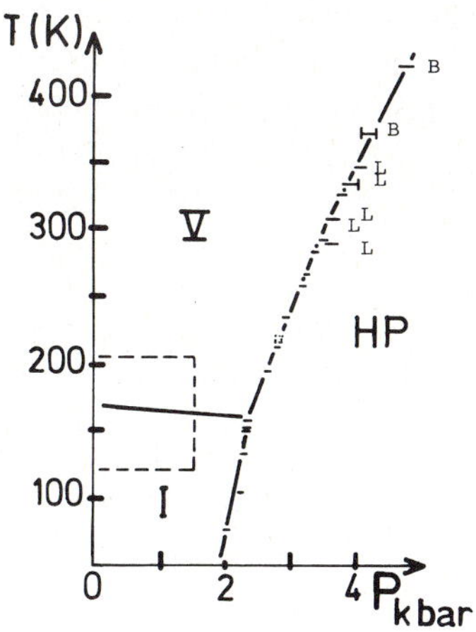

Fig. 6a. (P, T) phase diagram obtained by Raman spectroscopy; the region inside the dotted lines is shown in fig. 6b. (Figuière et al. 1975).

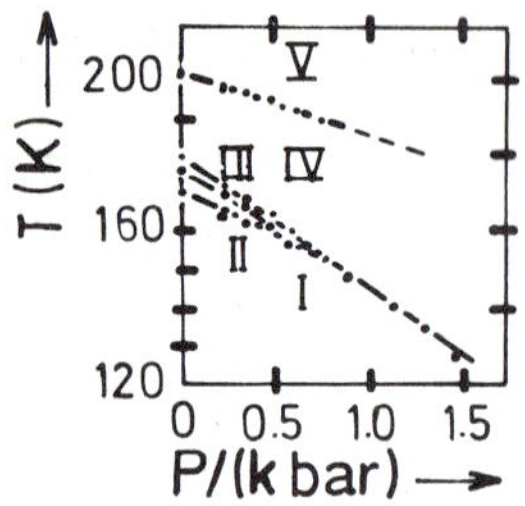

Fig. 6b. (P, T) phase diagram obtained from dielectric anomalies. (Gesi 1969.)

are that both T_i and T_c decrease with pressure and that some of the modulated phases, the so-called phases II and III, are not observed above 1 kbar.

As far as the structure of the high-temperature phase is concerned, the early X-ray powder diffraction data of Kabalkina (1961) could be indexed in terms of an orthorhombic structure with 12 molecules per cell, consistent with a 3b-superstructure. The identification of the high-pressure phase with a $\delta = \frac{1}{3}$ modulated structure was later confirmed by X-ray single crystal results (Moudden 1980, Moudden et al. 1980). In particular, the systematic extinction rules were found to be the same as for the commensurate 9b-phase, suggesting that the high-pressure 3b-phase belongs to the same class of lock-in phases.

3.2. *Neutron diffraction results*

One obvious question which arises is whether other new lock-in phases might be stabilized by the applied pressure, in the range from 0 to 2 kbar. In order to clarify this point an extensive set of neutron diffraction measurements was collected on a d-thiourea single crystal (Dénoyer et al. 1981, 1982). Figure 7 shows the isothermal variation of $\delta(T, P)$, for a few selected temperatures.

Generally speaking, δ increases with pressure while the modulation amplitude decreases. Above some finite pressure value $P_i(T)$, the "high temperature" paraelectric phase becomes stable; within experimental accuracy the transition at P_i remains second order. This type of behaviour is illustrated by the 213.6 K and 206.4 K isotherms in fig. 7. Below 205 K, the variation of δ with pressure follows the same qualitative trend as above, but additional features appear due to the existence of commensurate "lock-in" phases.

In particular, all isotherms below 205 K display a plateau at $\delta = 0.1428 \pm 2$. This result represents the *first evidence for a pressure-induced lock-in phase of period 7b*. One may notice the rapid variation of δ with P immediately below and above the $\frac{1}{7}$ plateau; continuous transitions with logarithmic singularities have been predicted theoretically (Aubry 1978). Note also the absence of a well defined plateau on the 157.6 K isotherm.

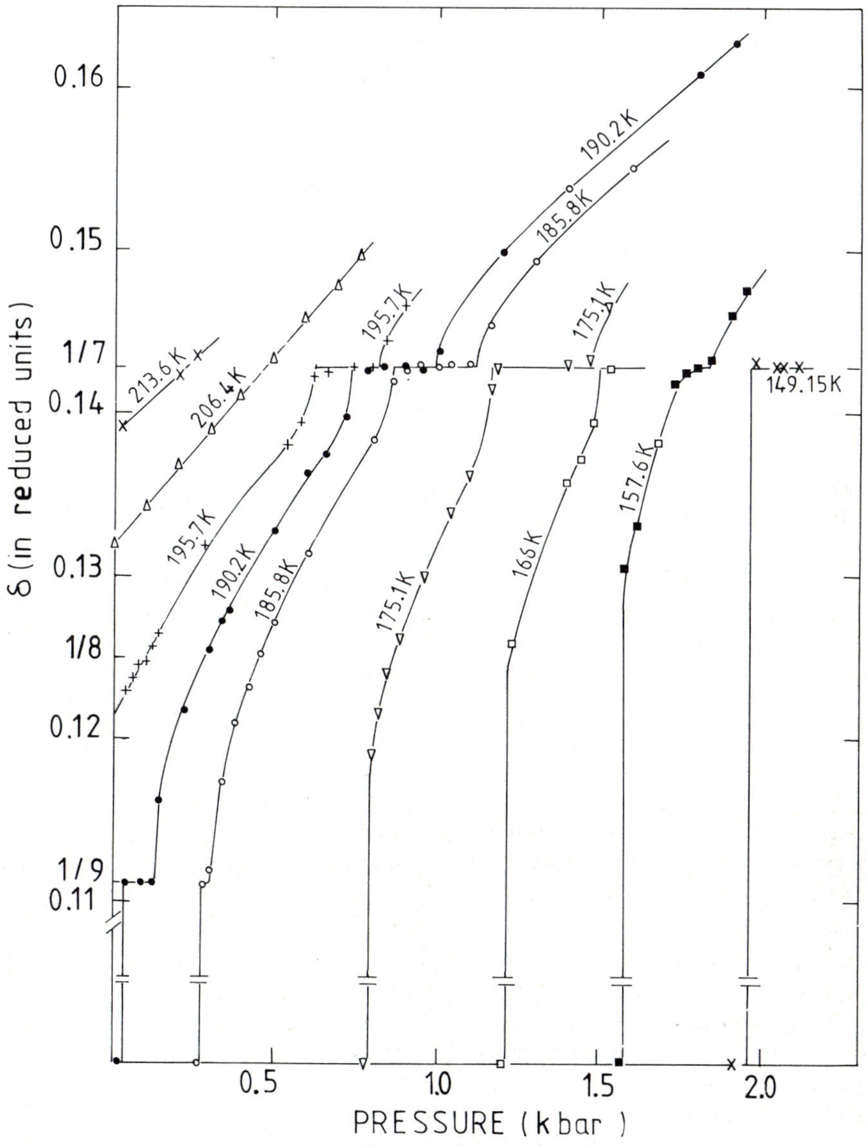

Fig. 7. Isothermal variations of the modulation wave vector $\delta(P, T)$ as a function of pressure: $P < 2200$ bar. (Dénoyer et al. 1982.)

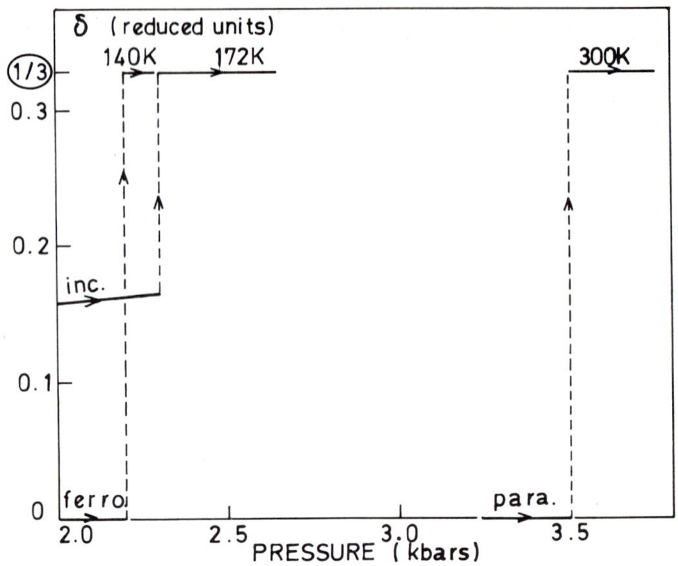

Fig. 8. Isothermal variations of δ as a function of pressure: $P > 2200$ bar.

Below 193 K, additional steps at $\delta = \frac{1}{9}$ and $\delta = 0$ appear, corresponding to the two zero-pressure commensurate lock-in phases. Below 180 K, the step at $\frac{1}{9}$ disappears altogether, while at still lower temperatures (ex. 149.15 isotherm), the system goes directly from the ferroelectric state into the 7b phase.

Figure 8 shows some additional $\delta(T, P)$-isotherms in the range from 2 to 3.5 kbar, obtained from the same neutron diffraction investigation. In all cases a discontinuous jump is observed at the onset of the $\delta = \frac{1}{3}$ phase. Note in particular, on the 172 K isotherm, that the $\delta = \frac{1}{3}$ lock-in occurs much before the value of the incommensurate wave vector approaches $\frac{1}{5}$, thus pre-empting a possible $\delta = \frac{1}{5}$ lock-in phase at high pressure.

3.3. *Summary*

Figure 9 shows the resulting $(P-T)$ phase diagram; The results may be summarized as follows:

(i) The stability region for the modulated phases is limited by a second order transition line $T_i(P)$ and by a first-order transition line $T_c(P)$. The values of δ along the $T_i(P)$ and $T_c(P)$ lines are shown as δ_i and δ_c respectively: (we ignore for the moment the 3b phase). δ_i varies between a value

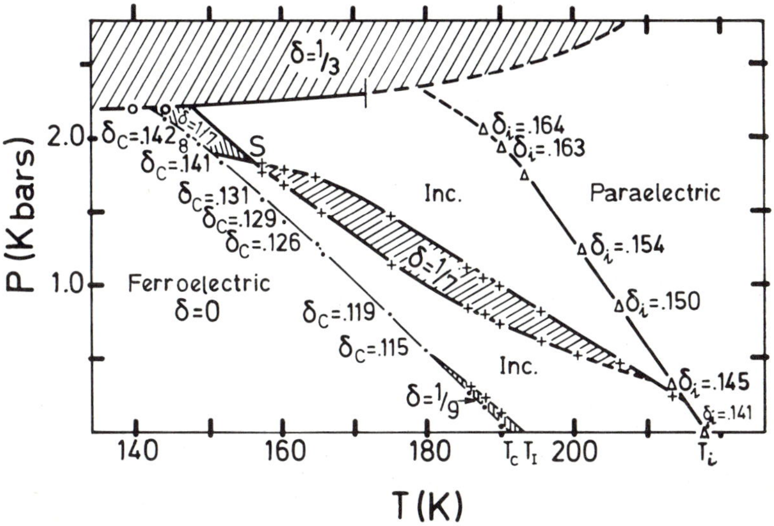

Fig. 9. (P, T) phase diagram. The hatched areas correspond to commensurate phases. (Dénoyer et al. 1982.)

somewhat smaller than $\frac{1}{7}$ to a value somewhat smaller than $\frac{1}{6}$, δ_c varying between $\frac{1}{9}$ and $\frac{1}{7}$. The values of the initial slopes, $(dT_i/dP)_{P=0} = -13.8 \pm 0.3$ K/kbar, $(dT_c/dP)_{P=0} = -20.2 \pm 0.5$ K/kbar, are somewhat lower than those obtained by other techniques (Gesi 1969, Klimowski et al. 1976, Benoit et al. 1983) on the hydrogenous compound.

(ii) The stability range of the 9b phase becomes progressively narrower with increasing pressure.

(iii) The 7b phase has a measurable stability range only below 205 K. Upon cooling, the stability range first increases, then saturates to a maximum value of about 300 bar, decreases to approximately zero at $T \simeq 156$ K, and subsequently increases again. This unusual behavior suggests the existence of two distinct 7b phases, with qualitatively different structures, an interpretation which is supported by satellite-intensity measurements in both phases. One can account for the two different 7b phases by postulating that along some curve passing through the point marked S in fig. 9 the $V^7(P, T)$ lock-in Umklapp term changes sign. The two lock-in phases would thus differ by a $\pi/7$ phase shift of the modulation wave with respect to the basic lattice. Such a phase shift would be presumably responsible for the observed change in satellite intensities.

4. Effect of applied electric field on the modulated phases

4.1. General features of the (E, T) phase diagram

The effect of an external electric field E applied along the ferroelectric ***a***-axis was first investigated by Futama (1962) by dielectric susceptibility measurements. The overall shape of the (E, T) phase diagram was later obtained by Jamet (1981), using the optical birefringence technique. His results are shown in fig. 10:

(i) The stability range of the modulated region decreases with field and eventually vanishes for $E > E_M \simeq 2300\,\mathrm{V/mm}$.

(ii) The first order transition line $T_c(E)$ has an initial slope

$$\left|\frac{\mathrm{d}T_c}{\mathrm{d}E}\right|^{-1} = 120\,\mathrm{V/mm\,K},$$

while the second-order transition line $T_i(E)$ starts with infinite slope.

(iii) The two lines join smoothly at a tricritical point with coordinates

$$E_{tr} = 2050\,\mathrm{V/mm}, \qquad T_{tr} = 212.5\,\mathrm{K}.$$

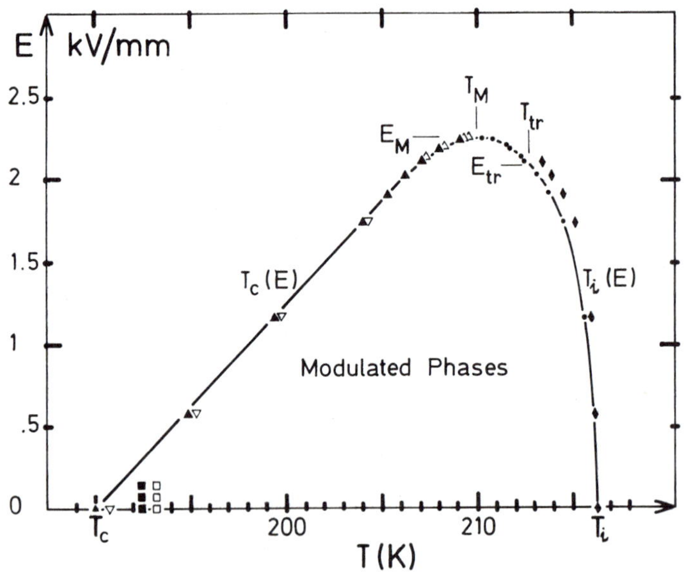

Fig. 10. Overall shape of the (E, T) phase diagram. (Jamet 1981.)

(iv) The point with a horizontal tangent (E_M, T_M) is often referred to as an isolated critical point, since from Clapeyron's relation it follows that the entropy jump changes sign at that particular point on the transition line.

The overall shape of the above phase diagram can be accounted for in terms of a continuum thermodynamic potential of the type developed in section 2.3.1, by simply adding a field term (cf. eq. 4)$-EP_0$. The (E, T) phase diagram is then derived by minimizing eq. (4) (including the field term) with respect to P_δ and P_0. This procedure has been used by Ishibashi and Shiba (1978), Coutinho-Filho and De Moura (1980), Yokoi et al. (1981), Lederer and Chaves (1981) and Cao et al. (1982), in various contexts.

Qualitatively, the way in which the applied field destabilizes the modulated state can be understood as follows:

(1) the field term $-EP_0$ stabilizes states with large homogeneous polarization P_0;

(2) the last term on the right-hand side of eq. (4), viz.

$$\tfrac{3}{2}(B_\delta + B)P_0^2P_\delta^2, \quad (>0)$$

penalizes states in which both P_0 and P_δ exist.

Thus, although there is no direct interaction between the applied field and the modulated polarization P_δ, there is an indirect one through the combined action of the field term and the anharmonic cross-term.

4.2. *Field-induced commensurate phases*

In section 2.1 we have pointed out that the large dielectric anomaly around 198.5 K in d-thiourea (so-called phase III) is associated with the passage of the modulation wave vector through the commensurate value $\delta(T)=\frac{1}{8}$. Although the corresponding commensurate phase is not stable in zero field [no observable plateau in the $\delta(T)$ curve], the large dielectric anomaly suggests an incipient lock-in phase of ferroelectric character.

This is confirmed by the following symmetry argument. Consider a mixed Umklapp term of the type

$$P_\delta^8 P_0, \tag{12a}$$

or, in the language of section 2.3.4,

$$Q^8(q_\delta, \tau_4) \quad Q(0, \tau_4) + \text{c.c.} \tag{12b}$$

This term is translationally invariant if $\delta=\frac{1}{8}$, and can be shown to satisfy the criterion for rotational invariance (eq. 10).

The presence of such a term in the free energy has two consequences:

(i) the commensurate $\delta = \frac{1}{8}$ phase is ferroelectric since the 8th-order harmonic wave $[Q(0, \tau_4)$, with wavevector $8\boldsymbol{q}_\delta = 0$ (modulo $\boldsymbol{b}^*$)] is polar (τ_4-symmetry).

(ii) the applied electric field will stabilize the ferroelectric $\delta = \frac{1}{8}$ phase over a finite temperature interval $\Delta T_8(E)$, at the expense of the adjacent incommensurate states.

The prediction for the occurrence of a field-induced 8b-lock-in phase, based on the above mechanism, was originally made by Dénoyer et al. (1982) and subsequently confirmed by dielectric measurements (Jamet et al. 1982, Gesi 1982) and neutron diffraction (Gesi and Iizumi 1982, Moudden et al. 1982).

Figure 11, from Gesi's work, shows the behaviour of the phase III-related dielectric anomaly for a few selected values of the d.c. bias field: the anomaly

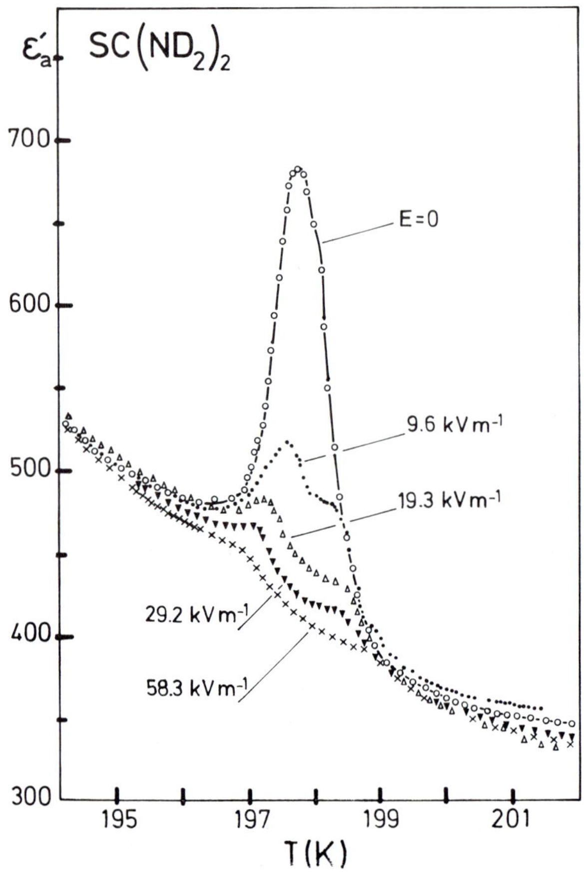

Fig. 11. The phase III-related dielectric anomalies as a function of bias field. (Gesi 1982.)

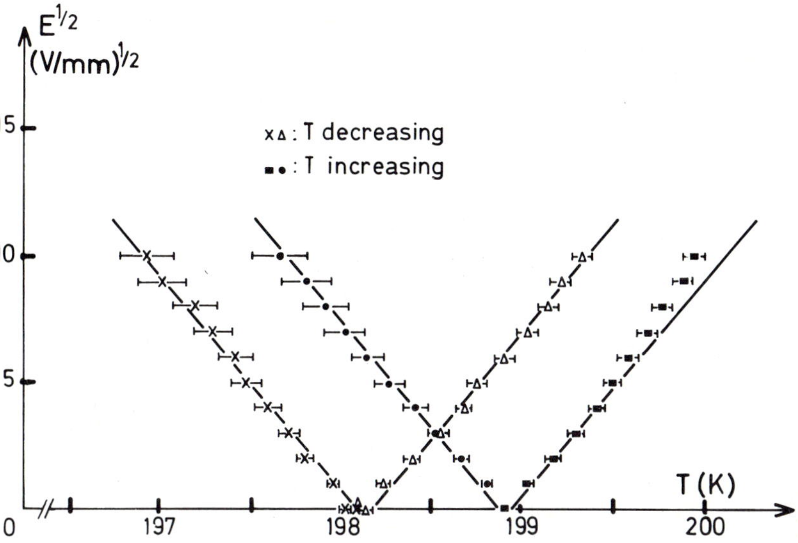

Fig. 12. Position of the susceptibility maxima as a function of $\sqrt{E}$ for small values of bias field. (Jamet et al. 1982.)

splits progressively into two separate peaks corresponding to the boundaries of the 8b-phase. The peak positions for various values of the field are plotted in fig. 12 (Jamet et al. 1982). The plot illustrates the relationship

$$\Delta T_8(E) \sim E^{1/2},$$

which is a direct consequence of the lock-in mechanism considered above.

The boundaries of the 8b-phase are also marked by weak optical birefringence anomalies. The complete (E, T) phase diagram, including the commensurate–incommensurate boundaries, has been recently mapped out (Barreto et al. 1983) using this latter technique. The results are summarized in fig. 13 which shows the overall shape of the 8b- and 9b-commensurate phases. The stability range $[\Delta T_8(E)]$ of the 8b-phase which at low fields grows as $\sqrt{E}$ is seen to saturate for field values of a few hundred V/mm and to decrease at still higher fields. This behaviour is well understood in terms of a lock-in potential of the type shown in eq. (12), where the equilibrium value of the modulated polarization P_δ is itself a decreasing function of the applied field (Jamet et al. 1982, Barreto et al. 1983, Durand et al. 1984).

Finally, we should mention the recent high-field *X-ray* diffraction results of Moudden et al. (1983), which conflict in several ways with the phase diagram shown in fig. 13: large plateaux are observed at high-order commensurate values, such as $\delta = \frac{2}{15}, \frac{4}{29}$ and $\frac{6}{43}$; furthermore the width of the latter plateau (at

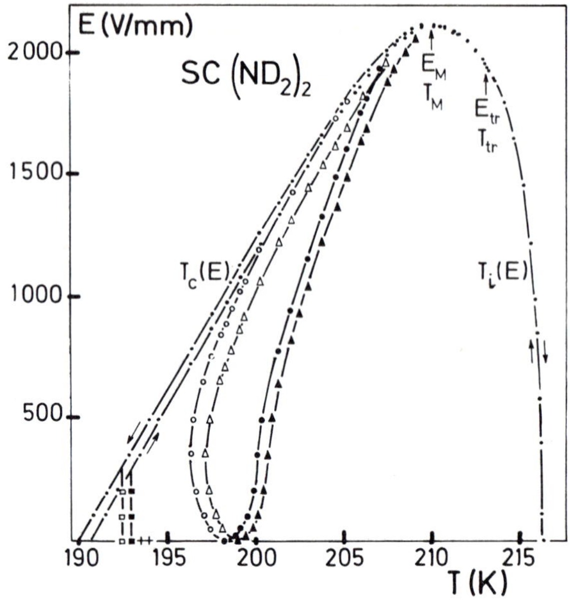

Fig. 13. (E, T) phase diagram for increasing and decreasing temperatures. The solid and open squares are related to the commensurate $(\delta = \frac{1}{9})$–incommensurate transitions for increasing and decreasing temperature. (Barreto et al. 1983.)

$E = 1800\,\mathrm{V/mm}$) appears to exceed the overall stability range of the modulated region as determined by the birefringence technique.

Since none of these surprising features can be found by neutron diffraction (fig. 14, Durand et al. 1984) the present authors believe that the plateaux reported by Moudden et al. are caused by the X-ray beam itself (radiation damage and defect-induced memory effects). Similar plateaux have been recently observed, by *neutron* diffraction in zero field, on a previously X-ray irradiated specimen (André et al. 1985b).

5. *Dynamical properties*

5.1. *General*

The current interest in thiourea is focused on the static behavior of the system in its modulated state, a behavior which is characterized by a highly sensitive response of the modulation periodicity to small changes in external physical parameters.

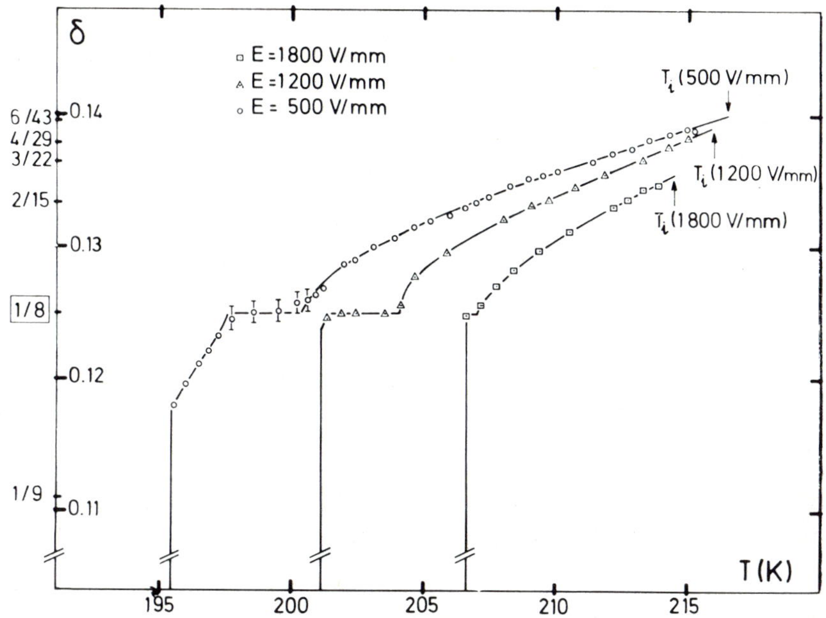

Fig. 14. Iso-electric field variations of the modulation wave vector $\delta(E, T)$ as a function of temperature. All the curves have been measured from right to left, i.e., going down in temperature from the paraelectric state. (Durand et al. 1984.)

In comparison the dynamical aspects associated with the observed phase transitions have received less attention. There are at least two obvious reasons for this:

(i) The low-frequency lattice-dynamics of thiourea in the normal state ($T > T_i$) is fairly complicated, with several interacting phonon branches below 1 THz, some of them heavily damped.

(ii) The situation is expected to be even more difficult in the modulated state, due to branch-splitting and zone-folding effects. There exist simpler model systems, where, for example, the sinusoidal approximation is expected to hold over a wide temperature range, which should be studied and understood first.

Nevertheless, there has been a number of optical studies by infrared, Raman and Brillouin spectroscopy as well as a few, largely unpublished, inelastic neutron scattering results, which all together provide some valuable insight into the microscopic nature of the structural instability at T_i. The main experimental results are reviewed below.

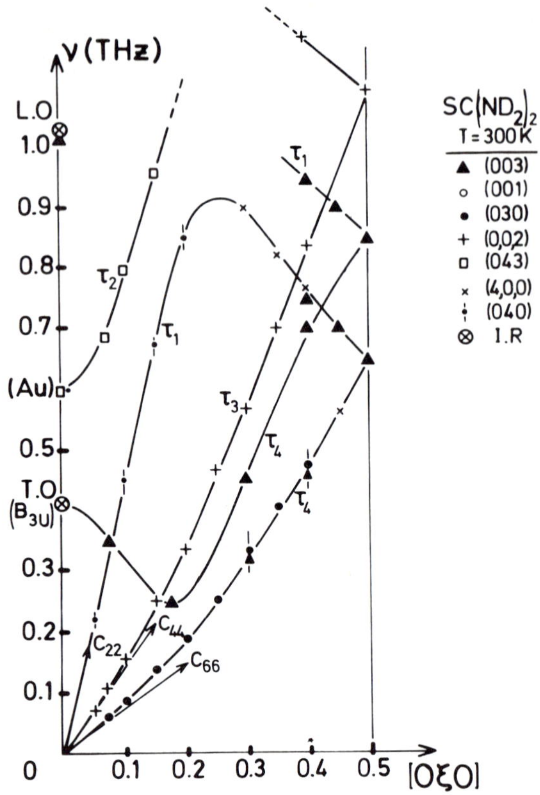

Fig. 15. Dispersion curves along the modulation direction at room temperature. (Moudden 1980.)

5.2. *Soft mode above T_i*

Between room temperature and T_i the softening of a polar optic mode (B_{3u}) is observed by infra-red spectroscopy (Bréhat et al. 1976, Khelifa et al. 1977, Siapkas, 1978, 1980) and inelastic neutron scattering (Moudden et al. 1978b, Moudden 1980). The position of the dispersion minimum is displaced by $\sim 0.15\boldsymbol{b}^*$ away from the Brillouin zone center. Hence the mode softening as seen by infra-red spectroscopy remains incomplete at T_i (cf. Petzelt 1981).

Figure 15 shows the low-frequency phonon branches propagating in the $\boldsymbol{b}^*$ direction, as obtained from neutron measurements at room temperature. The minimum in the soft-mode dispersion (upper τ_4-branch) is already well marked and cannot arise from a q-dependent interaction with the transverse acoustic mode (lower τ_4-branch), as originally postulated by Levanyuk and Sannikov (1976b). The interaction between the two τ_4-branches does exist however, and close to T_i, when the branches appear to cross, the experimental spectra can no longer be resolved into two separate response functions. In addition, the damping coefficient of the soft branch is substantial (a few hundred GHz at room temperature) and thus in order to extract the quasi-harmonic mode-frequencies a rather elaborate analysis of the experimental spectra is required. Such an analysis has not been attempted here, and thus the "dispersion curves" drawn in fig. 15 are only suggestive of the actual situation. More definite conclusions concerning the above-mentioned coupled-mode model should be based on a careful spectral analysis (currently in progress).

5.3. *The ferroelectric phase ($T < T_c$)*

The situation is much clearer below T_c. The soft optic branch is both Raman and infrared active at the zone center. While the LO-component is found to be approximately temperature independent (at ~1.05 THz), the TO-component, i.e. the $q \to 0$ limit of the soft-optic branch, renormalizes upwards on cooling (see for example Delahaigue et al. 1975 and Wada et al. 1978). The entire branch as well as the acoustic branches propagating along $\boldsymbol{b}^*$ have been observed by inelastic neutron scattering (Moudden 1980). Due to the lower temperature the phonon line shapes are sharper than in the paraelectric phase. In particular the zone-center TO mode is underdamped below ~180 K.

5.4. *The modulated phases ($T_c < T < T_i$)*

Raman scattering probes the amplitude mode frequency at the satellite position (Wada et al. 1978, Ishibashi and Takagi 1979), rather than at the parent structure's Brillouin zone center. This point is illustrated experimentally by the absence of angular dispersion on the observed frequency (no LO–TO splitting at the satellite position).

Attempts to observe a phason branch by inelastic neutron scattering have been largely inconclusive, due to the high damping of the low lying excitations below T_i. At lower temperature, a new acoustic-like phonon branch is observed, emanating from the satellite reflections. However, its slope is undistinguishable from that of a transverse acoustic branch, so that the nature of the new branch (true phason or "image" of the TA branch at the satellite position) remains in doubt.

6. *Concluding remarks*

In the preceding sections we have attempted to present a reasonably comprehensive account of recent experimental work on thiourea and of progress made in the understanding of its physical properties. Most of the current interest on this compound is associated with the response of the modulation to external parameters: temperature (section 2.1), hydrostatic pressure (section 3.2) and electric field (section 4.2).

Another active area of research, not covered here, is concerned with the interaction of the modulation with defects, frozen or mobile, giving rise to a variety of interesting phenomena (pinning, hysteresis, relaxations, memory effects, satellite broadening, etc.). These effects are not specific to thiourea [Unruh (1983), Schneck et al. (1984), see also the chapter by Hamano (1986) in Part III] although some have been discovered and analyzed first in thiourea (Jamet and Lederer 1983, 1984).

More quantitative information about the harmonic contents of the modulation itself would be useful in order to develop realistic models for the description of the modulated phases. Work in this direction is in progress (Simonson et al. 1985).

As far as the models themselves are concerned, there is an obvious need to bridge the gap between microscopic models on the one hand, which are not yet able to make quantitative predictions about phase diagrams, modulation wave vectors, etc., and on the other hand, phenomenological models based on Landau-type free energies with many unknown, adjustable parameters. Attempts to estimate some of these parameters starting from a microscopic approach should be encouraged.

References

André, G., D. Durand and F. Dénoyer, 1985a, to be published.

André, G., D. Durand, F. Dénoyer, F. Moussa and R. Currat, 1985b, to be published.

Andrews, S.R., and H. Mashiyama, 1983, J. Phys. **C16**, 4985.

Aslanyan, T.A., and A.P. Levanyuk, 1978, JETP Lett. **28**, 70.

Aslanyan, T.A., and A.P. Levanyuk, 1979, Solid State Commun. **31**, 547.

Aubry, S., 1978, in: Solitons and Condensed-Matter Physics, eds. A.R. Bishop and T. Schneider (Springer, New York).

Aubry, S., 1980, in: Seminar on the Riemann problem Spectral Theory and Complete Integrability, Lecture Notes in Mathematics, ed. D.G. Chudnovsky (Springer, Berlin).

Barreto, M., J.P. Jamet and P. Lederer, 1983, Phys. Rev. **B28**, 3994.

Benoit, J.P., M. Deniau and J.P. Chapelle, 1972, C.R. Acad. Sci. **275B**, 665.

Benoit, J.P., 1973, C.R. Acad. Sci. **277B**, 21.

Benoit, J.P., F. Thomas and J. Berger, 1983, J. Physique **44**, 841.

Bréhat, F., and A. Hadni, 1971, J. Physique **32**, 759.

Bréhat, F., J. Claudel, P. Strimer and A. Hadni, 1976, J. Phys. Lett. **37L**, 229.

Bridgman, P.W., 1938, Proc. Am. Acad. Arts Sci. **72**, 227.

Bridgman, P.W., 1948, Proc. Am. Acad. Arts Sci. **76**, 79.

Bruce, A.D., R.A. Cowley and A.F. Murray, 1978, J. Phys. **C11**, 3591.

Bruce, A.D., 1980, J. Phys. **C13**, 4615.

Calvo, C., 1960, J. Chem. Phys. **33**, 1721.

Cao, A.X., S. Krichene, G. Hauret, J.P. Benoit and J.P. Chapelle, 1982, Solid State Commun. **43**, 933.

Chang, S.S., and E.F. Westrum, 1963, Phys. Chem. Org. Solid State **1**, 126.
Comès, R., G. Shirane, S.M. Shapiro, A.F. Garito and A.J. Heeger, 1976, Phys. Rev. **B14**, 2376.
Coutinho-Filho, M.D., and M.A. De Moura, 1980, J. Magn. Magn. Mater. **15–18**, 433.
Cowley, R.A., and A.D. Bruce, 1978, J. Phys. **C11**, 3577.
Currat, R., 1984, in: Multicritical phenomena, Nato-ASI Series B Vol. 106, eds. R. Pynn and A.T. Skjeltorp (Plenum, New York).
Delahaigue, A., B. Khelifa and P. Jouve, 1975, Phys. Stat. Sol. (b) **72**, 585.
Dénoyer, F., A.H. Moudden and M. Lambert, 1980, Ferroelectrics **24**, 43.
Dénoyer, F., A.H. Moudden, A. Bellamy, R. Currat, C. Vettier and M. Lambert, 1981, C.R. Acad. Sci. Paris t 292 Serie II-13.
Dénoyer, F., A.H. Moudden, R. Currat, C. Vettier, A. Bellamy and M. Lambert, 1982, Phys. Rev. **B25**, 1697.
de Pater, C.J., J.D. Axe and R. Currat, 1979, Phys. Rev. **B19**, 4684.
Dolino, G., 1985, ch. 16, this volume.
Durand, D., F. Dénoyer, D. Lefur, R. Currat and L. Bernard, 1983, J. Physique Lett. **44**, L207.
Durand, D., F. Dénoyer, R. Currat and C. Vettier, 1984, Phys. Rev. **B30**, 1112.
Durand, D., F. Dénoyer and F. Moussa, 1985, to be published.
Dvoryankin, V.F., and B.K. Vainshtein, 1960 Kristallogr. **5**, 564.
Dzyaloshinskii, I.E., 1964, Zh. Eksp. Teor. Fiz. **47**, 992.
Elcombe, M.M., and J.C. Taylor, 1968, Acta Cryst. **A24**, 410.
Feigel'man, M.V., and I.F. Lyuksyutov, 1983, Solid State Commun. **48**, 397.
Figuière, P., M. Guelfenstein and H. Szwarc, 1975, Chem. Phys. Lett. **33**, 99.
Fisher, M.E., and D. Fisher, 1982, Phys. Rev. **B25**, 3192.
Futama, H., 1962, J. Phys. Soc. Jpn **17**, 434.
Futama, H., 1970, J. Phys. Soc. Jpn **28**, suppl., 295.
Futama, H., and A. Chiba, 1963, reported at the Annual Meeting of the Physical Society of Japan, Fukuoka.
Futama, H., Y. Shiozaki, A. Chiba, E. Tanaka, T. Mitsui and J. Furuichi, 1967, Phys. Lett. **25A**, 8.
Gesi, K., 1969, J. Phys. Soc. Jpn **26**, 107.
Gesi, K., 1982, J. Phys. Soc. Jpn **51**, 3.
Gesi, K., and M. Iizumi, 1982, J. Phys. Soc. Jpn **50**, 1047.
Goldsmith, G.J., and J.G. White, 1959, J. Chem. Phys. **31**, 1175.
Hamano, K., 1985, ch. 9, Part III.
Heine, V., and J.D.C. McConnell, 1981, Phys. Rev. Lett. **46**, 1092.
Hoegh-Jensen, M., and P. Bak, 1983, Phys. Rev. **B27**, 6853.
Iizumi, M., J.D. Axe, G. Shirane and K. Shimaoka, 1977, Phys. Rev. **B15**, 4392.
Ishibashi, Y., and H. Shiba, 1978, J. Phys. Soc. Jpn **45**, 409.
Ishibashi, Y., and Y. Takagi, 1979, J. Phys. Soc. Jpn **46**, 143.
Ishibashi, Y., W. Buchheit and J. Petersson, 1981, Solid State Commun. **38**, 1277.
Jamet, J.P., 1981, J. Physique Lett. **42**, L123.
Jamet, J.P., and P. Lederer, 1983, J. Physique Lett. **44**, L257.
Jamet, J.P., and P. Lederer, 1984, Ferroelectrics Lett. **1**, 139.
Jamet, J.P., A.M. Quittet and A.H. Moudden, 1980, EPS Conference, Antwerp, Belgium, Recent developments in Condensed Matter Physics, Vol. 9 (Plenum, New York) p. 61.
Jamet, J.P., P. Lederer and A.H. Moudden, 1982, Phys. Rev. Lett. **48**, 442.
Kabalkina, S.S., 1961, Zh. Fiz. Khim. **35**, 276.
Khelifa, B., A. Delahaigue and P. Jouve, 1977, Phys. Stat. Sol. (b) **83**, 139.
Klimowski, J., W. Wanarski and D. Ozgo, 1976, Phys. Stat. Sol. (a) **34**, 697.
Kovalev, O.V., 1965, Irreducible representation of the space groups (Gordon and Breach, New York).
Kunchur, N.R., and M.R. Truter, 1958, J. Chem. Soc., **517**, 2551.
Lederer, P., and C.M. Chaves, 1981, J. Physique Lett. **42**, L127.
Leonidova, G.G., 1964, Sov. Phys. Solid State **5**, 2519.

Levanyuk, A.P., and D.G. Sannikov, 1976a, Sov. Phys. Solid State **18**, 245.
Levanyuk, A.P., and D.G. Sannikov, 1976b, Sov. Phys. Solid State **18**, 1122.
MacKenzie, D.R., 1975a, J. Phys. **C8**, 1607.
MacKenzie, D.R., 1975b, J. Phys. **C8**, 2003.
MacKenzie, D.R., and J.S. Dryden, 1973, J. Phys. **C6**, 767.
MacKenzie, D.R., N.S. Ham and H.J. Whitfield, 1970, **8**, 2059.
McMillan, W.L., 1976, Phys. Rev. **B14**, 1496.
Michelson, A., 1977, Phys. Rev. **B16**, 577.
Mitsui, T., E. Nakamura, Y. Shiozaki, H. Motegi, T. Sekido, M. Ichikawa, T. Takama, M. Hosoya, K. Shibukawa, N. Wakaki and J. Furuichi, 1966, Proc. Int. Meeting on Ferroelectricity, Prague 1966, Vol. 1, p. 222.
Moncton, D.E., J.D. Axe and F.J. DiSalvo, 1975, Phys. Rev. Lett. **34**, 734.
Moudden, A.H., 1980, Thesis (Université Paris-Sud, Orsay) unpublished.
Moudden, A.H., and J. Als-Nielsen, 1980, unpublished.
Moudden, A.H., F. Dénoyer and M. Lambert, 1978a, J. Physique **39**, 1323.
Moudden, A.H., F. Dénoyer, J.P. Benoit and W. Fitzgerald, 1978b, Solid State Commun. **28**, 575.
Moudden, A.H., F. Dénoyer, M. Lambert and W. Fitzgerald, 1979, Solid State Commun. **32**, 933.
Moudden, A.H., L. Gatebois, F. Dénoyer and M. Lambert, 1980, Rev. Sci. Instrum. **51** (6), 836.
Moudden, A.H., E.C. Svensson and G. Shirane, 1982, Phys. Rev. Lett. **49**, 557.
Moudden, A.H., D.E. Moncton and J.D. Axe, 1983, Proc. A. Phys. Soc. **28**, 874; Phys. Rev. Lett. **51**, 2390.
Parlinski, K., 1984, Phys. Rev. **B29**, 410.
Parlinski, K., and K.H. Michel, 1984, Phys. Rev. **B29**, 396.
Petzelt, J., 1981, Phase Transitions, Vol. 2, Dielectric and light scattering spectroscopy of incommensurate phases in crystals (Gordon and Breach, New York).
Pokrovsky, V.L., and A.L. Talapov, 1979, Phys. Rev. Lett. **42**, 65.
Schneck, J., and F. Dénoyer, 1981, Phys. Rev. **B23**, 383.
Schneck, J., G. Calvarin and J.M. Kiat, 1984, Phys. Rev. **B29**, 1476.
Siapkas, D.I., 1978, in: Lattice Dynamics, ed. M. Balkanski (Flammarion, Paris) p. 692.
Siapkas, D.I., 1980, Ferroelectrics **29**, 29.
Shiozaki, Y., 1971, Ferroelectrics **2**, 245.
Shiozaki, Y., and M. Hosoya, 1970, J. Phys. Soc. Jpn **28** suppl., 290.
Simonson, T., F. Dénoyer, R. Currat and C. Vettier, 1985, to be published.
Solomon, L., 1956, Phys. Rev. **104**, 1191.
Takahashi, H.B., B. Schader, W. Meier and K. Gottlieb, 1967, J. Chem. Phys. **47**, 3842.
Tanisaki, S., and N. Nakamura, 1970, J. Phys. Soc. Jpn **28**, suppl., 293.
Truter, M.R., 1967, Acta Cryst. **22**, 556.
Unruh, H.G., 1983, J. Phys. **C16**, 3245.
Wada, M.A., A. Sawada, Y. Ishibashi and Y. Takagi, 1978, J. Phys. Soc. Jpn **45**, 1905.
Yokoi, C.S.O., M.D. Coutinho-Filho and S.R. Salinas, 1981, Phys. Rev. **B24**, 4047.

CHAPTER 15

Incommensurate Phase in β-$ThBr_4$

R. CURRAT and L. BERNARD

Institut Laue – Langevin
156X, 38042 Grenoble Cedex
France

and

P. DELAMOYE

Laboratoire de Radiochimie
Institut de Physique Nucléaire
B.P. 1, 91406 Orsay
France

Incommensurate Phases in Dielectrics 2
Edited by
R. Blinc and A.P. Levanyuk

Contents

1. Introduction . . . 164
2. The undistorted β-$ThBr_4$ structure . . . 165
   2.1. Polymorphism, chemical synthesis and crystal growth . . . 165
   2.2. Space group symmetry of the β-form . . . 167
   2.3. Normal modes at $q = 0$ . . . 168
3. The 95 K transition . . . 169
   3.1. Raman scattering below room temperature . . . 169
   3.2. Satellite reflections . . . 169
   3.3. Soft-mode behaviour above T_i . . . 170
   3.4. Discussion . . . 171
4. The modulated structure: static aspects . . . 173
   4.1. Form of the ionic displacements . . . 173
   4.2. Structural analysis . . . 176
      4.2.1. Elastic structure factor . . . 176
      4.2.2. Debye–Waller factors . . . 176
      4.2.3. Determination of η . . . 177
      4.2.4. Determination of θ . . . 180
      4.2.5. Determination of $\tilde{\eta}$ . . . 180
      4.2.6. Discussion . . . 181
5. The modulated structure: local properties . . . 181
   5.1. Thorium site symmetry . . . 181
   5.2. $ThBr_4$: U^{4+} optical spectra . . . 184
      5.2.1. Experimental absorption spectra . . . 184
      5.2.2. Crystal-field transitions . . . 185
      5.2.3. Discussion . . . 186
      5.2.4. Site-selective excitation experiments . . . 187
      5.2.5. Pinning of the modulation by the U^{4+} ions . . . 188
   5.3. Resonance spectroscopy . . . 189
      5.3.1. NQR study of β-$ThBr_4$ and β-$ThCl_4$ . . . 190
      5.3.2. EPR experiments on $ThBr_4 : Gd^{3+}$ . . . 191
6. The modulated structure: low frequency excitations . . . 194
   6.1. General . . . 194

6.2. Inelastic neutron scattering: experimental aspects . . . 196
6.3. Phase and amplitude modes in β-$ThBr_4$ at 81 K . . . 196
6.3.1. Experimental results . . . 196
6.3.2. Eigenvector analysis . . . 200
6.4. Experimental results at lower temperatures . . . 202
References . . . 203

1. Introduction

The criteria which determine the behaviour of crystalline solids near structural phase transitions have a universal character and are applicable to a broad range of physical systems, regardless of their microscopic structure. One particularly relevant variable is the dimensionality n of the order parameter, i.e. the number of degenerate degrees of freedom which display critical behaviour at the ordering temperature T_i.

Phase transitions leading to modulated ordered structures are described by order parameters with minimum dimensionality $n = 2$. The simplest situation corresponds to a single modulation characterized by an amplitude and a phase ($n = 2$). It is commonly found in systems having orthorhombic symmetry in the disordered state, with the modulation wavevector parallel to a two-fold rotation axis, e.g. K_2SeO_4, $(NH_4)_2BeF_4$, $SC(NH_2)_2$, $NaNO_2$, etc.

In cubic crystals on the other hand the minimum order parameter dimensionality leading to a modulated ordered state is $n = 6$. Below T_i, the number of ordered components is either 2, 4 or 6, corresponding, respectively, to single-q, double-q, or triple-q structures. Obviously, the high symmetry of the parent structure complicates rather than simplifies the description of the ordered state.

Nevertheless, it is convenient to select, within each universality class, a prototype system with the simplest possible structure and the highest degree of internal symmetry compatible with the class at hand. For example, one may look for the highest crystalline symmetry compatible with a $n = 2$ phase transition. As suggested by Cowley (1980), this particularly favourable instance is realized when a uniaxial crystal develops a modulation with wavevector parallel to the unique 3-, 4- or 6-fold rotation axis.

Indeed, this is the situation which pertains to the case of β-$ThBr_4$ (D_{4h}^{19}) and isomorphous β-$ThCl_4$. Although, in principle, the fact that these two compounds possess the full tetragonal point-group symmetry is of little consequence for the phase transition at T_i, we shall see that in practice the high symmetry simplifies the analysis of the modulated structure below T_i and allows more quantitative conclusions to be drawn.

In addition to the above symmetry arguments there are at least two other features which contribute to making thorium tetrabromide an attractive model system.

The first one concerns the nature of the modulation itself. Satellite intensity measurements indicate that the ionic displacements associated with the modulation remain small in amplitude and sinusoidal in character, down to the lowest temperatures. A related observation is the absence of lock-in transitions in $ThBr_4$ and $ThCl_4$, despite the fact that in both cases the modulation wavevector lies close to the commensurate value $\frac{1}{3}\boldsymbol{c}^*$. This feature is of theoretical importance since it allows the properties of the modulated phase to be discussed on the basis of a Landau free-energy, expanded in powers of a single Fourier component of the ionic displacements. This applies to the static properties and, within a quasi-harmonic approximation, to the excitation spectrum of the modulated lattice.

From an experimental viewpoint, however, the most favourable aspect of the structural phase transition in $ThBr_4$ is its typically displacive character, as evidenced by Raman and inelastic neutron scattering results (Hubert et al. 1981, Bernard et al. 1983). In both cases a well-defined soft optic phonon is observed whose quasi-harmonic frequency and damping coefficient can be extracted unambiguously from the experimental spectra. The exceptionally low value of the soft-mode damping coefficient ($\sim 70\,\mathrm{GHz}$ at T_i) greatly facilitates the study of the soft-branch renormalization in the modulated state. In particular it has made possible the identification of the phason branch by means of high-resolution inelastic neutron scattering (cf. section 6 below).

This chapter is organized as follows. In section 2 we review the structural and dynamical aspects of β-$ThBr_4$ in the undistorted phase ($T > T_\mathrm{i}$). Experimental results on the phase transition itself are presented in section 3, together with a discussion of the soft-mode symmetry properties. In section 4, the expression for the ionic displacements below T_i is derived and the values of the parameters entering this expression are obtained from neutron diffraction measurements. Some of these results are used in section 5 which deals with the local properties of the modulated phase as revealed by optical crystal-field spectroscopy and resonance techniques. Finally, inelastic neutron scattering results on the low-frequency fluctuation spectrum of the modulated phase are presented in section 6.

2. *The undistorted β-$ThBr_4$ structure*

2.1. *Polymorphism, chemical synthesis and crystal growth*

Thorium tetrabromide (chloride) exists in two crystalline forms with a $\beta \rightarrow \alpha$ transformation at about 400°C (Mason et al. 1973). Both forms have tetragonal symmetry but with different space groups, and can be easily identified by diffraction techniques. In principle the α-phase (C_{4h}^6) is the stable form at room temperature but the high-temperature β-phase (D_{4h}^{19}) is commonly found to be metastable at room temperature and below. In practice the form obtained depends crucially on the chemical synthesis and crystal growth techniques used (Brown 1968, Mason et al. 1974). For example, with the Bridgman technique (Hussonnois et al. 1981) the crystalline material obtained is primarily of the β-form. In what follows we shall be concerned exclusively with the properties of the β-form, at room temperature and below.

From a chemical standpoint $ThBr_4$ and $ThCl_4$, in the α- and β-forms, are characterized by a very high degree of hygroscopicity, a feature with important practical consequences, as far as specimen handling and storage are concerned.

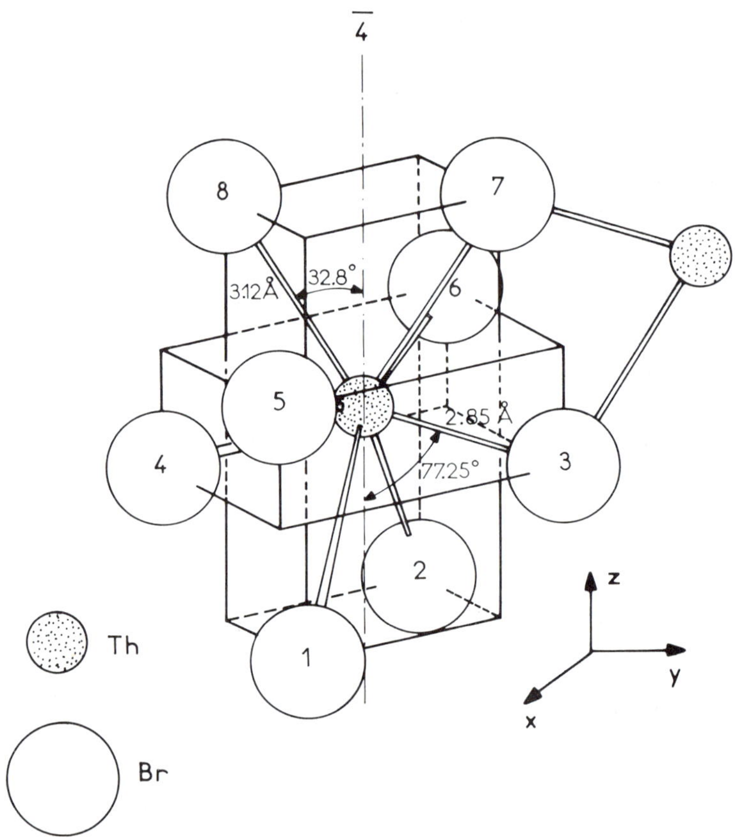

Fig. 1. Primitive unit cell of β-$ThBr_4$. The two Th^{4+} ions are located at (0,0,0) and (0, $\frac{1}{2}$, $\frac{1}{4}$). The positions of the Br^- ions are listed in table 1. (After Bernard et al. 1983.)

2.2. *Space group symmetry of the β-form*

The conventional body-centered tetragonal cell of β-$ThBr_4$ ($a = b = 8.93$ Å; $c = 7.96$ Å) contains 4 formula units. The primitive unit cell is defined by the basic vectors

$$\begin{aligned} \boldsymbol{t}_1 &= \tfrac{1}{2}(-\boldsymbol{a}+\boldsymbol{b}+\boldsymbol{c}), \\ \boldsymbol{t}_2 &= \tfrac{1}{2}(\boldsymbol{a}-\boldsymbol{b}+\boldsymbol{c}), \\ \boldsymbol{t}_3 &= \tfrac{1}{2}(\boldsymbol{a}+\boldsymbol{b}-\boldsymbol{c}), \end{aligned} \tag{1}$$

with only two formula units per cell. With the origin set at a thorium site ($\bar{4}2 = D_{2d}$) the second thorium ion is located at

$$\boldsymbol{t} = \tfrac{1}{2}\boldsymbol{b} + \tfrac{1}{4}\boldsymbol{c}. \tag{2}$$

The 8 bromine ions are located at equivalent (C_s) sites and can be chosen in such a way as to make up the first and second coordination shells of the Th^{4+} ion at the origin, as shown in fig. 1. The corresponding Br^- coordinates are listed in table 1, together with the symmetry operations in space group D_{4h}^{19}

Table 1
Symmetry operations in space group $D_{4h}^{19}(I_{4_1/amd})$. The fractional translation $\boldsymbol{t}$ is defined in eq. (2). The Br^- indices in the right-hand column are also shown in fig. 1. At room temperature Brown et al. (1973) give $x = 0.31$ and $z = -0.08$ (origin at $\bar{4}2m$).

Symmetry operations		Br^- positions	
(Kovalev)	(Seitz)	Coordinates	Index
$(h_1\|0)$	$(\varepsilon\|0)$	$0, x, z$	3
$(h_{14}\|\boldsymbol{t})$	$(C_{4z}\|\boldsymbol{t})$	$\frac{1}{2}-x, 0, z-\frac{1}{4}$	1
$(h_4\|0)$	$(C_{2z}\|0)$	$0, \bar{x}, z$	4
$(h_{15}\|\boldsymbol{t})$	$(C_{4z}^3\|\boldsymbol{t})$	$x-\frac{1}{2}, 0, z-\frac{1}{4}$	2
$(h_2\|\boldsymbol{t})$	$(C_{2x}\|\boldsymbol{t})$	$0, \frac{1}{2}-x, \frac{1}{4}-z$	7
$(h_3\|\boldsymbol{t})$	$(C_{2y}\|\boldsymbol{t})$	$0, x-\frac{1}{2}, \frac{1}{4}-z$	8
$(h_{13}\|0)$	$(C_2'\|0)$	$\bar{x}, 0, \bar{z}$	6
$(h_{16}\|0)$	$(C_2'\|0)$	$x, 0, \bar{z}$	5
$(h_{25}\|\boldsymbol{t})$	$(I\|\boldsymbol{t})$	$0, \frac{1}{2}-x, \frac{1}{4}-z$	7
$(h_{26}\|0)$	$(\sigma_{yz}\|0)$	$0, x, z$	3
$(h_{27}\|0)$	$(\sigma_{xz}\|0)$	$0, \bar{x}, z$	4
$(h_{28}\|\boldsymbol{t})$	$(\sigma_{xy}\|\boldsymbol{t})$	$0, x-\frac{1}{2}, \frac{1}{4}-z$	8
$(h_{37}\|\boldsymbol{t})$	$(\sigma_d\|\boldsymbol{t})$	$x-\frac{1}{2}, 0, z-\frac{1}{4}$	2
$(h_{38}\|0)$	$(S_4^3\|0)$	$x, 0, \bar{z}$	5
$(h_{39}\|0)$	$(S_4\|0)$	$\bar{x}, 0, \bar{z}$	6
$(h_{40}\|\boldsymbol{t})$	$(\sigma_d\|\boldsymbol{t})$	$\frac{1}{2}-x, 0, z-\frac{1}{4}$	1

($I_{4_1/amd}$). [Notice incidentally that the inversion center of the cell is not located on a thorium site, but half-way between the two thorium sites].

As seen from fig. 1 the bromine ions around the origin are located at the apices of two deformed tetrahedra: one tetrahedron is elongated along the c-axis (ions labelled 1,2,7,8) while the other is squashed (3,4,5,6). Each tetrahedron is separately invariant under the point-symmetry operations of the thorium site ($\bar{4}2m$).

2.3. *Normal modes at $q=0$*

Table 2 gives the normal-mode decompositions at the zone center and for wavevectors along $\boldsymbol{c}^*$. For the zone-center modes the Raman (R) and infrared

Table 2
Normal-mode decomposition along $[0,0,\zeta]$ and at the zone centre. The Raman and infrared frequencies are taken from Hubert et al. (1981).

$[0,0,\zeta]$	Zone Center	Raman and Infra-Red Activity	Frequency (THz)
$5\tau^1$	$2\tau^1$	(A_{1g})–R	3.27, 5.88
	$3\tau^4$	(A_{2u})–IR	(acoustic), 3.79, 4.70
$2\tau^2$	$1\tau^2$	(A_{1u})–S	
	$1\tau^3$	(A_{2g})–S	
$5\tau^3$	$3\tau^5$	(B_{1g})–R	2.00, 4.09, 5.55
	$2\tau^8$	(B_{2u})–S	
$2\tau^4$	$1\tau^6$	(B_{1u})–S	
	$1\tau^7$	(B_{2g})–R	1.36
$8\tau^5$	$4\tau^9$	(E_g)–R	1.15, 2.48, 3.70, 5.67
	$4\tau^{10}$	(E_u)–IR	(acoustic), 2.00, 2.94, 4.85

activity (IR) is indicated next to the representation symbol (in chemical notation). The corresponding frequencies as measured by Hubert et al. (1981), are shown in the right-hand column. For the infrared active modes (E_u and A_{2u}) only the TO frequencies are listed in table 2. The LO–TO splittings of the doubly-degenerate E_u-modes were also determined, via a Kramers–Kronig analysis of absorption and reflectivity spectra. The LO-frequencies of the (c-polarized) A_{2u}-modes could not be obtained in the same manner due to practical difficulties in preparing a cleaved surface of suitable orientation.

With this minor exception, the combination of infrared and polarized Raman spectroscopy has enabled all optically active frequencies to be identified, at room temperature. This result, which has proved quite useful (cf. section 3.3) is by no means trivial for a crystal structure with that many (30) degrees of freedom.

A similar study was carried out on $ThCl_4$ with comparable results. All observed frequencies are shifted upwards due to the lighter chlorine mass.

3. *The 95 K transition*

3.1. *Raman scattering below room temperature*

In the course of the same study Hubert et al. (1981) observed a splitting of the Raman lines, developing below $T_i = 92\,K$ in $ThBr_4$ (70 K in $ThCl_4$) and seemingly associated with a lowering of crystal symmetry.

The occurrence of a structural phase transition was further substantiated by the observation ($T < T_i$) of a new low-frequency line with characteristic soft-mode behaviour. The new mode is observed as a sharp feature with a width of a few cm^{-1} and a frequency ω_s varying approximately as $(T_i - T)^{1/3}$, over a wide temperature range (we shall comment on this last result in section 3.4).

Subsequent to this finding, neutron diffraction measurements were initiated in order to establish the nature of the low temperature distorted structure.

3.2. *Satellite reflections*

Powder diffraction (de Kouchkovsky et al. 1981) and single crystal studies (Bernard et al. 1983) on $ThBr_4$ below T_i have revealed the presence of additional (satellite) Bragg reflections located at reciprocal lattice positions of the type

$$\boldsymbol{Q} = \boldsymbol{G}_{hkl} \pm \boldsymbol{q}_s \quad \text{with} \quad \boldsymbol{q}_s = 0.310 \pm 0.005\boldsymbol{c}^*.$$

The satellite reflections grow in intensity below T_i, roughly in proportion to the square of the modulation amplitude. Even at 0 K however, their intensity is typically one to two orders of magnitude lower than that of a strong fundamental ($\boldsymbol{Q}=\boldsymbol{G}_{hkl}$) reflection.

At low temperatures, weak intensity maxima are observed at $\boldsymbol{Q}=\boldsymbol{G}_{hkl}\pm 2\boldsymbol{q}_s$. Their precise origin (second-order satellite reflections, or double scattering events involving two first-order satellite reflections) has not been established.

Thus, the emerging picture is that of a small-amplitude, sinusoidal, displacive modulation, with a temperature-independent wavevector.

One puzzling aspect of these results is the remarkable constancy of the modulation wavevector as a function of temperature. Depending upon experimental conditions, values for q_s have been found anywhere in the range from 0.305 to $0.315c^*$. However, in a specific experiment using a three-axis spectrometer at $\lambda = 2.36$ Å, where the specimen temperature was varied, all other experimental conditions remaining identical, the value of q_s/c^* was found to be 0.3115 ± 0.0015, between 80 K and 1.5 K. No tendency towards an increase or decrease of the reduced wavevector could be detected, while in the same temperature range measurable changes in the lattice parameters are observed.

This situation is reminiscent of the case of Rb_2ZnBr_4 where the modulation wavevector is found to be constant over a temperature interval of more than 140 K below the ordering temperature $T_i = 346$ K. Recently it has been pointed out (Hogervorst and de Wolff 1982, Iizumi and Gesi 1983) that the observed value ($0.293c^*$) was within experimental error of the commensurate value $\frac{5}{17}=0.2941$, and very probably identical to it.

In the case of $ThBr_4$, the simplest rational fraction compatible with observations would be $\frac{5}{16}=0.3125$. For $ThCl_4$ preliminary measurements (Delamoye et al. 1985) indicate $q_s/c^* = 0.360\pm0.003$ (possibly compatible with $\frac{5}{14}=0.357$), between 65 K and 10 K.

3.3. *Soft-mode behaviour above T_i*

The displacive nature of the 95 K transition, apparent in the Raman work of Hubert et al. (1981), is confirmed by the direct observation of a soft optic phonon branch by inelastic neutron scattering (cf. fig. 2). The soft branch is characterized by a frequency minimum at $\boldsymbol{q}_s$ $(0,0,q_s)$ with essentially isotropic dispersion around the minimum:

$$\omega^2(\boldsymbol{q}-\boldsymbol{q}_s)=\omega^2(\boldsymbol{q}_s)+\lambda_\perp\left(q_x^2+q_y^2\right)+\lambda_\parallel(q_z-q_s)^2+\mathrm{O}(|\boldsymbol{q}-\boldsymbol{q}_s|^3) \tag{3}$$

with ($T=150$ K):

$$\lambda_\perp \simeq \lambda_\parallel = 4.5\pm0.5\,\mathrm{THz^2\AA^2} \quad \text{and} \quad q_s = 0.285c^*.$$

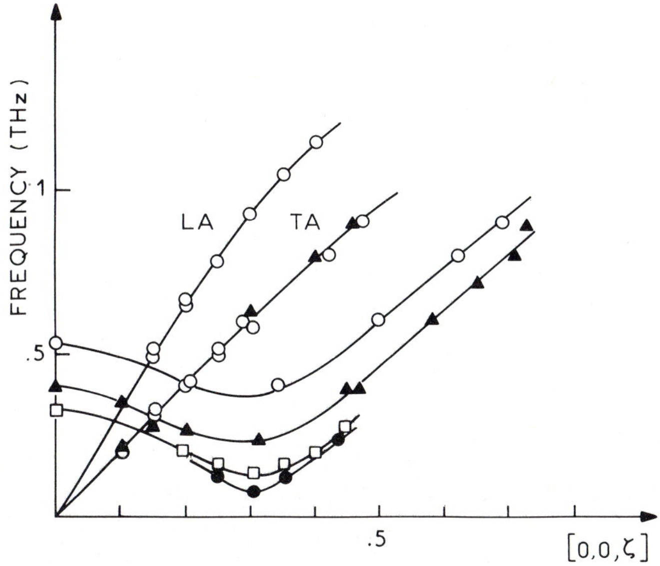

Fig. 2. Dispersion curves for the LA, TA and soft optic modes, propagating along $[0,0,\zeta]$: (○) 300 K; (▲) 150 K; (□) 120 K; (●) 101 K; $T_i = 95$ K. (After Bernard et al. 1983.)

The temperature dependence of $\lambda_\perp, \lambda_\parallel$ and q_s has not been determined accurately. The data in fig. 2, however, indicate that q_s increases between room temperature and T_i, in contrast with the behaviour of the satellite wavevector below T_i.

Close to T_i, the soft-mode response becomes overdamped. Simultaneously a central resonance develops (cf. fig. 3), which eventually dominates the critical dynamical response. The frequency width of the central peak is found to be resolution-limited (40 GHz).

Notice also that the zone-center frequency of the soft branch ($\simeq 0.55$ THz at room temperature) does not correspond to any of the observed infrared or Raman frequencies listed in table 2. This observation implies that the soft branch is optically silent at $\boldsymbol{q} = 0$, which rules out τ_1 and τ_5 as possible soft-mode irreducible representations.

The choice between the remaining 3 representations (τ_2, τ_3 and τ_4) can be made unambiguously on the basis of the observed satellite extinction rules, as discussed by Bernard et al. (1983). The correct choice turns out to be τ_4 which identifies the zone-center representation as B_{1u}.

3.4. *Discussion*

Little is known experimentally on the behaviour near T_i of macroscopic properties such as specific heat, thermal expansion coefficients, elastic compli-

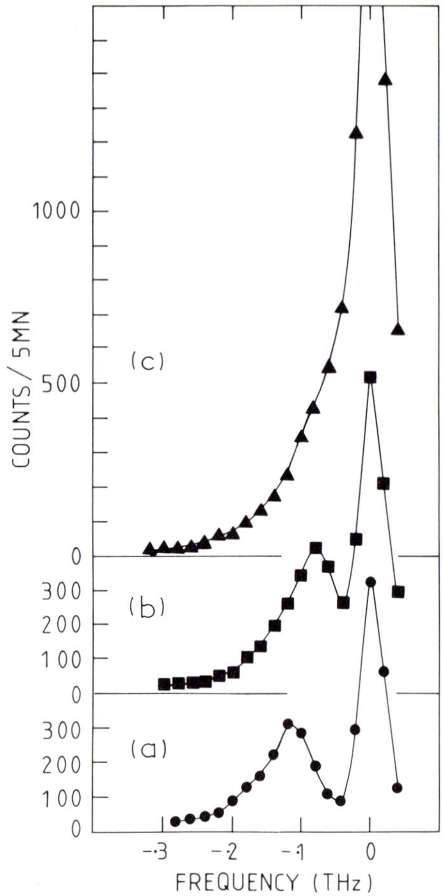

Fig. 3. Constant-Q scans at (2,3,0.69), showing the evolution of the soft-mode lineshape with temperature: (a) 108 K; (b) 101.5 K; (c) 97 K.

ances, optical indices, etc. Hence it is not possible to decide whether the transition is truly second-order or weakly first-order.

Presumably, if q_s is commensurate the transition at T_i should appear as weakly first-order, but one could also imagine an intermediate incommensurate phase, stable just below T_i, followed by a first-order or second-order lock-in transition. Reliable experimental information is obviously needed in order to decide which of the above conjectures, if any, is correct. Most techniques, however, are hampered by practical difficulties associated with the hygroscopic character of the samples. Satellite intensity measurements are clearly not suitable in this respect, due to the presence of critical scattering near T_i.

On the other hand, if q_s is incommensurate, which in effect we shall assume in what follows, one is left with the difficulty of explaining why it should remain so nearly constant, while, for instance, the position of the soft-mode minimum shows temperature dependence above T_i. One possible way out is to assume that the modulation wavevector is pinned by defects, as is observed in chemically disordered (Mashiyama et al. 1982) or radiation-damaged (André et al. 1985) systems. This type of argument is not fully satisfactory either, insofar as the nature of the defects has not been identified.

Another point which calls for comment is the unusual exponent value of $\sim \frac{1}{3}$ found for the temperature dependence of the Raman "soft-mode" below T_i (cf. section 3.1). On one hand mean-field (MF) theory predicts an exponent of $\frac{1}{2}$, as for the temperature dependence of the order-parameter ($\beta_{MF} = \frac{1}{2}$). On the other hand recent theoretical work (Meissner et al. 1981, Meissner and Blaschke 1984) indicates that in multicomponent order-parameter systems with weak anisotropy, the proportionality between the "longitudinal" soft-mode frequency and the order-parameter may extend beyond the range of validity of MF theory.

Thus close below T_i, in the non-MF regime one may expect both the Raman frequency and the order-parameter to vary as $(T_i - T)^{1/3}$, as in the case of the triply-degenerate R-point mode in cubic perovskites (Steigmeier and Auderset 1973, Jacoby et al. 1976). However, the reduced temperature range over which deviations from MF behaviour are observed is only a few percent in the perovskite case and is probably very similar here.

Hence there is no theoretical justification for the $\frac{1}{3}$ law, except possibly over a temperature range of a few K below T_i. Its apparent validity between T_i and 0 K must therefore result from a crossover between several successive regimes (critical, MF, saturated), as T decreases from T_i to 0 K.

4. *The modulated structure: static aspects*

4.1. *Form of the ionic displacements*

In the sinusoidal approximation the static ionic displacements, below T_i, are expressed as:

$$\boldsymbol{u}_{lk} = (m_k)^{-1/2} Q_{\boldsymbol{q}_s} \boldsymbol{e}_k(\boldsymbol{q}_s \tau_4) \exp(i\boldsymbol{q}_s \cdot \boldsymbol{r}_{lk}^0) + \text{c.c.}, \tag{4}$$

where $\boldsymbol{u}_{lk}$ is the average displacement of atom k in cell l away from its high-temperature equilibrium position $\boldsymbol{r}_{lk}^0$, $Q_{\boldsymbol{q}_s}$ is a two-component order-parameter written here as a complex normal coordinate; and $\boldsymbol{e}(\boldsymbol{q}_s \tau_4)$ is the 3×10-component soft-mode eigenvector.

As usual it is convenient to parametrize the soft-mode eigenvector in terms of symmetry-adapted eigenvectors with well-defined transformation properties with respect to the inversion:

$$e(q_s \tau_4) = \lambda e_1 + \mu e_2, \tag{5}$$

where e_1 and e_2 are normalized eigenvectors, compatible with τ_4-symmetry and with, respectively, odd and even character with respect to the inversion.

Since there are only two phonon branches of τ_4-symmetry, e_1 and e_2 are uniquely defined and must therefore be identical to the two τ_4-compatible zone center eigenvectors (cf. table 2):

$$e_1 \equiv e\{B_{1u}\}, \qquad e_2 \equiv e\{B_{2g}\}.$$

Making use of time-reversal symmetry (see for example Iizumi et al. 1977), eq. (5) above reduces to:

$$e(q_s \tau_4) = \cos\theta e_1 - \mathrm{i}\sin\theta e_2, \tag{6}$$

where θ is a real angle.

The ionic displacement patterns corresponding to the B_{1u} and B_{2g} modes are shown in fig. 4. Comparing figs. 1 and 4, we see that the B_{1u} mode consists in a rotation of the two Br^- tetrahedra about the $\bar{4}$-axis, while in the B_{2g} mode the tetrahedra experience a twist deformation about the same axis. In both cases the thorium positions are unaffected.

Combining eqs. (4) and (6) the real displacements are obtained as:

$$u_{lk} = \eta\left[\cos\theta e_1(k)\cos(q_s \cdot r_{lk}^0 + \phi) + \sin\theta e_2(k)\sin(q_s \cdot r_{lk}^0 + \phi)\right], \tag{7}$$

where we have set

$$2(m_k)^{-1/2} Q_{q_s} = \eta\, \mathrm{e}^{\mathrm{i}\phi}.$$

Equation (7) expresses the u_{lk}'s in a canonical form, i.e. as a superposition of two waves oscillating (spatially) in quadrature. Heine and McConnell (1981) have stressed that this was a general feature for all simple ($n = 2$) modulated structures, provided the high-temperature space group included at least one symmetry operation which changes q_s into $-q_s$. The above formulation has proved useful in several contexts, notably in the structural analysis of chemically modulated minerals (Heine and McConnell 1984, McConnell and Heine 1984a, b).

In the language of lattice dynamics, the existence of the two structural components oscillating in quadrature and the particular symmetry relationship between them, each having the symmetry of the other's gradient, appears as a direct consequence of time-reversal invariance.

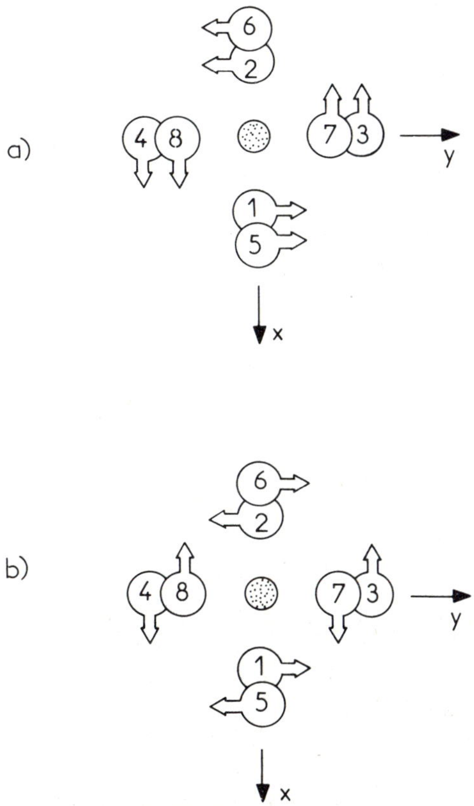

Fig. 4. Eigenvectors of (a) B_{1u} and (b) B_{2g} modes. All arrows have the same length, corresponding to equal amplitudes for all Br^- ions.

What is remarkable in the present case is that the two waves have *unique* symmetry properties and correspond to true eigenvectors of the $(q=0)$ dynamical matrix. As a consequence, not only does each component have the same symmetry as the gradient of the other but it *is* the gradient of the other. This is readily realized by choosing the phase variable ϕ in eq. (7) in such a way as to have a node at the origin of cell l, for the B_{2g}-component $(\boldsymbol{q}_s \cdot \boldsymbol{r}_l + \phi = 0)$: then, due to the remaining factor $\sin(\boldsymbol{q}_s \cdot \boldsymbol{r}_k^0)$ the ionic displacements generated by the B_{2g}-wave in the lth cell correspond to a rotation, rather than a twist, for the Br^- tetrahedra. Thus the B_{2g}-wave is "twist-like" near its extrema but "rotation-like" near its nodes, and conversely for the B_{1u}-wave. Note that since the two waves are phase-shifted by $\frac{1}{2}\pi$ their rotational and twist-like characters are precisely in phase.

4.2. Structural analysis

4.2.1. Elastic structure factor

The analysis of the modulated structure reduces to the determination of the parameters η and θ at each temperature. For a high-order commensurate structure the phase variable ϕ takes on a discrete set of possible values which must be determined as well. Here, however, we work under the assumption that the structure is incommensurate in which case the value of ϕ remains undetermined (except in the vicinity of defects). In the remainder we shall set it to zero.

In principle the task of evaluating θ and η is straightforward: we insert the expression for the $\boldsymbol{u}_{lk}$'s as given by eq. (7) into the elastic structure factor:

$$F(\boldsymbol{Q}) = \sum_{lk} b_k \mathrm{e}^{-W_k} \exp\{\mathrm{i}\boldsymbol{Q}\cdot(\boldsymbol{r}_{lk}^0 + \boldsymbol{u}_{lk})\}, \tag{8}$$

and calculate the structure factors $F(\boldsymbol{G})$ and $F(\boldsymbol{G} \pm \boldsymbol{q}_\mathrm{s})$ of the fundamental and first-order satellite reflections, in terms of θ and η. The latter are then determined by least-square fitting to the observed intensities. In order to do this we first note that

$$\boldsymbol{e}_2(k) = \varepsilon_k \boldsymbol{e}_1(k),$$

with

$$\varepsilon_k = +1\ (-1) \text{ for } k = 1\text{–}4\ (5\text{–}8)$$

which allows one to recast eq. (7) into the more convenient form

$$\boldsymbol{u}_{lk} = \eta \boldsymbol{e}_1(k) \cos(\boldsymbol{q}_\mathrm{s}\cdot\boldsymbol{r}_{lk}^0 - \varepsilon_k \theta). \tag{9}$$

Inserting eq. (9) into eq. (8) and using standard techniques we find

$$F(\boldsymbol{G}) = \sum_k b_k \exp(-W_k) J_0(\eta \boldsymbol{G}\cdot\boldsymbol{e}_1(k)) \exp(\mathrm{i}\boldsymbol{G}\cdot\boldsymbol{r}_k^0) \tag{10a}$$

and

$$F(\boldsymbol{G} \pm \boldsymbol{q}_\mathrm{s}) = \pm\mathrm{i} \sum_k b_k \exp(-W_k)\, J_1(\eta \boldsymbol{G}\cdot\boldsymbol{e}_1(k)) \exp(\pm\mathrm{i}\varepsilon_k\theta) \exp(\mathrm{i}\boldsymbol{G}\cdot\boldsymbol{r}_k^0). \tag{10b}$$

4.2.2. Debye–Waller factors

Equation (10) has been derived without considering the effects of those fluctuations which are specific to modulated structures and which, in loose terminology, may be referred to as fluctuations in the amplitude and phase of the displacement wave. In particular the Debye–Waller factors appearing in eqs. (10a) and (10b) only include the contribution of the "normal" phonons, in particular the long-wavelength acoustic modes, which are largely unaffected by the onset of the modulated state.

Overhauser (1971) has pointed out that the effect of random phase fluctuations ϕ_l, with a Gaussian distribution around a mean value (which we have chosen here to be zero) contributes an additional Debye–Waller factor $\exp\{-\frac{1}{2}n^2\langle\phi^2\rangle\}$ to $F(\boldsymbol{G} \pm n\boldsymbol{q}_s)$, where $\langle\phi^2\rangle$ is the cell-independent r.m.s. average of the fluctuations

$$\langle\phi^2\rangle \equiv \langle\phi_l^2\rangle.$$

Axe (1980) has shown that Overhauser's Gaussian phase approximation (GPA) is acceptable only for low momentum transfer (i.e. $Q\eta < 1$) and small phase fluctuations (i.e. $\langle\phi^2\rangle \ll 1$). Using Landau theory Axe has derived a more general approach, the so-called Gaussian displacement approximation (GDA), with a broader range of applicability; this includes the effect of both amplitude and phase fluctuations.

The main effect of fluctuations, and in fact the only effect relevant to the discussion of first-order satellite reflections, is the renormalization of the modulation amplitude. In the small momentum transfer, small fluctuation limit, the renormalized amplitude $\tilde{\eta}$ may be introduced by absorbing the "phason" Debye–Waller factor $\exp\{-\frac{1}{2}\langle\phi^2\rangle\}$ into the argument of the Bessel function in eq. (10b):

$$\mathrm{e}^{-(1/2)\langle\phi^2\rangle} J_1(\eta \boldsymbol{G}\cdot\boldsymbol{e}_1(k)) \simeq J_1(\tilde{\eta}\boldsymbol{G}\cdot\boldsymbol{e}_1(k)),$$

Thus eqs. (10b) and (10a) are formally unchanged except that the amplitudes η which appear in the Bessel function arguments are renormalized ($\tilde{\eta}$) for (10b) and unrenormalized (η) for (10a).

4.2.3. *Determination of* η

A large number of fundamental and first-order satellite reflections have been collected at several temperatures above and below T_i, using the four-circle hot-source neutron diffractometer D9 at ILL (Bernard et al. 1985b).

The fundamental reflections have been analyzed using the standard structure refinement program SHELX (Sheldrick 1976). The same procedure was followed above and below T_i, resulting in temperature-independent average Br^- positions (see parameters x and z in table 1).

Below T_i the modulation manifests itself by an additional "static Debye–Waller factor", as can be seen from eq. (10a):

$$J_0(\eta\boldsymbol{G}\cdot\boldsymbol{e}_1(k)) \simeq \exp\left\{-\tfrac{1}{4}(\eta\boldsymbol{G}\cdot\boldsymbol{e}_1(k))^2\right\}$$

More specifically, if we write

$$e^{-W_k} = \exp\left\{-\tfrac{1}{2}\boldsymbol{Q}\overline{\overline{U}}_k\boldsymbol{Q}\right\}$$

and define

$$U_\perp^{\mathrm{Br}}(k) \equiv \begin{cases} U_{xx}(k) & (k=3,4,7,8) \\ U_{yy}(k) & (k=1,2,5,6) \end{cases}$$

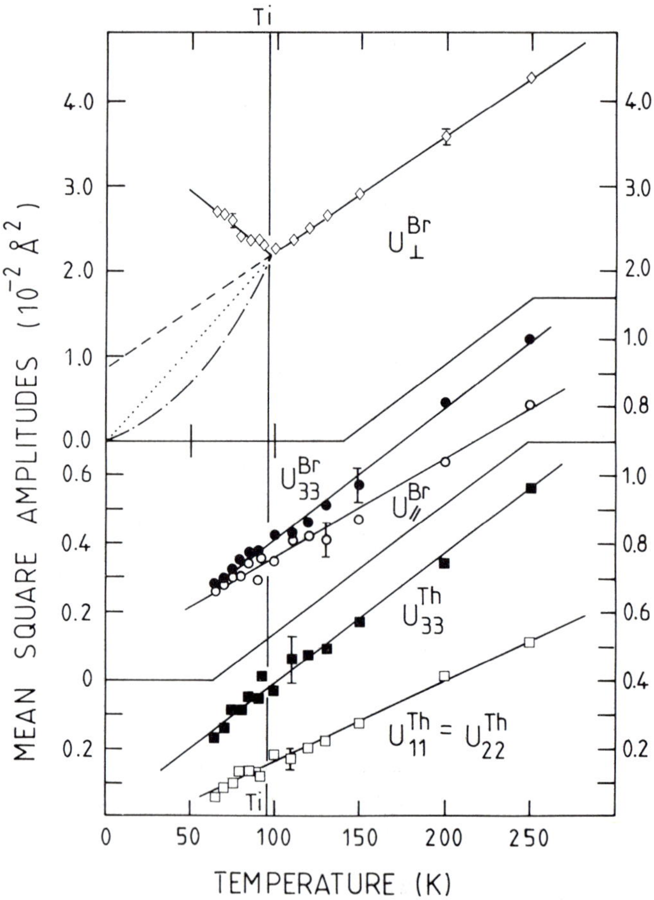

Fig. 5. Mean-square displacement amplitudes deduced from the analysis of the fundamental Bragg reflections. (After Bernard et al. 1985b.)

and

$$U_{\parallel}^{\mathrm{Br}}(k) \equiv \begin{cases} U_{yy}(k) & (k=3,4,7,8) \\ U_{xx}(k) & (k=1,2,5,6), \end{cases}$$

one then expects $U_{\perp}^{\mathrm{Br}}$ to be augmented by $\frac{1}{2}\eta^2$ below T_i:

$$U_{\perp}^{\mathrm{Br}} \rightarrow U_{\perp}^{\mathrm{Br}} + \tfrac{1}{2}\eta^2.$$

This is just what is observed, as shown in fig. 5, where the various components of the mean-square amplitude matrix are plotted as a function of temperature. All components display a normal monotonic behaviour with no apparent anomaly at T_i, except for $U_{\perp}^{\mathrm{Br}}$ which is much larger than any of the other components and increases below T_i, due to the static $\frac{1}{2}\eta^2$ contribution.

Attempts to extract a value for $\eta(T)$ from the behaviour of $U_{\perp}^{\mathrm{Br}}$ below T_i meet with the difficulty of properly subtracting the dynamical part of the Debye–Waller factor. If one follows a linear extrapolation scheme (dashed line in fig. 5) one ends up with a dynamical $U_{\perp}^{\mathrm{Br}}$ which is as large at 0 K as $U_{\parallel}^{\mathrm{Br}}$ at room-temperature (!). The resulting values for η can thus be safely taken as lower limits to the actual amplitudes. An alternate scheme consists in inter-

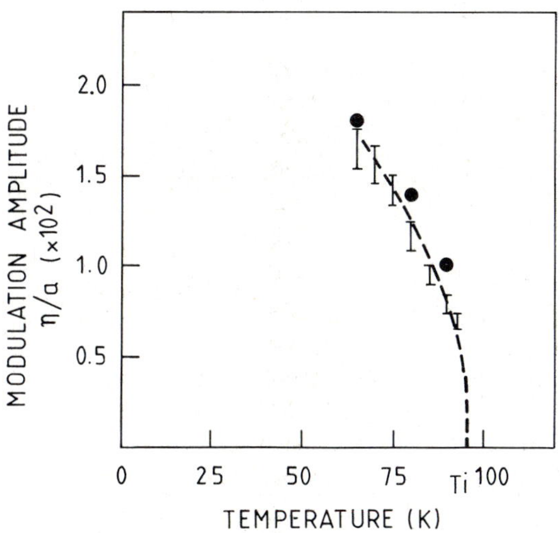

Fig. 6. Modulation amplitude as deduced from the analysis of the fundamental Bragg reflections (vertical bars) and of the first-order satellites (●). The upper (lower) end of each vertical bar corresponds to the dotted (dashed) line for the $U_{\perp}^{\mathrm{Br}}$ extrapolation scheme (cf. fig. 5). (After Bernard et al. 1985b.)

polating linearly between $U_{\perp}^{\mathrm{Br}}(T_{\mathrm{i}})$ and $U_{\perp}^{\mathrm{Br}}(T=0)=0$ (dotted line in fig. 5). This latter method would seem to yield a plausible upper limit for $\eta(T)$. The corresponding upper and lower values for η/a are shown by vertical bars in fig. 6.

4.2.4. Determination of θ

In principle θ and the renormalized amplitude $\tilde{\eta}$ can be obtained from a systematic analysis of the first-order satellite intensities. Such an analysis is currently in progress. As pointed out by Bernard et al. (1983), one can also obtain θ directly by considering pairs of satellite reflections $\boldsymbol{G}_{hkl}\pm\boldsymbol{q}_{\mathrm{s}}$. To the extent that the Bessel function J_1 appearing in eq. (10b) can be linearized (which is justified for low index reflections and values of η as shown in fig. 6), the ratio

$$\rho_{\mathrm{G}}=\frac{|F(\boldsymbol{G}+\boldsymbol{q}_{\mathrm{s}})|}{|F(\boldsymbol{G}-\boldsymbol{q}_{\mathrm{s}})|} \tag{11}$$

is a function of θ only. This statement becomes exact (i.e. no linearization needed) in the case of (h,k,l) reflections with $h=\pm k$. Twelve pairs of inequivalent satellite reflections collected at 55 K have been analyzed as indicated above. The values for θ thus obtained fall in the range from 15 to 25°. In particular, the (112), (224) and (332) satellite pairs give $\theta=21.2$, 19.1 and 21.5°, respectively. The value of θ at 55 K is thus estimated at $20\pm3°$.

Similar measurements performed at several temperatures did not detect any significant deviation from the above value, within the range of the measurements ($55\,\mathrm{K}<T<92\,\mathrm{K}$).

In a simple quasi-harmonic picture θ depends mainly on the value of the modulation wavevector q_{s}. Since the latter is found to be temperature independent, no particular temperature dependence is predicted for θ.

4.2.5. Determination of $\tilde{\eta}$

We are now in a position to estimate the value of the renormalized amplitude $\tilde{\eta}$ from the analysis of the first-order satellite intensities. We make use of eq. (10b) with the above-determined value for θ and values for $U_{\perp}^{\mathrm{Br}}$ obtained from the dashed-line extrapolation (cf. fig. 5). We wish to stress that the final result is quite insensitive as to which extrapolation scheme is used for $U_{\perp}^{\mathrm{Br}}$. In order to avoid having to linearize the Bessel function in eq. (10b) satellites of the type $\boldsymbol{G}_{hhl}\pm\boldsymbol{q}_{\mathrm{s}}$ were chosen. The corresponding values for $\tilde{\eta}/a$ are shown in table 3 for 3 temperatures and the average value $\langle\tilde{\eta}/a\rangle$ at each temperature is plotted as solid circles in fig. 6.

Contrary to expectation the amplitude values extracted from the first-order satellites lie above those obtained from the fundamental reflections.

Table 3
Modulation amplitude $\tilde{\eta}$ (in percent of cell edge a) as obtained from $\boldsymbol{G}_{hkl} \pm \boldsymbol{q}_s$ satellite intensities.

$\boldsymbol{G}_{hkl} \pm \boldsymbol{q}_s$	$\tilde{\eta}$		
	65 K	80 K	90 K
$(2,2,4+q_s)$	1.86	1.40	0.98
$(2,2,4-q_s)$	1.76	1.42	1.04
$(3,3,2+q_s)$	1.78	1.44	(not
$(3,3,2-q_s)$	1.83	1.34	available)
$\left\langle \frac{\tilde{\eta}}{a} \right\rangle_T \times 10^2$	1.80	1.40	1.01

4.2.6. Discussion

The major source of uncertainty in the procedure outlined above lies in the extrapolation scheme chosen to separate out the static from the dynamical part in the Debye–Waller coefficient $U_{\perp}^{\mathrm{Br}}$. This uncertainty affects directly the values of η extracted from the fundamental reflections.

These values have been underestimated if the dynamical part of $U_{\perp}^{\mathrm{Br}}$ displays a cusp at T_i as sketched in fig. 5 (dash-dotted line). Such type of behaviour has in fact been observed by Bulou et al. (1980) near the 200 K structural transition in the perovskite compound $RbCaF_3$. This latter transition is associated with the condensation of a triply-degenerate soft-mode at the R-point $(\frac{1}{2}, \frac{1}{2}, \frac{1}{2})$ of the simple cubic Brillouin zone. The cusp at the transition temperature is believed to arise from the rapid stiffening of the "longitudinal" soft-mode below $T_i (\equiv A_{1g}$-mode in the perovskite case; amplitude-mode in the modulated case), thus reducing the value of the relevant Debye–Waller coefficient.

One possible approach in order to discriminate between "static" and dynamical Debye–Waller contributions is to measure fundamental reflections at very large momentum transfers, where the Bessel function in eq. (10a) can be distinguished from a Gaussian-type behaviour. However, other difficulties arise at large momentum transfers, such as anharmonic Debye–Waller factors, and even with a suitable set of measured intensities, success is by no means certain.

5. The modulated structure: local properties

5.1. Thorium site symmetry

Much of the initial impetus for studying the physical properties of the thorium halides originated from crystal-field spectroscopists (Genet et al. 1977), whose interest was centered on the optical spectra of dilute actinide impurities such as U^{4+} (with $5f^2$ electronic configuration) substituted for Th^{4+}.

The unusual absorption and emission spectra recorded at helium temperature could not be interpreted on the basis of the accepted β-$ThBr_4$ structure, a situation which eventually led to the discovery of the 95 K structural transition (Hubert et al. 1981).

It is now realized (cf. next section) that the modification of the actinide environment caused by the sinusoidal bromine displacements provides the key to the interpretation of the observed absorption and emission spectra (Delamoye and Currat 1982, Delamoye et al. 1983a, b).

In order to calculate the crystal-field levels associated with the actinide 5f-electrons it is essential to know the local symmetry at the actinide site. This local symmetry, which in the absence of Jahn–Teller distorsion is entirely determined by the host matrix, is D_{2d} for the thorium site above T_i (cf. section 2.2 and fig. 1).

Below T_i, the concept of local symmetry breaks down, at least in a strict sense, since symmetry operations which map the crystal onto itself no longer exist. However, for the practical purpose of calculating crystal-field levels it is sufficient to define the symmetry of the thorium site in terms of the first two Br^- coordination shells.

With this operational definition, the site symmetry is expected to drop from D_{2d} ($T > T_i$) to C_2 ($T < T_i$), with C_2^z as the only remaining symmetry operation.

There are however special thorium planes, normal to the *c*-axis, for which the thorium site symmetry is higher than C_2. These are characterized by particular values of the local phase variable defined as

$$\phi_l = \boldsymbol{q}_s \cdot \boldsymbol{r}_l^0$$

or, more generally (cf. eq. 7):

$$\phi_l = \boldsymbol{q}_s \cdot \boldsymbol{r}_l^0 + \phi. \tag{12}$$

For an incommensurate wavevector $\boldsymbol{q}_s$, the variable ϕ_l (defined modulo 2π) has a *continuous* and uniform probability distribution over any chosen 2π interval. Thus it is not meaningful to talk about sites having a precise value of ϕ_l, say ϕ_0. One may nevertheless do so in an experimental context, when referring to sites for which ϕ_l is sufficiently close to ϕ_0 for the corresponding optical spectra to be undistinguishable. The situation is of course different for a commensurate modulation where ϕ_l takes on a discrete set of values.

For $\phi_l = 0$ or π the modulation is purely rotation-like (cf. section 4.1 and fig. 4a). Note in fig. 4a that all Br^- ions undergo the same *linear* displacement, corresponding, in general, to different *angular* rotations for the two Br^- tetrahedra. As a result the principal axes of the two tetrahedra no longer coincide and the symmetry elements σ_{xz}, σ_{yz} and C_2' ($\equiv C_2^{xy}, C_2^{x\bar{y}}$) are lost. The site symmetry is thus reduced to S_4.

For $\phi_l = \pm\frac{1}{2}\pi$ the modulation is purely twist-like and only C_2^z and C_2' survive. The resulting site symmetry is thus D_2.

Figures 4a and 4b show the displacement patterns associated with the $\boldsymbol{e}_1(k)$ and $\boldsymbol{e}_2(k)$ eigenvectors, respectively. This does not, however, give a true picture of the actual ionic displacements, simply because the latter are affected by additional phase factors $\cos(\phi_l + \boldsymbol{q}_s \cdot \boldsymbol{r}_k^0)$ and $\sin(\phi_l + \boldsymbol{q}_s \cdot \boldsymbol{r}_k^0)$, as seen in eq. (7). In particular, in the case $\phi_l = 0$, it is possible that the rotational contributions from the two component-waves combine in such a way that the two Br^- tetrahedra rotate by the *same* angle. This additional requirement may be written as:

$$\sum_{k=1,2,7,8} \frac{\boldsymbol{u}_{lk} \cdot \boldsymbol{e}_1(k)}{\rho_k} = \sum_{k=3,4,5,6} \frac{\boldsymbol{u}_{lk} \cdot \boldsymbol{e}_1(k)}{\rho_k}, \tag{13}$$

where ρ_k is the distance of ion k from the $\bar{4}$-axis. With the help of table 1 we find that condition (13) is satisfied, *simultaneously* for all values of l, if

$$\theta = \theta^* \equiv 2\pi\zeta_s z - \tan^{-1}\left\{ \frac{\frac{1}{2} - x - x\cos\left(\frac{1}{2}\pi\zeta_s\right)}{x\sin\left(\frac{1}{2}\pi\zeta_s\right)} \right\} \simeq 21.6°, \tag{14}$$

where

$$\zeta_s = q_s/c^* = 0.310$$

and x and z are given in table 1.

The value for θ obtained from the analysis of satellite reflections (cf. section 4.2.4) is not significantly different from the ideal value θ^* calculated above. This implies that for all practical purposes one can consider that the two Br^- tetrahedra rotate by the same amount, and this not only for $\phi_l = 0$ but for all values of ϕ_l. Thus, in $ThBr_4$ the thorium site symmetry is D_{2d} ($\phi_l = 0, \pi$) and D_2 ($\phi_l \neq 0, \pi$). For $\phi_l = \pm\frac{1}{2}\pi$ the symmetry is unchanged (D_2), since the rotational component of the Br^- displacements vanishes for these values of ϕ_l.

We do not see any convincing physical argument for the actual value of θ to be close to the ideal value θ^* which "approximately" preserves the highest-degree of local symmetry. Preliminary results (Delamoye et al. 1985) indicate that this apparent coincidence does not repeat itself in the case of $ThCl_4$, which in many other respects is similar to $ThBr_4$. There are however noticeable and systematic differences in the U^{4+} optical spectra from the two compounds (Delamoye et al. 1983a), but no causality relationship could be established between these differences and the departure of θ from the ideal value θ^*.

5.2. $ThBr_4 : U^{4+}$ optical spectra

5.2.1. Experimental absorption spectra

The timescale which characterizes photon-induced electronic transitions is fast compared to typical lattice-vibrational frequencies and hence, unlike resonance spectra, optical spectra are severely smeared out by thermal fluctuations. In the present context, this restricts the usefulness of the technique to temperatures below ~15 K. Even then, pure electronic transitions must be distinguished from vibronic transitions involving phonon creation.

Figure 7 shows examples of pure electronic absorption bands observed at 4.2 K in U^{4+}-doped (~1‰) $ThBr_4$ single crystals. Two distinct sets of absorption bands are obtained depending on the electric polarization of the incident radiation:

(1) When the incident light is polarized along the crystallographic *c*-axis ($E_{\parallel}$), the absorption spectrum consists of doublet lines with two edge-singularities. The frequency separation between the singularities (i.e. the bandwidth) is typically of the order of 8 cm^{-1}.

(2) When the incident light is polarized in the basal plane ($E_{\perp}$), the absorption bands are broader (40 to 80 cm^{-1}) and consist of three peaks with, as in the $E_{\parallel}$-case, two edge-singularities.

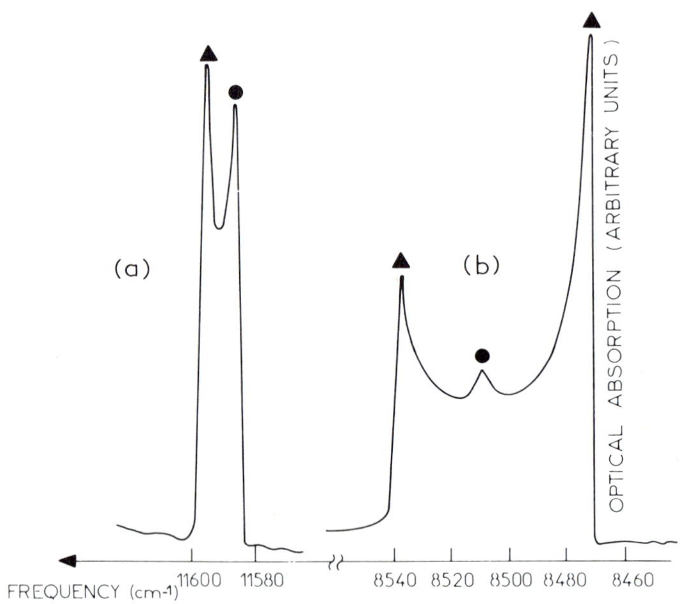

Fig. 7. Typical absorption bands from $ThBr_4 : U^{4+}$ at liquid helium temperature. The incident-light electric polarization is (a) parallel and (b) perpendicular to the tetragonal *c*-axis. (After Delamoye and Currat 1982.)

In general, single-ion crystal-field transitions at low temperatures give rise to sharp absorption lines with typical widths of $1\,cm^{-1}$ or less. The anomalous bandwidths shown in fig. 7 must therefore result from the modulation of the crystal-field energy levels due to the sinusoidal Br^- displacements.

5.2.2. *Crystal-field transitions*

The energy levels of an f^2 system in a D_{2d}-field correspond to singlet-states, labelled Γ_1, Γ_2, Γ_3 and Γ_4 and doubly-degenerate Γ_5-states.

Below T_i, the Br^- displacements $\boldsymbol{u}_{lk}$ give rise to a cell-dependent correction term $\Delta E_l^{(j)}$ to each energy level $E_0^{(j)}$. In addition, the lowering of the U^{4+} site symmetry lifts the degeneracy of the Γ_5-levels. The splitting between the corresponding sublevels is also spatially modulated and vanishes for D_{2d} (or S_4) sites ($\phi_l = 0, \pi$).

In lowest-order perturbation, the crystal-field energies $E_l^{(j)}$ below T_i may be written as

$$E_l^{(j)} = E_0^{(j)} + \Delta E_l^{(j)} = E_0^{(j)} + \eta^2\{\alpha_j + \beta_j \sin^2\phi_l\} \pm \gamma_j \eta \sin\phi_l, \tag{15}$$

where η is the modulation amplitude (cf. eq. 7) and α_j, β_j, γ_j are coefficients ($\gamma_j = 0$ for $(j) = \Gamma_1, \Gamma_2, \Gamma_3, \Gamma_4$). Equation (15) is readily established in the case of $ThBr_4:U^{4+}$ ($\theta \simeq \theta^*$), with D_{2d} ($\phi_l = 0, \pi$) and D_2 ($\phi_l \neq 0, \pi$) site symmetries. It appears to hold as well (Delamoye 1984) in the more general case of S_4 ($\phi_l = 0, \pi$), C_2 ($\phi_l \neq 0, \pi, \pm\frac{1}{2}\pi$) and D_2 ($\phi_l = \pm\frac{1}{2}\pi$) site symmetries ($\theta \neq \theta^*$).

Figure 8 shows examples of energy-level modulations as derived from eq. (15), for singlet and doublet states.

At each site l the possible transition frequencies $\nu_l^{jj'}$ are given by

$$\nu_l^{jj'} = E_l^{(j)} - E_l^{(j')}$$

and the spectral density $I^{jj'}(\nu)$ of the corresponding absorption or emission bands is proportional to

$$I^{jj'}(\nu) \propto |\mathrm{d}\nu_l^{jj'}/\mathrm{d}\phi_l|^{-1}.$$

Spectral singularities are observed whenever the condition

$$\frac{\mathrm{d}E_l^{(j)}}{\mathrm{d}\phi_l} = \frac{\mathrm{d}E_l^{(j')}}{\mathrm{d}\phi_l} \tag{16}$$

is fulfilled. As shown in fig. 8 this condition is satisfied when

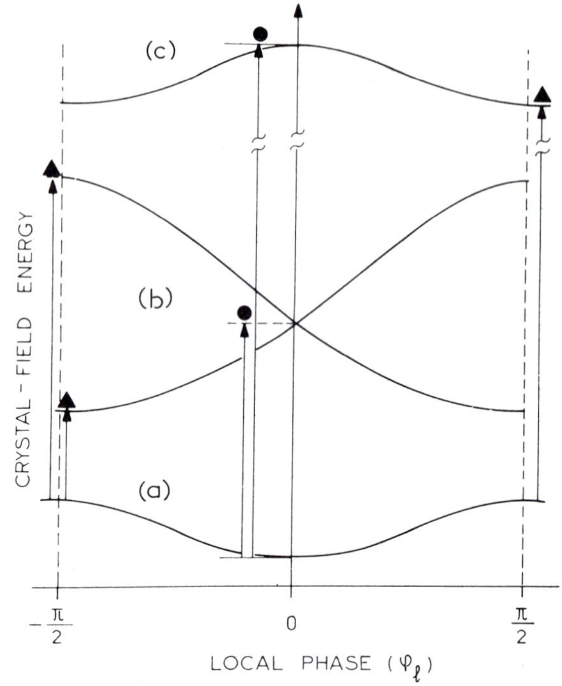

Fig. 8. Examples of U^{4+} crystal-field level modulations as obtained from eq. (15): (a) singlet state with $\beta_j > 0$; (b) doublet state; (c) singlet state with $\beta_j < 0$. The ground state of $ThBr_4:U^{4+}$ is believed to correspond to case (a) (see section 5.2.5). (After Delamoye and Currat 1982.)

(i) $\phi_l = 0, \pm\frac{1}{2}\pi$ for transitions involving singlet levels $[(j),(j') = \Gamma_1, \Gamma_2, \Gamma_3, \Gamma_4]$;

(ii) $\phi_l = \pm\frac{1}{2}\pi$ for transitions involving a Γ_5 sub-level.

5.2.3. Discussion

It is known from Zeeman spectroscopy that U^{4+} in a D_{2d} field has a singlet groundstate. In the case of $ThBr_4:U^{4+}$, detailed numerical analysis of the optical spectra (Delamoye et al. 1983b) has shown the groundstate to be of the Γ_4-symmetry type. Standard optical selection rules then indicate that the absorption of light with $E_{\parallel}$- and $E_{\perp}$-polarizations correspond to electronic transitions from the Γ_4-type groundstate to Γ_1- and Γ_5-type excited levels, respectively.

These two types of excited levels are affected by the modulated Br^- displacements in quite different ways, as seen in eq. (15) and fig. 8. This explains the qualitative differences in the absorption spectra obtained in the two polarizations. In particular the wider bandwidths of the $E_{\perp}$-lines can be correlated to the fact that the final state modulation is of order η for a Γ_5-type excited sublevel, while it is only of order η^2 ($\eta/a \ll 1$) for a Γ_1-type excited level and for the Γ_4-type groundstate.

The electronic transitions associated with the spectral singularities in fig. 7 have been indicated schematically in fig. 8. In the $E_{\parallel}$-case one edge singularity corresponds to $\phi_l \cong 0$ sites (solid circle) and the other to $\phi_l \cong \pm\frac{1}{2}\pi$ sites (solid triangle). In the $E_{\perp}$-case both edge singularities correspond, in general, to $\phi_l \cong \pm\frac{1}{2}\pi$ sites. The small peak in the middle of fig. 7b is not accounted for by the present model. Site selective excitation experiments (see next section) indicate that the corresponding transitions involve sites with $\phi_l \cong 0$ (solid circle). However, as clearly seen in fig. 8b, such transitions do not fulfill condition (16) and hence should not give rise to a peak in the line profile.

While the $\phi_l \cong 0$ peak is a systematic feature of the $E_{\perp}$-polarized absorption bands in $ThBr_4:U^{4+}$, a corresponding peak is never observed in $ThCl_4:U^{4+}$ which, otherwise, shows very similar spectra. Further comments on this point will be given in section 5.2.5.

As mentioned in section 5.1, the continuous character of the experimental absorption bands suggests a continuous distribution of ϕ_l values, consistent with a truly-incommensurate modulation wavevector ζ_s. For commensurate values of ζ_s, such as $\frac{5}{16}$, one expects a finite number of distinct ϕ_l values: $4 \times 16 = 64$, since there are 4 thorium ions per conventional cell.

Assuming the absorption spectrum from each individual site to be characterized by a Lorentzian shape-function $\mathscr{L}(E - E(\phi_l))$ with a finite, site-independent linewidth, γ, the overall absorption lineshape may be simulated via a simple summation procedure. Since the steepness of the spectral edges is directly controlled by the value of γ, the latter is readily determined by comparison to the experimental spectra.

The lineshapes generated by summing Lorentzian functions corresponding to 64 evenly-spaced values for ϕ_l, show a fine structure which is not apparent in the experimental spectra. The minimum number of distinct sites compatible with observations is about 80. This conclusion, which seemingly eliminates $\frac{5}{16}$ as a possible value for ζ_s, must be taken with some care, since it rests heavily on the assumption of a site-independent Lorentzian lineshape.

5.2.4. *Site-selective excitation experiments*

In addition to the absorption and broad-band fluorescence spectra described above, selective fluorescence spectra can be obtained using a narrow-band dye laser. By tuning the exciting laser to a specific frequency within a given absorption band, U^{4+} ions can be excited in a site-selective manner. The radiative decay of the excited ions gives rise to a number of fluorescence lines at frequencies which vary *continuously* with the exciting laser wavelength.

This unique feature can be used, for example, to determine which sites are associated with the different absorption singularities. With this technique,

Delamoye et al. (1983a) were able to show that all observed singularities can be assigned to only two types of sites, which, on the basis of the above model, can be identified as $\phi_l = 0$ and $\phi_l = \pm\frac{1}{2}\pi$ sites.

Selective excitation experiments can also be used for quantitative checks on the validity of eq. (15). The parameters which enter eq. (15), namely $E_0^{(j)} + \alpha_j\eta^2$, $\beta_j\eta^2$ and $\gamma_j\eta$, can all be determined from the position of the absorption singularities. Once this is done, the model can be applied to calculate the value of the local phase variable ϕ_l, corresponding to a particular absorption frequency, and from there to predict the values of the corresponding fluorescence frequencies. The agreement between predicted and observed frequencies is found to be excellent (Delamoye et al. 1983a). This provides a useful a posteriori vindication of the two main assumptions on which eq. (15) is based, namely the use of the sinusoidal approximation to describe the ionic displacements at low temperature and the use of lowest-order perturbation theory for the calculation of the crystal-field level modulations.

5.2.5. *Pinning of the modulation by the U^{4+} ions*

One outstanding experimental feature which does not fit into the above framework is the observation of a "third peak" in the $E_\perp$-polarized absorption spectra of $ThBr_4:U^{4+}$. As already stated the selective excitation technique allows one to assign the corresponding transitions to $\phi_l \simeq 0$ sites, but does not give any clue as to why these transitions should give rise to a peak in the line profile.

Delamoye and Currat (1982) have proposed an explanation based on the interaction of the U^{4+} impurities with the modulation. There is mounting evidence that defects play an essential role in the behaviour of modulated materials [cf. Hamano (1986), ch. 9, Part III, this volume]. Although in most cases the nature of the defects which couple to the modulation has not been identified, there are several examples of experimental studies in which the role of point defects such as chemical impurities or vacancies appears to be established (Mashiyama et al. 1982, Tolédano et al. (1985), ch. 17 in the present volume). Point defects are believed to interact with the modulation by pinning its phase, locally. For an initially random distribution of point defects the pinning interaction may lead to static phase distortions of the modulation or/and ordering of the defects through a thermally activated diffusion process. The relevant physical parameters are the pinning potential of a single isolated impurity, the phase compliance of the modulation, the concentration of impurities, and their activation energy for diffusion. The first two quantities vary with the modulation amplitude and are therefore strongly temperature dependent.

Thus even for a single type of impurity the problem is complex, and in general one expects to find the system in a metastable state where the impurities have only partially succeeded in minimizing their energies in the field of the modulation.

In the present context of $ThBr_4:U^{4+}$, one may first look for a plausible physical model for the interaction of the U^{4+} ions with the modulation. Since the main chemical difference between the U^{4+} ions and the Th^{4+} ions which they replace lies in the presence of a $5f^2$ outer electron shell on the former ions, it seems natural to assume that at least part of the U^{4+} pinning energy corresponds to the minimization of the $5f^2$ crystal-field energy.

The phase dependence of this energy is given by eq. (15):

$$E_l^{(0)} = E_0^{(0)} + \eta^2\left(\alpha_0 + \beta_0 \sin^2\phi_l\right), \tag{17}$$

where the subscript 0 refers to the (Γ_4-type) electronic ground-state. Two distinct situations occur depending on the sign of the coefficient β_0, which, unfortunately, cannot be determined by spectroscopic techniques:

(i) $\beta_0 > 0$: $E_l^{(0)}$ is minimized for $\phi_l \simeq 0$. This is the situation shown schematically in fig. 8a. Partial pinning of the modulation by the U^{4+} impurities will result in an *excess* of U^{4+} ions in $\phi_l \simeq 0$ sites. As a consequence, the $\phi_l \simeq 0$ singularity in the $E_{\|}$-absorption bands will be enhanced and an additional $\phi_l \simeq 0$ peak will appear in the $E_{\perp}$-spectra, as observed in $ThBr_4:U^{4+}$.

(ii) $\beta_0 < 0$: $E_l^{(0)}$ is minimized for $\phi_l = \pm\frac{1}{2}\pi$ and the corresponding singularities in the $E_{\|}$- and $E_{\perp}$-spectra are enhanced. No additional peak occurs in the $E_{\perp}$-spectra, as in the case of $ThCl_4:U^{4+}$.

The pinning mechanism proposed here is plausible but it is not necessarily unique. Its relevance to the present discussion must await confirmation by other techniques as, for example, magnetic circular dichroism (MCD). So far, MCD results (Briat et al. 1985) have not given any evidence for partial phase-pinning at $\phi_l \simeq 0$.

5.3. *Resonance spectroscopy*

Local information on modulated structures can also be obtained from resonance spectroscopy [see Blinc (1986), ch. 4, Part III, for a review]. Resonance lines are broadened inhomogeneously due to the distribution of local environments introduced by the modulation. The lineshape of the resonance spectrum mirrors the distribution of local fields, provided the latter fluctuates slowly on the characteristic timescale of the measurement (inverse inhomogeneous linewidth). This condition appears to be generally fulfilled except in the immediate vicinity of T_i where large-amplitude phase-fluctuations may give rise to motional narrowing (Blinc et al. 1983, Emery et al. 1984).

In the following sections we shall review some of the results obtained on β-$ThBr_4$ and β-$ThCl_4$ using ^{79}Br and ^{35}Cl NQR spectroscopy (Khan Malek et al. 1982) as well as the EPR work of Hubert et al. (1982) and Emery et al. (1984) on $ThBr_4:Gd^{3+}$.

5.3.1. NQR study of β-$ThBr_4$ and β-$ThCl_4$

Below T_i, broad NQR lineshapes have been observed in both compounds by Khan Malek et al. (1982). The line profiles are reminiscent of the U^{4+} absorption profiles, as shown in fig. 7a, although the spectral edge singularities are not as sharp in the resonance as in the optical spectra. Figure 9 shows the position of the two spectral maxima plotted as a function of temperature for the ^{79}Br NQR line in β-$ThBr_4$.

In a standard approach, the quadrupole coupling constant which determines the NQR frequency $\nu_Q(l, k)$ of an halogen nucleus at site (l, k) is expanded in powers of the (time-averaged) static displacement, $\boldsymbol{u}_{lk}$. Writing

$$\boldsymbol{u}_{lk} = \eta \boldsymbol{e}_1(k) \cos \phi_{lk},$$

where ϕ_{lk} is defined by comparison with eq. (9), and allowing for the initial C_s symmetry $(T > T_i)$ of the halogen site, the expression for the resonance frequency at site (l, k) is obtained as

$$\nu_Q(l, k) = \nu_0 + \beta_0 \eta^2 \cos^2 \phi_{lk} + \mathrm{O}(\eta^4), \tag{18}$$

where ν_0 is the resonance frequency in the undistorted phase. Close to T_i the above expression should be amended in order to allow for the effect of fluctuations.

Neglecting higher-order terms, eq. (18) leads to a continuous resonance spectrum bounded by two edge singularities at $\nu_Q = \nu_0$ and $\nu_Q = \nu_0 + \beta_0 \eta^2$, corresponding respectively to $\phi_{lk} = \pm \frac{1}{2}\pi$ and $\phi_{lk} = 0, \pi$.

The data in fig. 9 are not fully consistent with the above model. In particular the position of the lower singularity does not coincide with the extrapolated $\nu_0(T)$ line $(T > T_i)$. A better trial form might be:

$$\nu_Q(l, k) = \nu_0 + \eta^2(\alpha_0 + \beta_0 \cos^2 \phi_{lk}) + \mathrm{O}(\eta^4) \tag{19}$$

with

$$0 < \alpha_0 \ll \beta_0$$

and

$$\eta^2 \propto (T_i - T)^{2\beta} \quad (\beta \sim 0.35).$$

Similar results on $ThCl_4$ (also obtained by Khan Malek et al. 1982) can be interpreted in terms of eq. (19) with $\alpha_0 < 0$.

The physical origin of the additional term $\alpha_0 \eta^2$ in eq. (19) can be very diverse. For instance it may arise as a result of the order-parameter coupling to

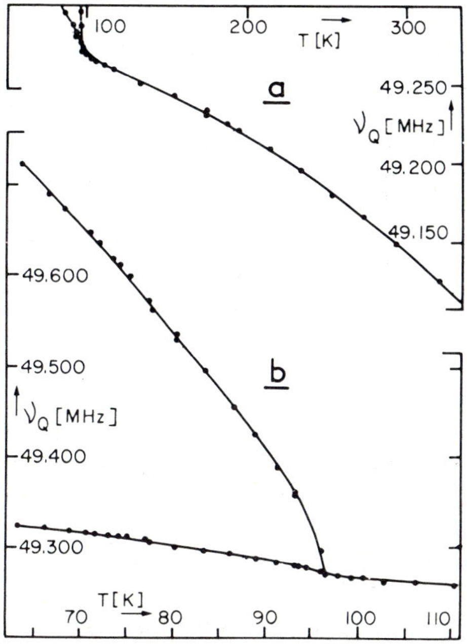

Fig. 9. [79]Br NQR spectrum in β-$ThBr_4$: (a) 77–333 K; (b) 63.15–110 K. Below T_i, the experimental points correspond to the positions of the maxima of a broad continuous spectrum. (After Khan Malek et al. 1982.)

elastic strains. Alternately it may reflect the inadequacy of an expansion of $\nu_Q(l, k)$ in powers of $\boldsymbol{u}_{lk}$ *alone*. The electric field gradient experienced by the (l, k) nucleus does not depend only on its own position $(\boldsymbol{r}^0_{lk} + \boldsymbol{u}_{lk})$ but also on the position of the neighbouring halogens $(\boldsymbol{r}^0_{l'k'} + \boldsymbol{u}_{l'k'})$. Thus in a detailed calculation a term of the type $\alpha_0\eta^2$ will necessarily appear due to the finite phase difference $(\phi_{lk} \neq \phi_{l'k'})$ between neighbouring halogen displacements.

5.3.2. *EPR measurements on $ThBr_4$: Gd^{3+}*

The charge unbalance caused by the substitution of Th^{4+} by Gd^{3+} is generally compensated by the presence of a Br^- vacancy on the nearest neighbour (nn) or next nn (nnn) shell of the paramagnetic ion. Although the presence of the vacancy should in principle give rise to orthorhombic terms in the spin-crystal field Hamiltonian these terms appear to be small, at least when the vacancy is located on the nnn shell (positions labelled 1, 2, 7 and 8 in fig. 1).

The Gd^{3+} $|\frac{5}{2}| \rightarrow |\frac{7}{2}|$ transition line associated with such centers has been studied by Hubert et al. (1982) and Emery et al. (1984). When the magnetic field is rotated in the basal plane

$$\alpha = (\boldsymbol{H}, \boldsymbol{a})$$

the resonance field $H(\alpha)$ varies approximately as:

$$H(\alpha) \cong H_0 + \lambda \cos 4\alpha, \tag{20}$$

where the coefficient λ is related to the fourth-order crystal-field parameter b_4^4. The latter has a large value in β-$ThBr_4$ due to the flat configuration of the nn Br^- tetrahedron (positions 3, 4, 5 and 6 in fig. 1).

The large anisotropy of the resonance field can thus be used to detect small static rotations of the local crystal axes, such as those associated with the rotational component of the modulated displacements below T_i. The sensitivity of the method ($\propto \mathrm{d}H(\alpha)/\mathrm{d}\alpha$) is optimized for field directions near $\alpha = \frac{1}{8}\pi$ (cf. eq. 20).

The corresponding EPR lineshapes are shown in fig. 10 for several temperatures.The experimental curves represent the derivative of the EPR signal with respect to H, near $H = H(\alpha)$ and for $\alpha = \frac{1}{8}\pi$.

At high temperatures the lineshape is Gaussian (static inhomogeneous broadening) and narrow, becoming progressively broader and Lorentzian in character (homogeneous broadening from critical fluctuations) as T_i is approached from above.

Below T_i, the EPR spectrum broadens further and develops side-peaks (cf. 93 K curve in fig. 10a), corresponding to the usual edge singularities. At still lower temperatures (cf. fig. 10b) additional structure develops on the low field end of the spectrum.

Emery et al. (1984) have interpreted the observed lineshapes using an expansion of the local resonance frequency (or field) in powers of the rotational and twist-like amplitudes of the Br^- displacements. The expansion includes linear and quadratic terms, the latter being responsible for the additional structure at low temperature. In contrast to the $ThBr_4:U^{4+}$ optical spectra, the $ThBr_4:Gd^{3+}$ resonance spectra give no indication of a tendency for the probe to pin the phase of the modulation.

Nevertheless the existence of phase-pinning centers, of unspecified nature, is implicit in the interpretation of resonance spectra from incommensurate structures: if the phase of the modulation could fluctuate freely the inhomogeneous line-broadening associated with the modulated ionic displacements would presumably disappear, through motional averaging.

Since pinning potentials vary as the modulation amplitude, and hence vanish in the limit $T \to T_i$, thermal energies should restore the "floating phase" regime, at least in a narrow temperature range below T_i. Evidence for a mixed regime with coexistence within the crystal bulk of floating and pinned regions,

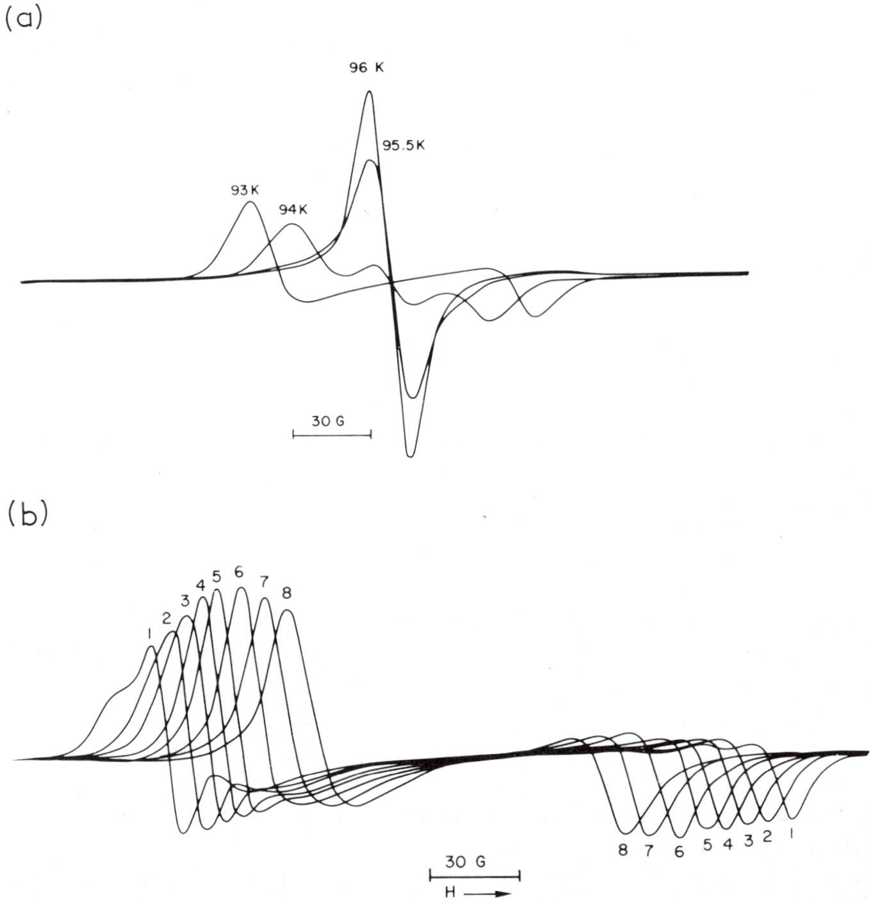

Fig. 10. EPR lineshape from the Gd^{3+} – nnn $V_{(Br^-)}$ centre in $ThBr_4$. The magnetic field orientation ($\alpha = \frac{1}{8}\pi$) is chosen in order to maximize the lineshape broadening from the modulation: (a) $T > 93$ K; (b) $T = 93$ K, 91.8 K, 90 K, 87 K, 86 K, 85 K, 84 K and 79.4 K for curves labelled from 8 to 1. (After Hubert et al. 1982.)

has been reported in Rb_2ZnBr_4 ($T_i - 10\,K < T < T_i = 346\,K$) by Blinc et al. (1983). Similar evidence is seen in fig. 10a, where the 94 K curve shows simultaneously, the side-peaks, corresponding to the edge singularities of the pinned spectrum, and the central resonance arising from motional narrowing in the floating regions. The effect is not due to gradients in T or T_i, since such gradients would smear out the side peaks.

Finally the EPR technique affords a direct and absolute measurement of the Br^- rotational amplitude and hence of $\tilde{\eta}$. By rotating the specimen around the

c-axis at constant resonant field, the two edge singularities [corresponding to $\phi_l = 0$ and π in eq. (12)], can be observed in turn. At $T_i - 7$ K, Hubert et al. (1982) find an angular separation of 4°20′ between the two peaks, corresponding to $\tilde{\eta}/a = 1.35\%$, in reasonable agreement with the value deduced (cf. fig. 6) from satellite intensity measurements: $1.1 \pm 0.1\%$.

6. The modulated structure: low frequency excitations

6.1. General

For a simple ($n = 2$) sinusoidally modulated solid, with displacive dynamics above T_i, the expressions for the phase- and amplitude-mode dispersions have been established by Axe (1976), using a simple Landau-type quasi-harmonic approach. Below T_i, the near-degeneracy between the soft-mode normal coordinates $Q(\pm \boldsymbol{q}_s + \boldsymbol{q})$ is lifted through the bilinear coupling term:

$$Q(\boldsymbol{q}_s)^2 Q(-\boldsymbol{q}_s + \boldsymbol{q}) Q^*(+\boldsymbol{q}_s + \boldsymbol{q}) + \text{c.c.},$$

where, as in eq. (4), $Q(\boldsymbol{q}_s)$ is a complex order-parameter which defines the amplitude and phase of the static modulation:

$$Q(\boldsymbol{q}_s) = Q^*(-\boldsymbol{q}_s) = \eta \mathrm{e}^{\mathrm{i}\phi}.$$

Starting from a soft-mode dispersion of the type given in eq. (3), namely

$$\omega^2(\pm \boldsymbol{q}_s + \boldsymbol{q}) = \omega_s^2 + \boldsymbol{q}\bar{\bar{\Lambda}}\boldsymbol{q} + \mathrm{O}(q^3) \tag{21}$$

with

$$\omega_s^2 \equiv \omega^2(\pm \boldsymbol{q}_s) \propto (T - T_i)^{\gamma} \tag{22}$$

and

$$\bar{\bar{\Lambda}} = \begin{pmatrix} \lambda_\perp & 0 & 0 \\ 0 & \lambda_\perp & 0 \\ 0 & 0 & \lambda_\parallel \end{pmatrix},$$

the phase- and amplitude-mode dispersions, $\omega_\phi^2(\boldsymbol{q})$ and $\omega_A^2(\boldsymbol{q})$ are obtained as:

$$\omega_\phi^2(\boldsymbol{q}) = \boldsymbol{q}\bar{\bar{\Lambda}}\boldsymbol{q} + \mathrm{O}(q^3), \tag{23a}$$

$$\omega_A^2(\boldsymbol{q}) = 2|\omega_s^2| + \boldsymbol{q}\bar{\bar{\Lambda}}\boldsymbol{q} + \mathrm{O}(q^3). \tag{23b}$$

The new normal coordinates ϕ_q and A_q are related to the original soft-mode coordinates, $Q(\pm \boldsymbol{q}_s + \boldsymbol{q})$, through:

$$\phi_{\boldsymbol{q}} = \frac{\mathrm{i}}{\sqrt{2}} \{ \mathrm{e}^{-\mathrm{i}\phi} Q(-\boldsymbol{q}_s + \boldsymbol{q}) - \mathrm{e}^{\mathrm{i}\phi} Q(\boldsymbol{q}_s + \boldsymbol{q}) \}, \tag{24a}$$

$$A_{\boldsymbol{q}} = \frac{1}{\sqrt{2}} \{ \mathrm{e}^{-\mathrm{i}\phi} Q(-\boldsymbol{q}_s + \boldsymbol{q}) + \mathrm{e}^{\mathrm{i}\phi} Q(\boldsymbol{q}_s + \boldsymbol{q}) \}. \tag{24b}$$

Using eqs. (24) above, one readily obtains the inelastic or "one-phonon" structure factor for the amplitude- and phase-fluctuations $A_{\boldsymbol{q}}$ and $\phi_{\boldsymbol{q}}$:

$$F_A^2(\boldsymbol{Q}\pm) = F_\phi^2(\boldsymbol{Q}\pm) = \tfrac{1}{2} F_s^2(\boldsymbol{Q}\pm) = \tfrac{1}{2} \left| \sum_k (m_k)^{-1/2} \boldsymbol{Q}_\pm \cdot \boldsymbol{e}_k(\pm \boldsymbol{q}_s) \mathrm{e}^{\mathrm{i}\boldsymbol{G}\cdot\boldsymbol{r}_k^0} \right|^2, \tag{25}$$

where $F_s(\boldsymbol{Q}_\pm)$ is the soft-mode inelastic structure factor at

$$\boldsymbol{Q}_\pm = \boldsymbol{G}_{hkl} \pm \boldsymbol{q}_s + \boldsymbol{q}.$$

The above description has been extended by Bruce and Cowley (1978) to allow for the effect of higher-order distorsion harmonics ($Q(2\boldsymbol{q}_s)$, etc...) and by Golovko and Levanyuk (1981, 1982) to include damping effects. To the extent that the soft-mode damping coefficient can be taken as uniform in the neighbourhood of $\pm \boldsymbol{q}_s$,

$$\Gamma(\pm \boldsymbol{q}_s + \boldsymbol{q}) \simeq \Gamma(\pm \boldsymbol{q}_s) = \Gamma_s, \tag{26}$$

then both the amplitude- and phase-modes are found to be affected by the same, q-independent, damping coefficient Γ_s.

Hence, unlike acoustic modes, phase-modes are necessarily overdamped in the long-wavelength limit, and, as such, do not qualify as true Goldstone excitations (Zeyher and Finger 1982). The damping arises from the fact that the atomic motions involved are inhomogeneous on a microscopic scale and are thus affected by ordinary anharmonic interactions.

In real solids, the presence of phase-pinning defects is likely to modify the phase-mode response, particularly in the long-wavelength limit ($\lambda \gtrsim$ average distance between defects). Although no detailed predictions have been made one can expect an enhancement of the phase-mode damping coefficient, or the appearance of a quasielastic "central" component in the phase-mode lineshape and/or the occurrence of a gap at $q = 2\pi/\lambda = 0$, depending whether the defects are able to relax and order themselves in the field of the modulation (e.g. by diffusion), and if so at which rate.

On the other hand for wavelengths which are short compared to the average distance between defects, the phase-mode response of the pure solid should be preserved. Most of the neutron scattering results which we shall review in the following sections correspond to this last regime and are thus representative of the pure solid.

6.2. *Inelastic neutron scattering: experimental aspects*

While eqs. (23b) and (22) predict a classic soft-mode behaviour for the amplitude mode, eqs. (23a) and (25) predict a temperature-independent dispersion and structure factor for the phase mode below T_i.

Experimentally, the phase mode is observed as an additional, acoustic-like, excitation branch emanating from the first-order satellite reflections.

One major practical difficulty is to distinguish the phason branch from the ordinary acoustic branches which are also observable near satellite peaks. The inelastic structure factors of the acoustic modes are proportional to the elastic structure factor of the relevant satellite peak, which in turn, varies as the modulation amplitude (cf. eq. 10b), and hence vanishes in the limit $T \rightarrow T_i^-$. Thus, sufficiently close below T_i the temperature-independent phase-mode response should dominate the inelastic neutron scattering spectra taken near a strong satellite reflection.

Unfortunately one then runs into the difficulty of separating the phase branch from the amplitude branch, which also contributes to the low-frequency spectra and whose response may overlap the phase-mode response due to damping effects and/or finite instrumental resolution.

In the case of K_2SeO_4, where the soft-mode damping is fairly large ($\Gamma_s \simeq$ 250 GHz) it has been possible to identify the phase branch, simply on the grounds that its response is overdamped (Quilichini and Currat 1983). In the case of biphenyl the acoustic slopes are sufficiently different from that of the phase mode that there can be no confusion between them, at least in some propagation directions (Cailleau et al. 1980). Neither argument can be applied convincingly to the case of $ThBr_4$. In particular the slope of the TA-branch propagating in the $\boldsymbol{c}^*$-direction is comparable to that expected for the phase branch in that direction. One must therefore rely on intensity arguments, as discussed by Bernard et al. (1983).

6.3. *Phase and amplitude modes in β-ThBr$_4$ at 81 K*

6.3.1. *Experimental results*

Figure 11 shows a set of constant-Q spectra taken near the strong (2, 3, 0.69) satellite reflection at $T = 81\,\mathrm{K} = T_i - 14\,\mathrm{K}$. The data were collected using the

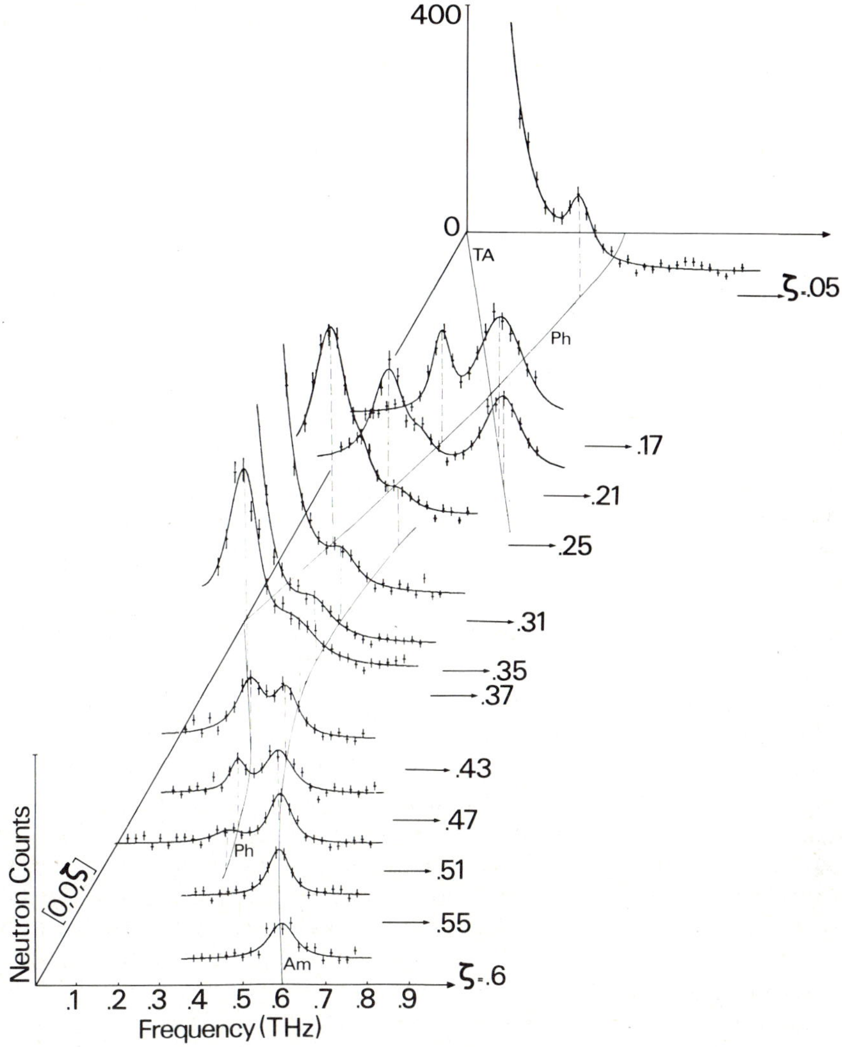

Fig. 11. Constant-Q spectra at $(2,3,1-\zeta)$ for $T=81$ K. For each scan the baseline is shown as a horizontal arrow pointing towards the relevant value of ζ. (After Bernard et al. 1985a.)

cold-source three-axis spectrometer IN12, operated in the up-scattering mode with a fixed incident neutron energy of 4.9 meV ($k_i = 1.54\,\text{Å}^{-1}$). The instrumental resolution was 25 GHz, as measured on a V-specimen.

Constant-Q scans taken at $(2,3,1-\zeta)$ are labelled according to the value of ζ. Hence the $(2,3,0.69)$ satellite position and the $(2,3,1)$ fundamental reflection correspond, respectively, to $\zeta = 0.31$ and $\zeta = 0.0$, in the figure.

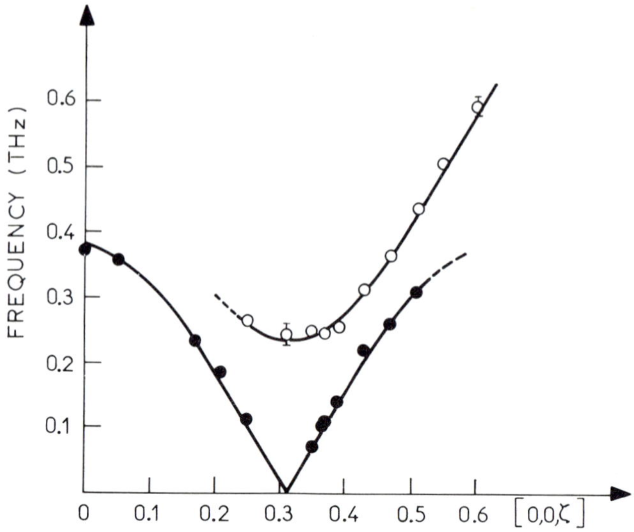

Fig. 12. Phase (●) and amplitude (○) mode dispersions at 81 K. The experimental frequencies are obtained from the analysis of the spectra shown in fig. 11. The solid lines are calculated from eq. (30).

Analysis of the inelastic spectra in fig. 11 gives evidence for the presence of three distinct excitation branches:

(i) an acoustic-like branch emanating from the (2, 3, 0.69) satellite position;
(ii) an optic-like branch with a frequency minimum at the satellite position;
(iii) a TA branch originating at the fundamental (2, 3, 1) reflection.

The first branch is identified with the phason branch, as it is much too intense to be of acoustic origin. Similarly the optic-like branch is identified with the amplitude-mode. Its frequency at $\zeta = 0.31$ is in good agreement with the Raman "soft-mode" frequency at that temperature, as deduced from the work of Hubert et al. (1981).

Figure 12 shows the quasi-harmonic frequencies of the phase and amplitude modes as obtained from a damped-harmonic-oscillator (DHO) fit to the spectra in fig. 11. Notice that the two branches are observed simultaneously only in the vicinity of the satellite peak ($0.25 < \zeta < 0.50$). For $\zeta < 0.25$ only the phase-mode is observed. For $\zeta > 0.50$ the phase-mode vanishes and the amplitude-mode is observed alone.

This type of assymmetrical behaviour has been predicted long ago (cf. fig. 7 in Axe 1976) and is observed here for the first time. The underlying physical picture will be discussed in more detail in the next section.

The experimental points in fig. 12 give no evidence for a gap in the phase-mode dispersion, at $q = 0$ ($\zeta = 0.31$). If such a gap exists, due to

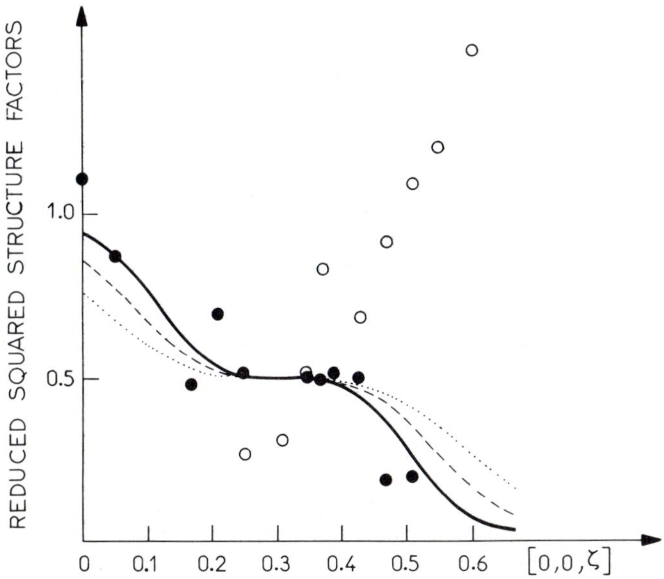

Fig. 13. Reduced squared structure factors $|F_{\phi,A}/F_s|^2$ for the phase (●) and amplitude (○) modes at 81 K. The experimental structure factors $|F_{\phi,A}|^2$ are obtained from the analysis of the spectra in fig. 11, in arbitrary units. The soft-mode structure factors $|F_s|^2$ are extracted from another series of inelastic measurements at $T = 110$ K, on a different specimen. The normalization between the two sets of data is performed on the basis of the observed (2,3,0.69) satellite intensities at $T_i - 14$ K. The calculated curves for $|F_\phi/F_s|^2$ at 81 K (solid line), 60 K (broken line) and 20 K (dotted line) are obtained from eqs. (31) and (33a).

high-order commensurability or defects its magnitude may not exceed ~ 40 GHz.

The fitting procedure also yield values for the mode damping coefficients. These values show a considerable amount of scatter due to the limited statistical accuracy of the experimental spectra. They lie in the range from 40 to 90 GHz for both modes, with no detectable trend as a function of wavevector.

The inelastic structure factors (ISF) for both modes deduced from the DHO fits, are shown in fig. 13. Apart from instrumental factors, the squared ISF's are related to the measured integrated intensities, J_A and J_ϕ, through:

$$J_{A,\phi} \propto \frac{kT}{\omega^2_{A,\phi}} |F_{A,\phi}|^2, \tag{27}$$

where the quasi-harmonic mode frequencies $\omega^2_{A,\phi}$ are shown in fig. 12. The

ISF's shown in fig. 13 are normalized to that of the soft mode measured at $T = 110\,\mathrm{K}$.

According to eq. (25) the normalized squared ISF's for both modes should be in the neighbourhood of 0.5. This prediction is consistent with the values shown in the central part of fig. 13 ($0.2 < \zeta < 0.4$). However, away from the satellite position, the experimental points indicate a strong wavevector-dependence for the mode ISF's, the amplitude mode increasing with ζ, while that of the phase modes decreases. This intensity reversal, already apparent on the raw spectra in fig. 11, is suggestive of an eigenvector exchange between the two branches. As we shall show, this effect is a simple consequence of the assymmetry of the soft-mode dispersion about its minimum. It can be readily accounted for by including odd-order terms in the expansion of the soft-mode frequency in powers of $\boldsymbol{q}$ (eq. 21) about its minimum at $\pm\boldsymbol{q}_s$.

6.3.2. Eigenvector analysis

Specializing to the case $\boldsymbol{q}\|\boldsymbol{c}^*$, eq. (21) may be rewritten as

$$\omega^2(q_s + q) = \omega_s^2 + G(q) + U(q), \tag{28}$$

where $G(q)$ and $U(q)$ are, respectively, even and odd in q:

$$G(q) = G(-q) = \alpha_2 q^2 + \alpha_4 q^4 + \cdots, \tag{29a}$$

$$U(q) = -U(-q) = \alpha_3 q^3 + \alpha_5 q^5 + \cdots, \tag{29b}$$

Similarly:

$$\omega^2(-q_s + q) = \omega^2(q_s - q) = \omega_s^2 + G(q) - U(q).$$

Following the same procedure as in section 6.1, we obtain the phase- and amplitude-mode frequencies as

$$\omega^2_{\left\{ A \atop \phi \right\}}(q) = |\omega_s^2| + G(q) \pm \left\{ |\omega_s^2|^2 + U^2(q) \right\}^{1/2}, \tag{30}$$

which reduces to eq. (23) in the limit $U(q) \to 0$.

Phase- and amplitude-mode dispersion curves calculated using eq. (30) are shown as solid lines in fig. 12. The expansion coefficients in $G(q)$ and $U(q)$ have been obtained from a fifth-order polynomial fit to the 150 K soft-mode dispersion curve (cf. fig. 2). The adjustable parameters are

(i) the value of ω_s^2 at 81 K, fitted to the $q = 0$ amplitude-mode frequency:

$$\omega_A^2(q = 0) = 2|\omega_s^2| = 0.055\,\mathrm{THz}^2;$$

(ii) the position of the soft-mode minimum which is constrained to coincide with the satellite position ($\zeta_s = 0.31$), whereas the fit to the 150 K data yields $\zeta_s = 0.288$.

Introducing the dimensionless ratio

$$z(q) = \frac{U(q)}{\left\{\left|\omega_s^2\right|^2 + U^2(q)\right\}^{1/2}} \tag{31}$$

the relationship between the (A_q, ϕ_q) variables and the original soft-mode coordinates $Q(\pm q_s + q)$, becomes

$$\phi_q = \frac{i}{\sqrt{2}}\left\{(1+z)^{1/2} e^{-i\phi} Q_{-q_s+q} - (1-z)^{1/2} e^{i\phi} Q_{q_s+q}\right\}, \tag{32a}$$

$$A_q = \frac{1}{\sqrt{2}}\left\{(1-z)^{1/2} e^{-i\phi} Q_{-q_s+q} + (1+z)^{1/2} e^{i\phi} Q_{q_s+q}\right\}. \tag{32b}$$

Below T_i, an inelastic up-scattering measurement at

$$\boldsymbol{Q}_{\pm} = \boldsymbol{G}_{hkl} \pm \boldsymbol{q}_s + \boldsymbol{q}$$

projects out the component of the displacement amplitudes proportional to $Q_{\mp q_s + q}$. Hence from eq. (32):

$$F_\phi^2(\boldsymbol{Q}_\pm) = \tfrac{1}{2}(1 \pm z) F_s^2(\boldsymbol{Q}_\pm), \tag{33a}$$

$$F_A^2(\boldsymbol{Q}_\pm) = \tfrac{1}{2}(1 \mp z) F_s^2(\boldsymbol{Q}_\pm). \tag{33b}$$

The data in fig. 11 have been measured at

$$\boldsymbol{Q}_- = \boldsymbol{G}_{231} - \boldsymbol{q}_s + \boldsymbol{q}$$

with

$$\boldsymbol{q} = (\zeta_s - \zeta)\boldsymbol{c}^*.$$

From eqs. (33) we get:

$$\left|\frac{F_\phi(Q_-)}{F_s(Q_-)}\right|^2 = \tfrac{1}{2}(1 - z(q)) \tag{34a}$$

$$\left|\frac{F_A(Q_-)}{F_s(Q_-)}\right|^2 = \tfrac{1}{2}(1 + z(q)). \tag{34b}$$

Figure 2 clearly shows that $dz/d\zeta$ is positive. Hence the phase-mode structure factor is expected to decrease with ζ, at the same rate as that of the amplitude mode increases.

The curves in fig. 13 illustrate the behaviour of the function $\frac{1}{2}(1-z(\zeta))$, for values of ω_s^2 corresponding to several temperatures below T_i. For $T=81$ K, the calculated curve for the phase-mode structure factor agrees rather well with the experimental points, while for the amplitude-mode, the observed wavevector-dependence appears to be more severe than predicted by eq. (34b).

6.4. *Experimental results at lower temperatures*

Figure 14 summarizes some recent inelastic results obtained near the $(2,3,1-\zeta_s)$ satellite reflection. Constant-Q scans at $1-\zeta=0.63$, 0.57 and 0.53 (corresponding to propagation wavevectors $q=-0.06, -0.12$ and $-0.16c^*$, respectively) have been performed for several temperatures between 83 K and 20 K. The vertical scales have been adjusted in such a way as to eliminate the effect of the temperature factor in the phonon cross-section (cf. eq. 27). Within each set of scans the peak heights can thus be directly compared.

The following features may be noticed:

(i) As expected, the amplitude-mode frequency increases on cooling. This is seen most clearly in fig. 14c, between 83 K and 60 K. At lower temperatures the amplitude-mode frequency lies outside the range of the scans.

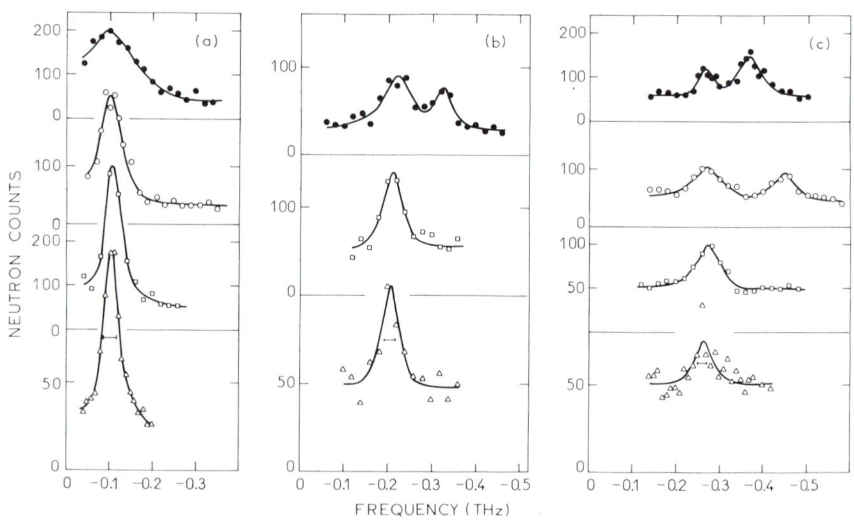

Fig. 14. Constant-Q scans at $\boldsymbol{Q}=\boldsymbol{G}_{231}-\boldsymbol{q}_s-\boldsymbol{q}$ for several temperatures between 83 K and 20 K: (a) $q/c^*=0.06$; (b) 0.12; (c) 0.16. (●) $T=83$ K; (○) 60 K; (□) 40 K; (△) 20 K.

(ii) In contrast, the phase-mode frequency, for all 3 wavevectors, remains very nearly constant between 83 K and 20 K.

(iii) At low q (cf. fig. 14a) the phase-mode lineshape becomes sharper on cooling while the normalized integrated intensity (J_ϕ/kT) remains very nearly constant. At 20 K the apparent linewidth is resolution-limited. These features are consistent with the behaviour expected from ordinary anharmonicity.

(iv) At large q, the statistical accuracy of the experimental spectra is not sufficient for a meaningful lineshape analysis. The data in fig. 14c seem to indicate an increase in integrated intensity (J_ϕ/kT) on cooling, a trend which is consistent with the prediction of the model developed in the preceding section (cf. fig. 13).

References

André, G., D. Durand, F. Dénoyer, F. Moussa and R. Currat, 1986, to be published.
Axe, J.D., 1976, Proc. of Gatlinburg Neutron Scattering Conf., CONF-760601-P1 (U.S. Dept. of Comm.) p. 353.
Axe, J.D., 1980, Phys. Rev. **B21**, 4181.
Axe, J.D., M. Iizumi and G. Shirane, 1980, Phys. Rev. **B22**, 3408.
Bernard, L., R. Currat, P. Delamoye, C.M.E. Zeyen, S. Hubert and R. de Kouchkovsky, 1983, J. Phys. **C16**, 433.
Bernard, L., R. Currat and P. Delamoye, 1986a, to be published.
Bernard, L., J. Pannetier, R. de Kouchkovsky, P. Delamoye, C.M.E. Zeyen and R. Currat, 1985b, to be published.
Blinc, R., 1986, ch. 4, Part III.
Blinc, R., D.C. Ailion, P. Prelovsek and V. Rutar, 1983, Phys. Rev. Lett. **50**, 67.
Briat, B., P. Delamoye, J.C. Rivoal, S. Hubert and P. Evesque, 1985, J. Phys. **46**, 1375.
Brown, D., 1968, Halides of Lanthanides and Actinides (Wiley, New York).
Brown, D., T.L. Hall and P.T. Moseley, 1973, J. Chem. Soc. Dalton Trans. **6**, 686.
Bruce, A.D., and R. Cowley, 1978, J. Phys. **C11**, 3609.
Bulou, A., C. Ridou, M. Rousseau, J. Nouet and A.W. Hewat, 1980, J. Physique **41**, 87.
Cailleau, H., F. Moussa, C.M.E. Zeyen and J. Bouillot, 1980, Solid State Commun. **33**, 407.
Cowley, R.A., 1980, Adv. Phys. **29**, 1.
de Kouchkovsky, R., M.F. Le Cloarec and P. Delamoye, 1981, Mater. Res. Bull. **16**, 1421.
Delamoye, P., 1984, private communication.
Delamoye, P., and R. Currat, 1982, J. Physique Lett. **43**, L-655.
Delamoye, P., J.C. Krupa, J.G. Conway and N. Edelstein, 1983a, Phys. Rev. **B28**, 4913.
Delamoye, P., K. Rajnak, M. Genet and N. Edelstein, 1983b, Phys. Rev. **B28**, 4923.
Delamoye, P., J.C. Krupa, L. Bernard, R. Currat and D. Petitgrand, 1986, to be published.
Emery, J., S. Hubert and J.C. Fayet, 1984, J. Physique Lett. **45**, L-693.
Genet, M., P. Delamoye, N. Edelstein and J.G. Conway, 1977, J. Chem. Phys. **67**, 1620.
Golovko, V.A., and A.P. Levanyuk, 1981, Ferroelectrics **36**, 306.
Golovko, V.A., and A.P. Levanyuk, 1982, Sov. Phys.-JETP **54**, 1217.
Hamano, K., 1985, ch. 9, Part III.
Heine, V., and J.D.C. McConnell, 1981, Phys. Rev. Lett. **46**, 1092.
Heine, V., and J.D.C. McConnell, 1984, J. Phys. **C17**, 1199.
Hogervorst, A.C.R., and P.M. de Wolff, 1982, Solid State Commun. **43**, 179.

Hubert, S., P. Delamoye, S. Lefrant, M. Lepostollec and M. Hussonnois, 1981, J. Solid State Chem. **36**, 36.
Hubert, S., J. Emery, J.J. Rousseau and J.C. Fayet, 1982, J. Physique Lett. **43**, L-815.
Hussonnois, M., J.C. Krupa, M. Genet, L. Brillard and R. Carlier, 1981, J. Cryst. Growth **51**, 11.
Iizumi, M., and K. Gesi, 1983, J. Phys. Soc. Jpn **52**, 2526.
Iizumi, M., J.D. Axe, G. Shirane and K. Shimaoka, 1977, Phys. Rev. **B15**, 4392.
Jacoby, Y., S. Yust and W.W. Kruhler, 1976, Ferroelectrics **12**, 117.
Khan Malek, C., A. Péneau and L. Guibé, 1982, J. Mol. Struct. **83**, 201.
McConnell, J.D.C., and V. Heine, 1984a, Act. Cryst. A., to appear.
McConnell, J.D.C., and V. Heine, 1984b, Phys. Rev. Lett., to appear.
Mashiyama, H., S. Tanisaki and K. Hamano, 1982, J. Phys. Soc. Jpn **51**, 2538.
Mason, J.T., M.C. Jha and P. Chiotti, 1973, J. Less Common Met. **34**, 143.
Mason, J.T., M.C. Jha, D.M. Bailey and P. Chiotti, 1974, J. Less Common Met. **35**, 331.
Meissner, G., and H. Blaschke, 1984, Ferroelectrics **53**, 199.
Meissner, G., N. Menyhárd and P. Śzepfalusy, 1981, Z. Phys. **B45**, 137.
Overhauser, A.W., 1971, Phys. Rev. **B3**, 3173.
Quilichini, M., and R. Currat, 1983, Solid State Commun. **48**, 1011.
Sheldrick, G.M., 1976, SHELX-Program for Crystal Structure Determination (University of Cambridge).
Steigmeier, E.F., and H. Auderset, 1973, Solid State Commun. **12**, 565.
Tolédano, J.C., J. Schneck and G. Errandonéa, 1985, ch. 17, this volume.
Zeyher, R., and W. Finger, 1982, Phys. Rev. Lett. **49**, 1883.

CHAPTER 16

The Incommensurate Phase of Quartz

G. DOLINO

Laboratoire de Spectrométrie Physique (LA n° 08)
Université Scientifique et Médicale de Grenoble
Boite postale no. 87, 38402 Saint-Martin-d'Hères
France

Incommensurate Phases in Dielectrics 2
Edited by
R. Blinc and A.P. Levanyuk

Contents

1. Introduction . . . 207
2. The properties of quartz and the discovery of the incommensurate phase . . . 207
3. Static properties of the INC phase of quartz . . . 209
   3.1. Macroscopic thermodynamical data . . . 210
   3.2. Experimental evidence of an incommensurate modulation . . . 212
      3.2.1. Scattering pattern in the incommensurate phase . . . 212
      3.2.2. Temperature variations of the scattering pattern . . . 213
      3.2.3. Discussion . . . 215
   3.3. Lock-in transition to the α phase . . . 216
      3.3.1. Electron microscopy observations . . . 216
      3.3.2. Observations on bulk samples . . . 218
      3.3.3. Conclusion . . . 218
4. Theoretical studies . . . 219
   4.1. Origin of the INC phase of quartz . . . 219
   4.2. Structure of the INC phase . . . 221
   4.3. Macroscopic symmetry of the INC phase . . . 223
   4.4. Comparison with experimental results . . . 225
5. Dynamical properties . . . 226
   5.1. Soft mode behavior in the α and β phases . . . 226
   5.2. Brillouin experiment . . . 227
   5.3. Raman experiment . . . 227
6. Conclusion . . . 229

References . . . 230

1. *Introduction*

As evidenced by this book, it has been realized in the last 10 years that a large number of dielectric crystals present intermediate phases characterized by incommensurate modulations, the origin of which may sometimes be related to the existence of a Lifshitz invariant in the expansion of the free energy as a function of the order parameter. For other cases, where, as in $NaNO_2$ or $SC(NH_2)_2$, the crystal symmetry does not allow the existence of a Lifshitz invariant, Levanyuk and Sannikov (1976) have shown that the origin of the incommensurate phase may be related to the existence of a Lifshitz-type invariant built by coupling the order parameter to another variable by a gradient interaction. Using this kind of argument, Aslanyan and Levanyuk (1978, 1979) predicted the possibility of the existence of an incommensurate phase near the $\alpha-\beta$ transition point of quartz, which was indeed subsequently discovered in a neutron scattering experiment by Dolino, Bachheimer and Zeyen (1982).

In this chapter, we will first recall the properties of the classical $\alpha-\beta$ transition of quartz and the steps which have led to the discovery of this new incommensurate (INC) phase of quartz. Then the static properties of this phase will be presented and compared to the theoretical results obtained from the gradient interaction model. Finally we shall review the dynamical properties with the aim of testing this model.

2. *The properties of quartz and the discovery of the incommensurate phase*

Under normal conditions, quartz is the stable crystalline form of SiO_2, which is one of the major constituents of the earth's crust. Due to the frequent occurrence of large and perfect natural crystals, this compound has played a great role in the development of crystal physics: piezoelectricity, Brillouin and Raman scattering in solids and nonlinear optical effects were discovered in quartz. It is also a material of great technological importance as a stable-frequency oscillator in electronics. This application comes from its piezoelectric properties as well as from its special elastic properties: in general elastic constants decrease with increasing temperature, but in quartz, one elastic constant (c_{66}) increases, which permits to build resonators with temperature-independent frequencies. This anomalous variation of c_{66} is related to the existence of a structural phase transition occurring at 846 K between the low-temperature α phase and the high-temperature β phase.

The results of the large number of experiments performed on this first order transition, since its discovery by Le Chatelier (1889), have been reviewed several times (Sosman 1965, Scott 1974, Dolino and Bachheimer 1982, Dolino et al. 1983b): the α-quartz structure is built from SiO_4 tetrahedra, interlinked by their corners and forming right or left helixes around the Z axis, which is an helicoidal axis of order 3, while the X axis is along one of the 3 rotation axes of order 2. The hexagonal elementary cell contains 3 silicium atoms in special position $(u, 0, 0)$ and 6 oxygen atoms in general position (x, y, z). Usually the $\alpha-\beta$ transition is considered to be displacive with the atoms shifting, in the β phase, to more symmetrical positions, $(\frac{1}{2}, 0, 0)$ for silicium and $(2x, x, \frac{1}{6})$ for oxygen. With these displacements the number of atoms in the elementary cell does not change but the symmetry of the β phase increases: the helicoidal axis is now of order 6, and a new rotation axis OY of order 2 perpendicular to OX and OZ appears. As the symmetry group (32) of the α phase is a subgroup of that (622) of the β phase, the $\alpha-\beta$ transition can be described in the frame of Landau theory: this is an example of a ferroic transition where the order parameter η is associated to the appearance in the α phase of the piezoelectric coefficient d_{xxx} and of the elastic constant c_{xxyz} (or c_{14}) (Aizu 1973). In a microscopic description, η is associated to the tilting angle of SiO_4 tetrahedra (Höchli and Scott 1971, Megaw 1973, Grimm and Dorner 1975). Two rotation senses are possible leading to a domain structure, known for a long time by mineralogists under the name of Dauphiné twins, because they were first detected in crystals coming from the Dauphiné, a province of France. The temperature variation of η measured by second harmonic generation of light is in good quantitative agreement with the predictions of Landau theory of first order transitions (Bachheimer and Dolino 1975).

Concerning dynamical properties, a soft mode has clearly been observed in the α phase by Raman scattering (Shapiro et al. 1967). But if Landau theory works quite well for the α phase, several discrepancies remain for the β phase and some puzzling features have been observed just at the transition itself:

Firstly, even the classical structure of the β phase is questioned: in a recent neutron investigation of this structure, Wright and Lehmann (1981) have concluded to the existence of a disordered structure, so that the $\alpha-\beta$ transition would present a mixed character, with features of order–disorder transitions due to the disordered structure of the β phase, and features of displacive transitions as the soft mode of the α phase.

In the β phase, several properties such as thermal expansion or elastic constants show large variations just above the transition, which cannot be explained by the usual Landau theory (Dolino and Bachheimer 1982).

Just at the $\alpha-\beta$ transition intense light scattering has been observed by Yakolev et al. (1956) and interpreted as a critical opalescence [these theories, developed by Ginzburg (1955) and by Ginzburg and Levanyuk (1958), correspond to the first treatment of the $\alpha-\beta$ transition of quartz according to Landau theory]. However it was shown later by Shapiro and Cummins (1968) that this light scattering occurs without any frequency shift, unambiguously excluding a thermal fluctuation origin. Rather, they attributed this phenomenon to the apparition of an extensive microtwinning of the Dauphiné kind, as already suggested by Young (1962) in his X-ray work.

Using electron microscopy, Van Tendeloo et al. (1975, 1976) and Malov and Sonyushkin (1975) have indeed observed a fine scale triangular structure between the α and the β phases, which they interpreted as microdomains of Dauphiné twins. A quite different interpretation was put forward later by Aslanyan and Levanyuk (1978, 1979): they suggested that this was a direct observation of an incommensurate phase, the existence of which they predicted from a coupling between the strains U_{ij} and the spatial derivatives of the order parameter η, given by

$$\tfrac{1}{2}a\left[\left(U_{xx}-U_{yy}\right)\partial\eta/\partial x-2U_{xy}\,\partial\eta/\partial y\right]. \qquad (2.1)$$

This expression introduces a Lifshitz-like invariant in the free energy of quartz which can lead to the existence of an incommensurate phase if the coupling coefficient a is large enough.

A study of the mechanical properties of quartz was independently started in Grenoble in order to specify the anomalous behaviour observed around the $\alpha-\beta$ transition. Results, coherent with Landau theory, were soon obtained for the α phase (Bachheimer and Dolino 1980) but the large variations of the β phase remain unclear. After improving the temperature resolution Bachheimer (1980) found some discontinuities in the variation of mechanical properties at a temperature T_i, 1.3 K above T_c, the transition temperature to the α phase upon cooling, disclosing the existence of an intermediate phase between T_i and T_c. The identification of the incommensurate character of this new phase was obtained by a neutron scattering experiment performed in december 1981 at the Institute Laue-Langevin, where satellite peaks were discovered along the Y axis, as predicted (Dolino et al. 1982, 1983a, Zeyen et al. 1983). We shall now describe the static properties of this new phase of quartz in more detail.

3. *Static properties of the INC phase of quartz*

We will first present some thermodynamical data in order to clarify the relations between the 3 phases: α–INC–β. Then we shall go on with more microscopic results obtained from neutron and X-ray scattering and we will end with some data about the lock-in transition from the INC to the β phase.

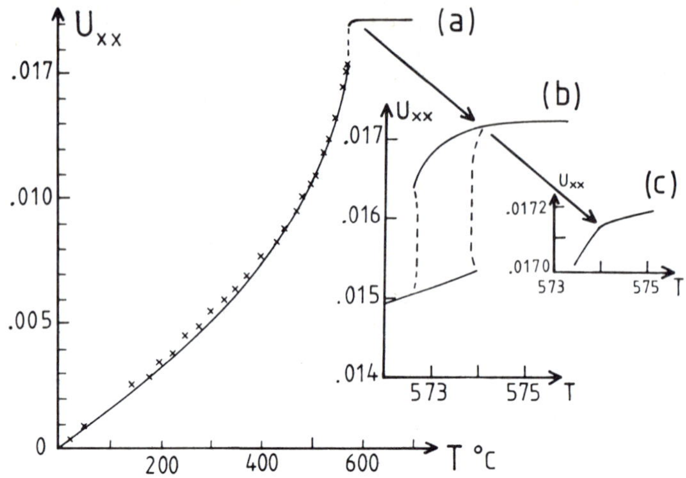

Fig. 1. Plot of the temperature variation of the thermal expansion U_{xx} of quartz along the X axis, presented with different magnifications: (a) from 0 to 700°C; (b) in the hysteresis region between the α and the β phases; (c) around the transition from the β to the INC phase. In (a), the crosses are experimental points and the full curve corresponds to the theoretical variation obtained from the Landau theory. In (b) and (c) the full curves correspond to experimental results in homogeneous phases, while the dashed curves correspond to coexistence states during the phase transition.

3.1. *Macroscopic thermodynamical data*

Most physical properties of quartz show quite distinct anomalies in the INC phase but these changes were unnoticed in most previous experiments, due to poor temperature resolution. However the presence of the INC phase is clearly displayed in the accurate thermal expansion measurements of Bachheimer (1980). These results, completed by older ones analyzed by Dolino and Bachheimer (1984) are presented with increasing resolution around the transition in Fig. 1a to 1c:

Figure 1a shows the general variation of the thermal expansion $U_{xx}(T)$ from room temperature to the transition: in the α phase $U_{xx}(T)$ is roughly proportional to the square of the order parameter $\eta(T)$ (the full curve of figure 1a is a parabola with a horizontal axis, representing the variation of $\eta^2(T)$ according to Landau theory). At higher temperature, in the β phase, the thermal expansion is small and even becomes negative.

On the magnified temperature-scale of figure 1b, the first-order nature of the transition is clearly revealed by the existence of discontinuities and by thermal hysteresis: upon heating the nucleation of the high temperature phase occurs at a temperature T_h while upon cooling the nucleation of the α phase occurs at a temperature T_c, about 1 K lower. One must also note the asymmetry of the

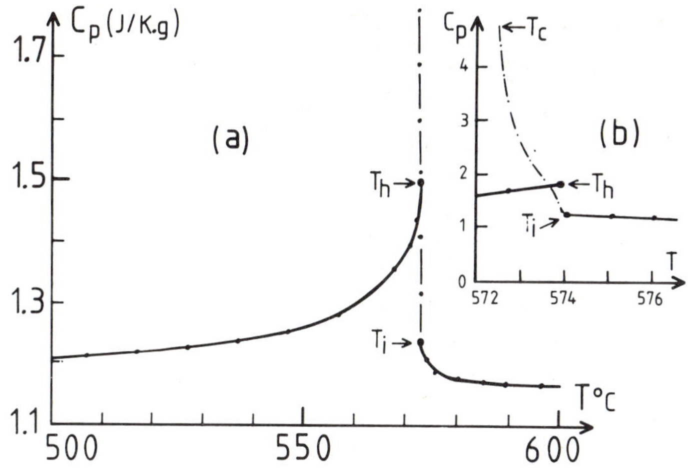

Fig. 2. Plot of the temperature variation of the heat capacity C_p of quartz: (a) from 500 to 600°C; (b) in the transition region. The dot-dashed curve corresponds to the INC phase, while full curves correspond to α and β phases.

hysteresis cycle: upon heating in the α phase, the $U_{xx}(T)$ curve is roughly linear while the upper curve upon cooling presents a strong downward curvature.

Finally with the further magnification of figure 1c, a slope discontinuity is observed at T_i, indicating the presence of a second-order transition and the occurrence of a new phase between T_i and T_c (Bachheimer, 1980). These results are confirmed by the measurements of the heat capacity C_p plotted in fig. 2 (Dolino et al. 1983a, Dolino and Bachheimer 1984), the variation of which is quite similar to that of the thermal expansion coefficient $\alpha_x = \partial U_{xx}(T)/\partial T$. In the α phase the increase of C_p is related to the order parameter variation, while in the β phase there is a small but distinct premonitory increase of C_p caused by the effect of fluctuations. Between T_i and T_c there is a large increase of C_p shown more clearly in figure 2b. (The quantitative relations between α_x and C_p will be developed in section 4). We only mention here that the anomaly of C_p integrated between T_i and T_c corresponds to $\frac{1}{4}$th of the transition enthalpy ΔH upon heating.

One must be cautious that the temperatures T_h, T_c and T_i have different thermodynamical characters: T_h and T_c are nucleation temperatures which are dependent upon experimental conditions (crystal shape and size, thermal conditions...). On the other hand the transition at T_i appears to be a second-order transition, so that T_i is determined by thermodynamical conditions. (In the following, temperatures are given relative to $T_i \simeq 847.2\,\mathrm{K}$.) In general T_h is a few tenths of a degree below T_i, so that the INC phase can also be observed upon heating, but experiments are easiest upon cooling. The variations of physical properties between T_i and T_c are nearly reversible (upon cooling or upon heating) as long as the nucleation of the α phase is avoided.

The effect of the INC phase on other physical properties has also been observed: elastic moduli S_{xx} and S_{zz} have variations similar to that of C_p (Bachheimer 1980) while the birefringence is proportional to U_{xx} (Berge et al. 1984a). [The observation of an intermediate phase has been claimed by Steinwehr (1938) from a birefringence measurement. But in fact he observed only steps in a state corresponding to the dashed part of figure 1b and not to the anomaly at T_i. These observations may come from temperature gradients.]

If the macroscopic measurements described above, show clearly the existence of an intermediate phase in a small temperature range of about 1.3 K inbetween the classical α and β phases, evidence for a modulated structure must be searched in neutron or X-ray scattering experiments.

3.2. *Experimental evidence of an incommensurate modulation*

The incommensurate nature of the intermediate phase of quartz has been verified by the observations of satellite peaks in scattering experiments with neutrons (Dolino et al. 1982, 1983a, 1984a) and with X-rays (Gouhara et al. 1983b) presented below.

3.2.1. *Scattering pattern in the incommensurate phase*

In the INC phase, more or less intense satellites are observed around most of the Bragg peaks. For example, fig. 3 (from Dolino et al. 1984a) shows a map of the scattered neutron intensities in the (001) plane around the (022) Bragg peak. (Coordinates are given relative to the hexagonal reciprocal axis system where OX is along [110], OY along [1$\bar{1}$0] and OZ along [001]). In figure 3, 6 satellites are observed along $\langle 100 \rangle$ directions at $\pm q_0 = 0.030a^*$ above a broad background of inelastic scattering. The presence of 6 satellites is a consequence of the hexagonal symmetry of the β phase. However as the scattering vector $\boldsymbol{Q}$ is not along the Z axis, one does not expect the scattering pattern to present an order 6 symmetry and indeed often the various satellites have different intensities: two extreme cases have been observed in the (001) plane:

For the Bragg peaks, along the $\langle 100 \rangle$ axis, only 4 satellites are observed, due to the absence of the two satellites along this axis. This can be related to the fact that atomic displacements U_i are expected to have a Σ_2 symmetry like the order parameter of the α–β transition (Dolino et al. 1984a).

On the other hand, around the (220) Bragg peak only 2 satellites are observed, at $(2+q_0, 2-q_0, 0)$ and at $(2-q_0, 2+q_0, 0)$ (Dolino et al. 1984c). This must come from a strongly anisotropic structure factor, probably related to the existence of the diffuse streaks discovered in the X-ray work of Arnold (1965) along the "4th hexagon" (from (220) to (400)).

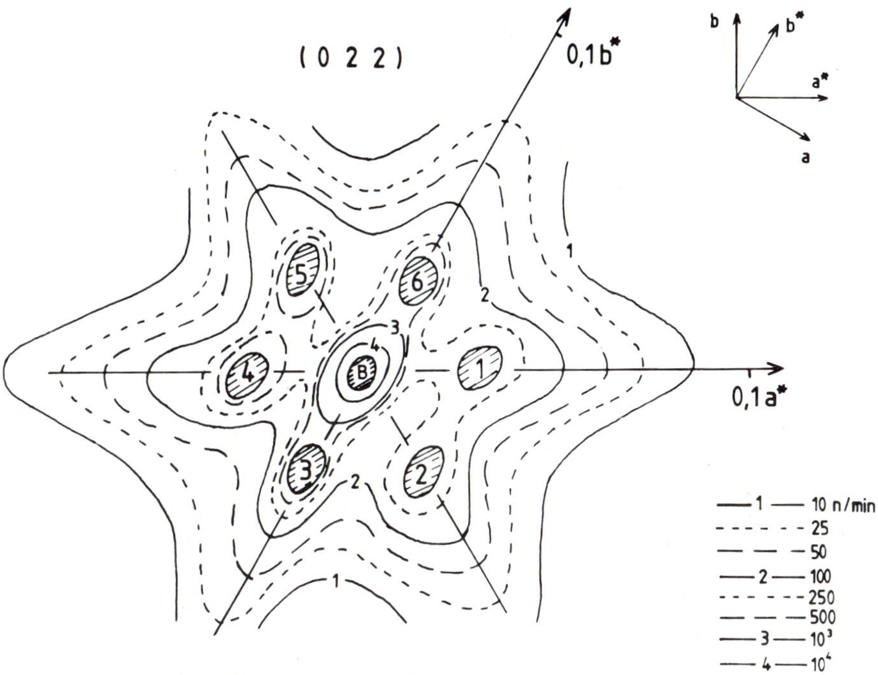

Fig. 3. Map of the neutron scattered intensity in a (001) plane around the (022) peak, showing six satellites around the Bragg peak. (The intensities of satellite peaks vary between 1900 and 2140 n/min while the intensity of the Bragg peak reaches 130000 n/min.) (From Dolino et al. 1984a.)

3.2.2. Temperature variations of the scattering pattern

We will now follow the temperature variations of the neutron scattering pattern around the (022) peak, obtained in the last experiment of Dolino et al. (1984c) and presented in fig. 4.

In the β phase a broad inelastic background is observed (curve 1 of fig. 4a) for a $(\xi, 2, 2)$ scan measured with an analyzer positioned at zero energy transfer with a resolution of 0.2 THz at FWHM (full width at half maximum). Upon cooling small humps appear above T_i (curve 2 of fig. 4a at $T_i + 2$ K). The curve 3 of figure 4a was measured just at T_i (within the experimental error of ± 0.1 K). These humps are observed along the 3 equivalent $\langle 100 \rangle$ directions. On the other hand for a $(0, 2, 2+\xi)$ scan a monotonously decreasing background is observed without much temperature variations.

Below T_i (in the INC phase), narrow satellites grow and shift towards the central Bragg peak as observed in fig. 4b and 4c at $T_i - 0.4$ K and $T_i - 1.1$ K respectively. Below $T_i - 0.8$ K, second order satellites are observed.

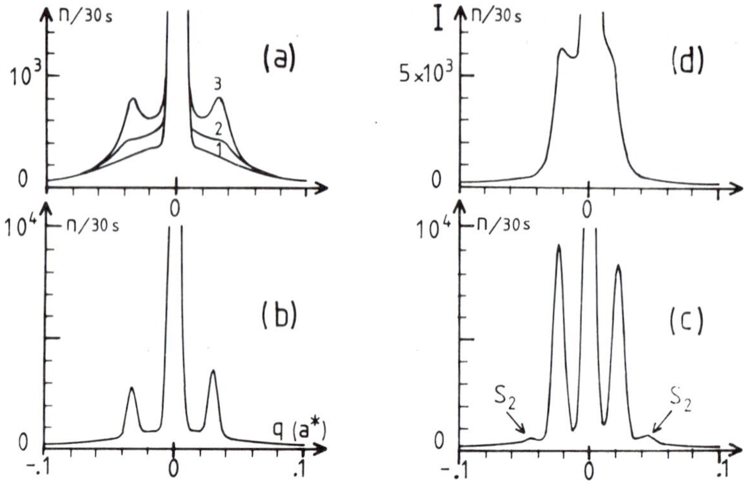

Fig. 4. Typical shape of satellite peaks for a $(\xi, 2, 2)$ scan at different temperatures: (a) curve 1 at $T_i + 7\,K$, curve 2 at $T_i + 2\,K$, curve 3 at T_i (within $\pm 0.1\,K$); (b) $T_i - 0.4\,K$; (c) $T_i - 1.1\,K$; (d) $T_i - 1.4\,K$ (just after nucleation of the α phase). (The q scale is in units of the reciprocal lattice).

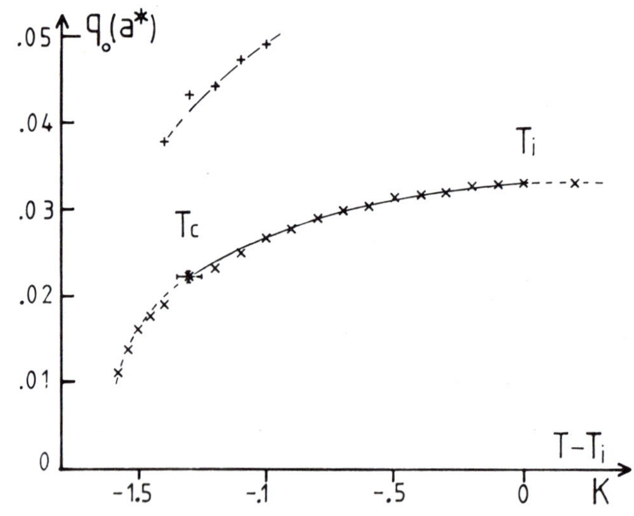

Fig. 5. Temperature variations of the distance between the Bragg peak and 1st (×) and 2nd order (+) satellites. The point above T_i corresponds to an inelastic hump. The narrow satellites of the INC phase are observed between T_i and T_c. Below T_c, broader satellites are observed after nucleation of the α phase.

After the nucleation of the α phase, which occurs at $T_i - 1.3\,K$, satellites shift closer to the Bragg peak and broaden, producing the ill-resolved structure of fig. 4d. The duration of this coexistence state depends on experimental conditions. After the full completion of the transition to the α phase, only the Bragg peak is observed.

The variations of the positions of first- and second-order satellite peaks, are plotted as a function of temperature in fig. 5. q_0 is observed to decrease from $0.034a^*$ at T_i to $0.022a^*$ at T_c. [This last value is lower than that previously obtained by Dolino et al. (1984a, b) due to improvements in temperature conditions.] The satellites of the INC phase have also been observed with X-rays (Gouhara et al. 1983b) and γ-rays (Bastie and Dolino 1985). Very recently Gouhara and Kato (1984) have observed a new feature: a splitting of the satellites produced by a rotation of $\pm\phi$ of q_0 from $\langle 100 \rangle$ directions, with ϕ increasing from 1° near T_i to about 7° around T_c. These rotations have been, predicted in the theoretical work of Aslanyan et al. (1983) which will be presented in section 4.

3.2.3. *Discussion*

The results of these scattering experiments clearly show the existence of the INC phase in a small temperature range between the α and the β phases of quartz.

The properties of the INC phase of quartz are somewhat similar to those of $NaNO_2$ (Durand et al. 1985) with an INC phase sandwiched in a small temperature range of 1.5 K between a paraelectric and a ferroelectric phase. In both cases, satellite peaks appear near the zone center (with $q_0 = 0.121a^*$ at T_i for $NaNO_2$) and get rapidly closer to the Bragg peak with decreasing temperature. But they are also significant differences: in quartz second-order satellites have been observed, as shown in fig. 5, and a global hysteresis in the variation of $q_0(T)$ has been recently found in the γ-ray scattering experiment of Bastie and Dolino (1985), while these features have not been detected in $NaNO_2$.

Another distinctive feature of quartz comes from its hexagonal symmetry. Such a symmetry which has not been found in other incommensurate dielectrics, occurs in charge density wave phenomena, for example in 2H-$TaSe_2$ or with the adsorbed phases on graphite. In these cases one can find either a single modulation wave ($1-q$ state) or a superposition of 3 modulation waves ($3-q$ state) (Bak et al. 1979). In a diffraction experiment, it is difficult to discriminate between these two possibilities as one expects that $1-q$ states with different orientations will occur in different regions of the sample producing a mean hexagonal diffraction pattern. However some information about this problem was obtained in quartz by direct electron microscopy observations, presented in the next part dealing with the lock-in transition from the INC to the α phase.

3.3. *Lock-in transition to the α phase*

The thermodynamical data presented in section 2, have clearly shown that this lock-in transition, occurring at the center of the Brillouin zone, is a first-order transition, which must proceed by nucleation and growth, with appearance of an interface separating the transformed part of the crystal from the untransformed matrix. Often in quartz, the nucleation of the new phase starts at a single point of the sample boundary, so that a single interface sweeps the whole sample. If the existence of an interface is a necessary consequence of the first-order nature of the transition, the presence of temperature gradients often stabilizes the coexistence of 2 (or even of 3) phases in a broader temperature range. This trend can be followed among different experiments where greater difficulties of temperature control lead to increasing temperature gradient (from 0.1 K/cm with neutrons, and X-rays to 0.1 K/μm with electron microscopy).

Since its first observation by Zubov (1956) by optical microscopy, the interface between the α and the INC phase (previously considered as the $\alpha-\beta$ interface) has been observed in numerous experiments. The most detailed observations, displaying at the same time the structure of the INC phase, are those by electron microscopy presented below. Other experiments, presented in section 3.4, have revealed the existence of larger size structures in the interface region which are at the origin of the light scattering discovered by Yakolev et al. (1956).

3.3.1. *Electron microscopy observations*

This technique gives a direct image of the modulated structure but only in a thin sample and with a large-temperature gradient of 0.1 to 1 K/μm, due to heating by the electron beam, so that the INC phase is always observed in the presence of at least the α phase. The first observations were made by Van Tendeloo et al. (1975, 1976) and Malov and Sonyushkin (1975) but they thought that they were observing the structure of the $\alpha-\beta$ interface. More recent observations of Van Tendeloo and coworkers have revealed a lot of new details (Dolino et al. 1984b). In fig. 6, which is a picture of the INC phase coexisting with the α phase, one observes:

The triangular pattern typical of the superposition of 3 modulation waves (3q-state).

An increase of the modulation period, in going from the higher temperature region (below) to the lower temperature region (upper left) where the α phase is present, with small Dauphiné twins. So one passes progressively from the INC to the α phase.

Near the α phase, two orientational directions tilted by about $\pm 10°$ from the $\langle 100 \rangle$ axis, probably related to the splitting of satellite peaks observed by X-rays reported in section 3.2.2.

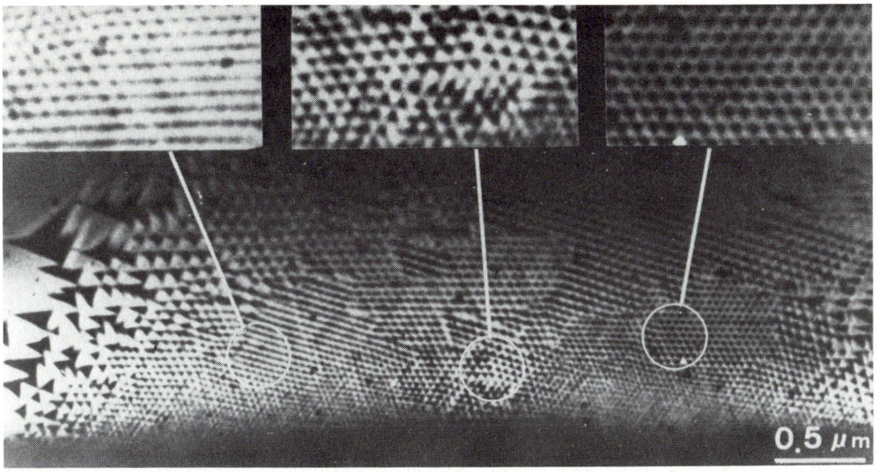

Fig. 6. Dark field micrograph of a thin quartz crystal during the transition: the INC phase covers much of the picture while the α phase can be observed in the upper left of the picture. The presence of rotated domains is clearly noted.

The insets are enlarged parts illustrating: on the right, a perfect hexagonal close packing, in the middle, a rotation grain boundary with a change in the direction of the black triangle, on the left, the presence of a dislocation in the modulated structure pattern. (Courtesy: G. Van Tendeloo, J. Van Landuyt and S. Amelinckx.)

The rightmost insert presents a region with a perfect hexagonal structure, while defects are observed in the two others. The central insert shows that the change of orientational direction occurs with a modification of the triangular structure: on the left the black triangles point downward, while on the right they point upward. Finally, the leftmost insert shows a dislocation in the periodic modulation pattern [similar observations of dislocations in discommensuration patterns have been made by electron microscopy in the orthorombic incommensurate phase of $2HTaSe_2$ (Fung et al. 1981, Chen et al. 1982). It appears quite plausible that the change of the modulation period of incommensurate structures occurs systematically via a dislocation mechanism (Janovec 1983), which can also be at the origin of the global hysteresis effects observed in so many incommensurate phases. [Hamano (1986), ch. 9, part III, this volume.]

These electron microscopy observations are in good agreement with a neutron scattering study performed during the coexistence of the INC and of the α phases. It was found that after the nucleation of the α phase, q_0 decreases further down to $0.012a^*$ and probably to lower values (Dolino et al. 1984b). Although the electron microscopy results are in agreement with those of neutron scattering, one must be cautious that some perturbations can be produced by the particular conditions of the experiment (very thin samples, strong temperature gradient, presence of elastic stresses and of electric fields...).

3.3.2. Observations on bulk samples

Observations on bulk samples have been made by X-ray topography or by optical techniques. Although the spatial resolution of these methods is far below that of electron microscopy a lot of information is obtained on bulk samples with good temperature conditions. Following the pioneering X-ray topography study of Inoue et al. (1974) recent real time observations have been made either with powerful X-ray sources (Gouhara et al. 1983a, b) or with synchrotron radiation (Zarka 1983), revealing elastic strains occurring during the transition as well as the behaviour of Dauphiné twins. These features are not easily observed by optical techniques, which are however quite sensitive to the heterogenous structures appearing at the α-INC boundary first revealed by the light scattering experiments of Yakolev et al. (1956). More recent studies have shown that these structures are quite complex, as two contiguous regions with different properties have been observed (Dolino and Bachheimer 1977, Shustin et al. 1978, 1981, Dolino 1980):

On the α phase side, there is a thin region with the right-angle light scattering discovered by Yakolev et al. (1956). This light scattering is quite anisotropic as it occurs only in directions where the scattering vector $\boldsymbol{\Delta Q}$ is perpendicular to the Z axis. The scattering objects are columns elongated along OZ, with cross-sections in the μm range.

On the incommensurate phase side, there is a broader region producing an intense small angle scattering. Here also the scattering objects are columns parallel to OZ but of larger cross-sections, in the tens of μm range so that they can be directly observed by optical microscopy.

It is quite possible that the smaller scattering objects correspond to the microdomains of the Dauphiné twin kind, observed in the electron microscope. A tentative explanation of the larger objects of the second region would be to identify them with the orientation domains of the INC phase present in fig. 6. These experiments with different spatial resolution show that the lock-in transition of quartz occurs in a complex way through inhomogeneous interfacial regions. It would be interesting to know whether these phenomena are particular to quartz or whether they can be found in other materials with incommensurate phases.

3.3.3. Conclusion

From these results it appears clearly that quartz presents a new incommensurate phase in a small temperature range of 1.3 K between the usual α and β phases.

A somewhat similar situation has recently been found in $AlPO_4$, a crystal isostructural to quartz with an $\alpha-\beta$ transition at 860 K (Ecolivet and Poignant

1981): in neutron scattering experiments Bachheimer et al. (1984) and Arnold et al. (1984) have observed satellite peaks with properties similar to those of quartz. The existence of this INC phase in $AlPO_4$ shows that quartz is not an isolated material, but that it is the prototype of a new family of incommensurate materials.

4. Theoretical studies

4.1. Origin of the INC phase of quartz

The main guide line to explain the origin of incommensurate phase in dielectrics is the existence of the Lifshitz invariant, which depends on crystal symmetry. This invariant can exist only if the order parameter is at least doubly degenerate at a high symmetry point of the Brillouin zone.

If the Lifshitz invariant does not exist the origin of the incommensurate phase is attributed to some subtle competition between different atomic interactions, leading to the existence of minima of the free energy at some arbitrary points of the Brillouin zone. However, Levanyuk and Sannikov (1976) have shown that the interaction of the order parameter with another variable introduces a Lifshitz-type invariant which can lead to the appearance of an incommensurate phase. Such effects can be expected quite generally near a structural phase transition induced by a soft mode at the zone center: as the frequency of the soft mode decreases the corresponding phonon branch must cross the acoustic branches and if they have the same symmetry they will interact: one may distinguish two cases depending on the type of the interaction:

If the interaction is allowed at the zone center, the limiting slope of the acoustic branch at $q = 0$ decreases and the transition is produced by an elastic instability as in the well known case of KH_2PO_4.

If the interaction is forbidden at $q = 0$, but allowed at $q \neq 0$, the coupling can be written as a gradient term proportional to q, a dip can appear in the low frequency branch at $q_0 \neq 0$ as shown in fig. 7. Then with decreasing temperature an instability can arise, first at $q = q_0$, leading to a transition to an incommensurate phase. In this case there is no change of the slope of the acoustic mode at $q = 0$.

The dip observed in the dispersion curves of $KTaO_3$ (Axe et al. 1970) and SbSI (Pouget et al. 1979) have been attributed to this kind of interactions, but in these materials, the coupling is too small to induce an incommensurate phase. As already stated in section 2, Aslanyan and Levanyuk (1978, 1979) have used this kind of argument to predict the existence of the INC phase of quartz. They made a development of the free energy F as a function of a

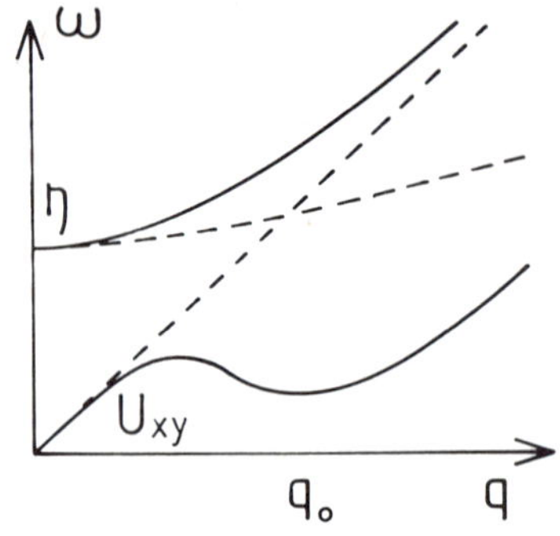

Fig. 7. Schematic representation of the coupling between the soft optical mode corresponding to the order parameter η and a transverse acoustic mode U_{xy}, due to a gradient interaction. (The dashed lines are the dispersion curves of the uncoupled mode, the full lines of the coupled mode.)

modulated order parameter $\eta(\boldsymbol{r})$, of its spatial derivatives, and of strains U_{ij} (to simplify the calculations we limit the spatial variations of η to the XY plane, in agreement with experimental results). The problem of the stability of the β phase can be discussed in retaining only quadratic terms in η and U_{ij}:

$$F=\frac{A}{2}(T-T_0)\eta^2+\frac{g}{2}(\nabla\eta)^2+\frac{h}{2}(\nabla^2\eta)^2+\tfrac{1}{2}C_{ijhk}U_{ij}U_{hk}$$
$$+\frac{a}{2}\left[(U_{xx}-U_{yy})\frac{\partial\eta}{\partial x}-2U_{xy}\frac{\partial\eta}{\partial y}\right]. \tag{4.1}$$

As usual in a Landau development the first term in η^2 is proportional to $T-T_0$. The second term is the usual gradient term favouring a homogeneous state, the higher-order spatial derivative of the third term is necessary to stabilize F at larger q values, the fourth term is the elastic energy, and the fifth term is the Lifshitz-type term, already given in eq. (2.1), which comes from a gradient interaction between the order parameter η and strains U_{ij}. Following Aslanyan et al. (1983) we make a Fourier transformation of eq. (4.1) expressing the strains U_{ij} as spatial derivatives of displacements u_{iq}. Furthermore we shift from cartesian to transverse and longitudinal coordinates relative to the wave vector $\boldsymbol{q}$ according to:

$$u_{\ell q}=u_{xq}\cos\phi+u_{yq}\sin\phi,\qquad u_{tq}=-u_{xq}\sin\phi+u_{yq}\cos\phi, \tag{4.2}$$

where ϕ is the angle between $\boldsymbol{q}$ and the X axis. With this choice of variables the elastic energy which is isotropic in the XY plane is diagonalized and the rotation symmetry of order 6 of the interaction term becomes conspicuous:

$$F=\sum_q \tfrac{1}{2}\left[\left(A(T-T_0)+gq^2+hq^4\right)\right]|\eta_q|^2$$
$$+\tfrac{1}{2}q^2\left[C_{11}|u_{\ell q}|^2+C_{66}|u_{tq}|^2+C_{44}|u_{zq}|^2\right]$$
$$+\tfrac{1}{2}aq^2\left[u_{\ell q}\cos 3\phi-u_{tq}\sin 3\phi\right]\eta_q. \tag{4.3}$$

To minimize F, we first eliminate the elastic variables using $\partial F/\partial u_{iq}=0$ so that:

$$F=\sum_q \frac{1}{2}\left[A(T-T_0)+\left[g-\frac{a^2}{4}\left(\frac{\cos^2 3\phi}{C_{11}}+\frac{\sin^2 3\phi}{C_{66}}\right)\right]q^2+hq^4\right]|\eta_q|^2. \tag{4.4}$$

One must note that the anisotropic terms in 3ϕ come from the bilinear gradient interaction with elastic strain. Without this coupling the soft mode is quite isotropic in the XY plane, as an angular dependence is introduced only when the development in q is pushed to the 6th power.

Since $C_{66}<C_{11}$, the minimum of F occurs for the transverse acoustic wave corresponding to $3\phi=(n+\frac{1}{2})\pi$. If $a^2>4gC_{66}$ there is a second-order transition at a temperature $T_i=T_0+(a^2/4C_{66}-g)^2/4Ah$ to a modulated phase characterized by a wave vector with modulus q_0

$$q_0=\left[\left(a^2/4C_{66}-g\right)/2h\right]^{1/2}.$$

We can transpose these results for vibration modes by using a quasiharmonic hamiltonian with an expression similar to that of the free energy from which one can obtain dispersion curves of the soft mode and of the acoustic mode similar to those of fig. 7 (Berge et al. 1984a).

4.2. *Structure of the INC phase*

As a consequence of the hexagonal symmetry of the β phase, the minima at $\pm\boldsymbol{q}_0$ discussed above occur simultaneously along the 3 equivalent $\langle 100\rangle$ directions: the star of $\boldsymbol{q}_0$ has 6 branches and the INC phase can present either a single modulation wave ($1-q$ state) or a superposition of several waves with wave vectors along different directions. Then the order parameter $\eta(\boldsymbol{r})$ can be written as a linear combination of sinusoidal waves of wavevector $\boldsymbol{q}_i$ and of amplitude $\rho_0 \exp(i\Psi_i)$. To discuss the structure of the INC phase one must take into account the effect of higher order terms in the development of the free energy:

There is a third-order term, expressed as

$$\eta^2\left(\frac{\partial^3\eta}{\partial x^3} - 3\frac{\partial^3\eta}{\partial x\partial y^2}\right)$$

in direct space, which produces interactions between 3 waves of wavevectors $\boldsymbol{q}_1, \boldsymbol{q}_2, \boldsymbol{q}_3$ satisfying $\boldsymbol{q}_1 + \boldsymbol{q}_2 + \boldsymbol{q}_3 = 0$, a situation allowed by the existence of a star of $\boldsymbol{q}_0$ with 6 branches.

There are several fourth-order terms, but we only keep the simplest one, η^4.

Furthermore we introduce a quadratic coupling between strains and η given by $\eta^2(r_1(U_{xx} + U_{yy}) + r_3 U_{zz})$ which is useful to explain the large thermal expansion of the INC phase. After Fourier transformation and elimination of the elastic variables we get the following expression for the development of the free energy:

$$F = \sum_{\boldsymbol{q}} \tfrac{1}{2}\left[A(T - T_0) + \left[g - \frac{a^2}{4}\left(\frac{\cos^2 3\phi}{C_{11}} + \frac{\sin^2 3\phi}{C_{66}}\right)\right]q^2 + hq^4\right]|\eta_q|^2$$

$$+ \frac{i}{3}\sum_{\boldsymbol{q}_{11}, \boldsymbol{q}_2, \boldsymbol{q}_3}\left[\frac{ar_1}{2C_{11}}((q_1\cos 3\phi_1 + q_2\cos 3\phi_2 + q_3\cos 3\phi_3)\right.$$

$$\left. + Gq_1q_2q_3\cos(\phi_1 + \phi_2 + \phi_3)\right]\eta_{q_1}\eta_{q_2}\eta_{q_3}\delta(\boldsymbol{q}_1 + \boldsymbol{q}_2 + \boldsymbol{q}_3)$$

$$+ \frac{B}{4}\sum_{\boldsymbol{q}_1, \boldsymbol{q}_2, \boldsymbol{q}_3, \boldsymbol{q}_4}\eta_{q_1}\eta_{q_2}\eta_{q_3}\eta_{q_4}\delta(\boldsymbol{q}_1 + \boldsymbol{q}_2 + \boldsymbol{q}_3 + \boldsymbol{q}_4). \tag{4.5}$$

One may note that the anisotropic terms in 3ϕ come from interaction with strains: for the η^2 terms, the bilinear gradient interaction leads to a q^2 dependence, while for the η^3 terms the quadratic coupling $U_{ii}\eta^2$ leads to a q dependence.

Due to the great number of coefficients, a complete minimization of the free energy given by eq. 4.5 has not been worked out and only the relative stabilities of a few simple cases have been compared:

$1q$ solution, $\eta(\boldsymbol{r}) = 2\rho\sin(\boldsymbol{q}\cdot\boldsymbol{r})$ with $\boldsymbol{q}$ along one of the $\langle 100\rangle$ directions. The third-order term is always nil, and the free energy reduces to $F = A_1\rho^2 + B_1\rho^4$, which depending on the sign of B_1 leads to a first or second-order transition.

Symmetrical $3-q$ state,

$$\eta(\boldsymbol{r})=2\rho_0\left[\exp i(\boldsymbol{q}_1\cdot\boldsymbol{r}+\Psi_1)+\exp i(\boldsymbol{q}_2\cdot\boldsymbol{r}+\Psi_2)+\exp i(\boldsymbol{q}_3\cdot\boldsymbol{r}+\Psi_3)\right]+\text{c.c.}$$

with $\boldsymbol{q}_1+\boldsymbol{q}_2+\boldsymbol{q}_3=0$ and $|\boldsymbol{q}_1|=|\boldsymbol{q}_2|=|\boldsymbol{q}_3|=q_0$.

The contribution of the third-order term, which has an imaginary coefficient, is maximized when $\Psi_1+\Psi_2+\Psi_3=(2n+1)\pi/2$, and favours the $3-q$ state relative to the $1-q$ state. As the coefficients of the quadratic and cubic terms in η have different angular dependence in ϕ, the minimum of F is obtained for a value $\phi_0\neq 0$, solution of $\partial F/\partial\phi=0$. One finds that $\cos 3\phi_0\sim\pm\rho_0/q_0$ so that the modulation waves are tilted away from the crystallographic $\langle 100\rangle$ axis. Such rotations have indeed been observed in X-ray experiments (Gouhara and Kato 1984) and in electron microscopy as shown in fig. 6. The morphology of the rotated domains has been studied in more details in the theoretical work of Walker (1983).

After eliminating $\cos 3\phi_0$ in eq. (4.5) one finds again a free energy given by $F=A_3\rho^2+B_3\rho^4$. As shown by Aslanyan et al. (1983), depending on the sign of the new coefficient B_3, the transition can be first or second order. One may note that the presence of a third-order term in the free energy does not necessarily lead to a first-order transition, when the coefficient of this term is anisotropic and vanishes for the directions along which the instability occurs. In conclusion, the most spectacular result of this theoretical analysis is the rotation of the symmetrical $3-q$ state away from the crystallographic directions.

4.3. *Macroscopic symmetry of the INC phase*

The general problem of the symmetry of incommensurate phase has been discussed by Dvořák et al. (1983): when the star of $\boldsymbol{q}_0$ has 2 branches the macroscopic symmetry of the incommensurate phase is the same as that of the high temperature phase, but when the star of $\boldsymbol{q}_0$ has more than 2 branches the symmetries of the 2 phases can be different. We will present the application of these ideas to quartz, made by Vallade (1984), following the simplified presentation of Berge (1984): one looks for invariants coupling macroscopic physical properties to powers of the order parameter and of its spatial derivatives:

For the $1-q$ state, an orthorombic distortion is obtained from a coupling term $(U_{xx}-U_{yy})q^2\rho^2$

For $3-q$ states, there are more possibilities: we consider only the superposition of 3 sinusoidal waves, with wave vectors of equal moduli, as in section 4.2, but with different amplitudes ρ_i.

Then one finds terms coupling η to strains U_{ij}, to a polarization P_z and to the homogeneous order parameter η_0 of the α phase, given respectively by:

$$q^2(2\rho_1^2-\rho_2^2-\rho_3^2)(U_{x'x'}-U_{y'y'})-2\sqrt{3}\,(\rho_2^2-\rho_3^2)U_{x'y'}, \tag{4.6}$$

$$q^3P_z\rho_1\rho_2\rho_3\sin\Psi\sin 3\phi, \tag{4.7}$$

$$\eta_0\rho_1\rho_2\rho_3\cos\Psi, \tag{4.8}$$

where $\Psi=\Psi_1+\Psi_2+\Psi_3$ is the relative phase of the three waves. [In eq. (4.6) the axes X' and Y' are rotated by an angle ϕ relative to X and Y]. In the symmetrical $3-q$ state where $\sin\Psi=\pm1$, eq. (4.7) predicts the appearance of an electric polarization along the Z axis. Corresponding to the two signs of $\sin\Psi$, there is a ferroelectric domain structure, as sketched in fig. 8a. One may note the close analogy of fig. 8a with the central insert of the electron microscopy picture of fig. 6. By changing the relative phases of the 3 waves, other structures are found; in particular, fig. 8b shows the structures corre-

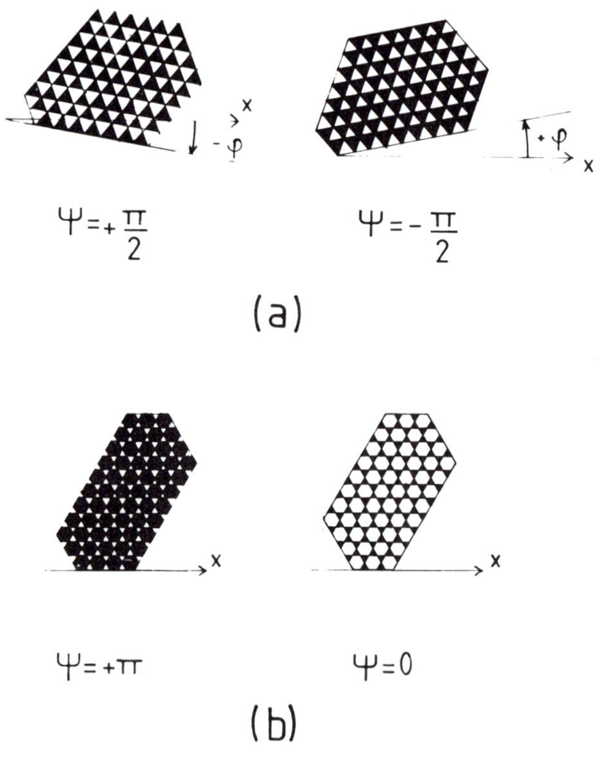

Fig. 8. Schematic representation of incommensurate $3-q$ states due to the superposition of three modulation waves with wavevectors forming an equilateral triangle. The black and white parts correspond to positive and negative displacements; Ψ is the relative phase between the 3 waves. (a) Stable rotated structure; (b) unstable non-rotated structure.

sponding to $\cos\Psi = \pm 1$. These structures present the order parameter η_0 of the α phase, also with a domain structure, now of the Dauphiné twin kind, corresponding to the two opposite signs of $\cos\Psi$. However these states do not produce minima of the free energy, but the stability of different states can be changed by the application of external fields: for example a uniaxial stress along the [100] axis favours a $1-q$ state relative to a $3-q$ state. Indeed, in a neutron experiment it was found that the application of a stress σ_{yy} of 60 MPa along [100] favours the development of a single modulation wave along this direction (Dolino et al. 1984c). In a similar way the apparition of a $3-q$ state with $\cos\Psi = \pm 1$ is expected from a nonlinear stress coupling of η_0 to $\sigma_{yz}(\sigma_{xx} - \sigma_{yy}) + 2\sigma_{xz}\sigma_{xy}$.

4.4. *Comparison with experimental results*

Even the Landau theory of the INC phase of quartz is quite complex, due to the great number of coefficients. Nevertheless we will discuss some points of the theoretical analysis in relation to experimental work:

The first question is the relevance of the mode coupling model proposed by Aslanyan and Levanyuk (1978, 1979). As will be discussed in section 5, the phonon dispersion curves of quartz around the Brillouin zone center are not known with enough accuracy to bring a definitive answer. However, mode coupling theory is probably the simplest explanation for the existence of the anisotropic instabilities of the phonon spectrum observed by neutron scattering at small q values.

Another problem is the structure of the INC phase: is it a $1-q$ or a $3-q$ state. The main evidence for a $3-q$ state comes from the direct electron microscopy pictures (fig. 6). A further confirmation comes from the rotation away of the $\langle 100\rangle$ direction predicted by the theory for the $3-q$ state and observed in the X-ray experiment of Gouhara and Kato (1984). The prediction that this state must present a ferroelectric polarization has not been tested yet.

Despite its predictive power, a phenomenological theory cannot determine the atomic displacements between the different phases. The experimental determination by Axe and Shirane (1970b) using neutron scattering, as well as the rigid tetrahedron model developed by Grimm and Dorner (1975) indicates that atomic displacements between the α and the β phases correspond mainly to the rotation of SiO_4 tetrahedra. It is quite plausible that the INC phase is produced by a modulated wave of SiO_4 rotations. The growth of higher-order harmonics near T_c indicates that these modulation waves become more rectangular. If one attributes the thermal expansion variations to a coupling quadratic in the order parameter η, one finds that near T_c, η reaches about 3° (which is about half the value of η in the α phase just below the transition temperature).

Finally, we will discuss the thermodynamical behaviour of the INC phase: the phenomenological theory is quite flexible about the order of the β-INC transition at T_i, which, depending on coefficient values, can be of first or second order, although experimentally the transition appears rather continuous. The nature of the lock-in transition to the α phase at T_c has not been investigated with the free energy of eq. (4.5) so that we can only refer to some other studies: for incommensurate phase, with dispersion modes even in q, Bruce et al. (1978) have shown by numerical calculations that one expects, at least for the $1-q$-state, a small range of existence of the incommensurate phase, little development of higher-order harmonics, and a first-order transition to the commensurate phase. Another reason for a first-order transition comes from the existence of a $3-q$ state which, following the argument of Bak et al. (1979), systematically leads to a first-order lock-in transition. Experimentally the INC-α transition is clearly first order, but nevertheless the variation of several properties accelerates near T_c, as shown for the curve $q_0(T)$ in fig. 5. Such a downward curvature has been also found in $NaNO_2$ and was interpreted with a coupling term $(\eta \nabla \eta)^2$ by Durand et al. (1983). However this treatment does not lead to a divergence of the heat capacity C_p near T_c, when only fundamental waves are considered. For the present time, these large variations of C_p, of the thermal expansion coefficient α_i, and of the elastic moduli S_{ii}, have been correlated by using Pippard relations, which for anisotropic materials can be written as (Garland, 1964)

$$\alpha_i = (\partial T_t / \partial \sigma_{ii}) C_p / vT + \alpha_i^0, \tag{4.9}$$

$$S_{ij} = (\partial T_t / \partial \sigma_{ii}) \alpha_j + S_{ij}^0, \tag{4.10}$$

where $\partial T_t / \partial \sigma_{ii}$ is the variation of the transition temperature T_t under the application of a stress σ_{ii}, v is the specific volume and α_i^0 and S_{ij}^0 are constants. These relations are well verified in the α and in the INC phase, while the test is more difficult in the β phase were the variations are smaller (Dolino and Bachheimer 1984).

5. *Dynamical properties*

One of the most interesting features of incommensurate phases is the existence of specific excitations (amplitudon and phason) which can be studied by various techniques (neutron or light scattering...). Furthermore in quartz a knowledge of phonon dispersion curves presents a special interest in order to test the gradient mode coupling model.

5.1. *Soft mode behavior in the α and the β phases*

One may recall that the soft mode of quartz has a long history: the frequency decrease of the $207\,\text{cm}^{-1}$ mode was already reported by Raman and Nedungadi

(1940) a long time before the introduction of the soft mode concept around 1960. More accurate data have been obtained by Shapiro et al. (1967), also by Raman scattering. Following the interpretation of these data by Scott (1968) one gets a soft mode with a mean field behaviour.

In the β phase the soft mode cannot be observed by Raman scattering, but in an inelastic neutron scattering experiment, Axe and Shirane (1970) found an overdamped soft mode around the zone center; its frequency ω_0 also followed the mean field behaviour given by $\omega_0^2 \sim T - T_0$. Furthermore they found a low frequency branch along $\langle 100 \rangle$ directions related to this soft mode. This branch which is at the origin of the diffuse streaks observed with X-Ray by Arnold (1965), has been investigated in more details by Bauer et al. (1971) and by Boysen et al. (1980).

Although their data (obtained only at $q_0 \simeq 0.05a^*$ and 13 K above T_i) are too scarce to conclude definitively, this low frequency branch corresponds probably to the dispersion curves sketched in fig. 7. [This is confirmed by a recent inelastic neutron investigation (Berge et al. 1985)].

5.2. *Brillouin experiment*

Although a Brillouin experiment gives information only on the behaviour of acoustic modes at small wavevector, the good energy resolution can produce some evidence of the coupling to the soft mode.

Indeed, in a recent experiment by Berge et al. (1984a), such effects have been found in measurements of transverse acoustic modes polarized in the XY plane and propagating either along the X or along the Y axis: between these two directions, there is a small frequency shift (at the limit of experimental uncertainty) as shown in fig. 9a, while a large damping Γ is unambiguously observed for the wave propagating along OY, in agreement with calculations of the mode coupling model. However larger effects are obtained for the longitudinal modes (along X, Y or Z axis) not related to the bilinear coupling $U_{ij}\partial\eta/\partial x_i$ but to the quadratic coupling $\eta^2(r_1(U_{xx}+U_{yy})+r_3U_{zz})$. These variations, shown in fig. 9b, were found previously by Shapiro and Cummins (1968) but their relation with the INC phase has been discussed only in the more recent work of Berge et al. (1984a). A large difference was found between these high frequency measurements (at 30 GHz) and those at low frequency (0 Hz) of Bachheimer (1980) pointing to important dispersion effects.

5.3. *Raman experiment*

A new investigation of the Raman spectrum of quartz has recently been performed. In addition to the soft mode present in the α phase and well studied by Shapiro et al. (1967) new low-frequency excitations were discovered in the three phases near the transition temperatures, as shown in fig. 10 (from

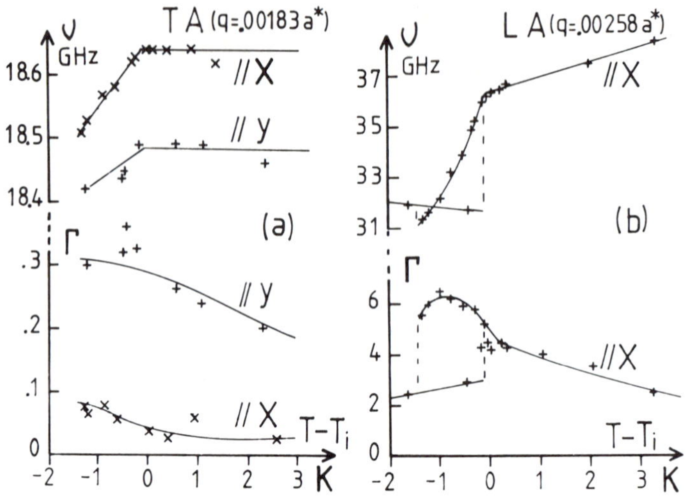

Fig. 9. Temperature variations of the frequency (ν) and damping (Γ) of acoustic modes of quartz in the transition region: (a) Transverse acoustic (TA) mode U_{xy}: a difference is observed between the mode propagating along the X axis ($\times$) or along the Y axis ($+$). (b) Longitudinal modes (LA): identical results are obtained for modes propagating along the X axis (shown here) and along the Y axis. (These spectra were obtained with an argon laser of wavelength 5145 Å at right angle scattering for TA modes and in backscattering for LA modes.)

the work of Berge et al. 1984b): a dynamic central peak is observed in a range of a few tens of K below and above T_i, with a narrow dip at about 50 GHz, resulting from an interference between this central peak and the longitudinal acoustic mode. In the INC phase narrow lines appear around 180 and 300 GHz. Their intensities increase while their frequencies decrease in going from T_i to T_c, so that they cannot be interpreted as amplitudons, the frequency of which is expected to vary in the opposite sense. Furthermore, from the large damping of the soft mode in the α and β phases, one expects that the amplitudon as well as the phason would be overdamped in the INC phase. A tentative explanation of these narrow lines is to attribute them to the acoustic modes made visible at q_0 by the multiplication of the Brillouin zone in the INC phase. In a similar way some higher-frequency modes (such as the 355 cm^{-1} line) allowed in the α phase and forbidden in the β phase, remain visible in the incommensurate phase. Finally an increase of the small Rayleigh line is observed in the INC phase and its maximum is found around T_i. Further investigations of dynamical properties are needed, from different techniques, to get a complete understanding of the phase transitions of quartz, but at the present time all observations appear to be coherent with the gradient mode coupling model.

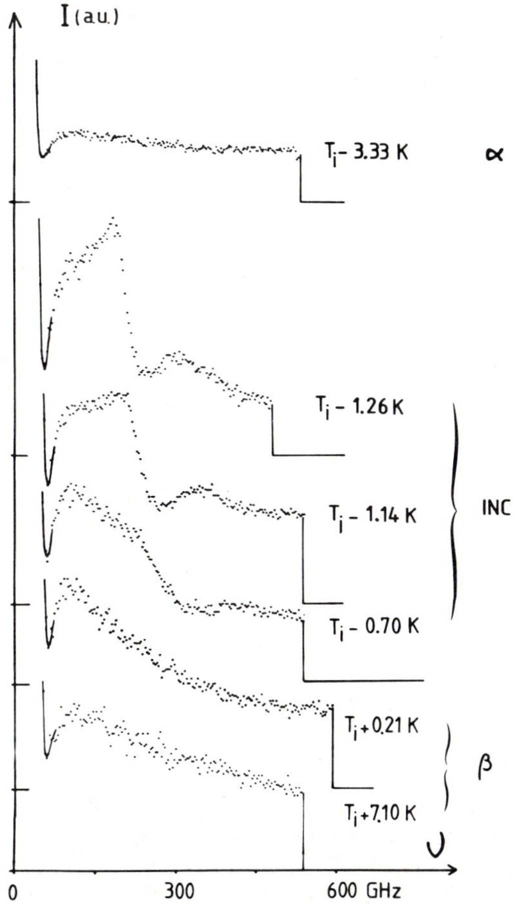

Fig. 10. Low frequency Raman scattering spectra of quartz in the transition region (for clarity the curves have been translated along the intensity axis). The full line on the left is the bottom of the longitudinal acoustic phonon, the dip is produced by an interference with a broad central peak observed in the α and in the β phases. The new excitations appearing in the INC phase may be attributed to acoustic modes made visible at the modulation wavevector position. (These spectra were obtained in the $x(zz)y$ configuration.)

6. *Conclusion*

It is now clear that quartz presents an intermediate incommensurate phase inbetween the classical α and β phases. The origin of this phase is attributed to a gradient interaction between the soft optical mode and a transverse acoustic mode. A phenomenological theory based on this coupling predicts that the most stable modulated structure is a $3-q$ state produced by the superposition of three waves with wavevectors rotated away from $\langle 100 \rangle$ directions.

However, some points are not well understood, such as the large variations of physical properties near T_c, or the heterogeneous structures, at the origin of the intense light scattering which appears during the INC$-\alpha$ transition. Further work is needed on these points as well as on the dynamical properties to test the validity of the gradient mode coupling model. The small value of the wavevector q_0, varying from 0.034 to $0.024a^*$, shows that a development as a function of q_0 can give quite valuable results.

Although the range of existence of the INC phase is rather small, quartz presents a great quality: the existence of large and perfect crystals permitting to perform many different experiments, with reproducible results. Furthermore, intrinsic properties are probably measured as crystals with very few impurities can be obtained. So one can hope that further studies will elucidate the properties of this INC phase, and that quartz will be a good example of structural transitions where the order parameter variation from the ordered to the disordered state occurs via an intermediate step presenting an incommensurate modulation.

Acknowledgments

Many results presented in this paper were obtained in collaboration with physicists of the Laboratoire de Spectrométrie Physique, Bachheimer, Bastie, Berge, Lajzerowicz and Vallade. We acknowledge the good cooperation of scientists from other laboratories, Zeyen from ILL and Martinez from SNCI, both at Grenoble, Boissier and Vacher from Montpellier University, Van Tendeloo, Van Landuyt and Amelinckx from Belgium and Aslanyan and Levanyuk from USSR. We thank prof. Kato for communicating his X-ray results prior to publication.

References

Aizu, K., 1973, J. Phys. Soc. Jpn **34**, 120.

Arnold, H., 1965, Z. Kristallogr. **121**, S145.

Arnold, H., J. Bethke, G. Eckold, Th. Hahn and J. Min Hua, 1984, Acta Cryst. **A40**, Suppl., C, 139.

Aslanyan, T.A., and A.P. Levanyuk, 1978, Zh. Eksp. Teor. Fiz. **28**, 76.

Aslanyan, T.A., and A.P. Levanyuk, 1979, Solid State Commun. **31**, 547.

Aslanyan, T.A., A.P. Levanyuk, M. Vallade and J. Lajzérowicz, 1983, J. Phys. **C16**, 6705.

Axe, J.P., and G. Shirane, 1970, Phys. Rev. **B1**, 342.

Axe, J.P., J. Harada and G. Shirane, 1970, Phys. Rev. **B1**, 1227.

Bachheimer, J.P., 1980, J. Physique Lett. **41**, L559.

Bachheimer, J.P., and G. Dolino, 1975, Phys. Rev. **B11**, 3195.

Bachheimer, J.P., and G. Dolino, 1980, Ferroelectrics **25**, 423.

Bachheimer, J.P., B. Berge, G. Dolino, P. St.-Grégoire and C.M.E. Zeyen, 1984, Solid State Commun. **51**, 55.

Bak, P., D. Mukamel, J. Villain and K. Wentowska, 1979, Phys. Rev. **B19**, 1610.

Bastie, P., and G. Dolino, 1985, Phys. Rev. **B31**, 2857.

Bauer, K.H.W., H. Jagodzinski, B. Dorner and H. Grimm, 1971, Phys. Stat. Sol. (b)**48**, 437.

Berge, B., 1984, thesis (Grenoble University) unpublished.

Berge, B., G. Dolino, M. Vallade, M. Boissier and R. Vacher, 1984a, J. Physique **45**, 715.

Berge, B., M. Vallade and G. Martinez, 1984b, J. Phys. **C17**, L167.

Berge, B., J.P. Bachheimer, G. Dolino, M. Vallade and C.M.E. Zeyen, 1985, Ferroelectrics, to be published.

Boysen, H., B. Dorner, F. Frey and H. Grimm, 1980, J. Phys. **C13**, 6127.

Bruce, A.D., R.A. Cowley and A.F. Murray, 1978, J. Phys. **C11**, 3591.

Chen, C.H., J.M. Gibson and R.M. Fleming, 1982, Phys. Rev. **B26**, 184.

Dolino, G., 1980, Phys. Stat. Sol. (a) **60**, 391.

Dolino, G., and J.P. Bachheimer, 1977, Phys. Stat. Sol. (a) **41**, 673.

Dolino, G., and J.P. Bachheimer, 1982, Ferroelectrics **43**, 77.

Dolino, G., and J.P. Bachheimer, 1984, in: Phase transformations in solids, ed. N. Tsakalatos (Mat. Res. Soc. Series, Elsevier, USA).

Dolino, G., J.P. Bachheimer and C.M.E. Zeyen, 1982, European condensed matter conference, Europhysics conference abstract, 6A, 257; ILL report, 1982, 121.

Dolino, G., J.P. Bachheimer and C.M.E. Zeyen, 1983a, Solid State Commun. **45**, 295.

Dolino, G., J.P. Bachheimer, F. Gervais and A.F. Wright, 1983b, Bull. Mineral. **106**, 267.

Dolino, G., J.P. Bachheimer, B. Berge and C.M.E. Zeyen, 1984a, J. Physique **45**, 361.

Dolino, G., J.P. Bachheimer, B. Berge, C.M.E. Zeyen, G. Van Tendeloo, J. Van Landuyt and S. Amelinckx, 1984b, J. Physique **45**, 901.

Dolino, G., J.P. Bachheimer, B. Berge, P. St.-Grégoire and C.M.E. Zeyen, 1984c, ILL report.

Durand, D., F. Dénoyer, D. Le Fur, R. Currat and L. Bernard, 1983, J. Physique. Lett. **44**, L207.

Durand, D., F. Dénoyer, R. Currat and M. Lambert, 1985, ch. 13, this volume.

Dvořák, V., V. Janovec and Y. Ishibashi, 1983, J. Phys. Soc. Jpn **52**, 2053.

Ecolivet, C., and H. Poignant, 1981, Phys. Stat. Sol. **63**, K107.

Fung, K.K., S. McKernan, J.W. Steeds and J.A. Wilson, 1981, J. Phys. **C14**, 5417.

Garland, C.W., 1964, J. Chem. Phys. **41**, 1005.

Ginzburg, V.L., 1955, Dokl. Akad. Nauk SSSR **105**, 240.

Ginzburg, V.L., and A.P. Levanyuk, 1958, J. Phys. Chem. Solids **6**, 51.

Gouhara, K., and N. Kato, 1984, J. Phys. Soc. Jpn **53**, 2177.

Gouhara, K., Y. Hao Li and N. Kato, 1983a, J. Phys. Soc. Jpn **52**, 3821.

Gouhara, K., Y. Hao Li and N. Kato, 1983b, J. Phys. Soc. Jpn, **52**, 3697.

Grimm, H., and B. Dorner, 1975, J. Phys. Chem. Solids **36**, 407.

Hamano, K., 1985, ch. 9, Part III, this volume.

Höchli, U.T., and J.F. Scott, 1971, Phys. Rev. Lett. **26**, 1627.

Inoue, N., A. Iida and K. Kohra, 1974, J. Phys. Soc. Jpn **37**, 742.

Janovec, V., 1983, Phys. Lett. **99A**, 384.

Le Chatelier, H., 1889, C. R. Acad. Sci. (Paris) **108**, 1046.

Levanyuk, A.P., and D.G. Sannikov, 1976, Fiz. Tverd. Tela **18**, 1927 [Sov. Phys. Solid State **18**, 1122].

Malov, Y.U., and V.E. Sonyushkin, 1975, Kristallogr. **20**, 1054 [Sov. Phys. Cryst. **20**, 644].

Megaw, H., 1973, Crystal structures: a working approach (Saunders, Philadelphia).

Pouget, J.P., S.M. Shapiro and K. Nassau, 1979, J. Phys. Chem. Solids **40**, 267.

Raman, C.V., and T.M.K. Nedungadi, 1940, Nature **145**, 147.
Scott, J.F., 1968, Phys. Rev. Lett. **21**, 907.
Scott, J.F., 1974, Rev. Mod. Phys. **46**, 83.
Shapiro, S.M., D.C. O'Shea and H.Z. Cummins, 1967, Phys. Rev. Lett. **19**, 361.
Shapiro, S.M., and H.Z. Cummins, 1968, Phys. Rev. Lett. **21**, 1578.
Shustin, O.A., J.G. Chernevich, S.A. Ivanov and I.A. Yakolev, 1978, Pis'ma Zh. Eksp. Teor. Fiz. **27**, 349 [Sov. Phys. JETP Lett. **27**, 330].
Shustin, O.A., J.G. Chernevich, S.A. Ivanov and I.A. Yakolev, 1981, Solid State Commun. **37**, 65.
Sosman, R.B., 1965, The Phases of Silica (Rutgers University, New Brunswick).
Steinwehr, E.H., 1938, Z. Krist. **99**, 292.
Vallade, M., 1984, Private communication.
Van Tendeloo, G., J. Van Landuyt and S. Amelinckx, 1975, Phys. State Sol. (a) **30**, K 11.
Van Tendeloo, G., J. Van Landuyt and S. Amelinckx, 1976, Phys. Stat. Sol. (a) **33**, 723.
Walker, M.B., 1983, Phys. Rev. **B28**, 6407.
Wright, A.F. and M.J. Lehmann, 1981, J. Solid State Chem. **36**, 371.
Yakolev, I.A., L.F. Mikheeva and T.S. Velichkina, 1956, Kristallografiya **1**, 123 [Sov. Phys. Cryst. **1**, 91].
Young, R.A., 1962, U.S. Air Force (Office of Scientific Research) Report n.2569.
Zarka, A., 1983, J. Appl. Cryst. **16**, 354.
Zeyen, C.M.E., G. Dolino and J.P. Bachheimer, 1983, Physica **120B**, 280.
Zubov, V.G., 1956, Kristallografiya **1**, 243 [Sov. Phys. Cryst. **1**, 188].

CHAPTER 17

Incommensurate Phase of Barium Sodium Niobate

J.C. TOLÉDANO, J. SCHNECK and G. ERRANDONÉA

Laboratoire de Bagneux, C.N.E.T.
196 rue de Paris, 92220 Bagneux
France

Incommensurate Phases in Dielectrics 2
Edited by
R. Blinc and A.P. Levanyuk

Contents

1. Introduction . . . 235
2. Characteristics of the incommensurate phase in BSN . . . 236
3. Macroscopic symmetry breaking and related anomalies . . . 242
4. Thermal hysteresis and relaxation effects . . . 244
5. Conclusions . . . 249
References . . . 250

1. Introduction

Among insulating incommensurate materials, barium sodium niobate, $Ba_2NaNb_5O_{15}$ (BSN), possesses two important peculiarities. On the one hand, it is the only confirmed example displaying a point symmetry change between the normal phase ($T > T_I$) and the incommensurate phase (INCP) ($T_L < T < T_I$) while such a change occurs below the lock-in transition in other known systems. As a consequence, the macroscopic anomalies induced by the INCP are expected to be of a different nature than those usually encountered. This characteristic of BSN is itself a consequence of an unusual, but not unique, circumstance: the four-dimensionality of the order parameter associated to the INCP. On the other hand, BSN shows a set of phenomena related to the presence of the INCP but departing from the standard behaviour expected for incommensurate systems. These phenomena, which include a large thermal hysteresis of a specific type, slow time-relaxation processes, and a "memory effect", are currently assigned to the occurrence of an interaction between the incommensurate modulation and mobile point defects. Similar phenomena have been pointed out in other substances, but in BSN their amplitude is larger by an order of magnitude.

BSN belongs to the family of mixed oxide compositions which crystallizes with the tetragonal tungsten bronze structure (Jamieson et al. 1969). An intricate pattern of structural transitions has been observed in it. A standard ferroelectric transition (P4/mbm P4bm) occurs at about 580°C (Singh et al. 1970). Two transformations separated by ≈ 30°C are located near 300°C and surround an incommensurate phase possessing an average orthorhombic mm2 point symmetry (Schneck and Dénoyer 1981). This point symmetry persists down to -160°C at which another transformation restores a tetragonal symmetry (Schneck et al. 1977). The latter transition has been noted to be very sensitive to a deviation from the stoichiometric composition (Schneck et al. 1980a).

The phase changes near 300°C have been the focus of most of the experimental and theoretical efforts in the past years. Three stages can be distinguished in the progressive clarification of their characteristics. In the first place, it was assumed that a single transition occurred in this range. Noting that this transition was accompanied by a softening of elastic constants,

Yamada et al. (1970) associated it with an elastic instability, the order parameter of the transition being identified to the shear strain e_{ab}. In this approach, the transition was implied to preserve the translational periodicity of the crystal.

This interpretation was subsequently shown by Tolédano (1975) to be incompatible with the temperature dependence of the elastic anisotropy ($C_{22} - C_{11}$) (Tolédano et al. 1976). The onset of the spontaneous strain e_{ab} was suggested to be a secondary effect of the transition resulting from a quadratic coupling of e_{ab} to another degree of freedom related to a breakdown of the translational periodicity along the *c*-direction. This inference was experimentally confirmed by Burgeat and Tolédano (1976) who established the occurrence of a period doubling along *c* below 300°C. Raman scattering measurements (Boudou and Sapriel 1980) consistently disclosed extra lines in the spectra. However a series of X-ray and neutron scattering experiments, set up to check the details of the preceding mechanism, eventually revealed a more complex situation where two transitions exist, separated by the range of stability of an incommensurate phase (Schneck and Dénoyer 1981, Schneck et al. 1982).

2. *Characteristics of the incommensurate phase in BSN*

As will be discussed in section 4, the precise range of stability of the INCP varies widely with the conditions of the measurements. Its approximate range is between ~ 270°C and ~ 300°C. X-ray precession measurements (Schneck and Dénoyer 1981), elastic neutron scattering (Schneck et al. 1982) and transmission electron microscopy (Manolikas 1981) have shown that the satellite reflections associated to the modulation are specified by the modulation vector:

$$\boldsymbol{k}_{\mathrm{c}} = (1+\delta)\left(\frac{\boldsymbol{a}_{\mathrm{t}}^{*} + \boldsymbol{b}_{\mathrm{t}}^{*}}{4}\right) + \frac{\boldsymbol{c}^{*}}{2}, \tag{1}$$

$\boldsymbol{a}_{\mathrm{t}}^{*}$, $\boldsymbol{b}_{\mathrm{t}}^{*}$, and $\boldsymbol{c}^{*}$ being the reciprocal lattice vectors of the tetragonal "normal" crystalline phase stable above T_{I}. The modulation is therefore a simple twofold superstructure along $\boldsymbol{c}$, as previously detected in the early X-ray measurements mentioned above (Burgeat and Tolédano 1976). The incommensurate part of the modulation is located in the *ab* plane and directed along one bisectrix of the tetragonal unit cell. This modulation destroys the macroscopic equivalence between the two bisectrices and sets in the system an "average" mm2 point symmetry in which the orthorhombic axes ($\boldsymbol{a}_0$, $\boldsymbol{b}_0$) are rotated by 45° with respect to the tetragonal one ($\boldsymbol{a}_{\mathrm{t}}$, $\boldsymbol{b}_{\mathrm{t}}$). Accordingly, in the INCP, macroscopic

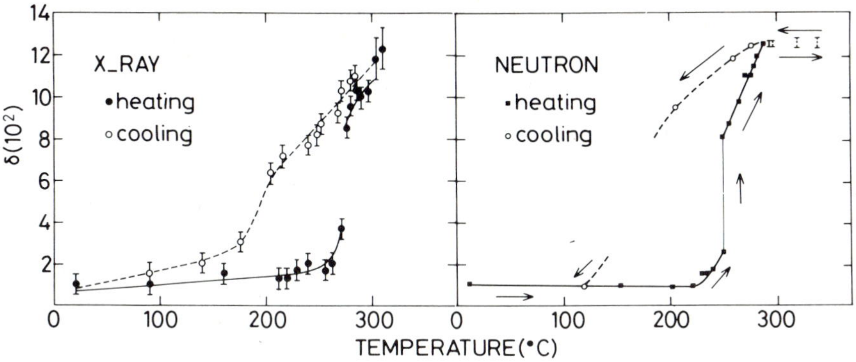

Fig. 1. Temperature dependence of the incommensurability parameter δ as determined from an X-ray precession experiment (left) and from an elastic neutron scattering experiment (right). (Schneck and Dénoyer 1981, Schneck et al. 1982.)

domains can be observed in the samples (Schneck and Dénoyer 1981, Schneck et al. 1982, Manolikas 1981). They correspond to mutually perpendicular directions of the incommensurate modulation, in the *ab* plane. In a single domain sample, the "modulated" orthorhombic axis is the "short" $\boldsymbol{a}_0$ axis.

Up to now, the most complete and accurate information on the value and variations of the δ parameter has been obtained from neutron scattering data (Schneck et al. 1982). Some very accurate but less complete results have become recently available from X-ray goniometry (Schneck et al. 1984). In the range of the INCP, δ varies on heating (fig. 1), between 0.08 at T_L and 0.125 at T_I, with a linear variation in this range. Both the neutron and the X-ray measurements have clearly established that, on heating, the T_L transition is discontinuous. Figure 1 shows the corresponding discontinuity in the value of δ. It also shows that the phase stable below T_L is not perfectly commensurate: unlike the situation in most other substances, the lower limit of the INCP does not correspond to a complete lock-in of the modulation on a simple superstructure. δ has been checked to keep a non-zero value of the order of 1% down to ~15 K. This value is sample dependent (Schneck et al. 1982). The phase below T_L which possesses a small residual incommensurability has been designated as "quasi-commensurate" (QCP) by analogy with the similar phase existing in the two-dimensional conductor 1T-TaS_2 (DiSalvo et al. 1975). The first order of the T_L transition gives rise to a latent heat of $\approx 0.02\,\text{cal/g}$ (Tolédano and Pateau 1974), and to a small discontinuity in the variation of the intensity of the incommensurate satellites (Schneck et al. 1982).

On the basis of a variety of measurements (Schneck et al. 1982, Schneck 1982) the incommensurate–normal transition at T_I has been assigned a continuous character. However, it must be stressed that this continuity is of a peculiar type, the vanishing of the relevant quantities (Tolédano and Schneck 1975, Schneck et al. 1982) occurring with a horizontal tangent instead of the expected vertical one. The origin of this "rounding", which is dependent on the measuring conditions, is not yet identified with certainty (especially in regard of the fact that no rounding is observed at T_L).

Higher-order satellites, which are often detected in the lower limit of an INCP, have been searched by means of X-ray (Schneck and Dénoyer 1981, Schneck 1982) and neutron scattering (Schneck et al. 1982) techniques. Weak second-order satellites have been observed in the QCP, while no such features could be detected in the INCP. This is another puzzling result, since the temperature dependence of δ (which is large in BSN), and the non-sinusoidal shape of the modulation (which seems unsignificant in BSN, judging from the absence of higher order satellites) are usually given a common origin, in the framework of the current phenomenological theories (Levanyuk and Sannikov 1976, McMillan 1976, Ishibashi 1978).

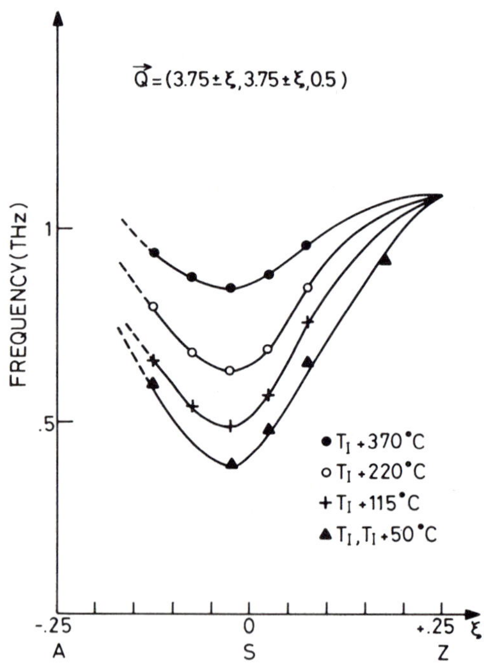

Fig. 2. Dispersion of the soft phonon branch along the modulation direction for $T > T_i$ near a satellite location. (Schneck et al. 1982.)

The INCP induces a widening of the satellites with respect to their width in the QCP. This anomalous width, which has an interesting behaviour when the measuring conditions are varied, will be analyzed in section 4.

The onset of the INCP gives rise to precursor effects above T_I. Thus a soft mode has been detected (Schneck et al. 1982) at the critical wavevector $\boldsymbol{k}_c$. This mode is underdamped above 400°C, and becomes overdamped between 400°C and T_I. Far above the transition its frequency is already low: 30 cm^{-1} at 650°C. In conformity with the behaviour recorded at a number of structural transitions (Fleury and Lyons 1981), the soft mode in BSN coexists in the neutron energy scans with a "central peak" whose intensity diverges at T_I (Schneck et al. 1982). Below T_I, the scans only reveal a wide central component with a decreasing intensity on cooling. Up to now, the restoration of an underdamped regime at $\boldsymbol{k}_c$ has not been detected, down to 2 K, either by means of neutron scattering or by Raman scattering (Boudou and Sapriel 1980). In the normal phase, the dispersion of the soft phonon branch along [110] (i.e. parallel to the modulation) shows a minimum at $\boldsymbol{k}_c$ (fig. 2) which appears to be independent of temperature and corresponds to $\delta \approx 0.125$. Along the perpendicular direction [110], the soft branch is very flat (Schneck et al. 1980b, 1982). The soft phonon dispersion in the *ab* plane is represented by a valley-shaped surface whose bottom is along [110] (Schneck et al. 1981a). This shape has been related to the observation of diffuse scattering in X-ray precession photographs of $(hk\frac{1}{2})$ reciprocal planes (Schneck and Dénoyer 1981). One observes diffuse scattering rods which have been identified with the bottom of the valley of the dispersion surface of soft modes (fig. 3). These rods can be clearly observed up to 600°C, while below T_I their intensity vanishes rapidly. A careful examination of their shape shows that they are constituted

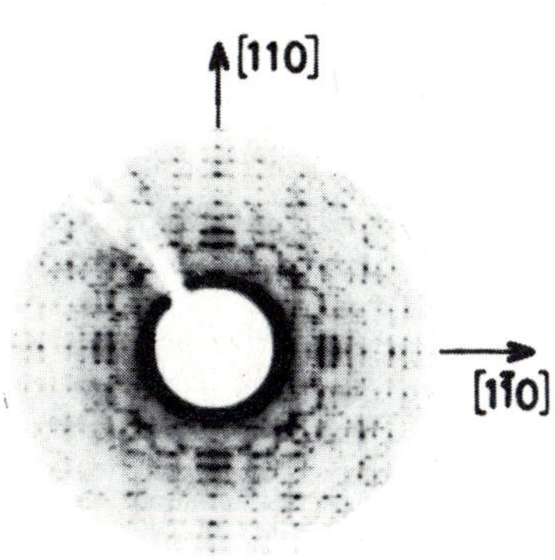

Fig. 3. X-ray precession photograph of a $(hk\frac{1}{2})$ reciprocal plane showing the diffuse scattering rods joining the satellites above T_I. (Schneck and Dénoyer 1981.)

by portions of curved lines joining neighbouring satellites (fig. 3). This peculiar shape has not yet been given an interpretation.

The existence of a lattice vibration mode, softening at T_I, has been taken as an evidence that the modulation present in the INCP or in the QCP is of a displacive type. The nature of the corresponding collective atomic displacement has been deduced (Schneck et al. 1982) from the room-temperature structural data. The data for BSN are incomplete, at present, since they were determined (Jamieson et al. 1969) without taking into account the satellite reflections. Nevertheless, these data, which are relative to the "average" structure of BSN, contain valuable information on the characteristics of the modulation (Schneck et al. 1982). Figure 4 shows the structure of BSN inferred on this basis, for $\delta = 0$, i.e. for a modulation assumed to be perfectly locked-in below T_L. The basic tungsten bronze structure consists of a skeleton of oxygen octahedra sharing corners and forming two types of large tunnels in which the barium and the sodium ions are respectively located. The modulation (incommensurate or locked) mainly consists in a collective shearing of the oxygen octahedra in which the basis of each octahedron in the *ab* plane is fixed, while the upper and lower vertices move parallel to this plane in opposite directions. This alternation along the *c*-axis corresponds to the doubling of the periodicity along this axis, while the amplitude of the displacements is modulated periodically along $\boldsymbol{a}_0$. Half of the barium ions are also affected by the modulation, the sodium ions remaining unaffected. In first approximation (i.e. neglecting the coupling to secondary degrees of freedom) the soft mode detected above T_I should involve the same collective shearing of oxygen octahedra.

The INCP is associated with an order parameter (OP) which has been specified by Tolédano et al. (1981). This OP has four components ($\eta+, \eta-$; $\xi+, \xi-$) respectively corresponding to the wavevectors $^{\pm}\boldsymbol{k}_c$ and $^{\pm}\boldsymbol{k}'_c$, with

$$\boldsymbol{k}'_c = (1+\delta)\left(\frac{\boldsymbol{a}^*_t - \boldsymbol{b}^*_t}{4}\right) + \frac{\boldsymbol{c}^*}{2}. \tag{2}$$

For incommensurate dielectrics, the usual situation (Levanyuk and Sannikov 1976) is that of a two-component OP, and aside from BSN, four-component OP's have been found in the cases of $BaMnF_4$ and biphenyl. The symmetry characteristics of the OP are used to build the free-energy F which is the basis of a phenomenological description of the standard behaviour of an incommensurate system (Levanyuk and Sannikov 1976, Ishibashi 1978). If we take, in BSN, the reference wavevector $\boldsymbol{k}_0$ which is implicitly taken into account in such theories, to coincide with $\boldsymbol{k}_c(\delta = 0)$, and put as variables ρ_i and ϕ_i such that $\eta_{\pm} = \rho_1\exp(\pm i\phi_1)$ and $\xi_{\pm} = \rho_2\exp(\pm i\phi_2)$, the free energy F has the form

Fig. 4. Schematic structure of BSN projected along $\boldsymbol{c}$. The modulation is assumed to be locked ($\delta = 0$). The shaded areas represent the displacements involved in the modulation. (Schneck 1982.)

$F = \int f(x, y)\,\mathrm{d}x\,\mathrm{d}y$, with:

$$f(x,y) = \frac{\alpha}{2}\left(\rho_1^2+\rho_2^2\right) + \frac{\beta_1}{4}\left(\rho_1^4+\rho_2^4\right) + \frac{\beta_2}{4}\left[\rho_1^4\cos(4\phi_1)+\rho_2^4\cos(4\phi_2)\right]$$
$$+\frac{\beta_3}{2}\rho_1^2\rho_2^2 + \lambda\left(\rho_1^2\frac{\partial\phi_1}{\partial x}+\rho_2^2\frac{\partial\phi_2}{\partial y}\right)$$
$$+\kappa_1\left[\rho_1^2\left(\frac{\partial\phi_1}{\partial x}\right)^2+\left(\frac{\partial\rho_1}{\partial x}\right)^2+\rho_2^2\left(\frac{\partial\phi_2}{\partial y}\right)^2+\left(\frac{\partial\rho_2}{\partial y}\right)^2\right]$$
$$+\kappa_2\left[\rho_1^2\left(\frac{\partial\phi_1}{\partial y}\right)^2+\left(\frac{\partial\rho_1}{\partial y}\right)^2+\rho_2^2\left(\frac{\partial\phi_2}{\partial x}\right)^2+\left(\frac{\partial\rho_2}{\partial x}\right)^2\right]. \tag{3}$$

Among the various solutions minimizing F, the case realized in BSN (Tolédano et al. 1981) corresponds to the freezing, in the INCP, of one pair of OP components, i.e. to either of the conditions ($\rho_1 \neq 0$; $\rho_2 = 0$) or ($\rho_1 = 0$; $\rho_2 \neq 0$) which respectively define two domains with mutually perpendicular modulation directions. Assuming as usual (Ishibashi 1978) "phase modulation only" (i.e. ρ_i spatially homogeneous), eq. (3) determines the phase ϕ_1 through the same sine-Gordon equation as for a two components OP:

$$\frac{\mathrm{d}^2\phi_1}{\mathrm{d}x^2} = -\frac{\beta_2}{2\kappa_1}\rho_1^2\sin[4\phi_1(x)]. \tag{4}$$

The standard sequence of events predicted for an INCP with a two-component order parameter is therefore expected to remain valid for BSN (variations of δ, characteristics of the lock-in, soliton shape of the modulation, etc...).

However specific consequences of the four-dimensionality of the OP are to be seen in the behaviour of the macroscopic quantities which we will now examine.

3. Macroscopic symmetry breaking and related anomalies

The anomalies of macroscopic quantities induced by the INCP can be studied on the basis of the coupling scheme between these quantities and the OP. For BSN the prominent anomalies are those of the mechanical quantities (strain, elastic constants), and one is therefore led to consider the coupling of the OP to the components of the strain tensor e_i. The corresponding contributions to the free-energy density have the form:

$$f' = \frac{C_{11}}{2}(e_1^2 + e_2^2) + C_{12}e_1e_2 + C_{13}(e_1 + e_2)e_3 + \frac{C_{33}}{2}e_3^2$$
$$+ [\mu_1(e_1 + e_2) + \mu_3 e_3](\rho_1^2 + \rho_2^2) + \nu(e_1 - e_2)(\rho_1^2 - \rho_2^2), \quad (5)$$

where the tensorial quantities refer to the orthorhombic axes.

The striking characteristic of eq. (5) is that the coupling terms do not depend on the phase part of the parameter ϕ_i. For $\rho_i \neq 0$, $\rho_j = 0$, the macroscopic

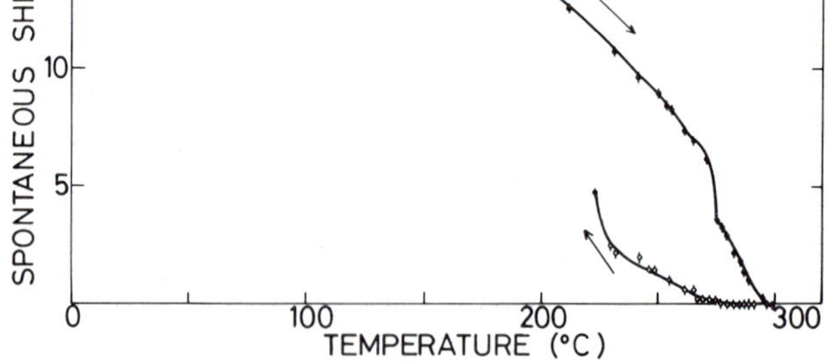

Fig. 5. Temperature dependence of the spontaneous strain in BSN as determined from γ-ray diffraction. (Schneck 1982.)

symmetry is broken through the onset of a spontaneous value for $(e_1 - e_2) \sim \rho_1^2$, and an orthorhombic symmetry is established both in the incommensurate ($\delta \neq 0$) and in the commensurate phase ($\delta = 0$). No change of the macroscopic symmetry is therefore expected at T_L. As pointed out in the introduction, this situation contrasts with the usual case for incommensurate systems. This peculiarity of BSN relies partly on the four-dimensionality of its OP. Thus for the common two-component case, the modulus of the OP, ρ^2, is totally symmetric, and the symmetry breaking macroscopic components are necessarily coupled to a phase-dependent term whose space average is zero in the INCP. The INCP of BSN can be considered as a special case of an improper ferroelastic (Tolédano 1974).

Equation (5) determines a temperature dependence of the shear strain $(e_1 - e_2) = e_{ab}$ identical to that of ρ^2. This dependence has been deduced from γ-ray diffractometry measurements (fig. 5). It shows a slight discontinuity at T_L and an approximately linear dependence within the INCP. The birefringence $(n_1 - n_2)$ has the same qualitative variations (Tolédano and Schneck 1975); this will be discussed in greater detail in the next section. For this quantity, the

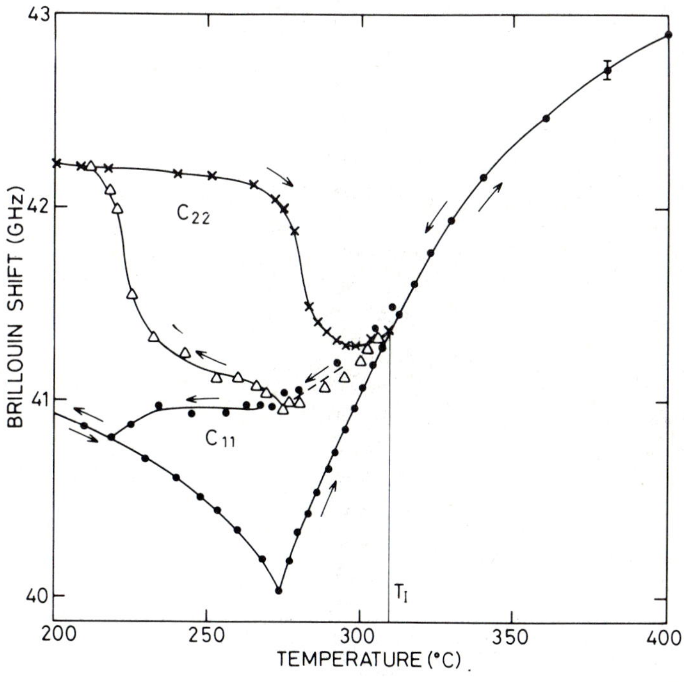

Fig. 6. Temperature dependence of the elastic constants C_{11} and C_{22} determined by Brillouin scattering measurements. (Hebbache and Errandonéa 1984.)

respective shares of direct coupling of the electronic polarizability to the OP, and of photoelastic coupling to the shear $(e_1 - e_2)$ have not yet been sorted out.

Figure 6 shows the behaviour of the elastic constants C_{11} and C_{22} which are inverse susceptibilities associated to e_1 and e_2. In the usual case of a phase-dependent coupling between a macroscopic quantity and the OP, the phenomenological theory (Levanyuk and Sannikov 1976) predicts that the susceptibility will be very slightly affected at T_I and that it will diverge at the lock-in, T_L. This divergence is related to the onset of the spontaneous quantity below T_L, due to the setting of the ϕ_i phase to a non-modulated value. As shown above, the situation is different in BSN, for which macroscopic symmetry breaking occurs at T_I. Errandonéa and Hebbache (1984) have recently analyzed this situation. Their calculations show that one expects at T_I the same downward step variation of the C_{ii} as encountered in improper ferroelastic transitions (Tolédano 1974), and a limited softening of these constants at T_L. However, interpretation of the experimental data (fig. 6) reveals a more complex situation which requires invoking both an important role of the fluctuations of the OP (Tolédano et al. 1976, Errandonéa and Hebbache 1984) and the occurrence of a frequency dispersion in the INCP (Scott 1983, Errandonéa and Hebbache 1984). In addition, the results show the occurrence of a large thermal hysteresis whose characteristics and origin are analyzed in the next section. Another consequence of the coupling of the strain components with the modulation is the evolution of the INCP with an applied pressure. Recent measurements (Savary and Tolédano 1985) show a narrowing of the temperature domain of the INCP. Actually T_L increases with pressure $(+14.2 \pm 0.8\ °C/kbar)$ when T_I appears less affected.

4. *Thermal hysteresis and relaxation effects*

Phenomena departing from the standard behaviour of incommensurate systems have been studied in BSN mainly by combining birefringence- (Errandonéa et al. 1984), and X-ray goniometric measurements (Schneck et al. 1984) as a function of time and of temperature. In the temperature dependence of $(n_1 - n_2)$, T_L is indicated by a convexity change, while T_I corresponds to the vanishing point of this quantity. In the X-ray data, T_L is marked by a discontinuity of δ (fig. 1) and also by a significant broadening (fig. 7) of the satellites along the direction of the modulation a_0 (the width increases at T_L from $2.4 \times 10^{-2} a_0^*$ up to $3.45 \times 10^{-2} a_0^*$). A similar broadening has been observed in Rb_2ZnCl_4 (Mashiyama et al. 1982), but it is confined to the vicinity of the lock-in transition, while in BSN it persists in the whole range of existence of the INCP.

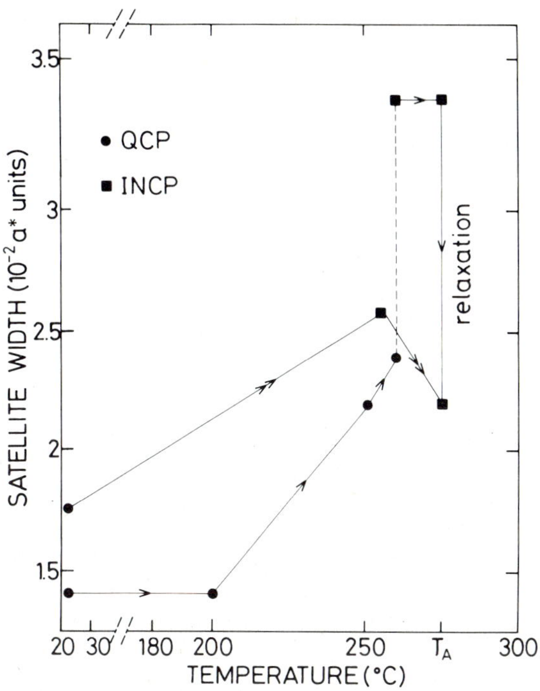

Fig. 7. Evolution with temperature and time of the linewidth of an incommensurate satellite, showing the "memory effect" which follows a relaxation and a cooling at room temperature. (Schneck et al. 1984.)

The various anomalous phenomena revealed by the preceding experiments are the following:

(i) Within the INCP, when the sample is maintained at a constant temperature T_A, a slow increase of the birefringence occurs, which saturates after a time which is typically two to four days (the birefringence can be multiplied by a factor as large as 8). In a similar way the intensity of the satellites increases and their width decreases markedly (fig. 8). These various relaxations have been fitted by a sum of two exponential decays, with one relaxation time in the range 0.5–3 h, and the other in the range 10–50 h (fig. 9). A similar relaxation exists in the normal phase ($T > T_I$), as shown by measurements of the dielectric permittivity ε_a. This relaxation is observed only after heating the sample across the INCP. Moreover, if after the end of a relaxation in the normal phase the sample is brought to another temperature within this phase, no further relaxation is recorded.

(ii) By waiting until the end of the relaxation at a temperature T_A (which can be considered as a kind of "annealing" at T_A) and then quenching the crystal

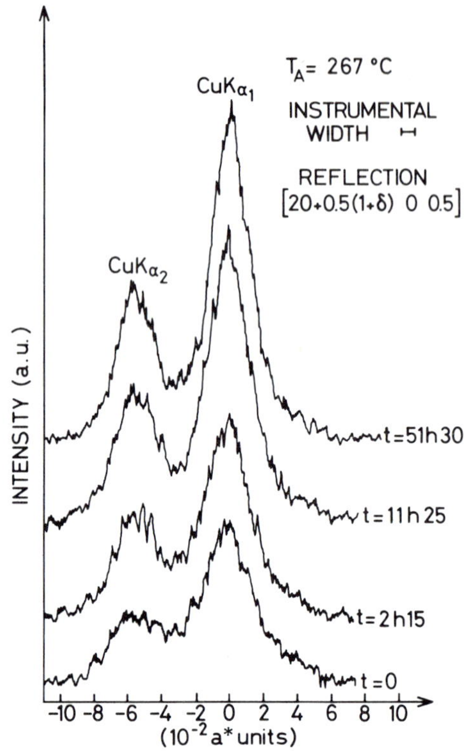

Fig. 8. Time dependence of an incommensurate satellite during a relaxation within the incommensurate phase. (Schneck et al. 1984.)

down to room-temperature, one is able to modify in a major way the events occurring during the subsequent heating run: one has either an anticipated "lock-out transition" (i.e. an anticipated onset of the INCP on heating, figs. 7, 10) at $T_L \approx 260°C$, if the annealing temperature is within the INCP, or a retarded "lock-out" at $T_L \approx 300°C$, if the annealing is performed below the range of the INCP (in the QCP). X-ray experiments further show that the "narrow" satellite width reached at the end of an annealing is preserved during the subsequent heating run (fig. 7). These anticipation–delaying phenomena constitute a "memory" effect, in the sense that the heating run bears a memory of the annealing temperature which has preceded it. Similar memory effects have first been pointed out in thiourea (Jamet and Lederer 1983) and later in Rb_2ZnCl_4 (Unruh 1983). In BSN, the memory can be erased by thermally cycling the sample several times between the room temperature and the normal phase. A wide satellite is then recovered in the INCP.

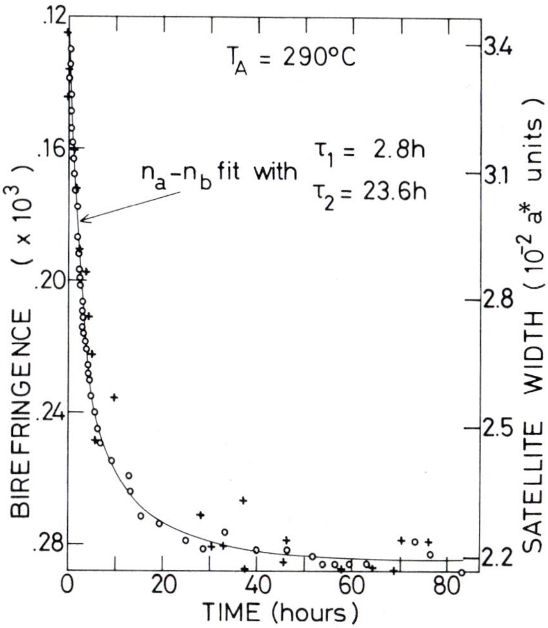

Fig. 9. Fitting of the relaxation of the birefringence (circles) and of the satellite linewidth (crosses) with a sum of two exponential decays. (Errandonéa et al. 1984.)

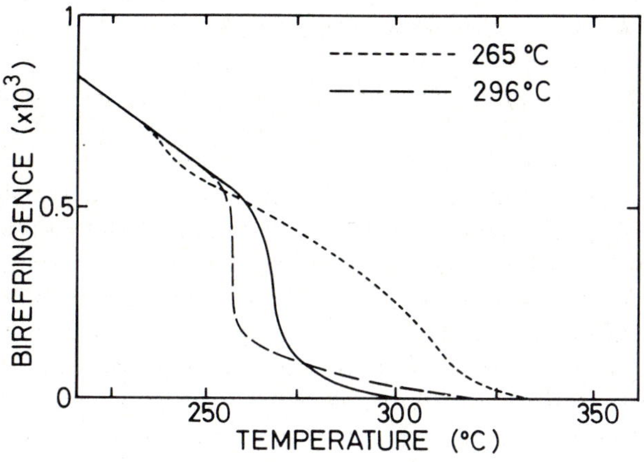

Fig. 10. Plots of the birefringence following respectively an annealing in the QCP (265°C) and in the INCP (296°C). The reference solid curve is obtained by repeatedly cycling the sample at 2.5°C/min between 20°C and 350°C. (Errandonéa et al. 1984.)

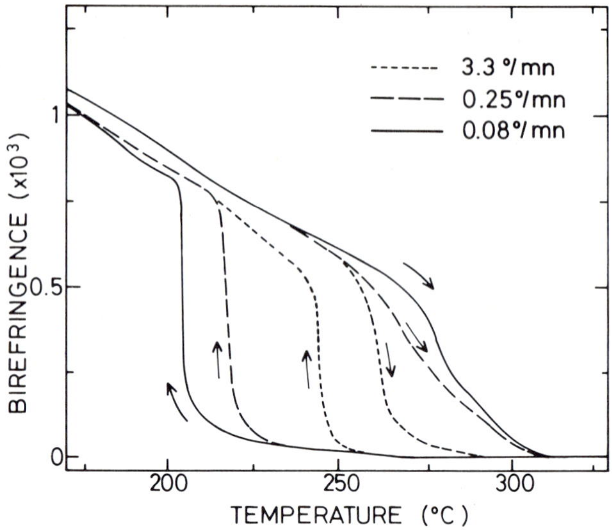

Fig. 11. Birefringence plots obtained by thermally cycling a sample at various heating rates. For each rate a reproducible plot is obtained after two to four cycles. (Errandonéa et al. 1984.)

(iii) If a sample is thermally cycled between 20°C and 350°C (above T_I) at a constant rate, one observes after a few cycles an accurately reproducible plot of $(n_1 - n_2)$, which does not depend on the previous thermal history of the sample. This plot displays a large thermal hysteresis (fig. 11) as do the variations of all the physical properties of BSN (fig. 1, 5, 6). This plot has a striking dependence on the heating/cooling rate (fig. 11). For increasing rates, there is in particular a considerable decrease of the magnitude of the thermal hysteresis.

An interpretation of the various effects has been given (Errandonéa et al. 1984, Schneck et al. 1984) on the basis of the following mechanism: point-defects are assumed to be present in BSN and are able to diffuse through the structure; in the normal phase they have, at equilibrium, a uniform spatial distribution. In the INCP, or in the QCP, they interact with the incommensurate modulation (pinning of the phase): this induces their diffusion towards structural sites where definite values of the phase of the modulation are realized; as a consequence a non-uniform distribution of defects occurs, patterned after the periodicity of the modulation; this evolution of the defect pattern is accompanied by evolution of the modulation towards an extension of its long-range order, initially limited by its pinning to an unadjusted distribution of the defects.

Evidence for this complex mechanism involving mobility (i.e. diffusion) of the defects is the persistence of relaxations in the normal phase, after each crossing of the INCP. In the assumed framework, these relaxations reflect the restoration of a uniform pattern of defects through thermally activated diffusion (since no modulation is present in this phase). If the temperature is varied within the normal phase one does not expect any remodelling of the defect distribution, in agreement with the experimental observation that a single complete relaxation can be observed in this phase.

Other evidence is provided by the "memory effect", for which Jamet and Lederer (1983) first stressed the probable importance of mobile defects in this phenomenon. In the case of BSN, the annealing at T_A sets in the sample a spatial distribution of defects patterned according to the state of the modulation at T_A. This pattern is almost frozen during the rapid cooling, and the subsequent run reveals the defect distribution through its influence on the temperature dependence of the measured quantity (e.g. the birefringence). Consistently, annealing in the INCP or in the QCP have respectively the effects of inducing the anticipated onset or the retarded onset of the INCP in the subsequent heating run (fig. 10). X-ray data disclose the extension of the long-range order of the modulation (fig. 7, 8) during a relaxation, and show that the memory effect is related to a recovery of the long-range order at T_A, probably due to the freezing of the defect distribution reached at the end of the annealing at T_A.

The assumed mechanism has also been shown (Errandonéa et al. 1984) to account satisfactorily for the dependence of the thermal hysteresis on the speed of the heating/cooling runs (fig. 11).

The nature of the incriminated mobile defects in BSN has not yet been identified with certainty, nor has their diffusion been established directly. However, it has been noted (Schneck et al. 1980a) that the composition of all real samples of BSN departs from the stoichiometric formula $Ba_2NaNb_5O_{15}$, the main structural defect introduced by the non-stoichiometry being the occurrence of vacancies at the sodium sites. On this basis, preliminary studies (Schneck 1982) show that the sodium vacancies have a suitable concentration (e.g. 10% of the sodium sites), and a possibility to migrate easily along the modulation direction (due to the low packing of atoms in the relevant lattice planes as seen in fig. 4) which makes it plausible to consider these as the mobile defects in cause. Consistently with this assumption, recent measurements have shown that the temperature range of the INCP, and the qualitative characteristics of the anomalous effects, depend strongly on the sodium stoichiometry of the samples.

5. *Conclusions*

Several important aspects remain to be clarified in BSN which have been pointed out in the preceding sections, both in its intrinsic characteristics as an incommensurate system and in the influence of defects. In addition to the preceding points it would be important to elucidate the nature of the low-temperature phase transition at $\sim -160°C$ (Schneck et al. 1977). This transition, which seems to have the same order parameter as the INCP, displays a similar anomalous thermal hysteresis and a striking sensitivity to stoechiometry defects (Schneck et al. 1980). On the other hand, we can note that BSN does not have an isolated behaviour in its structural family. $Sr_2KNb_2O_{15}$ and $Ba_{0.29}Sr_{0.71}Nb_2O_5$ which are structurally isomorphous, possess modulated phases at room temperature (Schneck et al. 1981). In these cases, there are two directions of modulation associated with $\boldsymbol{k}_c$ and $\boldsymbol{k}'_c$, and larger values of the δ parameter (resp. $\delta = 0.16 \pm 0.05$, and $\delta = 0.26 \pm 0.05$). Few data are known at present for these substances.

References

Boudou, A., and J. Sapriel, 1980, Phys. Rev. **B21**, 61.
Burgeat, J., and J.C. Tolédano, 1976, Solid State Commun. **20**, 281.
DiSalvo, F.J., J.A. Wilson, B.G. Bagley and J.V. Waszczak, 1975, Phys. Rev. **B12**, 2220.
Errandonéa, G., and M. Hebbache, 1984, unpublished.
Errandonéa, G., J.C. Tolédano, J. Schneck, H. Savary and J. Aubrée, 1984, J. Physique Lett. **45**, L329.
Fleury, P.A., and K. Lyons, 1981, Structural Phase Transitions vol. 1, eds. K.A. Muller and H. Thomas (Springer, Berlin) p. 9.
Hebbache, M., and G. Errandonéa, 1984, Ferroelectrics **55**, 39.
Ishibashi, Y., 1978, Ferroelectrics **20**, 103.
Jamet, J.P., and P. Lederer, 1983, J. Physique Lett. **44**, L257.
Jamieson, P.B., S.C. Abrahams and J.L. Bernstein, 1969, J. Chem. Phys. **50**, 4352.
Levanyuk, A.P., and G.D. Sannikov, 1976, Sov. Phys. Solid State **18**, 423.
Manolikas, C., 1981, Phys. Stat. Sol. (a) **68**, 653.
Mashiyama, M., S. Tanizahi and K. Hamano, 1982, J. Phys. Soc. Jpn **51**, 2538.
McMillan, W.L., 1976, Phys. Rev. **B14**, 1496.
Savary, H., and J.C. Tolédano, 1985, Phys. Rev. **B31**, 3134.
Schneck, J., 1982, Thesis (Université Paris VI) unpublished.
Schneck, J., and F. Dénoyer, 1981, Phys. Rev. **B23**, 383.
Schneck, J., J. Primot, J. Ravez and R. Von Der Muhl, 1977, Solid State Commun. **21**, 57.
Schneck, J., B. Joukoff and R. Mellet, 1980a, Ferroelectrics **26**, 775.
Schneck, J., J.C. Tolédano, B. Joukoff, F. Dénoyer and C. Joffrin, 1980b, Ferroelectrics **26**, 661.
Schneck, J., J.C. Tolédano, J. Aubrée, B. Joukoff and C. Joffrin, 1981a, J. Physique **C6**, 734.
Schneck, J., J.C. Tolédano, R. Whatmore and W. Ainger, 1981b, Ferroelectrics **36**, 327.
Schneck, J., J.C. Tolédano, C. Joffrin, J. Aubrée, B. Joukoff and A. Gabelotaud, 1982, Phys. Rev. **B25**, 1766.
Schneck. J., G. Calvarin and J.M. Kiat, 1984, Phys. Rev. **B29**, 1476.
Scott, J.F., 1983, Ferroelectrics **47**, 33.
Singh, S., D.A. Draegert and J.E. Geusic, 1970, Phys. Rev. **B2**, 2709.
Tolédano, J.C., 1974, Ann. Telecomm. **29**, 249.
Tolédano, J.C., 1975, Phys. Rev. **B12**, 943.
Tolédano, J.C., and L. Pateau, 1974, J. Appl. Phys. **45**, 1611.

Tolédano, J.C., and J. Schneck, 1975, Solid State Commun. **20**, 281.
Tolédano, J.C., M. Busch and J. Schneck, 1976, Ferroelectrics **26**, 775.
Tolédano, J.C., J. Schneck and C. Lamborelle, 1981, Symmetries and Broken symmetries in condensed matter physics, ed. N. Boccara (IDSET, Paris) p. 217.
Unruh, H.G., 1983, J. Phys. **C16**, 3245.
Yamada, T., H. Iwasaki and R. Niizeki, 1970, J. Appl. Phys. **41**, 4141.

CHAPTER 18

The Incommensurate Phase of Na_2CO_3

P.M. DE WOLFF and F. TUINSTRA
Vakgroep Vaste Stof — Laboratorium voor Technische Natuurkunde
Technische Hogeschool Delft
Delft
The Netherlands

Incommensurate Phases in Dielectrics 2
Edited by
R. Blinc and A.P. Levanyuk

Contents

1. Phase transitions in alkali carbonates ... 255
   1.1. Introduction ... 255
   1.2. Transition points and unit cell parameters ... 255
   1.3. Symmetry and structure type of the α, β and γ phases ... 258
2. The crystal structure of α- and β-Na_2CO_3 ... 262
   2.1. The crystal structure of α-Na_2CO_3 ... 262
   2.2. The crystal structure of β-Na_2CO_3 ... 263
3. The crystal structure of γ-Na_2CO_3 ... 266
   3.1. Type of modulation and crystal symmetry ... 266
   3.2. Crystal structure and interpretation ... 268
4. Physical properties of Na_2CO_3 ... 270
   4.1. Introduction ... 270
   4.2. Diffuse scattering ... 273
   4.3. Unit cell parameters as a function of temperature ... 274
   4.4. Raman and infrared spectroscopy ... 275
   4.5. Other physical properties ... 276
      4.5.1. Crystal growth ... 276
      4.5.2. Ferroelasticity ... 277
      4.5.3. Birefringence ... 277
      4.5.4. NMR ... 278
      4.5.5. Electric conductivity ... 279
5. Evidence for a lock-in transition at low temperatures ... 279

References ... 280

1. *Phase transitions in alkali carbonates*

1.1. *Introduction*

The anhydrous alkali carbonates are well-crystallized, brittle substances. Since they are very hygroscopic, experimental investigation of their properties is not easy, the more so because good single crystals—especially of Na_2CO_3— are hard to obtain. These facts may account for the comparative scarcity of accurate crystallographic data for such simple compounds. They have received increasing attention, however, since the discovery (Brouns et al. 1964) of the incommensurate character of Na_2CO_3 at room temperature.

This γ-Na_2CO_3 phase is the only incommensurate one among the known alkali carbonate phases. Nevertheless we shall deal in this section with the rubidium and potassium carbonate phases as well, because their structures are very closely related to that of Na_2CO_3. [Li_2CO_3 has a rather different type of structure; for Cs_2CO_3, too few data are known. Binary systems such as $(Na, K)_2CO_3$ (Christmann and Papin 1979) will not be treated either.]

In a wider context, Na_2CO_3 shows a remarkable analogy with the A_2BX_4-compounds treated in chapters 6 and 7. Hexagonal α-Na_2CO_3 may even be considered as an idealized prototype of these compounds in their various modifications (Eysel 1971).

The crystalline phases of Na, K and Rb carbonate are called α, β and γ in order of decreasing temperature. All three γ phases are stable at room temperature. The possible existence of a δ-Na_2CO_3 phase below 130 K will be touched upon in section 5; for the other two carbonates, no low-temperature data have been published.

1.2. *Transition points and unit cell parameters*

Experimental data obtained by DTA or diffraction methods for the temperatures of the transitions $\alpha-\beta$ and $\beta-\gamma$ are summarized in table 1. The caloric effects (all endothermic) are small, so it is not amazing that not all investigators have found both transitions. Data derived from other properties, such as birefringence, are in accordance with those listed. Only for Na_2CO_3 a weak additional caloric effect at 620°C has formerly been mentioned. Though it was later ascribed to contamination (Reisman 1958), an anomaly at the same

Table 1
Transition points of the α-β and β-γ transitions of three alkali carbonates.

Carbonate	T (°C)		Method	Reference
	α−β	β−γ		
Na_2CO_3	489	361	DTA	Reisman 1959
	479	349	DTA	Papin 1973
	485	356	DTA	Klement and Cohen 1975
	486	358	DSC	Cantor and Heatherly 1977
K_2CO_3	428	250	DTA	Marakov and Shulgina 1940
	422	253	DTA	Reisman 1958
	420	340	DTA, XRD	Papin 1973
	420	—	DSC, XRD	Schneider and Levin 1973
	421	—	DTA	Klement and Cohen 1975
	—	250	XRD	Becht and Struikmans 1976
	421	—	DSC	Cantor and Heatherly 1977
Rb_2CO_3	570	310	DTA, XRD	Karataeva and Lepeshkov 1970
	550	300	XRD	Hertogs 1977

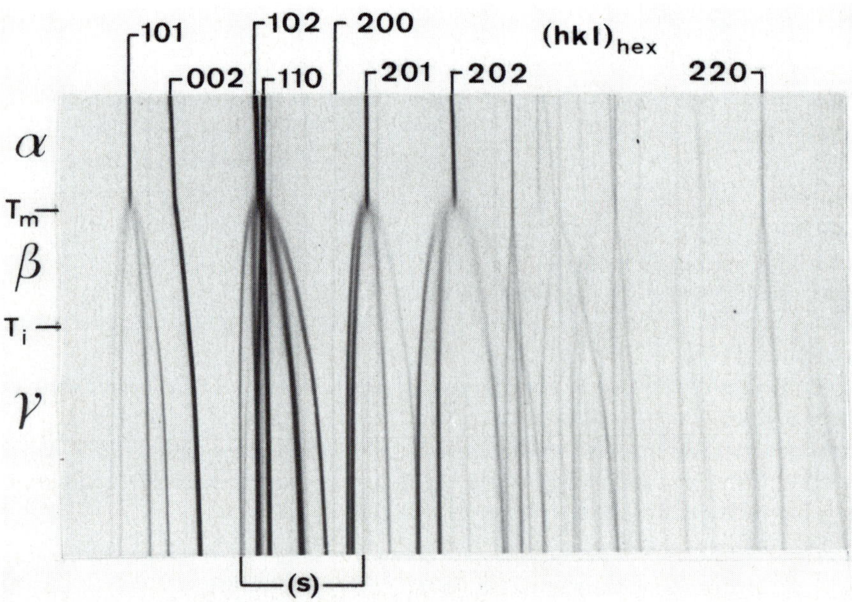

Fig. 1. X-ray powder diffraction pattern of Na_2CO_3 taken with a Guinier–Lenné camera, as a continuous function of the temperature in the range 130°C (bottom) to 590°C (top). Two strong satellite reflections are marked (S). Vertical smearing is caused by 3 mm slit height.

temperature has been found recently by Cerisier and Roux (1977) in the electric conductivity, using very pure samples. A change in rotational disorder, as they suppose, seems possible (free rotation is disproved by diffraction evidence up to 700°C, cf. section 2.1).

All effects are without detectable hysteresis and are of the type characteristic for second-order transitions. This is in accordance with the continuity observed in X-ray diffraction patterns, especially when these are obtained as a continuous function of temperature as in fig. 1 for Na_2CO_3.

The α phase of Na_2CO_3 has a hexagonal point group, hence it follows from the continuity of the transitions that its β and γ phases have pseuso-hexagonal symmetry. The same is true for the potassium and rubidium carbonates: their α phases are hexagonal as well. The lattices of the three β phases, though all pseudo-hexagonal, belong to different Bravais classes. Of the γ phases, the lattice types are again different; only γ-K_2CO_3 and γ-Rb_2CO_3 possess lattices of the same type.

In order to facilitate comparison, each of the nine phases will be described by axes defined as follows: ***c*** is the axis of (pseudo-) sixfold symmetry, whereas

Table 2
Unit cell parameters and space groups of the α, β and γ phases of three alkali carbonates. The parameters refer in each case to the (pseudo-)orthohexagonal cell with $a \simeq b\sqrt{3}$. They are given in two lines: a, b, c on the upper line and $\alpha, \beta, a/b\sqrt{3}$ on the lower line. Numbers are entered only when not fixed by symmetry.

Na_2CO_3					K_2CO_3					Rb_2CO_3				
T(°C)	Space group; parameters			Ref.[a]	T(°C)	Space group; parameters			Ref.[a]	T(°C)	Space group; parameters			Ref.[a]
	α:$P6_3/mmc$					α:$P6_3/mmc$					α:$P6_3/mmc$			
500	—	5.20	6.50	[1]	500	—	5.66	7.10	[6]	520	—	5.80	7.78	[10]
	—	—	—			—	—	—			—	—	—	
500	—	5.21	6.62	[2]	500	—	5.66	7.30	[2]	565	—	5.895	7.811	[7]
	—	—	—			—	—	—			—	—	—	
565	—	5.215	6.584	[3]	550	—	5.70	7.31	[7]	600	—	5.909	7.843	[11]
	—	—	—			—	—	—			—	—	—	
	β: C12/m1					β: C2/c11					β: Pcmn			
440	9.00	5.24	6.31	[1]	310	9.920	5.675	7.018	[8]	400	10.126	5.914	7.641	[11]
	—	96.7	0.9916			96.8	—	1.0092			—	—	0.9985	
385	9.005	5.253	6.290	[4]										
	—	97.3	0.9897											
	γ: C12/m1 (av.str.)					γ: $P2_1/c11$					γ: $P2_1/c11$			
20	8.904	5.239	6.042	[5]	23	9.841	5.638	6.876	[9]	20	10.116	5.873	7.344	[11]
	—	101.35	0.9812			98.66	—	1.0078			97.36	—	0.9945	

[a] References:

[1] Brouns et al. (1964).
[2] Christmann and Papin (1979).
[3] Visser (1975).
[4] de Jong (1979).
[5] van Aalst et al. (1976).
[6] Papin (1973).
[7] Hertogs (1977).
[8] Becht and Struikmans (1976).
[9] Visser (1966).
[10] Eysel (1971) p. 49.
[11] Visser (1984).

b and ***a*** correspond to [100] and [120] of the hexagonal lattice ("orthohexagonal description") so that $a \approx b\sqrt{3}$. Since in the hexagonal lattice [100], [010] and [110] are equivalent there are three possible choices; in the present cases we always choose the one for which ***a*** and ***b*** are mutually perpendicular by (monoclinic or orthohombic) symmetry. In this way the conventional unit cells for these systems are obtained. Only the labels a, b, c may depart from conventions.

Figure 1 also shows that the diffraction patterns—and therefore the unit cell parameters—change considerably with temperature even within a single-phase region. At one transition point ($\alpha-\beta$ for Na- and K-, $\beta-\gamma$ for Rb-carbonate) which we call T_m (m = monoclinic), there is an exceptionally dramatical change. In each case it involves an axial angle φ becoming obtuse such that roughly $(\varphi - 90°)^2 = C(T_m - T)$. In itself this is a familiar kind of ferrodistortive phenomenon; but the constant C has an order of magnitude parallelled by very few other compounds. As far as we know, H_3BO_3 is the only other example of ferroelasticity with such a huge spontaneous strain (Wadhawan 1978).

Upon analysis of the patterns, another common feature is found, viz. a very prominent thermal expansion in the direction of the (pseudo-)hexagonal axis c as compared to the a- and b-axes.

In table 2, unit cell parameters are listed for the three phases of each carbonate. From the above it follows that the temperature is crucial. This is not so important for the γ phase data at room temperature, among which the most reliable ones have been chosen. For the β phases only few data are known. Several data exist for the α phases and the choice is difficult because little is known of the temperature precision. Therefore most of these data are listed, only those obviously in error having been omitted.

1.3. Symmetry and structure type of the α, β and γ phases

In table 2 a space group is listed for each of the nine phases. This symmetry assignment is based on more or less complete structure analyses except for β-Rb_2CO_3, as discussed in more detail below.

The three α phases have been investigated mainly by powder diffraction. Early unpublished structure analyses by Bijk and Visser (1968) for Na_2CO_3, Hertogs (1977) for K_2CO_3 and Bijk (1967) for Rb_2CO_3 point to a complete isomorphism of these α phases, in accordance with the complete miscibility in the phase diagrams, cf. Reisman (1958) and Karataeva and Lepeshkov (1970).

For α-Na_2CO_3, the structure found in this early work agrees with the neutron powder diffraction analysis by de Pater (1978). The α-phase structure can be described as follows:

The alkali ions are equally divided over two topologically different sites I and II. Site I is the centre of an AO_6 octahedron with all A–O distances equal

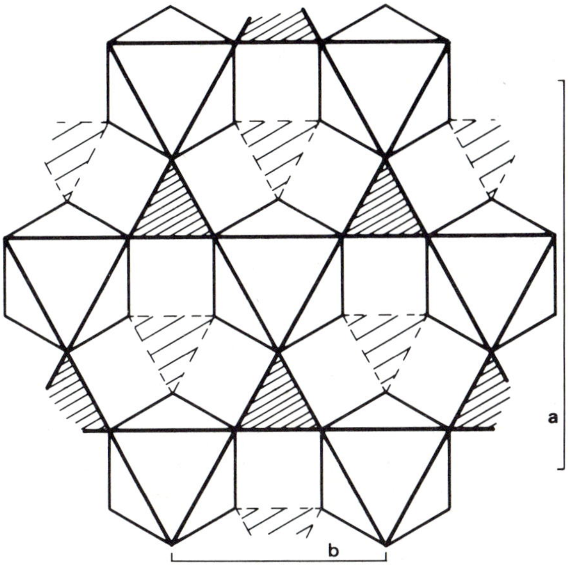

Fig. 2. Structure of α-$(Na,K,Rb)_2CO_3$ viewed along $\boldsymbol{c}$. The oxygen atoms are at the vertices of hatched CO_3 triangles. The thinly hatched triangles differ in height from the densely hatched ones by $\frac{1}{2}c$. Carbon and alkali atoms are not shown; a and b are orthohexagonal axes.

(Na–O = 2.37 Å) and with two opposite faces perpendicular to $\boldsymbol{c}$ (A = alkali). These faces are shared with neighbouring octahedra, thus giving rise to an infinite string of octahedra parallel to $\boldsymbol{c}$; in this direction the octahedra are elongated because of the A–A repulsion. The strings are arranged in a hexagonal array as shown in a projection along $\boldsymbol{c}$ in fig. 2, where just one octahedron of each string is drawn. Its neighbours in the string, though they coincide with it in this projection, are differently oriented in space. Only the second neighbours have the same orientation, so the repeat distance c covers two octahedra. Their two centres are crystallographically equivalent in the α phase, but not in all other phases, hence the roman numeral is used to avoid confusion with the usual arabic numeral of an atom in different crystallographic sites.

Site II for the remaining half of alkali ions is situated midway between three adjacent strings of octahedra. Its projection in fig. 2 coincides with the centre (not shown) of each hatched triangle representing a CO_3 ion. This anion of course shares all oxygen atoms with the AO_6 octahedra. In the direction of $\boldsymbol{c}$, the CO_3 ions alternate with the AII ions, as seen in the $\boldsymbol{b}$-projection of fig. 3a; both C and AII are at the z-level of the O atoms. Hence AII is surrounded by a flat hexagon of 6 O's at the same z-level, clearly visible in fig. 2 (NaII–O =

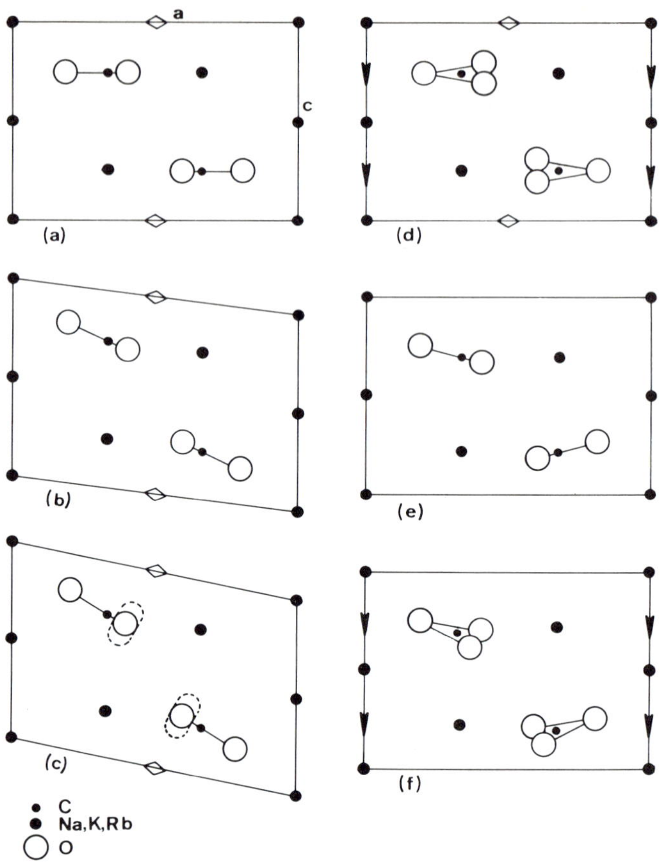

Fig. 3. Projection along b of a slice of A_2CO_3 containing A and CO_3 ions at the same y level: (a) α-Na_2CO_3, α-K_2CO_3, α-Rb_2CO_3; (b) β-Na_2CO_3; (c) γ-Na_2CO_3; (d) β-K_2CO_3; (e) β-Rb_2CO_3; (f) γ-K_2CO_3, γRb_2CO_3. Lines with small rhomb: projection of a centred face. Lines with arrows: c-axis sloping towards observer in arrow direction. The modulation is schematically shown in (c).

2.62 Å). This unusual coordination is completed by 3 + 3 more oxygen atoms of the neighbouring CO_3 ions in the row CO_3–AII–CO_3–AII parallel to $\boldsymbol{c}$; for these, the distance NaII–O is as large as 3.50 Å. The CO_3 ions in all phases are equilateral triangles with C (not shown in fig. 2) at the centre.

The above description of the structure in terms of a columnar array is of course not the only one possible. In contrast to other descriptions, however, it explains to a considerable extent the distortions occurring at the transition points: Each column of octahedra can be regarded as a rigid bar. Each CO_3 ion in itself is rigid as well, so it can be regarded as a stiff triangular plate. Its three corners constitute the joints with the three surrounding bars; and these joints

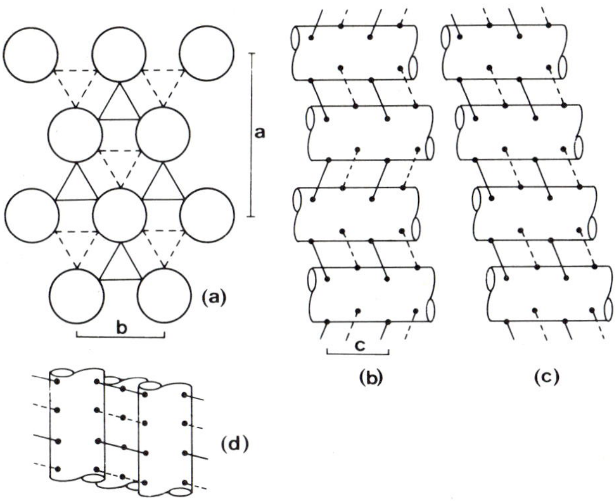

Fig. 4. (a) Schematized projection of A_2CO_3 along c, strings of octahedra (fig. 2) symbolized by cilinders; (b,c) S_5 fold and S_5 shear seen along b; (d) S_4 shear seen along a.

are flexible. It is this flexibility which introduces a peculiar aptitude towards shear deformation of the α phase. The three types of deformation which actually occur are illustrated by fig. 4; they correspond to the β-phases of Na, K and Rb carbonate.

The distortions in the former two phases are homogeneous shear strains: S_5 for Na_2CO_3, and S_4 for K_2CO_3. In a mechanical model in which the joints would be ideally flexible for all rotations, such a shear would not meet with any resistance at all. In terms of elasticity the model has vanishing torsion stiffness for these two shear situations ($C_{44} = C_{55} = 0$). Elasticity is hardly relevant for the model, however, since the torque remains zero also for finite strains of the type just mentioned.

In the actual crystal this is of course not true. There must be a restoring force which keeps the strain zero in the α phase. The α–β transition occurs when this force vanishes. The ensuing distortion can be expected to be stable for a finite strain value depending upon the temperature, and to be limited—at decreasing temperature—mainly by the AII-ions. These have so far been ignored, because their very loose and symmetric coordination is not likely to resist a small shear of any kind. At larger deformations, however, the environment of AII is distorted and this will lead to close AII–O contacts.

Whereas Na and K carbonate undergo a homogeneous and macroscopic shear strain in the α–β transition, this is not so for β-Rb_2CO_3. Here the S_5 shear occurs domain-wise, successive slabs parallel to b, c and $\frac{1}{2}a$ thick being sheared in equal but opposite amounts; we shall call this the 'S_5 fold'.

All this can be verified in more detail in fig. 3. These schematic pictures are based on single crystal structure analysis for β-Na_2CO_3 (de Jong and Tuinstra 1979) and for β-K_2CO_3 (Becht and Struikmans 1976). For β-Rb_2CO_3 only an indexed powder diffraction pattern is available (Visser 1984), but the systematic extinctions are sufficiently obvious to determine the space group and, at least schematically, the kind of distortion permitted by that group.

As regards the γ phases, the data for γ-K_2CO_3 (Gatehouse and Lloyd 1973) and for γ-Rb_2CO_3 (Ehrhardt et al. 1980) reveal that these two phases are isomorphous. Indeed from the corresponding schematic picture in fig. 3 it is seen that the distortion is a superposition of the S_4-shear and the S_5-fold. At decreasing temperature, K_2CO_3 undergoes these two distortions in that order, whereas for Rb_2CO_3 the order is reversed. The 'hinging' model does indeed allow superposition of both distortions, without any mechanical resistance. Evidently neither of the two in itself leads to a stable configuration at low temperatures.

Using these facts, one may be tempted to predict what will happen in the case of Na_2CO_3. For the spontaneous S_5 shear strain in β-Na_2CO_3, superposition with an S_4 shear might be expected to occur as a transition mechanism. Actually, however, the $\beta-\gamma$ transition in this compound follows a modification of that course; this will be examined in section 3.

2. *The crystal structures of α- and of β-Na₂CO₃*

2.1. *The crystal structure of α-Na₂CO₃*

The hexagonal unit cell accommodates two Na_2CO_3 formula units. The powder diffraction pattern (Visser 1975) shows a systematic extinction (in hexagonal indices: *hhl* present for l = even only) which points to the space group P 6_3/mmc. One obtains qualitative agreement when the atoms are placed in the special positions, in hexagonal coordinates

2NaI	in (a)	$0,0,0$;	$0,0,\frac{1}{2}$,
2NaII	in (d)	$\frac{1}{3},\frac{2}{3},\frac{3}{4}$;	$\frac{2}{3},\frac{1}{3},\frac{1}{4}$,
2C	in (c)	$\frac{1}{3},\frac{2}{3},\frac{1}{4}$;	$\frac{2}{3},\frac{1}{3},\frac{3}{4}$,
6O	in (h)	$x,2x,\frac{1}{4}$	etc.

of that space group. The above positions constitute the ideal structure first given by Bijk and Visser (1968) and later cited by Eysel (1971) and by Dubbeldam and de Wolff (1969). In this ideal structure (figs. 2 and 3a) there is only one free parameter; it can be expressed either as x in the above O-position, or as the distance r = C–O.

Very strong anisotropy can be expected in the thermal motion, on account of the anistropic thermal expansion. From the various attempts to introduce anisotropy we mention the analysis by de Pater (1978) of the neutron powder diffraction pattern at 520°C, using the Rietveld method. Here the thermal motion is simulated by splitting each atom in two half atoms removed a distance u (parallel to c) from its ideal position, and further splitting each half oxygen atom in two quarter atoms removed a distance v (perpendicular to c and to the vector C–O) from the former position. In this way a reliability index value of 8.5%, based on intensities, was obtained. The values of $u = 0.20$ for NaI, $u = 0.27$ for C, $u = 0.40$ for O and for NaII, and $v = 0.27$ for O (all in Å), superimposed upon an isotropic root mean square amplitude of 0.39 Å, do indeed demonstrate the expected large and strongly directional effect.

Apart from the strong vibrations parallel to c, there is also an obvious libration of the anions, expressed by v. Although considerable, the value of v can in no way be interpreted as a free rotation of the anions in their plane. Moreover, such a free rotation would make reflections with $h - k = 3n$ and $l =$ odd vanish, whereas the reflection 301 is found to have a non-zero observed intensity at least up to 700°C.

The distance C–O in this refinement is found to be $r = 1.23$ Å (corrected for libration), and it is even less in the earlier refinements from X-ray data. Now CO_3 is known to be a very rigid element of any structure in which it appears, with a C–O distance seldom exceeding the range $r = 1.29 \pm 0.01$ Å. The discrepancy can be caused only by strong anharmonicity of the thermal motion.

As regards the space group it can be remarked that several symmetries lower than $P6_3/mmc$ are not excluded by diffraction evidence. Nevertheless these groups are unlikely to be valid for the α phase, because they do not include as subgroups the symmetries of the β and γ phases. Indeed, $P6_3/mmc$ is the only group which does include them for all three carbonates. In using this argument we assume, of course, that the continuity of all transitions implies a group–subgroup relationship of the two space groups involved in each transition.

2.2. *The crystal structure of β-Na_2CO_3*

Below the α–β transition temperature at $T_m = 489$°C, the structure is monoclinic. The spacegroup of the β-modification is C2/m with 4 Na_2CO_3 in the centred unit cell given in table 2, and positions occupied as follows:

2Na(1)	in	(a)	0,0,0.	etc.	4C	in	(i)	$x,0,z$	etc.
2Na(2)	in	(c)	$0,0,\frac{1}{2}$	etc.	4O(2)	in	(i)	$x,0,z$	etc.
4Na(3)	in	(i)	$x,0,z$	etc.	8O(1)	in	(j)	x,y,z	etc.

The NaI atoms are those numbered (1) and (2), whereas Na(3) represents the NaII atoms of section 1. The structure has been determined by de Jong and Tuinstra (1979) using single-crystal X-ray diffraction. The detailed data in the present section are cited from the full report of that investigation (de Jong 1979) which was carried out at 380°C. [The data obtained by de Pater (1978) from the neutron powder diffraction pattern are compatible with de Jong's but are much less accurate.]

We shall now discuss the question whether the β-phase structure confirms the mechanism of the α–β transition presented in section 1. Three main aspects of that model can be distinguished:

Rigidity of the CO_3 ion. This can be judged by looking at its internal interatomic distances. The experimental findings are C–O(2) = 1.27 Å; C–O(1) = 1.22 Å. The latter value occurs twice because C and O(2) lie in a mirror plane of the structure. For the same reason there are just two independent O–O distances, viz. O(2)–O(1) = 2.14 Å and O(1)–O(1) = 2.13 Å. None of these values has been corrected for libration. Just as in the α phase—though to a smaller extent—such a correction is uncertain because of the very strong thermal motion (the rms amplitudes are 0.3–0.4 Å). As it was found that O(1) has by far the largest thermal amplitude, the fact that C–O(1) is too short can

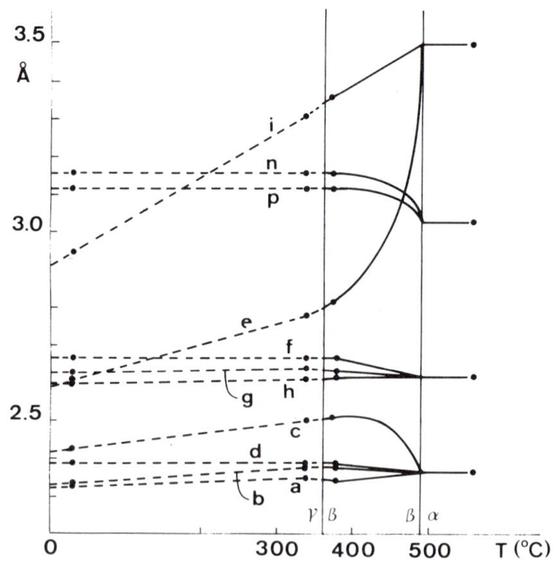

Fig. 5. Length of (a,...,i) Na–O bonds and (n,p)O–O bonds, the letters corresponding to those in fig. 6, as a function of temperature. Experimental values at 30, 340, 385 and 550°C are shown by dots. In the γ-phase, both the dots and the dashed interpolated lines refer to average values of a range due to modulation which is often quite large, cf. fig. 8.

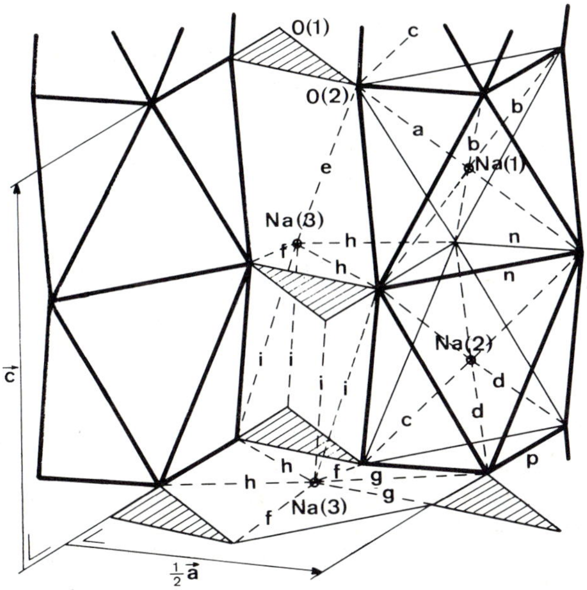

Fig. 6. Perspective view of the structure of β-Na_2CO_3, showing two strings of NaO_6 octahedra and some CO_3 triangles (hatched). The two Na(3) atoms shown are related by a 2_1-axis.

be ascribed to thermal libration of the anion about C–O(2). So there is no reason to doubt that the CO_3 ion is again an equilateral triangle with C in the centre. On the other hand, the discrepancy for C–O(1) means that interatomic distances derived from this structure determination cannot be trusted to within better than a few hundredths of an Å.

Rigidity of the NaO_6 octahedra. This feature can be judged by looking at the plot (fig. 5) of distances versus temperature, and at the perspective view of the structure (fig. 6) in which each distance is marked by the same letter as in the plot (corresponding to those used by van Aalst et al. 1976). The relevant distances are Na–O (a, b, c, d) and O–O (n, p), the latter being edges of the face shared by two octahedra. The Na–O bond lengths are within the range corresponding to the sum 2.40 of the ionic radii, with the exception of distance $c = 2.52$ Å. Even this deviation, though significant, is small. Very remarkable is the fact that the average Na–O distance in the octahedra equals the value for the α phase. It means that the strong thermal expansion parallel to c corresponds not only to an elongation of the octahedra in that direction, but simultaneously to a lateral compression. This is borne out also by the behaviour of the relevant O–O distances n and p in fig. 5, both significantly larger than in the α phase.

Environment of Na(3). For the six coplanar Na(3)–O bonds (f, g, h), all equal in the α phase but equal in pairs here, the plot shows that not much change or spread occurs. The hexagon becomes somewhat buckled but it is still recognizable in fig. 6, where it is shown in full for the lower Na(3). Evidence for the dramatic character of the $\alpha-\beta$ transition is obviously not to be found in this planar coordination.

The one and only nearest-neighbour bond which changes spectacularly is the distance e between Na(3) and the O(2) of one of the CO_3 ions situated about $\frac{1}{2}c$ from it. The drop of e from 3.50 to 2.82 Å is the single major consequence of the S_5 shear (fig. 4) for the bonding scheme. This drop is so large that the bond e to Na(3) begins to compete with the six bonds of the latter's plane coordination. It also unbalances the symmetrical distribution of forces in the α phase. In particular the above-mentioned lengthening of the Na-O distance c can now be seen as a consequence of the attraction along bond e, acting on O(2) where it joins the bond c to Na(2); cf. fig. 6, top centre. Obviously the distance $e = 2.82$ is still to large to put a stop to the S_5 shear mechanism; as we shall see, another mechanism, leading to the γ phase, sets in before a seventh close Na(3)–O contact is reached by further decrease of e.

3. The crystal structure of γ-Na_2CO_3

3.1. Type of modulation and crystal symmetry*

The diffraction pattern of the γ phase shows all reflections of the β phase. For these 'main reflections' there is at the transition no discontinuity, either in position or in intensity (fig. 1). Besides, however, there appear a great many satellites some of which are clearly visible even in the powder pattern. Single-crystal diffraction intensities show that the 2/m point group symmetry of the β phase is still valid in γ-Na_2CO_3. The positions of the satellites are given by

$$H = h\boldsymbol{a}^* + k\boldsymbol{b}^* + l\boldsymbol{c}^* + m\boldsymbol{q} \quad (h, k, l, m \text{ integer})$$

based on the same axes $\boldsymbol{a}, \boldsymbol{b}, \boldsymbol{c}$ as in the β phase, so that main reflections have $m = 0$ and satellites have $m \neq 0$. The vector $\boldsymbol{q}$ is perpendicular to $\boldsymbol{b}$:

$$\boldsymbol{q} = \alpha\boldsymbol{a}^* + \gamma\boldsymbol{c}^*$$

Values of α and γ for some temperatures are plotted in fig. 7. These coefficients appear to be continuous and mutually independent functions of the temperature at least down to -100°C (cf. also section 5). This means that

*This section is based largely on the paper by van Aalst et al. (1976).

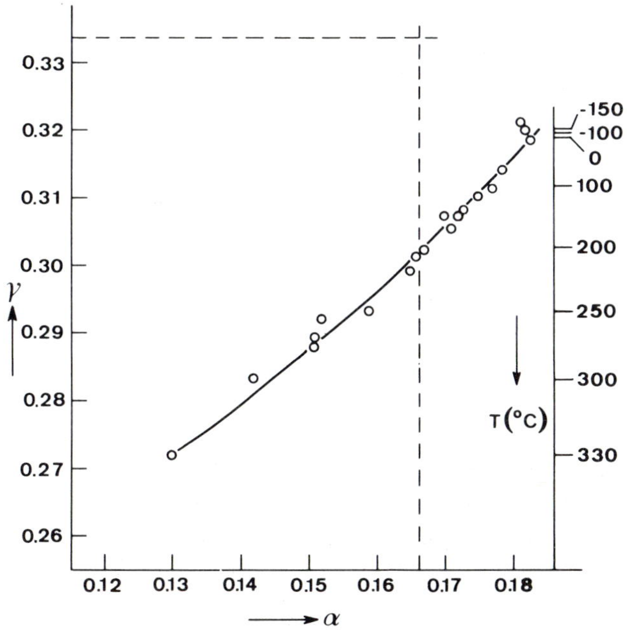

Fig. 7. The components of the modulation vector $\boldsymbol{q} = \alpha \boldsymbol{a}^* + \gamma \boldsymbol{c}^*$ at various temperatures. Dashed lines represent $\alpha = \frac{1}{6}$, $\gamma = \frac{1}{3}$.

γ-Na_2CO_3 is a typical example of an incommensurably modulated structure*. It has been solved on the assumption that the modulation is a displacive one. In order to describe such a distortion, a separate wave ("modulation function") has to be assigned to each atom of the unit cell of β-Na_2CO_3 for each of its three coordinates. Even assuming harmonic modulation, this yields six modulation parameters per atom.

Fortunately there are many symmetry relations between these parameters. Rather than following the original symmetry description in four dimensions, one can see these relations as the symmetry of the pattern formed by all modulation waves in three-dimensional space. For instance, it turns out that if in the basic structure (the β-phase structure) two atoms are related by a reflection in a mirror plane of the C2/m symmetry, then their modulation waves are related by a glide reflection in the same plane. The glide component is half the modulation wavelength; it causes a systematic extinction among satellites, the discovery of which was crucial for the structure analysis. Of the latter's result, three main aspects will now be described.

*In one microscopic theory cited in Ch. 2 (Fortuin 1977), Na_2CO_3 is used as an example and its modulation is explained at least qualitatively.

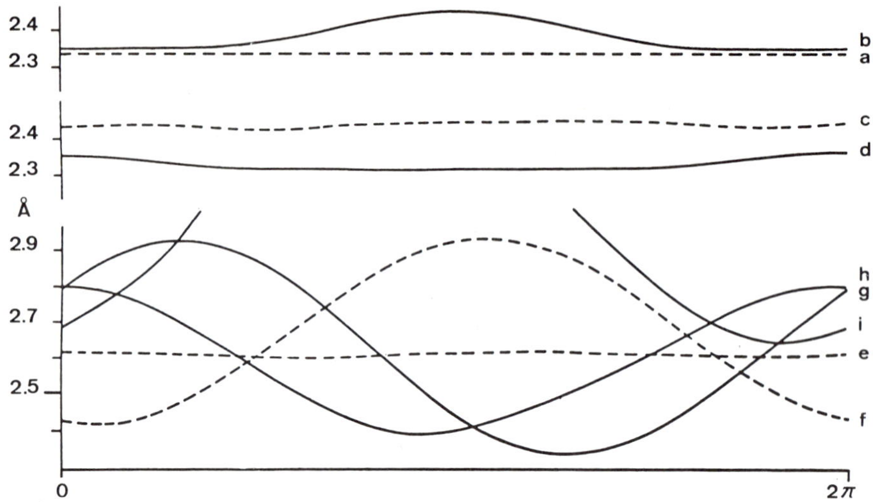

Fig. 8. Length of ($a,\ldots,i$) Na–O bonds, the letters corresponding to those in fig 6, as a function of the phase of the modulation wave, at room temperature.

3.2. *Crystal structure and interpretation*

A detailed insight in the bonding scheme can be obtained by studying interatomic distances in relation to the phase of the modulation, that is, the behaviour of each distance (between atoms situated in a fixed unit cell) when the entire modulation 'wave' is imagined to propagate through the crystal. Then each bond undergoes a periodic variation. This is shown graphically in fig. 8 for the bonds $a \ldots i$ indicated in fig. 6. The main conclusions are:

–Internal bonds in the CO_3 anion and in the NaO_6-octahedra vary only by little amounts, in accordance with the rigidity ascribed to these structural units.

–Na(3) alternately comes in close contact with three out of the six surrounding oxygens (bonds f, g and h) at distances around 2.40 Å.

–The seventh Na(3)–O bond e formed in the β phase does not reach low values, but it does not vary from its mean, 2.61 Å. A similar bond length is reached by the two bonds i, though only in alternating deep minima.

It can be concluded that the limit of the distortion is set by the low minimum of bonds f, g and h which at last seem to establish a stable configuration around Na(3).

The modulation functions are smooth periodic functions. The supposition of a square-wave type of distortion made formerly by Dubbeldam and de Wolff (1969) has been disproved, but Hogervorst and de Wolff (1979) have shown that the higher harmonics in the modulation functions are not negligible. The wave form of these functions is something in between a sinusoid and a square

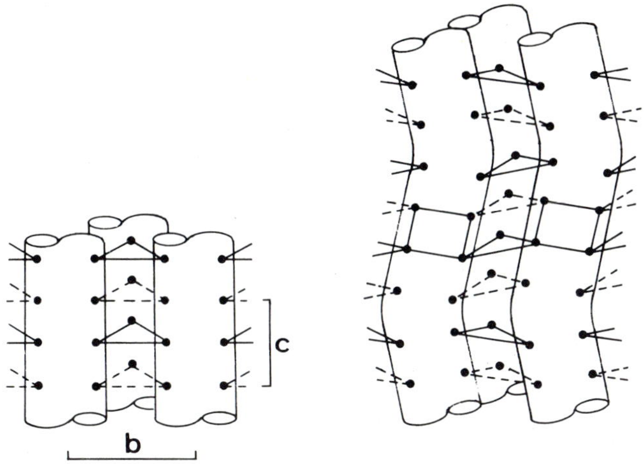

Fig. 9. (left)β-Na_2CO_3 in the symbolized S_5 shear version of fig. 4c, but now seen along the normal to the b, c-plane; (right) γ-Na_2CO_3 seen in the same direction. The rectangles indicate the NaO_6 octahedra.

wave. Displacements are mainly parallel to $\boldsymbol{b}$, and the average amplitude is 0.3 Å.

Disregarding deviations from this average amplitude, the *overall picture* of the modulation is, therefore, a structure derived from the β-phase structure by a shear deformation with a magnitude alternating periodically along the direction of $\boldsymbol{q}$, but strictly homogeneous in planes perpendicular to $\boldsymbol{q}$.

In terms of our columnar model, such a deformation requires periodic shear and bending of the columns as well. If this deformation would follow the quasi-homogenous overall picture, the interatomic distances within the columns would undergo very considerable changes. Such changes are in contradiction with the above experimental findings and also with expectations, in particular for the strong Na(1)–O and Na(2)–O bonds.

Therefore the detailed behaviour of the columns is expected to be rather as in fig. 9: In the more or less straight parts of the column between successive bends, the column is merely tilted in the b, c plane without internal strains. This requires a considerable wiggle of the rigid CO_3 ions which again act as freely hinging connections between neighbouring columns. (Indeed the actual wiggle amounts to as much as 20° to each side). Thus we are led to examine the deformations allowed by the β-phase structure assuming rigid columns and rigid anions. Upon close analysis (de Wolff and Tuinstra 1985) such a shear mechanism (neglecting a possible rotation of columns about their axes) is found to yield a surprising result: it can happen only in such a way that there is no macroscopic distortion in the average plane of the CO_3 ions.

Since, on the other hand, the wave vector $\boldsymbol{q}$ of the modulation has to be perpendicular to such an undistorted plane this result forces $\boldsymbol{q}$ to be parallel to the normal $\boldsymbol{n}$ of the average CO_3 plane. Now the actual azimuth values of $\boldsymbol{q}$ and $\boldsymbol{n}$ in the a, c plane are indeed found to be equal within the error of measurement (1°).

The incommensurability of the modulation can be explained as follows: If a modulation of the β-phase structure is assumed to occur, such a compulsary parallelism of $\boldsymbol{q}$ and the CO_3-normal [verified also at temperatures (290 and 340°C) close to T_i by de Jong (1979)] forces the modulation to be incommensurate: In the β phase the CO_3-plane is not parallel to a net plane. Therefore at T_i, $\boldsymbol{q}$ is already trapped in an incommensurate direction. At lower temperatures, the average CO_3-orientation has no reason to become that of a net plane either, hence the components of $\boldsymbol{q}$ remain irrational.

In the above mentioned analysis, attention has also been given to a possible rotation of the columns about their axis. It is found that if a reasonable amount of rotation is assumed, the parallelism of $\boldsymbol{q}$ and $\boldsymbol{n}$ is hardly affected. Hence we may conclude that the columnar model adequately explains the direction of the modulation vector.

Finally, the S_4 shear strain expected in γ-Na_2CO_3, (section 1.3) indeed turns out to resemble the actual situation. It would have been realized if $\boldsymbol{q}$ had been parallel to $\boldsymbol{c}$ instead of being at a considerable angle with it, viz. 31° at room temperature. Since the distortion is an alternating strain, it might be called a pseudo-S_4 fold analogous to the S_5 fold in K- and Rb carbonate.

The main feature of the other transitions is evident for β-γ Na_2CO_3 as well. It is the very large shear strain allowed by the hinge function of the anions, represented here by their wiggle rotation. Though in the diffraction pattern the change does not look as drastic as the $\alpha-\beta$ transition, the actual amount of strain is of the same order, viz. about 10% in the regions of maximal distortion.

4. *Physical properties of Na_2CO_3*

4.1. *Introduction*

In the previous sections of this chapter a relationship between the $\alpha-\beta$ transition and the $\beta-\gamma$ transition is suggested by the structural changes. In the present section we will see that this relationship is also shown by other physical properties.

The $\alpha-\beta$ phase transition is of second order or nearly so and occurs at the centre of the Brillouin zone: $\boldsymbol{q} = 0$. The crystal point symmetry is lowered from $6/mmm$ (D_{6h}) to $2/m(C_{2h})$ with the hexagonal a-axis along the two fold axis of $2/m$. The representations which can induce such a transition should

Table 3
Symmetry of lattice modes at $\boldsymbol{q} = 0$ of α-Na_2CO_3.

Type	Number					Modes					
Total	36	A_{1g}	A_{2g}	$3B_{2g}$	E_{1g}	$4E_{2g}$	$4A_{2u}$	$2B_{1u}$	B_{2u}	$5E_{1u}$	$2E_{2u}$
acoustical	3						1			1	
translational	15			2		2	2	1		2	1
rotational	6		1		1				1		1
internal	12	1		1		2	1	1		2	

conserve the 2/m symmetry and break the symmetry of all higher order subgroups of 6/mmm. The representations of 6/mmm which fulfill the former requirement are A_{1g}, B_{2g}, E_{1g} and E_{2g}. Of these A_{1g} leaves 6/mmm untouched while B_{2g} leaves $\bar{3}$m (D_{3d}) invariant and thus would lead to a trigonal rather than a monoclinic β phase. E_{2g} leads to mmm (D_{2h}) symmetry. Only E_{1g} leads, with one of its two components, to the appropriately oriented 2/m point group. Since E_{1g} does not allow a third order term in the free energy expansion, this representation is active in the sense of the Landau theory. Also the Lifshitz condition is fulfilled. E_{1g} thus would be able to induce a second-order $\alpha \rightarrow \beta$ transition.

A group analysis (see also de Pater and Helmholdt 1979, and Maciel et al. 1981) of the modes of the α-structure at $\boldsymbol{q} = 0$ is given in table 3. The E_{1g} mode is a rotational mode of the CO_3-ion; the doubly degenerate mode is depicted in fig. 10 resolved in two components. If at the phase transition this lattice mode gets soft it does so for both components. In the case of Na_2CO_3 apparently the left hand component in fig. 10 is chosen; K_2CO_3 just chooses the other

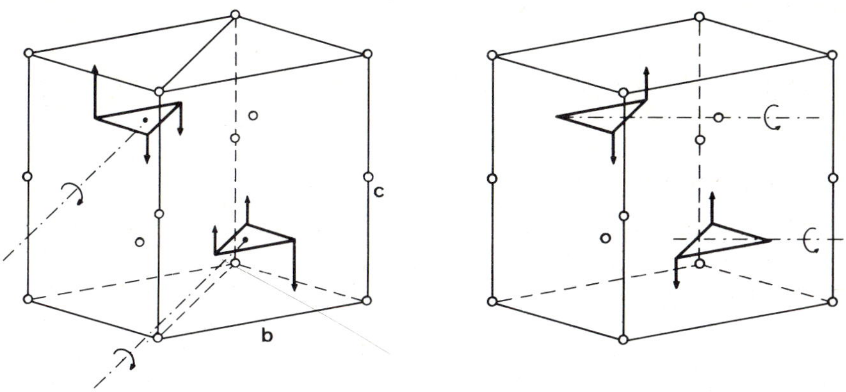

Fig. 10. The two components of the E_{1g}-mode of the α phase. The mode is a rotational mode of the CO_3 ions; the arrows indicate the atomic amplitudes, circles are Na.

component. The preference for one or the other must be determined by nonlinear effects, i.e. properties described by tensors of order higher than four. In view of the thermal motions observed in X-ray analysis and the diffuse scattering occurrence of nonlinear effects is not amazing.

The above analysis fits in the structural picture developed in the previous sections. For the $\alpha-\beta$ transition the shear and the observed rotation of the CO_3 group, both about the *b*-axis, transform according to E_{1g}. The shear, however, cannot be characterized as a lattice mode at $\boldsymbol{q}=0$ but belongs to the acoustic branch with $\boldsymbol{q}=(0,0,q_3)$. This branch originates from the doubly degenerate E_{1u} representation at $q=0$. One of its components induces the S_5-shear. Along the direction of $\boldsymbol{c}$ the branches originating at E_{1u} and E_{1g} couple since both belong to the same representation of the group of $\boldsymbol{q}=(0,0,\gamma)$.

The main distortions accompanying the $\beta-\gamma$ transition are a *b*-polarized periodic shear and a rotation of the CO_3 ion about the C–O(2) bond with the same periodicity. The acoustic mode and the rotational mode from which they seem to condense belong to branches containing exactly the two components

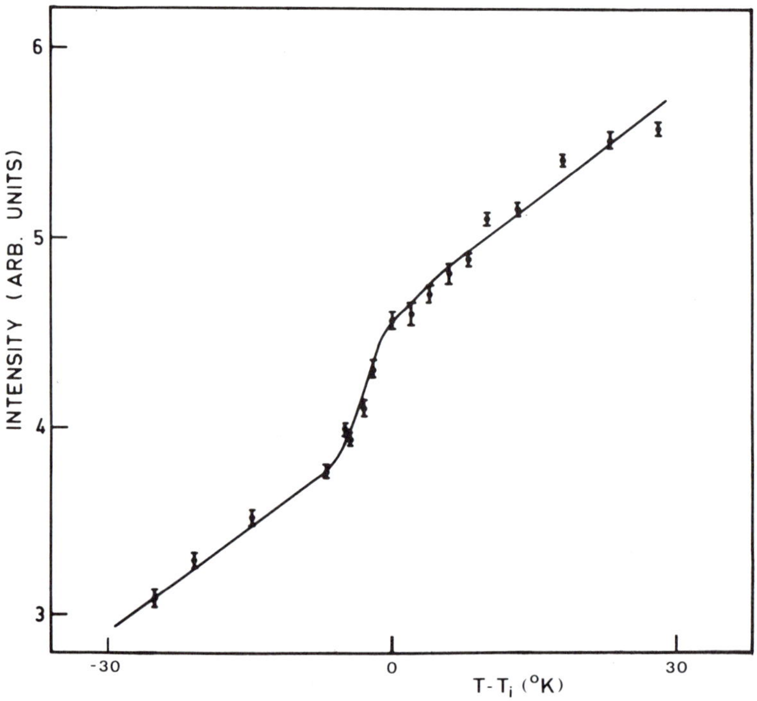

Fig. 11. Temperature dependence of the intensity of the diffuse scattering at a point close to the (040) reflection, in the direction of $\boldsymbol{q}$. (de Pater 1978.)

of E_{1g} and E_{1u} left over in the $\alpha-\beta$ transition. In the monoclinic phase these representations are called B_g and B_u respectively.

In short, at the $\alpha-\beta$ transition two modes couple at $\boldsymbol{q} \approx 0$; one is a component of E_{1g} the other from E_{1u}, at 489°C they get soft. At the $\beta-\gamma$ transition two branches containing the components of E_{1g} and E_{1u} conjugate to those of the $\alpha-\beta$ transition condense at a point in the $\boldsymbol{a}^*, \boldsymbol{c}^*$ plane of reciprocal space.

4.2. *Diffuse scattering*

Diffuse neutron as well as X-ray scattering was extensively studied by de Pater (de Pater 1978, de Pater and Helmholdt 1979).

Because of the second-order character of both phase transitions, diffuse scattering is to be expected. For the $\gamma-\beta$ transition fluctuations of the order parameter should give rise to diffuse intensity in the β phase at positions where in the γ phase the satellite reflections occur. Around the $\alpha-\beta$ transition diffuse intensity should be present around the Bragg-reflections.

de Pater actually observed diffuse intensity peaking at the satellite positions in a temperature range extending from the $\gamma-\beta$ transition up to some 15°C above it.

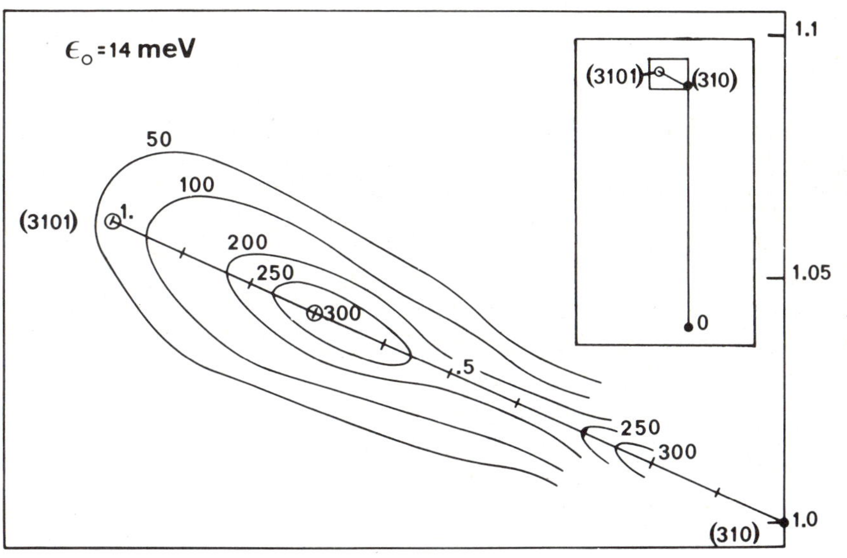

Fig. 12. Diffuse neutron scattering in the neighbourhood of the (310) reflection at 17.5 K above T_i. The rod of intense diffuse scattering extends from the main reflection (below right) towards the position where the (3101) satellite was observed below T_i. The anisotropic intensity distribution close to (310) is also visible. (de Pater 1978.)

The diffuse intensity around the main reflections is also observed. de Pater and Helmholdt (1979) noted some remarkable features:

(i) It is already observable 60°C below the $\gamma-\beta$ transition and is present throughout the whole range of the β phase. Above 370°C it acquires more the character of a broadening of the Bragg reflections.

(ii) Its intensity increases quite abruptly on passing the $\gamma-\beta$ transition from below (fig. 11).

(iii) The intensity is not isotropically distributed around the main reflections. The diffuse scattering is confined to a rod-like region extending from the main reflection in the direction of the modulation wave vector $\boldsymbol{q}$ (fig. 12).

(iv) The rod mentioned under (iii) is longer, the higher the k-index of the corresponding Bragg reflection. The reflections ($h0l$) show no diffuse rod.

The last property indicates that the atomic displacements causing the diffuse scattering are in the $\boldsymbol{b}$-direction. The acoustic b-polarized mode is apparently soft throughout the β phase. Its a-polarized companion already condensed at the $\alpha-\beta$ transition. Approaching the $\alpha-\beta$ transition from higher temperatures the acoustic branch softens making a trough in the dispersion surface in the direction of $\boldsymbol{q}$. It couples with the wiggle mode of the CO_3 ions at $\boldsymbol{q}$ and hardens in the γ phase.

4.3. *Unit cell parameters as a function of temperature*

As stated before, the unit cell parameters change continuously and no hysteresis as a function of temperature is observed.

Thermal expansion of a meltgrown crystal has been measured by Midorikawa et al. (1978). They present a plot of the thermal expansion along the $\boldsymbol{c}$-direction and along an unspecified direction perpendicular to $\boldsymbol{c}$.

Crooijmans et al. (1983) reported coefficients of thermal expansion based on X-ray powder diagrams. Their results are shown in table 4. In the regions I and III the expansion is linear. In region III the expansion coefficient in the $\boldsymbol{c}$ direction is more than three times the value in range I. It should be noted that the transition from region I to region III occurs at 280°C, which temperature is well below the $\beta-\gamma$ phase transition occurring at 360°C.

The anomalous thermal expansion in region III can to a certain extent be understood with the model introduced in section 1.3. The c-axis is expanding at a high rate. If now the NaI–O distance is taken to be constant the elongation of the c-axis will make the shared triangular faces of the oxygen octahedra in the columns shrink. As the CO_3 ion is not expected to be affected by thermal expansion, the final effect will be shrinking cell dimensions perpendicular to the c-axis. This explains the negative expansion coefficient in the $\boldsymbol{b}$ direction. The a-axis may still expand due to the change of the angle β.

Table 4
Cell parameters of Na_2CO_3 at four temperatures. The linear expansion coefficients along the crystallographic axes are given as α_a, α_b, and α_c for the ranges I and III. Range II is a transition range with nonlinear expansion.

Quantity	Value						
		I		II		III	
T(°C)	−180		230		330		485
a(Å)	8.868(2)		8.942(2)		8.977(1)		9.028(3)
b(Å)	5.220(2)		5.251(2)		5.252(1)		5.210(1)
c(Å)	5.971(2)		6.120(2)		6.220(1)		6.465(3)
β(°)	101.87(3)		100.43(3)		99.13(1)		90.69(4)
$\alpha_a(10^{-4}\,K^{-1})$		0.22(2)				0.42(2)	
$\alpha_b(10^{-4}\,K^{-1})$		0.13(2)				−0.6(2)	
$\alpha_c(10^{-4}\,K^{-1})$		0.62(2)				2.2(1)	

4.4. *Raman and infrared spectroscopy*

Raman and infrared studies of the incommensurate phase of Na_2CO_3 were published by Brooker and Bates (1971). Later Maciel and Ryan (1981) made an investigation of the Raman spectra as a function of temperature.

Of the higher frequency spectra, i.e. internal vibrations of the CO_3-ion, the most striking feature is a splitting of the spectral lines which are assigned to nondegenerate vibrational modes. The symmetric breathing mode for instance, which is Raman active and observed close to the free ion frequency of CO_3 at $1063\,cm^{-1}$, is split into two lines a few cm^{-1} apart. The splitting, which is observed in the γ phase only, was measured as a function of temperature. The distance Δf of the two lines was found to be proportional to $T_i - T$.

In both papers mentioned the splitting of this mode was assigned to a two-site effect. The sites used in the argument are however crystallographic equivalent positions, because the modulation wave function has a glide plane as a symmetry element. This would yield the same frequency for both sites.

As shown for NMR (Blinc 1981), the modulation does all the same cause a splitting of nondegenerate spectral lines.* Travelling along the direction of the modulation wave vector through the crystal we would meet CO_3 ions at continuously changing sites. In a harmonic approximation $u = a\sin 2\pi tq$ the population density n of sites would diverge at $u = \pm a$: $dn/du = (a^2 - u^2)^{1/2}$. Because of the symmetry (the glide plane) the frequencies of vibrational modes should be the same at $+u$ and $-u$. Therefore the frequency f must be an even function of u: $f = b + ca^2\sin^2 2\pi tq$. This causes dn/df not only to diverge at

*The validity of this argument for the Raman scattering was suggested to us by Dr. A.C.R. Hogervorst.

$f = b + a^2c^2$ ($u = \pm a$), but also at $f = b$ ($u = 0$). The separation of the two lines is $\Delta f = ca^2$. In a harmonic approximation we would expect two spectral lines of about the same intensity.

In this way the appearance of two spectral lines with comparable intensities is explained. An intensity difference will be caused by anharmonic effects which occur at temperatures away from T_i. The observed linear temperature dependence of Δf causes the order parameter a to depend on temperature as $(T_i - T)^{1/2}$ over a wide temperature range down to about 100 K (Maciel and Ryan 1981); this is in accordance with the classical Landau theory.

The assignment of the lattice modes at low frequencies is speculative since no polarized Raman data are available. This is due to the lack of single crystals of appropriate size. Nevertheless two modes are observed to get soft on approaching the $\alpha - \beta$ transition. They are believed to be librational modes of the CO_3 ion. Two additional lines observed in the γ phase are assigned to the coupling of the acoustic librational branches at $\boldsymbol{q}$.

4.5. Other physical properties

4.5.1. Crystal growth

Work on the physical properties of Na_2CO_3 is severely hampered by the lack of single crystals of appropriate size. Many investigations have only been done with powder samples, with which anisotropic crystal properties are hard to detect.

Crystal growth from an aqueous solution has been reported by van Aalst et al. (1976). Crystals in the form of thin platelets up to 0.3 mm thick and a few mm across were grown at vapour pressure of 1.5 bar. The solution boils at 120° at this pressure, which temperature is well above the point where the anhydrous carbonate and the monohydrate are in equilibrium as solid phases in a saturated solution.

At atmospheric pressure crystals can be grown from the ternary system Na_2CO_3–NaCl–H_2O. Anhydrous Na_2CO_3 crystals are deposited at temperatures between 97°C and the boiling point. A.J. van den Berg was acknowledged by de Pater and Helmholdt (1979) for growing their crystals in this way. Crystals with dimensions of a few mm in all directions were obtained.

The most successful growth has been reported by Midorikawa et al. (1978). They grew crystals from the melt by a modified Bridgman method. The recipe is described in detail in their publication. The maximum size reported is 3×3×2 mm. Above 20°C a small compressive stress seems enough to transform the multidomain crystal into a single domain one (Midorikawa et al. 1980). Later on, crystal growth from the melt was reported by Maciel and Ryan (1981) using Bridgman and Czochralski techniques. Though their crystals were relatively large (up to 2 cm^3) they were heavily microtwinned and opaque.

The microtwinning, which causes the opaqueness of melt-grown crystals, is related to the ferroelastic behaviour of Na_2CO_3 in the β-phase.

4.5.2. *Ferroelasticity*

The reduction of the crystal pointgroup from 6/mmm to 2/m at the continuous α–β transition makes a crystal of Na_2CO_3 vulnerable to ferroelasticity in the β phase. According to Aizu (1970) the symmetry reduction makes Na_2CO_3 a complete ferroelastic in the β phase; 6/mmm F 2/m(S) in Aizu's nomenclature. This implies that the number of equivalent orientationally different crystals in which a single crystal from the α phase can be transformed in the β phase is six.

Ferroelastic behaviour was indeed observed in Na_2CO_3 by Midorikawa et al. (1980). Unfortunately they were able to look with the polarizing microscope down the *c*-axis only and thus could observe only three of the six domain orientations. The orientations of the domains could be seen to be 30° apart; the domain walls were parallel to (110) and (310). Above 470 K (that is, even in the γ phase) the domain walls could be made to move. The direction of the motion was such as to make domains grow of which the *a*-axis was parallel to the applied stress. This is in agreement with the observed ratio a/b which is smaller than $\sqrt{3}$ in the β and γ phases. The coercive stress wad estimated to be 30–40 N/cm^2 at 470 K, being lower at higher temperatures.

Since not $(a/b-\sqrt{3})$ but rather $(\beta-90)$ is the primary order parameter, it would be interesting to know the response to a shear stress about the *b*-axis. In order to observe this part of the ferroelasticity one should have crystal plates cut perpendicular to the *b* axis at one's disposal.

From our mechanical model one would expect small coercive stresses, however the lattice distortions are quite large; the *c*-axis or the *a*-axis must rotate through $2\times(\beta-90)°$. It is supposedly the multidomain situation of this type which makes the crystals opaque after the α–β transition, especially when viewing perpendicular to *c*.

4.5.3. *Birefringence*

In all phases of Na_2CO_3 the planar CO_3 ions are packed more or less parallel to each other. This causes the birefringent properties of Na_2CO_3 to be determined predominantly by the optical anisotropy of the CO_3 ion which also causes the well known birefringence of calcite.

Experimentally, Waldeck et al. (1932) observed a large birefringence. At room temperature Na_2CO_3 is optically biaxial: $n_x=1.410$; $n_y=1.537$ and $n_z=1.544$ (±0.001); the biaxial angle $2V=28\pm2°$.

The monoclinic symmetry requires one axis, the *y* axis, of the indicatrix to be parallel to the crystallographic *b*-axis. The *x* and the *z*-axis of the indicatrix are coplanar with the *a* and the *c* axis but their directions are not specified by symmetry.

Lindeyer (1969) determined the orientation of these axes with respect to the crystallographic c and a axes. At room temperature the x axis of the indicatrix makes an angle of 30° with the crystallographic c axis and lies in the obtuse angle between the a and c axis. At room temperature the direction of the x axis thus coincides with the normal to the plane of the CO_3 ion.

We assume at all temperatures the orientation of the indicatrix to be fully determined by the orientation of the CO_3 ion, its shorter axis being normal to the plane of the anion and the other two virtually equal.

At T_m the x axis of the indicatrix coincides with the crystallographic c axis; the difference between n_y and n_z has vanished. As the temperature is raised from T_i and approaches T_m this situation is reached in a continuous manner by a combination of a rotation of the indicatrix about the b axis and a change of its form.

In our mechanical model the angle ψ between the plane of the CO_3 ion and the a axis depends linearly on $(\beta - 90)$ or nearly so. A mere rotation of the indicatrix at T_m as indicated would change the birefringence Δn by an amount proportional to ψ^2 i.e. $(\beta - 90)^2$.

The difference between n_y and n_z is caused by a difference in packing density of the CO_3 ions along the y and the z direction respectively. A projection of the structure onto the z axis gives a denser packing than a projection onto the y $(= b)$ axis. This difference, which vanishes at T_m, can from our model be found also to be proportional to ψ^2 i.e. $(\beta - 90)^2$.

In all, Δn is expected to change as $(\beta - 90)^2$. Now thermal expansion data show that close to T_m, $(\beta - 90°)^2$ is proportional to $T_m - T$, as predicted also by classical Laue theory. Accordingly, Δn as well should vary as $T_m - T$. This result was derived by Midorikawa et al. (1980) using a more elaborate calculation essentially along the same lines. The latter authors moreover established an experimental verification by measuring the birefringence in a crystal plate cut perpendicular to the c axis; it did show the expected linear dependence upon T.

4.5.4. NMR

Recently Jäger and Heide (1983) studied the temperature dependence of the ^{23}Na-central lineshape in Na_2CO_3 as a function of temperature.

At the phase transition temperatures T_i and T_m no change in the lineshape was observed. Apparently the environment of the sodium nuclei is not altered at these temperatures. From previous sections it is clear that the octahedral surroundings of the sodium ions NaI change slowly and gradually as a function of temperature. It is rather the relative position of the rather stiff octahedra that changes at the phase transition. The NMR lineshape does change in the temperature region of 240°C to 300°C. In this region the environment of Na(3) is changed quite drastically, which change is possibly responsible for the deformation of the lineshape.

Another observation of Jäger and Heide is a linear increase of 6% in the quadrupole frequency as the temperature is raised from 320°C to 450°C. The authors attribute this to a decrease in the NaO distance. From fig. 5 one can see that indeed the average NaO distances shrink a little in the temperature region between T_i and T_m.

4.5.5. *Electric conductivity*

The lack of single crystals of good quality inhibits experimental studies of these properties. The electric conductivity has been measured by Cerisier and Roux (1977) on powder samples. The conductivity γ, though reproducible, depends strongly on the atmosphere under which the experiments are performed. All experiments show the phase transition points T_i and T_m as changes in the slope of $\log \gamma$ against $1/T$. A change in slope at 620°C has been discussed in section 1.2.

5. *Evidence for a lock-in transition at low temperatures*

In their study of the low temperature behaviour of γ-Na_2CO_3 de Pater and Helmholdt (1979) reported a peculiar lock-in transition at -143°C. In a single crystal the modulation wave vector was observed to follow a line $\gamma = \alpha + 0.140$ in the reciprocal unit cell as the temperature is lowered from T_i to room temperature. At room temperature the endpoint of the modulation wave vector changes it trajectory along another line $\alpha + \gamma = 0.5$ and reaches a point $(\frac{1}{6}, \frac{1}{3})$ at -143°C.

The manner in which the alleged locked-in situation is reached is quite unique. Unfortunately the accuracy in the low-temperature range is very low.

Later Crooijmans et al. (1983) tried to reproduce these results using a very accurate powder X-ray camera. In their temperature runs down to -180°C no deviation of the modulation wave vector from a smooth line (roughly $\gamma = \alpha + 0.140$) could be found. The X-ray measurements exclude a lock-in at $\frac{1}{6}, \frac{1}{3}$ in powder samples above -180°C, see fig. 7.

Most of the evidence produced by de Pater and Helmholdt (1979) is on single crystals. It could be possible that in larger crystals a lock-in occurs due to inhomogeneities, internal stress or impurities. In a powder the ratio of crystal surface area to bulk volume is so much larger that the inhomogeneities etc. can easily diffuse to the surface. Nevertheless de Pater and Helmholdt reported also to have found the lock-in transition in a neutron powder experiment.

The lock-in transition has also attracted attention from scientists using other methods.

Midorikawa et al. (1980) measured the specific heat and the dielectric constant. They did not observe any anomaly in either property in a temperature range around -143°C.

Also in the Raman spectra of Maciel and Ryan (1981) there is barely any evidence for a transition at low temperatures. One low frequency mode is only detected below $-110°C$; its origin is uncertain.

In summary: at present the occurrence of the peculiar lock-in transition at $-143°C$ is controversial and needs further investigation.

References

Aizu, K., 1970, Phys. Rev. **B2**, 754–772.

Becht, H.Y., and R. Struikmans, 1976, Acta Cryst. **B32**, 3344–3346.

Blinc, R. 1981, Phys. Reports **79**, (5)331–398.

Brooker, M.H., and J.B. Bates, 1971, J. Chem. Phys. **54**, 4788–4796.

Brouns, E., J.W. Visser and P.M. de Wolff, 1964, Acta Cryst. **17**, 614.

Bijk, P., 1967, Internal report (T.H. Delft).

Bijk, P., and J.W. Visser, 1968, Internal report (T.H. Delft).

Cantor, S., and D.E. Heatherly, 1977, Oak Ridge Report 5297.

Cerisier, P., and F. Roux, 1977, J. Solid State Chem. **22**, 245–251.

Christmann, M., and G. Papin, 1979, Rev. Chim. Miner. **16**, 485–489.

Crooijmans, P.C.B., G.H. Kolkman and F. Tuinstra, 1983, Abstr. Eighth European Crystallography Meeting, Liège, 243.

de Jong, F., 1979, Internal report (T.H. Delft).

de Jong, F., and F. Tuinstra, 1979, Structural changes at the incommensurate phase transition of Na_2CO_3, in: Modulated Structures, Hawai, 1979, eds. J.M. Cowley, J.B. Cohen, M.B. Salamon and B.J. Wuensch (AIP Conf. Proc. No. 53, New York) p. 220–222.

de Pater, C.J., 1978, Doctoral Thesis (T.H. Delft).

de Pater, C.J., 1979, Physica **96B**, 89–95.

de Pater, C.J., and R.B. Helmholdt, 1979, Phys. Rev. **B19**, 5735–5745.

de Wolff, P.M., and F. Tuinstra, 1985, to be published.

Dubbeldam, G.C., and P.M. de Wolff, 1969, Acta Cryst. **B25**, 2665–2667.

Ehrhardt, H., H. Schweer and H. Seidel, 1980, Z. Anorg. Allg. Chem. **462**, 185–198.

Eysel, W., 1971, Strukturelle und Kristallchemische Verwandschaft bei Verbindungen $A_2(BX_4)$ und $A_2(BX_3)$. Habilitationsschrift (RWTH, Aachen).

Fortuin, C.M., 1977, Physica **86A**, 224–256.

Gatehouse, B.M., and D.J. Lloyd, 1973, J. Chem. Soc. Dalton **1973**, 70–72.

Hertogs, P.H.S., 1977, Internal report (T.H. Delft).

Hogervorst, A.J., and P.M. de Wolff, 1979, Modulation wave form for Na_2CO_3, in: Modulated Structures, Hawai, 1979, eds. J.M. Cowley, J.B. Cohen.

M.B. Salamon and B.J. Wuensch (AIP Conf. Proc. No. 53, New York) p. 217–219.

Jäger, C., and K. Heide, 1983, J. Phys. **C16**, L159–L163.

Karataeva, I.M., and I.L. Lepeshkov, 1970, J. Inorg. Chem. **15**(12), 1751–1754.

Klement Jr, W., and L.H. Cohen, 1975, Ber. Bunsenges. Phys. Chem. **79**, 327–334.

Lindeyer, G., 1969, Internal report (T.H. Delft).

Maciel, A., J.F. Ryan and P.J. Walker, 1981, J. Phys. **C14**, 1611–1618.

Maciel, A., and J.F. Ryan, 1981, J. Phys. **C14**, L509–L514.

Marakov, S.Z., and M.P. Shulgina, 1940, Izv. Akad. Nauk SSSR, Otd. Khim. Nauk, 691–703.

Midorikawa, M., T. Hashimoto, Y. Ishibashi and Y. Takagi, 1978, J. Crystal Growth **44**, 505–506.

Midorikawa, M., K. Tsunoda and Y. Ishibashi, 1980, J. Phys. Soc. Jpn **49**(1), 242–245.

Papin, G., 1973, C.R. Acad. Sci. Paris **277**, 691–694.

Reisman, A., 1958, J. Am. Chem. Soc. **80**, 3558–3561.

Reisman, A., 1959, J. Am. Chem. Soc. **81**, 807–811.

Schneider, S.J., and E.M. Levin, 1973, J. Am. Cer. Soc. **56**, 218–219.

van Aalst, W., J. den Hollander, W.J.A.M. Peterse and P.M. de Wolff, 1976, Acta Cryst. **B32**, 47–58.

Visser, J.W., 1966, Card 16-820 of the Powder Diffraction File of JCPDS/CDD, Swarthmore.

Visser, J.W., 1975, Card 25-815 of the Powder Diffraction File of JCPDS/CDD, Swarthmore.

Visser, J.W., 1984, Cards to appear in future sets of the Powder Diffraction File of JCPDS/CDD, Swarthmore.

Wadhawan, V.K. 1978, Acta Cryst. **A34**, S 363.

Waldeck, W.F., G. Lynn and A.E. Hill, 1932, J. Am. Chem. Soc. **54**, 928.

CHAPTER 19

Statics and Dynamics of Incommensurate $BaMnF_4$

J. F. SCOTT

Department of Physics
University of Colorado
Boulder, CO 80309
USA

Incommensurate Phases in Dielectrics 2
Edited by
R. Blinc and A.P. Levanyuk

Contents

1. Introduction 285
2. Measurements near T_I 286
3. Linear birefringence 288
4. Evidence for a second transition at T_{II} 289
5. Extended Ising model calculations for incommensurates 291
6. Evidence for defect-stabilized structure in incommensurate $BaMnF_4$ 293
7. Magnetoelectric phenomena 294
References 298

1. Introduction

$BaMnF_4$ is a member of a family of ferroelectrics which includes $BaFeF_4$, $BaNiF_4$, $BaZnF_4$, $BaMgF_4$ and $BaCoF_4$. These crystals are isomorphic at ambient temperatures, crystallizing in space group $A2_1am$ (C_{2v}^{12}). They are ferroelectric at all temperatures; a prototypic paraelectric D_{2h} phase is approached at high temperatures, but the crystals melt before reaching T_C. Of the six members of the family, the Fe, Mn, Ni and Co crystals order antiferromagnetically with three-dimensional Néel temperatures of order 25 to 50 K and in-plane, two-dimensional spin ordering at temperatures near $2T_N$. At low temperatures all of the family members except $BaMnF_4$ retain their C_{2v} structure. For purely steric reasons (Mn^{++} is the smallest divalent transition metal ion) $BaMnF_4$ undergoes a low-temperature structural phase transition. Above $T_I = 250$ K there is enough thermal rms motion of fluorine ions to stabilize the C_{2v} structure shown in fig. 1; whereas below T_I the fluorine octahedra buckle around the undersized Mn-ions. The room temperature structure is known from the X-ray study of Keve et al. (1969), whereas the

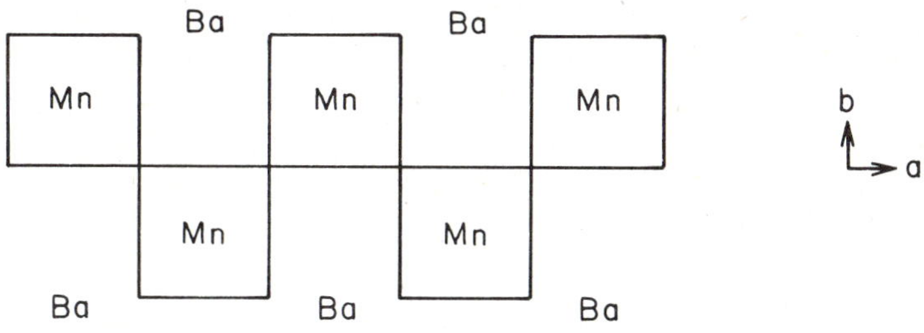

Fig. 1. Schematic diagram of $BaMnF_6$ in the prototype paraelectric D_{2h} phase. In the actual ambient C_{2v} structure the MnF_6 octahedra rotate about the c-axis, out of phase for adjacent octahedra. Below T_I the structure becomes incommensurate with modulation vector q_0 slightly less than $2a^*/5$. At low temperatures ($T_{II} = 170$ K upon warming) there is a pseudo-lock-in transition to a structure with repeat length given along a by five MnF_6 octahedra; that is, the structure shown above constitutes one primitive unit cell below T_{II} (since the primitive cell at room temperature consists of two diagonally adjacent octahedra, this is a net change in translation vector of $\frac{5}{2}$). MnF_6 displacements are along the c-axis in the incommensurate phase.

structure below T_I was shown to involve doubling of the primitive cell in the *bc*-plane (Ryan and Scott 1974) and a modulated, incommensurate distortion along ***a*** (Cox et al. 1979, 1981, 1983). The first of the Brookhaven reports (Cox et al. 1979) showed an incommensurate periodicity of $0.392a^*$ that is temperature independent from $T_I = 250\,\mathrm{K}$ down to 4 K. However, subsequent studies revealed a small temperature dependence (Cox et al. 1981, 1983), and recent data from Grenoble (Barthes-Regis et al. 1983) show that the temperature dependence of the incommensurate modulation is highly sample dependent and varies from $0.389a^*$ to $0.399a^*$ in the "best" samples. We show below that this is in accord with the extended Ising model calculations of Yamada and Hamaya (1983) for incommensurates, which predicts one possible trajectory in phase space which leads to $2a^*/5 > q > 5a^*/13$, i.e., $0.400 > q > 0.385a^*$. The final contribution regarding low-temperature structure was provided by the birefringence and optical activity studies of Pisarev et al. (1983); their work showed that the point group symmetry lowered at T_I from C_{2v} to C_2, with a monoclinic distortion angle of 4′ of arc observed below T_I. This value of 4′ is compatible with subsequent measurements via neutron scattering (Barthes-Regis et al. 1983) and X-ray scattering (Nelmes 1983).

2. *Measurements near* T_I

The first evidence of a phase transition near $T_I = 255\,\mathrm{K}$ in $BaMnF_4$ was the observation of divergence in the ultrasonic attenuation (Spencer et al. 1970). However these authors were tentative in ascribing that observation to a phase transition, since they could find no independent confirmation through optical absorption or magnetic resonance; and in fact their work was not cited even in subsequent work from the same laboratory. Subsequently, Ryan and Scott (1974) rediscovered the transition via Raman spectroscopy. Their work showed that the transition was primarily displacive, with a soft optical phonon decreasing in energy from approximately $40\,\mathrm{cm}^{-1}$ to about $10\,\mathrm{cm}^{-1}$ before becoming overdamped. The Raman study also allowed the explicit prediction of a lambda-shaped divergence in the dielectric constant along the twofold *a*-axis, a prediction confirmed by Samara and Richards (1976). Neither the Raman studies nor the dielectric measurements were able to ascertain whether the phase transition is truly continuous, but they showed any discontinuities to be extremely small, if present. The clearest evidence that the transition is slightly first order is provided by the birefringence studies (Schafer et al. 1983, Pisarev et al. 1983). They are compatible with the predictions from renormalization group theory (Cox et al. 1979) that the transition must be at least slightly first order. These results are also in complete accord with the specific heat results (Scott et al. 1982), which yielded an exponent $\alpha = 0.54$ above T_I for the specific heat divergence; mean-field theory without fluctuations predicts a critical exponent $\alpha = 0.5$ for systems very near tricritical points (i.e., where transitions go from first order to second order) but only for the low-temperature phase. However, fluctuations, especially in the case of a multicomponent order parameter, give $\alpha = 0.5$ in both phases, even far from tricritical points. The specific heat results yielded experimentally even larger values of α' describing the divergence below T_I; these were approximately 1.0 and compatible with the defect theory of Levanyuk et al. (1979). Note that this result implies that the defects which dominate specific heat below T_I disappear above T_I. This immediately suggests the role of discommensurations.

A number of early measurements showed strong sample-dependent effects. The soft mode data of Ryan and Scott (1974) could not be quantitatively reproduced by other laboratories with other specimens. A comparison of several such data sets is given by Scott (1983a). Typically the data differ in the degree of softening observed for the soft mode. In some specimens the totally symmetric soft mode decreases to only $\sim 20\,\mathrm{cm}^{-1}$ and remains underdamped down to T_I. This suggests that some samples are more order–disorder in character than others. There are also minor variations in T_I measured from about 250 K to 254 K. Lavrencic and Scott (1981) showed that this variation in T_I could be explained by fluorine vacancy density, through an experiment in which they cycled specimens to temperatures near the melting point and measured T_I after each cycle. Changes in T_I as great as 9 K were observed.

More dramatic and more troubling sample dependences were discovered by Levstik et al. (1975, 1976). In contrast to the dielectric results of Samara and Richards (1976), they found rather small increases in the *a*-axis dielectric constant at T_I with no lambda shape (Levstik et al. 1976) and, in some specimens, two *a*-axis dielectric anomalies 4 K apart (Levstik et al. 1975). The reasons for this variation were quite unclear, and the only hypothesis advanced —of fluorine vacancies as the culprit—was speculative, although the magnitude, temperature dependence, and anisotropy of ionic conductivity in $BaMnF_4$ were all compatible with hopping of fluorine vacancies as the dominant mechanism (Samara, private communication, quoted in Scott 1979).

The indication mentioned above that some specimens exhibit two phase transitions near 250 K, rather than one, became a source of considerable controversy. Occasionally this was extremely unfriendly. Repeated measurements at Brookhaven confirmed the original results that there exists only a single incommensurate transition near 250 K, with nearly temperature independent structure below that temperature (Cox et al. 1979, 1981, 1983). However, we know now that such a conclusion cannot be drawn from diverse data on other specimens. The evidence for two transitions is reviewed in the following sections.

3. Linear birefringence

In contrast to cubic crystals or the case of light propagating along the optic axis in uniaxial crystals, the phase transition in $BaMnF_4$ produces changes in linear birefringence proportional to the square of the order parameter (Fousek and Petzelt 1979):

$$\Delta n_{ij}(T) = \left[\frac{\rho_{ii}}{2n_i^0} - \frac{\rho_{jj}}{2n_j^0}\right]\langle \xi^2 + \eta \rangle, \tag{1}$$

where ρ_{ii} are coefficients coupling the order parameter (two-dimensional, with components ξ and η) and electric field, and n_i^0, n_j^0 are indices of refraction far from T_I. This birefringence was studied in $BaMnF_4$ by Regis et al. (1980, 1981), by Schafer et al. (1983), and by Pisarev et al. (1983). In addition to fitting the observed birefringence behavior below T_I, the theory used by Pisarev et al. predicts a rotation angle γ of the optical index ellipsoid given by

$$\gamma = \tfrac{1}{2}\tan^{-1}\left[2(\xi^2 - \eta^2)\rho_{bc}/(\varepsilon_b - \varepsilon_c)\right], \tag{2}$$

where ε_b, ε_c are the diagonal values of the optical permeability. For small γ, γ is proportional to $\xi^2 - \eta^2$. Pisarev et al. find that eq. (1) is satisfied and yields critical exponent $\beta = 0.28$ for the order parameter temperature dependence below T_I. Schafer et al. independently obtain 0.28 ± 0.005. Regis et al. obtained an earlier but less accurate 0.30. These are in agreement with the proximity to a tricritical point, for which $\beta = 0.25$, and with the specific heat data discussed below, which give $\alpha = 0.54$. However, eq. (2) is not verified, in that γ is found to be $\sim 4'$ of arc and independent of temperature near T_I. The neutron scattering experiments show only one kind of domain (i.e., $\xi \neq 0$, $\eta \neq 0$), so we would expect γ to be proportional to Δn and each to vary as reduced temperature $t^{2\beta} = t^{0.56}$. This is not the case experimentally, and the lack of temperature dependence for the monoclinic distortion angle γ_0 ($\gamma_0 \approx \gamma$) is not understood. A possible explanation in terms of "frozen-in" chiral strains is suggested below.

The data for $\Delta n_{ji}(T)$ above T_I are also in reasonable agreement with theory. Experimentally these go as reduced temperature $t^{-0.48 \pm 0.04}$ in agreement with a theoretical value of $\frac{1}{2}$ (Courtens 1976, Kleeman et al. 1979). Schafer et al. conclude that the long "tails" in the birefringence data are due to phase-like fluctuations and not to fluctuations in amplitude mode population. They also note that the experimental relationship

$$|\Delta n_{ab}|_{\mathrm{f1}} \ll |\Delta n_{ac}|_{\mathrm{f1}} \approx |\Delta n_{bc}|_{\mathrm{f1}} \tag{3}$$

due to fluctuations, implies, according to the theory of Fousek and Petzelt (1979), a strong anisotropy in fluctuation contributions to the electric susceptibility. It is not clear whether this is compatible with the anisotropy observed by Lyons et al. (1982) via quasielastic "central mode" scattering. It was emphasized by both Pisarev et al. (1983) and Schafer et al. (1983) that birefringence and inelastic neutron scattering data probe fluctuations which last for different time scales: $\Delta t \gg 1/f$, where f is the probe frequency. Pisarev et al. show that the optical phase lag is proportional to the neutron Bragg peak intensity at (0.39, $\frac{1}{2}$, $\frac{1}{2}$) from $T_{\mathrm{I}} - 100\,\mathrm{K}$ to $T_{\mathrm{I}} - 10\,\mathrm{K}$, but is much larger than predicted within 10 K of T_{I} due to the different range (in energy and space) of fluctuations sampled via the two techniques.

The linear birefringence data yield no evidence for a second structural phase transition below T_{I}. Although the early birefringence data of Regis et al. (Regis et al. 1980, 1981, Barthes-Regis et al. 1983) had some suggestive "kinks" below T_{I}, these were not reproduced in more detailed subsequent work (Schafer et al. 1983, Pisarev et al. 1983). It is important to keep this in mind in reviewing the material summarized in the following section.

4. *Evidence for a second transition at* T_{II}

In addition to the early dielectric study of Levstik et al. (1975), which evidenced two transitions 4 K apart in one specimen of $BaMnF_4$, both specific heat and piezoelectric resonance data were published (Scott et al. 1982) which showed clearly two transitions at 247 K and 254 K. The specific heat evidence was reproducible although not dramatic: a small anomaly in C_p is observed at 247 K of magnitude 2.0 ± 0.7 mcal/mol K (entropy change), compared with the larger peak at 254 K ($\Delta S = 25 \pm 2$ mcal/mol K). The piezoelectric resonance data, reproduced in fig. 2, were more spectacular, showing two sharp anomalies at 247 K and 254 K. It is important to note that both the specific heat and piezoelectric resonance data were taken with increasing temperatures. They were *not* made on the same samples.

Additional evidence for two phase transitions came from the dynamic central mode light scattering measurements of Lyons et al. (1982). These authors found at 241 K the onset of dynamical central mode scattering; at 247 K this scattering intensity exhibits a maximum; and at 254 or 255 K the scattering abruptly vanishes. These results provide independent confirmation that some phase transition-like anomalies occur at the same temperatures (247 and 255 K) in different samples.

Very recently Barthes-Regis et al. (1983) have provided new inelastic neutron data that conflict with the Brookhaven results (Cox et al. 1979, 1981, 1983) and which appear to reconcile the diverse, sample-dependent data. They

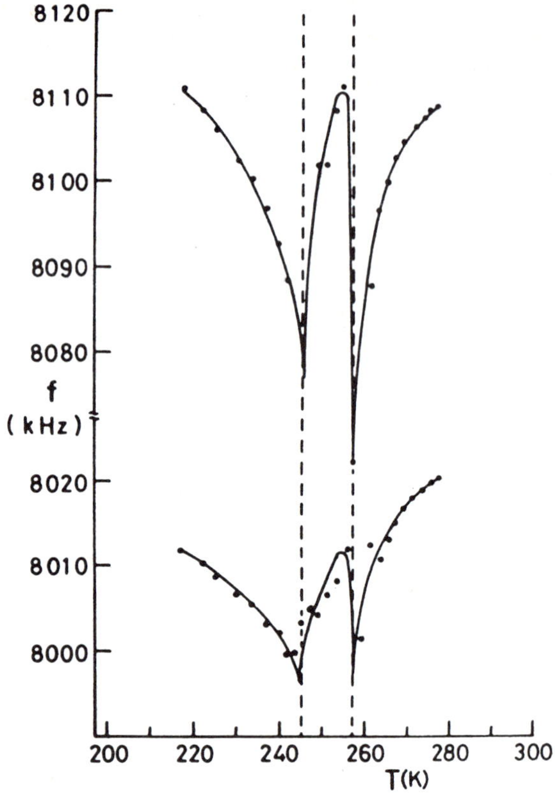

Fig. 2. Piezoelectric resonance frequencies near 9 MHz in $BaMnF_4$. (M. Hidaka, K. Inoue, S. Yamashita and J.F. Scott, unpublished; cited in Scott et al. 1982.)

find that the incommensurate translation vector is strongly temperature dependent in some samples, varying from $0.399a^*$ to $0.389a^*$. The temperature dependence exactly at T_I is quite steep, so that $0.389a^*$ is in fact an upper estimate. In addition, they find that there is something approximating a lock-in transition at 60 K (cooling) or 170 K (heating) to a nearly commensurate structure with $q_0 = 0.399a^*$, a value which is commensurate ($0.400a^*$) within experimental uncertainty. This situation is strongly reminiscent of that in $Ba_2NaNb_5O_{15}$, where there is a phase transition from a highly incommensurate modulation (0.12 to 0.08) to a nearly commensurate structure ($q_0 = 0.02 \pm 0.01$). In the nearly commensurate phase of both $Ba_2NaNb_5O_{15}$ and $BaMnF_4$ the modulation is within $0.01a^*$ of being perfectly commensurate. The present author believes that such structures are defect-stabilized and not intrinsically stable. Evidence for such a conclusion is discussed in the following section. However, first we would like to offer an explanation for the strange values $0.399a^*$ and $0.389a^*$ observed as the end-point values in the high and low ends of the incommensurate phase in $BaMnF_4$.

5. *Extended Ising model calculations for incommensurates*

Early studies of incommensurates such as K_2SeO_4 were described by free energies of Landau–Ginzburg form, with key interaction terms of, for example, Q^3P, where Q is the order parameter and P is the spontaneous electric polarization (Iizumi et al. 1977). Whereas such a third-order term seemed vaguely plausible to explain the cell-tripling lock-in transition ($q_0 = a^*/3$) in K_2SeO_4, the subsequent discovery of lock-in structures for incommensurates at, say, $4a^*/13$, or, in the case of Rb_2ZnBr_4, $5a^*/17$ (Gesi and Iizumi 1978, de Pater et al. 1979, Ueda et al. 1982) made it clear that such a Landau approach, requiring the key coupling in the 17th-order term in the free energy expansion, was quite unphysical [or, as Yamada and Hamaya (1983) put it: "Ridiculous"]. As an alternative to this approach, following Bak and von Boehm (1980), Villain and Gordon (1980), and Axel and Aubry (1981), Yamada and Hamaya (1983) have developed an extended Ising model to describe the sequence of phase transitions which can be reached with decreasing temperature in the A_2BX_4 family. They allow for third-nearest neighbor interactions. The nearest neighbor interaction is antiferroelectric ($J_1 < 0$); the second neighbor interaction is also antiferroelectric and of the same general magnitude as J_1 (i.e., $J_2 = 0.2J_1$ to $1.0J_1$); the third neighbor interaction is small and positive: $J_3 = -0.1J_1$. Their results are shown in fig. 3, which graphs normalized temperature versus J_2/J_1 with J_3 fixed at $-0.1J_1$. The phase transition trajectories are drawn as slightly non-vertical lines to permit the temperature dependence of the exchange constants. This diagram suffices to explain the observed sequence of phases in about a dozen A_2BX_4 incommensurates. The trajectories for K_2ZnCl_4, Rb_2ZnCl_4 and Rb_2ZnBr_4 are marked in. In the diagram the shaded zones are incommensurate, and the lock-in phases are marked not with the experimental value q_0 of modulation vector, but with q_{mod}, a quantity defined from a reduced-Brillouin zone scheme as:

$$q_{\text{mod}} = (1 - q_0)/2. \tag{4}$$

For the other dozen known A_2BX_4 incommensurates, it is necessary to modify the Ising Hamiltonian to account for the observed sequence of phases: A coupling term $a_0 u \Sigma_i \sigma_i$ must be added to the Hamiltonian, where u is the uniform strain. Such a term is known by direct measurement to be present in tetramethyl ammonium $CuCl_4$ (Sugiyama et al. 1980). This term has the

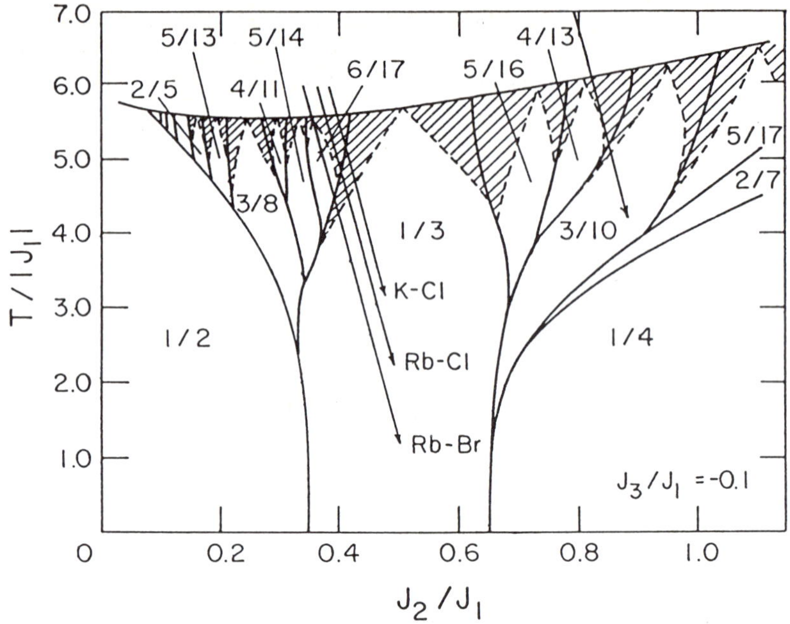

Fig. 3. Global phase diagram for A_2BX_4 incommensurates. Reduced temperature is plotted versus the ratio of nearest neighbor and next-nearest neighbor exchange constants (both assumed negative). Nearly vertical lines represent possible trajectories for sequences of phase transitions. $J_3 = -0.01J_1$. The line entering the shaded incommensurate region near $q_{\text{mod}} = \frac{4}{13}$ (i.e., $q_0 = \frac{5}{13}$) represents $BaMnF_4$. This predicts $2a^*/5 > q_0 > 5a^*/13$ for the incommensurate phase of $BaMnF_4$, in agreement with experiment. It suggests that $BaMnF_4$, which is an incommensurate of form $A^{2+}BX_4$, can be treated together with A_2BX_4 family members. (From Yamada and Hamaya 1983.)

theoretical effect of increasing the width of the $\frac{1}{3}$ phase and of making certain phase transition sequences accessible. A coupling constant a_0 of 3.6 suffices to describe the phase sequences in all known tetramethyl ammonium A_2BX_4 incommensurates.

What is interesting about this diagram in the case of $BaMnF_4$ is that, without modification, it permits a phase sequence (marked as a solid line) from the high temperature prototype parent phase into an incommensurate phase bounded by $q_{\text{mod}} = \frac{4}{13}$ and $q_{\text{mod}} = \frac{3}{10}$; that is, by $q_0 = \frac{5}{13}$ and $q_0 = \frac{2}{5}$. Such a trajectory would therefore yield an incommensurate structure with q_0 slightly greater than $2a^*/5$ at the lowest temperatures. This is, of course, precisely the range of values observed by Barthes-Regis et al. (1983) for $BaMnF_4$! Thus, we see that the extended Ising model of Yamada and Hamaya (1983) fits $BaMnF_4$ as well as the A_2BX_4 incommensurates, and with the same value of J_3/J_1 and the same antiferroelectric J_1 and J_2. This seems quite remarkable.

An additional result from this theory is that the incommensurate phase should not be the ground state at $T=0$. This suggests that in $BaMnF_4$ the incommensurate phase is not intrinsically stable but is stabilized by defects.* Independent evidence for this is summarized in the next section. Before leaving the extended Ising model theory, however, let us note that it seems equivalent in predictions to chiral clock models. In particular, the phase region near $q_0=\frac{1}{3}$ in fig. 3 is precisely equivalent to that generated by the three-state Potts model (Ostlund 1981, Huse 1981).

6. *Evidence for defect-stabilized structure in incommensurate* $BaMnF_4$

In addition to the indirect arguments sketched above in relation to the Ising model theory of Yamada and Hamaya (1983), there are more direct data which show that the incommensurate phase in $BaMnF_4$ is not one of thermal equilibrium over much of the temperature range for which measurements have been carried out. The clearest evidence is the extreme hysteresis reported for q_0 in the neutron scattering experiments of Barthes-Regis et al. (1983). They show that very strong hysteresis effects occur for heating/cooling cycles even if the cycles stay within the commensurate phase. This result was first established for other incommensurate crystals via dielectric measurements by Hamano et al. (1980). In the case of $BaMnF_4$ and $RbH_3(SeO_3)_2$ there is another piece of evidence for nonequilibrium states and defect stabilization: Those are the ultrasonic observations that the transverse sound velocities V_{ij} are not equal to V_{ji} in the incommensurate phases (Fritz 1975, Esayan et al. 1981). Here V_{ij} denotes a shear wave velocity with propagation along i and polarization along j.

The observation that $V_{ij} \neq V_{ji}$ in incommensurates poses an apparent paradox: For we naively assume that

$$V_{ij}=(C_{ij}/d)^{1/2}, \tag{5}$$

where C_{ij} is an elastic constant and d is the density; and it is well known that thermodynamic arguments alone suffice to prove that the elastic constants are symmetric with interchange of the subscripts ij. Therefore, if $C_{ij}=C_{ji}$, how can $V_{ij} \neq V_{ji}$? The present author was the first to propose that this occurs because of finite frequency, finite wave vector effects (Scott 1983b). Since $C_{ij}=C_{ji}$ is true at zero frequency, whereas V_{ij} are measured in the MHz regime, perhaps

*Randa and Scott (1985) show that the spontaneous polarization can also stabilize the $2a^*/5$ phase at $T=0$ in $BaMnF_4$.

this is due to finite wave vector interaction terms. This suggestion was presented as a developed theory by Dvořák and Esayan (1982) for $RbH_3(SeO_3)_2$. However, their theory predicts that the difference $V_{ab} - V_{ba}$ for $BaMnF_4$, or $V_{zy} - V_{yz}$ for $RbH_3(SeO_3)_2$, should become much bigger in the GHz regime; in preliminary Brillouin studies we could not confirm that prediction. That led us to suggest that perhaps $C_{ij} \neq C_{ji}$. This is a radical suggestion, but we believe that in incommensurate $BaMnF_4$ and $RbH_3(SeO_3)_2$ it is true.

It is important to note that both $BaMnF_4$ and $RbH_3(SeO_3)_2$ have screw axes. Recall also that the thermodynamic proof that $C_{ij} = C_{ji}$ assumes thermal equilibrium (Auld 1973). I propose that in these two incommensurate crystals there are chiral strain fields around the screw axes; these chiral strains could cause (see Nelson and Lax 1970) $C_{ij} \neq C_{ji}$, and at the same time stabilize the incommensurate structure, preventing in the case of $BaMnF_4$ the lock-in to the $2a^*/5$ phase. An explicit test of these predictions can be made: The two samples of $BaMnF_4$ studied at Grenoble by Barthes-Regis et al. (1983) should have rather different behavior of $V_{ab}(T) - V_{ba}(T)$.

In summarizing this section, let us argue that $BaMnF_4$ may provide the best system in which to study the role of defects in stabilizing incommensurate structures. It is clear that the extended Ising model theory applies to it, and the q_0 in the incommensurate phase is bounded by $\frac{2}{5}$ and $\frac{5}{13}$ a^* values. It would be desirable to study the family $BaMn_{x-1}Zn_xF_4$, for which a continuous range of solid solutions can be grown. This should permit further application of the global phase diagram in fig. 3. Finally, it seems that the $V_{ij} \neq V_{ji}$ transverse sound velocities observed may be a good way to measure nonequilibrium properties and helical (or chiral) defect strain fields. The presence of these "frozen-in" chiral strains could explain the total lack of temperature dependence observed by Pisarev et al. (1983) for the optical activity below T_I in $BaMnF_4$. The optical activity is theoretically expected to scale with the linear birefringence; however, whereas the birefringence varies as the square of the order parameter, as expected, the optical activity appears temperature independent. This line of argument might also explain the temperature independence for the a-axis lattice constant (Cox et al. 1979).

It seems appropriate to end this section with a quote from Professor L. Keldysh at the Varna Conference on Solid State Physics (1982): "Maybe those incommensurate structures aren't in thermal equilibrium."

7. *Magnetoelectric phenomena*

$BaMnF_4$ is ferroelectric in its incommensurate phase. This is unusual, if not unique. Since it orders magnetically as temperature is lowered within the incommensurate phase, the possibility exists for it to be simultaneously ferroelectric and (weakly) ferreomagnetic, magnetoelectric and incommensurate! I know of no other crystal which displays these characteristics simultaneously. In this section I will try to summarize the magnetic studies of $BaMnF_4$ with emphasis on the most recent work.

Associated with the two-dimensional magnetic ordering, for which evidence exists as high as 70 K, is a broad anomaly in the b-axis dielectric constant, first measured by Samara and Richards (1976) and Samara and Scott (1977). It was shown (Scott 1977 Glass et al. 1977) that the magnitude of this anomaly is proportional to the nearest neighbor average $\langle S_j^z S_{j+1}^z \rangle$. Below $T_N = 26\,K$, magnetic susceptibility measurements (Holmes et al. 1969) show three-dimensional magnetic ordering. Due to single-ion anisotropy, the spin orientation is tilted at 9° of arc from the b-axis (Cox et al. 1979). In addition to this tilting, weak canting (weak ferromagnetism) has been reported (Venturini and Morgenthaler 1975) with a canting angle of 3 mrad at 4.2 K. The analysis leading to this conclusion that spin canting exists is not wholly unambiguous, due to the tilted orientation of the antiferromagnetic axis, however; and in fact rather recent attempts to measure the weak ferromagnetism directly (Kizhaev 1984) gave a null result. However, on the assumption that the spin canting inferred by Venturini and Morgenthaler is correct, the magnetic point group symmetry is 2′ (Dvořák 1975, Fox and Scott 1977).

A series of papers have dealt with the interesting a-axis dielectric anomaly which occurs below T_N (Samara and Scott 1977; Scott 1979, Albuquerque and Tilley 1978, Fox and Scott 1977, Fox et al. 1980, Tilley and Scott 1982, Scott et al. 1984). The temperature dependence and sign of the static dielectric constant for the a-axis can be accounted for by an interaction of form $(b_0 + b_i p + b_2 p^2) M_x L_z$ in the free energy, where $L = M_1 - M_2$ and $M = M_1 + M_2$ are the usual variables describing a canted antiferromagnet with sublattices 1 and 2, and where p is the part of the spontaneous electric polarization induced by magnetic ordering. Recently predictions of frequency dependence for the a-axis dielectric constant below T_N were made (Tilley and Scott 1982) and subsequently found not to be in agreement with experimental data (Kozlov et al. 1985). The reason is now understood to be the need for an additional term of form $(b_0' + b_2' p + b_2' p^2) L^2$ in the free energy. This gives rise to a frequency-independent contribution to $\varepsilon_a(T)$ which is much larger than the frequency-dependent term arising from the expression in $M_x L_z$ given below:

$$\Phi = \tfrac{1}{2}(A + a)L^2 - \tfrac{1}{2}aL_z^2 + \tfrac{1}{4}GL^4 + \tfrac{1}{2}BM^2 + \tfrac{1}{2}D(\boldsymbol{L}\cdot\boldsymbol{M}) + (\beta_0 + \beta_1 p + \beta_2 p^2) M_x L_z - \gamma M_z L_x + \tfrac{1}{2}Kp^2 - \boldsymbol{p}\cdot\boldsymbol{E} - \boldsymbol{M}\cdot\boldsymbol{H}. \tag{6}$$

This free energy was used to calculate both static and dynamic properties in the magnetoelectric phase, in particular, of the a-axis electric susceptibility

$\chi_{yy}^{ee}(\omega, T)$. A Landau–Khalatnikov equation of motion approach was utilized for the dynamics. This calculation, together with the free energy given in eq. (5), was sufficient to explain all published data for $BaMnF_4$. Some attention was devoted to the justification for neglect of additional terms in eq. (5). In addition to explaining the sign and temperature dependence of the dielectric anomaly below T_N, this work made an explicit prediction of the frequency dependence of $\Delta\varepsilon a$ below T_N:

$$\Delta\varepsilon_a(\omega, T) = \left[\frac{\beta_1^2\omega_m^2}{\omega_m^2 - \omega^2} + \beta_0\beta_2\right]\frac{\langle|\boldsymbol{L}\times\boldsymbol{M}|\rangle F}{\left(\omega_{TO}^2 - \omega^2\right)^2}, \tag{7}$$

where ω_m is the $q = 0$ spin wave energy (about $3\,\text{cm}^{-1} = 90\,\text{GHz}$); ω_{TO} is the frequency of the lowest $q = 0$ optical phonon of totally symmetric representation; and F is a constant independent of T and ω.

At frequencies $\omega \ll \omega_m$ eq. (7) predicts a $\Delta\varepsilon_a$ which can be negative if $\beta_0\beta_2$ is < 0 and greater in magnitude than the positive definite term in β_1^2. This is in accord with experiment and was a principal aim of the early theoretical work. It also predicts a temperature dependence given by the thermal average $\langle|\boldsymbol{L}\times\boldsymbol{M}|\rangle$; note that this three-dimensional expectation value vanishes abruptly at T_N despite the strong two-dimensional spin ordering above T_N. This T-dependence also seemed to agree with experiment, although insufficient data were obtained near T_N to make a conclusive test.

In the last year two new experiments were reported on magnetoelectric $BaMnF_4$. The first out was a measurement of $\Delta\varepsilon_a(\omega)$ for ω from 55 to 500 GHz (1.8 to $17\,\text{cm}^{-1}$) (Kozlov, Volkov and Scott, 1985); this was done to test the prediction of eq. (6). It was assumed that for ω near ω_m, $\Delta\varepsilon_a(T \ll T_N)$ would change sign from -0.10 to something measurably large and positive. However, the data show that $\Delta\varepsilon_a$ is independent of frequency within a measuring uncertainly of $\pm 0.5\%$ over this frequency range. This could be due to an accidentally small value of β_1^2, since $\Delta\varepsilon_a(\omega)$ was measured for discrete values of $\omega = 1.8\,\text{cm}^{-1}$, $5\,\text{cm}^{-1}$, and higher; small β_1^2 would yield no measurable frequency dependence except at probe frequencies closer to $\omega_m = 3\,\text{cm}^{-1}$. However, we believe that it is due to neglect of additional terms in the free energy Φ, discussed below. The second set of new data are those of Schafer, Kleeman and Tsuboi (1983). These data show a discrepancy below T_N: the spin wave theory does not account for the temperature dependence of Δn below T_N, an effect qualitatively ascribed to magnetoelectricity but quantitatively incompatible with a $\langle|\boldsymbol{L}\times\boldsymbol{M}|\rangle$ thermal average, which would follow from eq. (5).

Let us add to the free energy Φ in eq. (5) a term

$$\Phi' = \left(\beta_1' p + \beta_2' p^2\right) L^2. \tag{8}$$

Such a term is allowed by symmetry but was not included in our original analysis because there was no experimental evidence for it. If we incorporate it in the Landau–Khalatnikov equations of motion and solve for $\Delta\varepsilon_a(\omega,T)$ as before, we get an additional contribution to eq. (6). For $T \ll T_N$ the complete expression can be written as

$$\Delta\varepsilon_a(\omega,0)=\left[\frac{(\beta_1+4B\beta_1'/\beta_0)^2\omega_m^2}{\omega_m^2-\omega^2}+\beta_0\beta_2+4B\beta_2'\right]L_0^2F^2, \tag{9}$$

where F_2 is a constant and L_0 is twice the magnitude of sublattice magnetization. We see from eq. (9) that there are two new effects in this low-temperature limit: first, there is an extra frequency independent term $4B\beta_2'$. If we assume that β_2 and β_2' are of roughly the same order of magnitude, then this term is much larger than that $(\beta_0\beta_2)$ originally included, since $4B/\beta_0=1.34\times10^3$ from Fox et al. (1980); there is a similar modification of the frequency-dependent terms; β_1 in eq. (7) is replaced by $\beta_1+4B\beta_1'/\beta_0$. If $\beta_1\approx\beta_1'$ this would increase this term by a factor similar to that for the frequency-independent term. However, we see that β_1 and $4B\beta_1'/\beta_0$ may be of opposite signs, making the net frequency-dependent contribution to $\Delta\varepsilon_a(\omega,T)$ quite small.

When we look at expressions for $\Delta\varepsilon_a(\omega,T)$ closer to T_N, we find a second new effect arising from Φ': Whereas Φ given in eq. (6) predicts $\Delta\varepsilon_a$ proportional to $\langle|\boldsymbol{L}\times\boldsymbol{M}|\rangle$, Φ' yields terms proportional to $\langle L^2\rangle$, which changes very slowly at T_N, due in part to the two-dimensional spin ordering above T_N. Including Φ' in the free energy therefore predicts that $\Delta\varepsilon_a(T)$ should vary as $\langle L^2\rangle$ near T_N, which does not vanish abruptly at T_N.

Efforts to find direct evidence for magnetoelectric phenomena in $BaMnF_4$ have been generally unsuccessful, although a qualitative effect was reported by Al'Shin et al. (1972). In 1983 Iio et al. (1983) and Schafer et al. (1983) showed that the linear birefringence is directly proportional to the magnetoelectric tensor (α_{13} for $BaMnF_4$). However the only experimental evidence for such an effect was the observation that below T_N the temperature derivative of the linear birefringence is not proportional to the magnetic specific heat $C_m(T)$, contrary to expectations (Schafer et al. 1983). A difference between the two curves is observed which increases as T is lowered below T_N; it vanishes smoothly as T_N is approached from below. This smooth disappearance as T_N is approached from below is in contrast with the temperature dependence predicted for the magnetoelectric constant α_{13} by Fox and Scott (1977) and Fox et al. 1980; they predicted that it would vary as the sublattice magnetization. However, more detailed theories of magnetoelectric off-diagonal tensor elements (Englman and Yatom 1975) show that in general this difference need not vanish as $\langle S_z\rangle$ does; it can decrease more slowly or more rapidly as $T\to T_N$. A slower decrease is possible for cases where Dzyaloshinskii anisotropic exchange is present, and in $BaMnF_4$ two sources of spin canting and magnetoelectricity are expected: the first is the usual Dzyaloshinskii–Moriya anisotropic exchange (Tilley and Scott 1982); the second is canting induced by ferroelectric moment (Fox and Scott 1977). These are expected to have different temperature dependences. Therefore it seems that the discrepancy between linear birefringence temperature derivatives in $BaMnF_4$ below T_N and $C_m(T)$ is probably due to magnetoelectric contribution, as suggested by Schafer et al. (1983), and that the exact temperature dependence observed by them can be explained by more sophisticated calculation of magnetoelectric properties.

Several explicit predictions have been made from the models of magnetoelectricity in $BaMnF_4$. Scott (1979) predicted that the *a*-axis dielectric anomaly in $BaMnF_4$ would vanish in $BaMn_{0.99}Co_{0.01}F_4$, a crystal in which the antiferromagnetic spin orientation is rotated from near $\boldsymbol{b}$ to along $\boldsymbol{a}$. This was verified by Fox et al. (1980). A second prediction was that the *a*-axis anomaly would vanish for applied magnetic fields above the spin flip field of 1*T*. This should occur for the same symmetry reasons, since the effects of Co-doping and large applied magnetic fields upon the spins are the same: to flip the antiferromagnetic alignment from near $\boldsymbol{b}$ to along $\boldsymbol{a}$. However, very recent work at fields above 1.0T (Kizhaev 1984) shows that this is false: the *a*-axis dielectric constant below T_N is totally unaffected by such applied magnetic fields. This is unexplained, and it is important to note that Kizhaev's sample contained 0.1% Co and that his $\Delta\varepsilon_a(T)$ curve is qualitatively different from that of Samara and Richards (1976), even for $H=0$; 0.5% Co is known to change the magnetic symmetry of the crystal. If Kizhaev's result cannot be explained by the cobalt concentration, it seems that one of two effects left out of the theoretical models (Tilley and Scott 1982) must be put into the calculations: either the spin tilting of 9° from *b*, or the incommensurate lattice structure (ignored in the magnetic calculations) must be reintroduced. If any of these magnetic phenomena are explicitly due to the incommensurate lattice modulation, the calculations will be formidable.

References

Albuquerque, E.L., and D.R. Tilley, 1978, Solid State Commun. **26**, 817.

Al'Shin, B.I., D.N. Astrov and R.V. Zorin, 1972, Zh. Eksp. Teor. Fiz. **63**, 2198 [1973, Sov. Phys. JETP **36**, 1161].

Auld, B.A., 1973, Acoustic Fields and Waves in Solids (Wiley, New York).

Axel, F., and S. Aubry, 1981, J. Phys. **C14**, 5433.

Bak, P., and J. von Boehm, 1980, Phys. Rev. **B21**, 5297.

Barthes-Regis, M., R. Almairac, P. St.-Gregoire, C. Filippini, U. Steigenberger, J. Nouet, and J.-Y. Gesland, 1983, J. Physique Lett. **19**, L-829.

Courtens, E., 1976, Proc. Int. School of Physics Enrico Fermi, course LIX (North-Holland, Amsterdam) p. 293.
Cox, D.E., S.M. Shapiro, R.A. Cowley, M. Eibschutz and H.J. Guggenheim, 1979, Phys. Rev. **B19**, 5754,
Cox, D.E., S.M. Shapiro, R.J. Nelmes, T.W. Ryan, H. Bleif, M. Eibschutz and H.J. Guggenheim, 1981, Bull. Am. Phys. Soc. **26**, 303.
Cox, D.E., S.M. Shapiro, R.J. Nelmes, T.W. Ryan, H. Bleif, M. Eibschutz and H.J. Guggenheim, 1983, Phys. Rev. **B28**, 1640.
de Pater, C.J., J.D. Axe and R. Currat, 1979, Phys. Rev. **B19**, 4684.
Dvořák, V., 1975, Phys. Stat. Sol. (b) **71**, 269.
Dvořák, V., and J. Fousek, 1980, Phys. Stat. Sol. (a) **61**, 99.
Dvořák, V., and S. Kh. Esayan, 1982, Solid State Commun. **44**, 901.
Englman, R., and H. Yatom, 1975, in: Magnetoelectric Interaction Phenomena in Crystals, eds. A.J. Freeman and H. Schmid (Gordon and Breach, New York) p. 17.
Esayan, S. Kh., V.V. Lemanov, N.Mamatkulov and L.A. Shuvalov, 1981, Kristallografiya **26**, 1086 [Sov. Phys. Crystallogr. **26**, 619].
Fousek, J., and J. Petzelt, 1979, Phys. Stat. Sol. (a) **55**, 11.
Fox, D.L., and J.F. Scott, 1977, J. Phys. **C10**, L329.
Fox, D.L., D.R. Tilley, J.F. Scott and H.J. Guggenheim, 1980, Phys. Rev. **B21**, 2926.
Fritz, I.J., 1975, Phys. Lett. **51A**, 219.
Gesi, K., and M. Iizumi, 1978, J. Phys. Soc. Jpn **45**, 1777.
Glass, A.M., M.E. Lines, M. Eibschutz, F.S.L. Hsu and H.J. Guggenheim, 1977, Commun. Phys. **2**, 103.
Hamano, K., Y. Ikeda, T. Fujimoto, K. Ema and S. Hirotsu, 1980, J. Phys. Soc. Jpn **49**, 2278.
Holmes, L.M., M. Eibschutz and H.J. Guggenheim, 1969, Solid State Commun. **7**, 973.
Huse, D.A., 1981, Phys. Rev. **B24**, 5180.
Iio, K., T. Hoshino and K. Nagata, 1983, J. Magn. Magn. Mater. **31**, 587.
Iizumi, M., J.D. Axe, G. Shirane and K. Shimaoka, 1977, Phys. Rev. **B15**, 4392.
Keve, E.T., S.C. Abrahams and J.L. Bernstein, 1969, J. Chem. Phys. **51**, 4928.
Kizhaev, A., 1984, unpublished; submitted to Sov. Phys. Solid State.
Kleeman, W., F.J. Schafer and J. Nouet, 1979, Physica **B97**, 145.
Kozlov, G.V., A.A. Volkov and J.F. Scott, 1985, to be published.
Lavrencic, B.B., and J.F. Scott, 1981, Phys. Rev. **B24**, 2711.
Levanyuk, A.P., V.V. Osipov, A.S. Sigov and A.A. Sobyanin, 1979, Zh. Eksp. Teor. Fiz. **76**, 345 [Sov. Phys. JETP **49**, 176].
Levstik, A., R. Blinc, P. Kadaba and S. Cizikov, 1975, Bull. Am. Phys. Soc. **20**, 558.
Levstik, A., R. Blinc, P. Kadaba, S. Cizikov, I. Levstik and C. Filipic, 1976, Ferroelectrics **14**, 703.
Lyons, K.B., T.J. Negran and H.J. Guggenheim, 1982, Phys. Rev. **B25**, 1791.
Nelmes, R.J., 1983, private communication.
Nelson, D.F., and M. Lax, 1970, Phys. Rev. Lett. **24**, 379.
Ostlund, S., 1981, Phys. Rev. **B24**, 398.
Pisarev, R.V., B.B. Krichevtzov, P.A. Markovin, O. Yu. Korshunov and J.F. Scott, 1983, Phys. Rev. **B28**, 2677.
Randa, J., and J.F. Scott, 1985, Bull. Am. Phys. Soc. **30**, 273.
Regis, M., M. Candille and P. St.-Gregoire, 1980, J. Phys. Lett. **41**, L-423.
Regis, M., M. Candille and P. St.-Gregoire, 1981, in: Recent Developments in Condensed Matter Physics, vol. **4**, eds. J.T. Devreese et al. (Plenum, New York) p. 107.
Ryan, J.F., and J.F. Scott, 1974, Solid State Commun. **14**, 5.
Samara, G.A., and J.F. Scott, 1977, Solid State Commun. **21**, 167.
Samara, G.A., and P.M. Richards, 1976, Phys. Rev. **B14**, 5073.
Schafer, F.J., W. Kleeman and T. Tsuboi, 1983, J. Phys. **C16**, 3987.

Scott, J.F., 1977, Phys. Rev. **B16**, 2329.
Scott, J.F., 1979, Rep. Prog. Phys. **42**, 1055.
Scott, J.F., 1983a, in: Light Scattering Near Phase Transitions, eds. H.Z. Cummins and A.P. Levanyuk (North-Holland, Amsterdam) p. 261.
Scott, J.F., 1983b, Ferroelectrics **47**, 33.
Scott, J.F., F. Habbal and M. Hidaka, 1982, Phys. Rev. **B25**, 1805.
Scott, J.F., D.R. Tilley and M. Seehra, 1984, Ferroelectrics **56**, 451.
Spencer, E.G., H.J. Guggenheim and G.J. Kominiak, 1970, Appl. Phys. Lett. **17**, 300.
Sugiyama, J., M. Wada, A. Sawada and Y. Ishibashi, 1980, J. Phys. Soc. Jpn **49**, 1405.
Tilley, D.R., and J.F. Scott, 1982, Phys. Rev. **B25**, 3251.
Ueda, T., S. Iida and H. Terauchi, 1982, J. Phys. Soc. Jpn **51**, 3953.
Venturini, E.L., and F.R. Morgenthaler, 1975, Proc. 20th Conf. on Magnetism and Magnetic Materials, eds. C.D. Graham and J.J. Rhyne (AIP, New York) p. 168.
Villain, J., and M.B. Gordon, 1980, J. Phys. **C13**, 3117.
Yamada, Y., and N. Hamaya, 1983, J. Phys. Soc. Jpn **52**, 3466.

CHAPTER 20

Unique Incommensurate–Commensurate Phase Transitions in a Layer-Structure Perovskite

R. KIND and P. MURALT

Laboratory of Solid State Physics
Swiss Federal Institute of Technology ETH
Hönggerberg, CH 8093 Zürich,
Switzerland

Incommensurate Phases in Dielectrics 2
Edited by
R. Blinc and A.P. Levanyuk

Contents

1. Introduction 303
2. The $\beta-\gamma-\delta$ sequence 306
  2.1. ^{2}H- and ^{35}Cl-NMR–NQR 307
  2.2. Theory of the $\beta-\gamma-\delta$ sequence 310
3. The $\delta-\varepsilon-\zeta$ sequence 315
  3.1. ^{2}H- and ^{35}Cl-NMR–NQR 315
  3.2. Neutron scattering experiments 316
4. Conclusions 317
References 318

1. *Introduction*

The incommensurate γ-phase of the perovskite type layer structure compound $(C_3H_7NH_3)_2MnCl_4$ (PAMC) is sandwiched between two phases which have the same structure. In contrast to the usual lock-in transition, where the amplitude of the modulation wave continues to increase at T_C but the incommensurate part of the average wave vector of the modulation vanishes, in PAMC the amplitude of the incommensurate modulation wave vanishes on both sides of the γ-phase, whereas the wave vector is not critical.

Perovskite type layer structures of the formula $(C_nH_{2n+1}NH_3)_2MCl_4$ with short hydrocarbon chains ($n < 5$) are known to exhibit several structural phase transitions which are connected with reorientational jumps of the alkylammonium chains (Kind 1980 and refs. cited therein). The structure of these compounds consists of infinite sheets of MCL_6 octahedra sharing corners. Such an arrangement closely resembles a plane of the perovskite structure with the metallic ions (M = Mn, Cd, Fe) occupying the B sites in the centers of the octahedra, whereas the A sites in the cavities between the octahedra are occupied by the NH_3 groups of the alkylammonium ions. $(C_nH_{2n+1}NH_3)$ chains are thus attached by weak N–H..Cl bonds on both sides of the layers (fig. 1). These neutral strata are staked upon each other to form the three-dimensional structure. Owing to the essentially quadric form of the cavities the alkylammonium chains can have four different equilibrium orientations according to the four possible ways the NH_3 head can couple to the layer. At high temperature there exists a dynamical disorder of the chains among the four equilibrium positions, whereas this motion is "frozen in" at low temperature.

Within this family the compound $(C_3H_7NH_3)_2MnCl_4$ (PAMC) is a special case because of its complicated phase sequence with two incommensurate phases (Depmeier 1979, 1981, Depmeier and Mason 1982). The six different phases in PAMC are denoted as α, β, γ, δ, ε and ζ, where γ and δ are incommensurate phases (fig. 2). In contrast to PAMC the isomorphous Cd-salt (PACC) exhibits only four phases: I4/mmm–Cmca–Pbca–unknown low temperature phase, i.e. the two compounds show the same behaviour at high temperatures but their low temperature behaviour is different (Chapuis 1978).

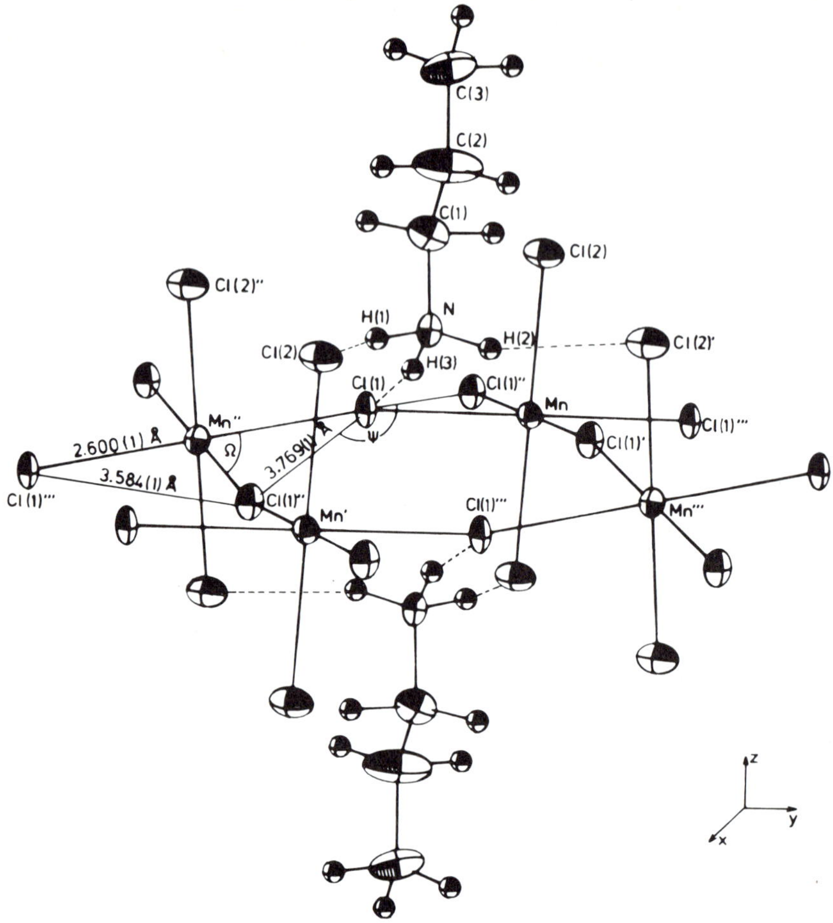

Fig. 1. Part of the structure of $\delta-(C_3H_7NH_3)_2MnCl_4$ PAMC showing the attachment of two propylammonium ions to the perovskite-type layer via hydrogen bonds. On each side of the layer only one propylammonium group is shown. (After Depmeier 1979.)

The behaviour of PACC can be described by a rigid lattice model just taking order–disorder-transitions of the basic reorientable units, i.e. the propylammonium (PA) chains, into account. In the tetragonal high temperature phase the four possible orientations α of each molecule are occupied with equal probability n_α in time: $n_1 = n_2 = n_3 = n_4$. At the transition to the orthorhombic room temperature phase Cmca the potential wells become inequivalent in pairs: $n_1 = n_2 \neq n_3 = n_4$, or $n_1 = n_4 \neq n_2 = n_3$. For two 90°-domains

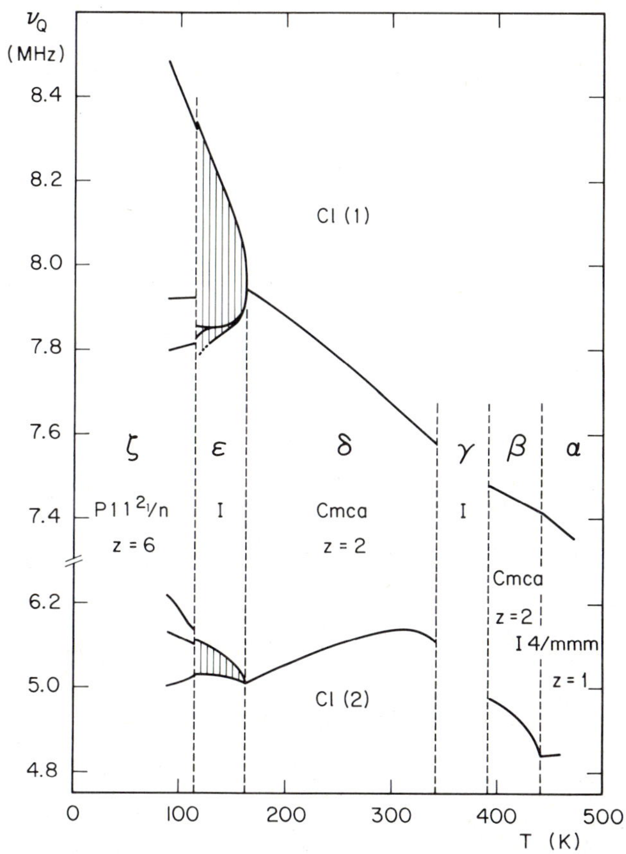

Fig. 2. Phase sequence of partially deuterated PAMC as observed by means of pure ^{35}Cl-NQR. In the α-, β- and δ-phase there are only two chemically inequivalent Cl-atoms, the equatorial Cl(1)-atoms and the axial Cl(2)-atoms according to fig. 1. In the incommensurate γ-phase the ^{35}Cl-lines are smeared out completely, whereas one can observe the edge singularities of the spectrum in the incommensurate ε-phase. The z-values are given for the primitive unit cell. (After Muralt 1984.)

which are obtained in this way one can define order parameters:

$$p_1 = n_1 + n_2 - n_3 - n_4,$$
$$p_2 = n_1 + n_4 - n_2 - n_3,$$
$$-1 < p_{1,2} < +1.$$

So one has either $p_1 \neq 0$ and $p_2 = 0$ with $n_1 = n_2$ or $p_1 = 0$ and $p_2 \neq 0$ with $n_1 = n_4$. For the sake of simplicity we continue with the first domain only. On lowering the temperature the order parameter p_1 becomes saturated and at the transition to the orthorhombic low temperature phase n_2 becomes different from zero too. In the ground state one has e.g. $p_1 = 1$, $p_2 = 1$, i.e. $n_1 = 1$, $n_2 = n_3 = n_4 = 0$.

The same model can be applied for the $\alpha - \beta$ transition in PAMC, but it does not describe the system below the β-phase although the ordering of the PA-groups into the ground state is probably the driving mechanism of the $\varepsilon - \zeta$ transition.

The most interesting feature of PAMC is the reentrant behaviour of the $\gamma - \delta$ transition. It was shown by means of NMR–NQR that the β- and the δ-phase have indeed the same structure (Muralt et al. 1982). They differ only in the saturation of the order parameter p_1 of the $\alpha - \beta$ transition.

In the incommensurate γ phase the modulation has a nearly temperature independent wave vector

$$k_\gamma = (\tfrac{1}{6} + \delta)\boldsymbol{c}^* + \boldsymbol{b}^*,$$

i.e. the corresponding mode becomes soft on the H-line (between the Y- and T-point) of the orthorhombic base centered Brillouin zone (BZ). (Notation after Bradley and Cracknell 1972). Instead of a lock-in of the wave vector at the $\gamma - \delta$ transition the order parameter η of the γ-phase just decreases to zero and thus the β-phase is restored.

The transition from δ to the incommensurate ε-phase is of second order. The incommensurate part δ of the wave vector $\boldsymbol{k}_\varepsilon = (\tfrac{1}{3} + \delta)\boldsymbol{a}^*$ decreases with temperature but instead of a lock-in at $\boldsymbol{k} = \tfrac{1}{3}\boldsymbol{a}^*$ the wave vector changes direction at the $\varepsilon - \zeta$ transition to $\boldsymbol{k}_\zeta = (\boldsymbol{a}^* \pm \boldsymbol{b}^*)/3$. (Depmeier and Mason 1982, 1983). Therefore the $\varepsilon - \zeta$ transition must be of first order.

In the following sections the partial phase sequences $\beta - \gamma - \delta$ and $\delta - \varepsilon - \zeta$ will be treated individually since both are starting from the same parent phase with space group Cmca.

2. *The $\beta - \gamma - \delta$ sequence*

From the structure determination of the δ-phase of PAMC (Peterson and Willet 1972, Depmeier and Mason 1978) it is known that the molecular axes of the PA-chains are essentially perpendicular to the layers. The order parameter p_1 of the $\alpha - \beta$ transition, which is almost saturated in the δ-phase, has therefore almost no influence on the orientation of this axis. However, the octahedra in the layers exhibit a washboard-like tilting around the a-direction of about 8°. This tilt angle vanishes in the tetragonal α-phase.

The structure determination of the incommensurate γ-phase (Depmeier 1981) revealed a modulation of the layers with an amplitude in the y-direction

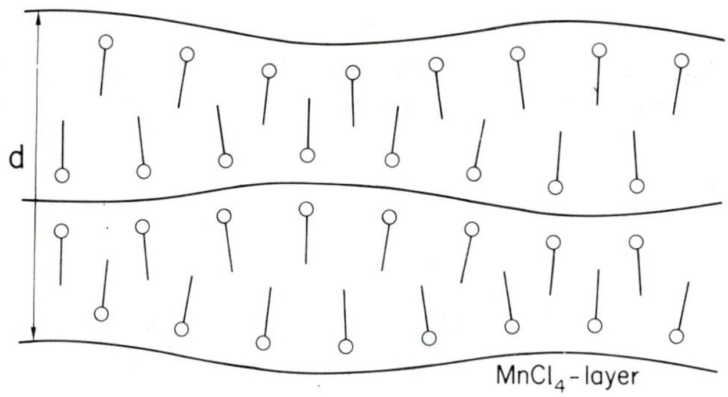

Fig. 3. Schematic structure of γ-PAMC showing the incommensurate modulation of the layers and the propylammonium tilt angles. (After Muralt et al. 1982.)

and a wave vector in the c-direction $\boldsymbol{k} \simeq \boldsymbol{c}^*/6$. A closer analysis shows, however, that the wave vector has also a commensurate component $\boldsymbol{b}^*$ which destroys the base centering of the unit cell. This leads to a modulation where adjacent layers have opposite modulation amplitudes, so that the interlayer distance is modulated (fig. 3). The modulation wavelength is about 40 Å and the amplitude $\simeq 0.5$ Å. No information was obtained about the behaviour of the PA-chains. Their axes are drawn tilted in a way that they remain essentially perpendicular to the local layer tangent. This was one of the main results of the NMR–NQR investigations discussed in the following section. Another aim of these investigations was to see whether the driving mechanism could be assigned to the interactions between the chains.

2.1. ^{2}H- and ^{35}Cl-NMR–NQR

In order to obtain detailed information about the behaviour of the PA-chains, the measurements were performed on partially deuterated crystals. Since the chains are always in the all-trans configuration, it was sufficient to deuterate only the NH_3-groups. It is known from measurements on similar compounds (Kind et al. 1978) that the isotope effect is very small and that the phase sequence should remain unchanged besides a slight shift of the transition temperatures. This could indeed be proved by means of pure ^{35}CL-NQR (fig. 2) (Caravatti 1981). Down into the ζ-phase the deuterons within the NH_3-groups exhibit a rapid exchange so that the electrical field gradient (efg) tensor of the three individual sites is time averaged to a single efg tensor with its principal z-axis being nearly parallel to the N–C bond of the molecule (Muralt 1984). By measuring the quadrupolar splitting of the DNMR-line as a function of the crystal orientation in the external magnetic field, one can derive the orientation of the efg tensor axes and thus the direction of the N–C bonds.

From the N–C(1)–C(2) bond angle of 113° it follows that the N–C bonds are tilted by an angle $\theta_0 = 33.5^0$ with respect to the layer normal. In the disordered phases an additional averaging process takes place—due to the reorientational jumps of the PA-chains—so that only the time averaged direction of the C–N bonds can be obtained from the NMR–NQR measurements. The time averaged efg tensor $\overline{V}$ becomes

$$\overline{V} = n_1 V_1 + n_2 V_2 + n_3 V_3 + n_4 V_4. \tag{2.1}$$

Since in the β- and δ-phases we have $n_1 = n_2 \neq n_3 = n_4$ the time averaged N–C bond as well as the time averaged z-direction of the efg tensor are located for any temperature in the mirror plane m_x. The order parameter p_1 can be derived from the tilt angle θ of the z-axis of the efg tensor with respect to the layer normal. In the rigid lattice approximation it becomes

$$\eta_1 = \frac{\tan 2\theta}{\tan 2\theta_0'}, \qquad \theta_0' = \arctan\left(\sin\theta_0 \cos\tfrac{1}{4}\pi\right). \tag{2.2}$$

A more accurate treatment is given in Muralt (1984). The results of this treatment are shown in fig. 4 for the β, γ and δ phase. One can see that this order parameter is not much affected by the modulation in the γ-phase.

A modulation of the tilt angle ϕ of the molecules in the γ-phase as shown in fig. 3 is transferred directly into a modulation of the angles θ and θ_0'. This

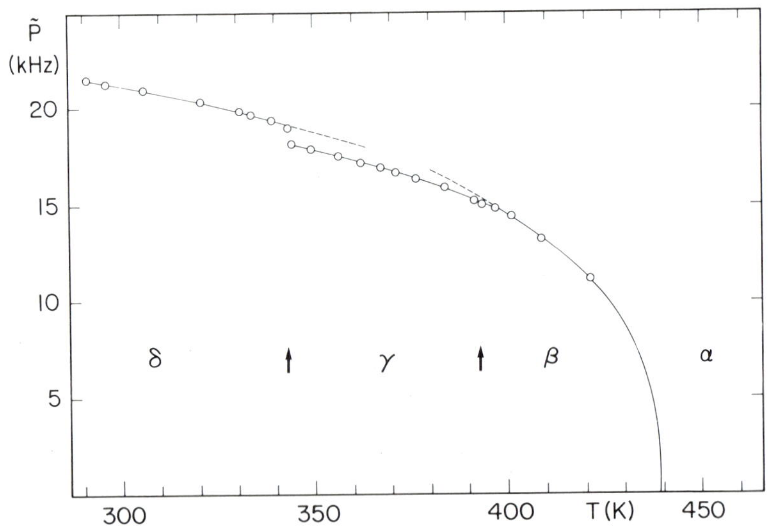

Fig. 4. Order parameter $\tilde{p} \propto p_1$ of the α–β transition versus temperature. (After Muralt 1984.)

leads to a characteristic powder-like frequency distribution with sharp edge singularities for each NMR-line (for a review see Blinc 1981). From the separation of these edge singularities the amplitude of the tilt angle ϕ can be obtained. This amplitude reaches 3.8° in the middle of the γ-phase and drops down to zero at the transitions to the β- and δ-phase, respectively (fig. 5) (Muralt 1984).

In a similar way the modulation amplitude of the octahedra tilting was measured by means of ^{35}Cl NMR–NQR. This was measured on the $-\frac{1}{2}+\frac{1}{2}$ transition in a magnetic field of 5 Tesla. As one can see from fig. 2 no pure NQR-signal of ^{35}Cl could be obtained in the γ-phase. At $T=361$ K the amplitude is measured to be 2.8°. This value is in good agreement with the value obtained from the modulation wavelength (40 Å) and amplitude (0.5 Å). From the fact that the tilt amplitude of the molecules is about 25% higher than that of the octahedra one can conclude that the driving force for the modulated phase originates from the direct interaction of the PA-chains, i.e. the modulation of the layer is a consequence of the chain modulation.

According to the symmetry description of the γ-phase (superspace group $M^{Cmca}_{1\,\bar{1}\bar{1}}$ given by Depmeier (1981), the soft mode of this phase can be identified as one of the 15 modes of H_3-symmetry, which conserve the mirror plane m_x

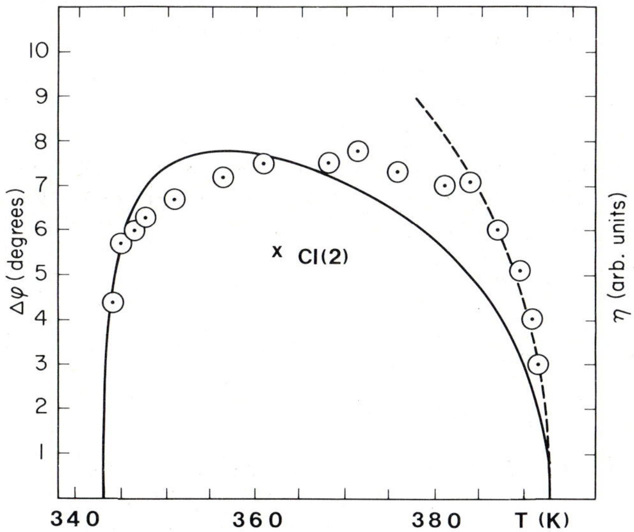

Fig. 5. Peak-to-peak values of the propylammonium modulation tilt angles $\Delta\phi$ versus temperature as obtained from deuteron NMR–NQR. The solid line and the dashed line are two different scales for the tilt angle resulting from the fit of the dilatation shown in fig. 6. The cross indicates the peak-to-peak value of the octahedra modulation tilt angle as measured by means of high field chain NQR–NMR. (After Muralt 1984.)

and the glide mirror plane m_y of Cmca (Muralt 1984). From the deuteron NMR-rotation pattern around the a and b axes one can see that this is indeed the case, i.e. the disorder of the PA-chains $n_1 = n_2 \neq n_3 = n_4$ is strictly maintained in the γ-phase.

2.2. *Theory of the $\beta-\gamma-\delta$ sequence*

The problem encountered with γ-PAMC is, that no lock-in transition takes place, but the high symmetry phase re-enters. Thus the usual treatment, to start with a thermodynamical potential where the order parameter is expanded around the critical wave vector $\boldsymbol{k}_c$ ($\boldsymbol{k}_c \equiv$wave vector of the lock-in phase), cannot be applied. A direct transition from the β- to the δ-phase would be caused by a non symmetry breaking order parameter. There is no obvious reason why such a transition should be connected with an instability with respect to an incommensurate modulation, unless one assumes a strong coupling between the order parameter η of the modulation and this totally symmetric order parameter e. Such a coupling leads but to a renormalization of some initial coefficients and to a spontaneous contribution to the equilibrium value of e. If, however, the equilibrium value of e exhibits an explicit temperature dependence, the coupling can quench the amplitude of the modulation completely, as we shall see later on.

In view of our assumption that the wave vector in γ-PAMC is not critical (Muralt et al. 1982) we choose a critical wave vector as close as possible to the measured incommensurate wave vector, i.e. $\boldsymbol{k}_c = \boldsymbol{b}^* + \boldsymbol{c}^*/6$ which corresponds to a general point on the H-line at the BZ-boundary.

The irreducible representations of the space group Cmca on the H-line are two-dimensional and conjugate complex. Their basis vectors Q_k, Q_{-k} are linear combinations of eigenvectors of normal modes belonging to the same irreducible representation:

$$Q_k = \alpha_k \eta_k + i\beta_k s_k, \qquad Q_{-k} = \alpha_{-k}\eta_{-k} + i\beta_{-k}s_{-k},$$

where

$$\alpha_k = \alpha_{-k}, \qquad \beta_k = \beta_{-k}, \qquad \eta_k = \eta^*_{-k}, \qquad s_k = s^*_{-k}. \tag{2.3}$$

For the H_3-symmetry the modes η_0 and s_0 transform according to the Y_2^- and Y_4^- at the Y-point, i.e. they are orthogonal. On the H-line, however, the η_k and s_k modes are coupled by the Lifshitz-invariant term (Landau and Lifshitz 1958, Levanyuk and Sannikov 1976a)

$$g_{\eta s} = i\gamma \left(Q_k \frac{dQ_k^*}{dz} - Q_k^* \frac{dQ_k}{dz} \right). \tag{2.4}$$

Assuming that the η-mode is dominant in the admixture exhibiting the lowest energy one can minimize the free energy density with respect to s by taking into account only the harmonic potential $g_s = \frac{1}{2}A'ss^*$ of the s-mode. The coupling term $g_{\eta s}$ then reads, with $\alpha = \alpha(k_c)$ and $\beta = \beta(k_c)$:

$$g_{\eta s} = -\frac{4\alpha^2\beta^2\gamma^2}{A'}\frac{\mathrm{d}\eta_k}{\mathrm{d}z}\frac{\mathrm{d}\eta_k^*}{\mathrm{d}z} + \mathrm{i}\gamma\alpha^2\left(\eta_k\frac{\mathrm{d}\eta_k^*}{\mathrm{d}z} - \eta_k^*\frac{\mathrm{d}\eta_k}{\mathrm{d}z}\right) + \text{higher order terms.} \tag{2.5}$$

The free energy expansion of the η-mode contains of course the harmonic term $\eta_k\eta_k{}^*$ and higher orders of it. The terms $\eta_k^2 \pm \eta_k^{*2}$ and $\eta_k^3 \pm \eta_k^{*3}$ fail to be invariants for any wave vector on the H-line. The lowest order Umklapp-term is $\eta_k^4 + \eta_k^{*4}$ for $\boldsymbol{k}_c = \boldsymbol{b}^* + \frac{1}{2}\boldsymbol{c}^*$. Since the k-vector of the modulation is not near the T-point, this term can be omitted. Up to the fourth order the free energy density reads

$$g_\eta = \frac{1}{2}A\eta\eta^* + \frac{1}{4}B(\eta\eta^*)^2 - \frac{1}{2}(\kappa - \delta)\frac{d\eta}{dz}\frac{d\eta^*}{dz} + \mathrm{i}\gamma\alpha^2\left(\eta\frac{\mathrm{d}\eta^*}{\mathrm{d}z} - \eta^*\frac{\mathrm{d}\eta}{\mathrm{d}z}\right) + \frac{1}{2}\lambda\frac{\mathrm{d}^2\eta}{\mathrm{d}z^2}\frac{\mathrm{d}^2\eta^*}{\mathrm{d}z^2}, \tag{2.6}$$

where

$$\kappa = \frac{8\alpha^2\beta^2\gamma^2}{A'} > 0, \qquad \delta, \lambda > 0, \qquad A = a(T - T_0).$$

The δ term corresponds to the elastic energy which competes with the κ term (Levanyuk and Sannikov 1976b). If the s-mode has a sufficient low energy (small A') $\kappa - \delta$ can become positive, so that the λ term is needed to stabilize the modulation. Attributing to η a single Fourier component $\eta_q = \eta \mathrm{e}^{-\mathrm{i}qz}$, $\boldsymbol{q} = \boldsymbol{k} - \boldsymbol{k}_c$, one obtains:

$$g_\eta = \frac{1}{2}A_q\eta_q\eta_q^* + \frac{1}{4}B(\eta_q\eta_q^*)^2$$

with

$$A_q = a(T - T_0) + \lambda q^4 - (\kappa - \delta)q^2 - 4\gamma\alpha^2 q. \tag{2.7}$$

By minimizing A_q one obtains the wave vector for which the β-phase becomes unstable with respect to η_q.

We are now introducing a phenomenological model describing the reentrant behaviour of γ-PAMC. As mentioned in the introduction of this section our hypothesis is that a coupling of the order parameter η of the modulation with a non-symmetry breaking order parameter of the β-phase is able to quench the amplitude of the modulation completely. The specific feature of our substance is an unusually high thermal expansion along the y-axis, i.e. perpendicular to the layers (see fig. 6). By extrapolation from the β-phase one can distinguish two contributions to the thermal expansion: a fairly linear background which amounts to 2.5% over the whole range of the γ-phase, and a spontaneous contribution due to a coupling to the order parameter η of the γ-phase, with a maximum value of about 1%. A microscopic reason for this coupling can be found in the special symmetry of the eigenvector of η. $|\eta|$ is the amplitude of a frozen out zone boundary mode with a transverse antiphase modulation of adjacent layers as shown exaggerated in fig. 3. As the minimal interlayer distance is restricted for sterical reasons (intercalation of PA-chains is not possible), the spontaneous contribution to the mean interlayer distance must be positive and must increase with rising η. As the spontaneous strain component $e_{yy}(\eta)-e_{yy}(0)$ is by far the largest macroscopic consequence of the incommensurate modulation in the γ-phase, and as the total homogeneous

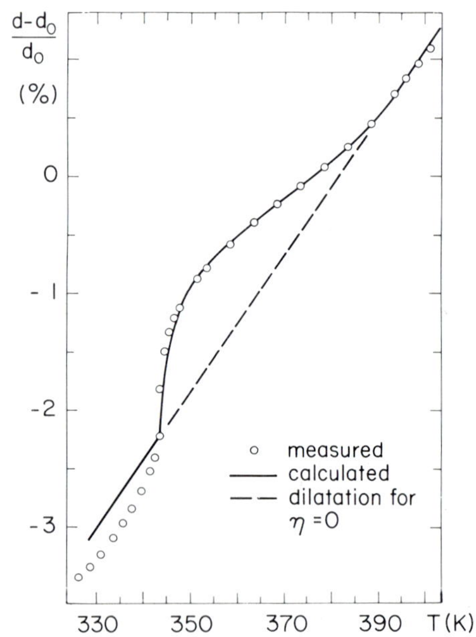

Fig. 6. Thermal expansion $(d-d_0)/d_0$ along the b-axis of PAMC for the $\beta-\gamma-\delta$ phase sequence. The solid curve is a fit to the Landau theory described in section 2.2. (After Muralt et al. 1982.)

strain $e_{yy} \equiv e_2$ transforms according to the identity representation of the β-phase, we postulate that the coupling of η and e_2 is the essential reason for the reentrant behaviour of γ-PAMC.

Therefore we have to introduce two additional terms to our free energy density (2.6): g_e and $g_{e\eta}$, describing the additional elastic energy due to the dilatation of the layers and coupling between the dilatation and the modulation wave:

$$g_e = C(T - T_0)e + \tfrac{1}{2}De^2 + \cdots, \qquad D > 0, \quad C < 0. \tag{2.8a}$$

Minimizing g_e with respect to e yields

$$e_0(\eta = 0) = -\frac{C}{D}(T - T_0),$$

where $-C/D$ is the linear expansion coefficient. Since the temperature dependence of $e_{yy}(0)$ is fairly linear over the whole range of the γ-phase we can neglect higher orders in e.

The coupling term $g_{e\eta}$ up to the 4th order in η reads

$$g_{e\eta} = \left(Ee + \tfrac{1}{2}Fe^2\right)\eta\eta^* + \left(E'e + \tfrac{1}{2}F'e^2\right)(\eta\eta^*)^2 + \cdots,$$
$$E, E' < 0, \quad F, F' > 0. \tag{2.8b}$$

This term describes in a simplified way the microscopic properties of the system discussed above. Minimizing the total free energy density $g = g_\eta + g_e + g_{e\eta}$ with respect to e one obtains

$$e_0 = \frac{C(T - T_0) + E|\eta|^2 + E'|\eta|^4}{D + F|\eta|^2 + F'|\eta|^4} \tag{2.9}$$

for the equilibrium value of e. Substituting this back into the equation for g, and expanding up to the fourth order in η, we find

$$g = g_0 + \tfrac{1}{2}\tilde{A}\eta\eta^* + \tfrac{1}{4}\tilde{B}(\eta\eta^*)^2 - \tfrac{1}{2}(\kappa - \delta)\frac{\mathrm{d}\eta}{\mathrm{d}z}\frac{\mathrm{d}\eta^*}{\mathrm{d}z}$$
$$+ \mathrm{i}\gamma\alpha^2\left(\eta\frac{\mathrm{d}\eta^*}{\mathrm{d}z} - \eta^*\frac{\mathrm{d}\eta}{\mathrm{d}z}\right) + \tfrac{1}{2}\lambda\frac{\mathrm{d}^2\eta}{\mathrm{d}z^2}\frac{\mathrm{d}^2\eta^*}{\mathrm{d}z^2}, \tag{2.10}$$

with

$$g_0 = -\frac{C^2}{2D}(T-T_0)^2,$$

$$\tilde{A} = \tilde{a}(T-T_0) + \frac{C^2F}{D^2}(T-T_0)^2, \quad \tilde{a} = a - \frac{2EC}{D},$$

$$\tilde{B} = B - \frac{2E^2}{D} + \frac{4C}{D}\left(\frac{EF}{D} - E'\right)(T-T_0) - \frac{2C^2}{D^2}\left(\frac{F^2}{D} - F'\right)(T-T_0)^2.$$

The coupling of η with the temperature dependent strain e_2 thus results in a temperature dependence of the renormalized Landau expansion coefficients. Assuming that the incommensurate soft mode is resulting from a coupling between two low lying modes with small dispersion, one can neglect the Lifshitz-invariant γ-term. In the constant-amplitude approximation the Euler equation for the phase angle $\phi(z)$ of the order parameter $\eta = r\exp(\mathrm{i}\phi)$ becomes

$$\lambda\phi''' - 2\lambda\phi'^3 + (\kappa - \delta)\phi' = 0, \qquad \phi' = \mathrm{d}\phi/\mathrm{d}z. \tag{2.11}$$

It was shown by Muralt and Kind (1983) that the plane-wave solution $\phi(z) = q_0 z$ has the lowest total free energy in the constant-amplitude approximation $[r(z) = r(0)]$. As the Euler equation for the phase does not contain the amplitude r of the order parameter but only essentially temperature independent coefficients, the wave vector is temperature independent too: $q = q_0 = ((\kappa - \delta)/2\lambda)^{1/2}$. This behaviour predicted by Muralt et al. (1982) was confirmed by measurements of Depmeier (1983).

Inserting the plane wave solution into eq. (2.10) yields

$$g = g_0 + \tfrac{1}{2}\left(\tilde{A} - \frac{(\kappa - \delta)^2}{4\lambda}\right)r^2 + \tfrac{1}{4}\tilde{B}r^4. \tag{2.12}$$

From the expression for $\tilde{A}$ it is evident that $\tilde{A} - (\kappa - \delta)^2/4\lambda$ becomes negative within a certain temperature region $T_{c1} > T > T_{c2}$. For this region, which corresponds to the γ-phase, the equilibrium value of r is different from zero, but equal to zero outside. From the structure of eq. (2.12) it follows that both transitions must be of second order.

In reality the reentrant phase transition at T_{c2} is slightly discontinuous, which is due to an invariant third-order term of e_2 neglected in eq. (2.8). By choosing appropriate values for the coefficients it is possible to obtain a very good fit for the observed thermal expansion of γ-PAMC as shown in fig. 6. In fig. 5 the equilibrium value of the order parameter η obtained for the same coefficients is shown. The discrepancy between measured and calculated temperature dependence of η is due to neglected higher-order terms in eq. (2.10).

3. The $\delta-\varepsilon-\zeta$ sequence

At $T_I = 165.2\,K$ partially deuterated PAMC undergoes a second-order phase transition to an incommensurate phase ε with a quasi triplication of the unit cell along the a-axis. The lock-in transition at $T_c = 114.5\,K$ does not lead to a structure with a modulation wave vector of $\boldsymbol{k}_c = \boldsymbol{a}^*/3$, but to a structure with $k_c = (a^* \pm b^*)/3$. Hence the unit cell in the ζ-phase is triplicated along the primitive basis vectors $\boldsymbol{t}_1$ or $\boldsymbol{t}_2$ of the C-centered unit cell. The two possibilities correspond to two different domains.

The two phase transitions were discovered by Depmeier et al. (1977) by means of differential scanning calorimetry and X-ray scattering. Elastic neutron scattering performed by Depmeier and Mason (1982, 1983) revealed the incommensurability of the ε-phase and the switching of the wave vector direction at the $\varepsilon-\zeta$ transition. The superspace group $P^{Cmca}_{\bar{1}1s}$ for ε-PAMC was derived from X-ray scattering (Brunskill and Depmeier 1982) and from elastic neutron scattering (Depmeier and Mason 1982). According to the same authors the symmetry of ζ-PAMC belongs to the Bravais-class 2/m. Further details of the structure have not been obtained so far by means of scattering experiments. Group theoretical considerations in combination with our NQR-results allow, however, to determine the space group unambiguously to the P11 $2_1/n$ (Muralt 1984).

3.1. ^2H- and ^{35}Cl-NMR–NQR

It was possible to detect the pure chlorine NQR signals with the pulse-technique also in the ε-phase, where the spectral lines are anomalously broadened over a range of up to 500 kHz in case of the Cl(1)-sites, and up to 80 kHz in case of the Cl(2)-sites (fig. 2). On lowering the temperature additional structure is observed in the Cl(1) spectrum besides the usual edge singularities. This is probably due to the occurrence of soliton-like distortions of the incommensurate modulation wave. Below T_c the number of NQR lines is triplicated with respect to δ-PAMC, demonstrating the triplication of the unit cell in the ζ-phase. The measurements clearly reveal the continuity of the $\delta-\varepsilon$ transition and the discontinuity of the $\varepsilon-\zeta$ transition. From the temperature dependence of the edge singularities the critical exponent β of the order parameter of the $\delta-\varepsilon$ transition was derived to be $\beta = 0.37 \pm 0.03$ (Muralt 1984).

The DMR-rotation patterns in the ε-phase revealed that the modulation wave is of transversal nature. The tilting of the PA-groups is modulated mainly around the c-axis with an amplitude of 11° at 131 K. This modulation destroys the mirror plane m_x. Around the a-axis the modulation amplitude is measured to be 2.3°, whereas no rotation around the b-axis (layer normal) was observed (Muralt 1984).

In the ζ-phase the DMR-rotation patterns clearly show the glide mirror plane m_z and also the connection of the lines from the two different domains by the lost mirror plane m_x. The amplitude of the PA tilt angle around the c-axis amounts now up to 20°, whereas the amplitude around the a-axis is reduced to 1.2°. This can well be understood by taking the gradual freezing-in of the PA-motion into account. In the δ-phase we still have a disordered state with $n_1 = n_2 \neq n_3 \neq n_4$ (required by the mirror plane m_x). In the ε-phase the mirror plane m_x is destroyed by the modulation and n_1 becomes different from n_2. At the lower end of the ε-phase the reorientational jumps start to freeze in, a fact which can be observed in an increase of the second moment of the proton NMR-lines around 120 K. The lock-in transition at 114.5 K is thus connected with this freeze-in (Muralt 1984).

3.2. *Neutron scattering experiments*

In order to obtain more information about the soft mode and the lock-in transition of the ε-phase elastic neutron scattering measurements were performed on a partially deuterated single crystal (Muralt 1984). The experiments confirmed previous data on normal PAMC (Depmeier and Mason 1982, 1983)

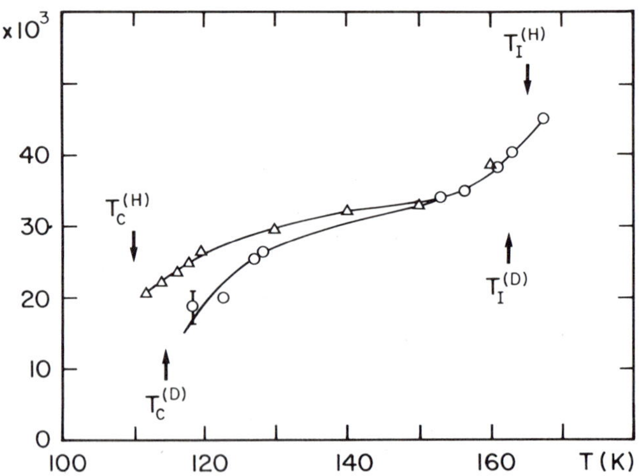

Fig. 7. Incommensurate part δ of the wave vector $k_0 = (1/3 + \delta)a^*$ of ε-PAMC as a function of temperature. Triangles, normal PAMC (after Depmeier and Mason 1982, 1983); open circles, partially deuterated PAMC (after Muralt 1984). The solid curves are guides to the eye.

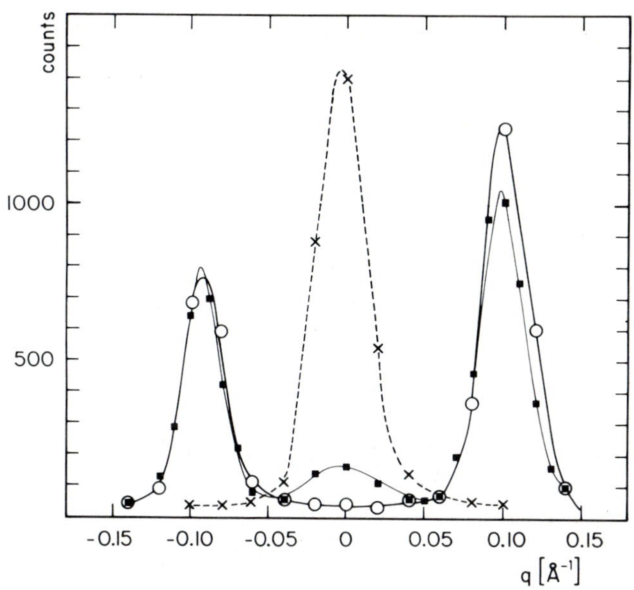

Fig. 8. Intensity of the satellite reflexions versus wave vector q for different temperatures, demonstrating the switching of the wave vector at the $\varepsilon-\zeta$ transition. The scattering vector of the neutrons is $(-3.333a^*,\ 13b^*+q,\ 0$. Crosses, $T=115\,\mathrm{K}$ (ε-phase); full squares, $T=114.1\,\mathrm{K}$ (ζ-phase); open circles, $T=112.7\,\mathrm{K}$ (ζ-phase). (After Muralt 1984.)

concerning modulation and lock-in wave vectors. In addition, strong critical scattering of some satellite reflexions of the ε-phase could be observed already in the δ-phase which allowed to determine the wave vector also above T_I (fig. 7). In fig. 8 the intensity of the satellite reflexions is plotted as a function of the wave vector q for different temperatures. For the chosen scattering vector a lock-in on the Σ-line would not give rise to an intensity at $q=\pm 0.1\,\text{Å}^{-1}$ but only at $q=0$. Thus the figure clearly shows the switching of the wave vector at T_c.

4. *Conclusions*

We have shown that the two incommensurate phases in PAMC exhibit unusual transitions at the low temperature side. For the γ-phase it is a reentrant high symmetry transition which is due to a coupling of the order parameter with a homogeneous deformation of the crystal. This behaviour can well be described by a Landau-type thermodynamical potential. For the ε-phase it is the switching of the wave vector direction at T_c. This behaviour is related to the freezing-in of the reorientational motion of the propylammonium chains. A Landau theory explaining the wave vector switching and moreover also the values of the observed modulation vectors is given in Muralt (1984).

References

Blinc, R., 1981, Phys. Rep. **79**, 331.

Bradley, C.J., and A.P. Cracknell, 1972, Theory of Symmetry in Solids (Clarendon Press, Oxford).

Brunskill, I.H., and W. Depmeier, 1982, Acta Crystallogr. **A38**, 132.

Caravatti, P., 1981, Diploma Thesis (Department of Physics, Swiss Federal Institute of Technology, Zürich) unpublished.

Chapuis, G., 1978, Acta Crystallogr. **B34**, 1506.

Depmeier, W., 1979, J. Solid State Chem. **29**, 15.

Depmeier, W., 1981, Acta Crystallogr. **B37**, 330.

Depmeier, W., 1983, Solid State Commun. **45**, 1089.

Depmeier, W., and S.A. Mason, 1978, Acta Crystallogr. **B34**, 920.

Depmeier, W., and S.A. Mason, 1982, Solid State Commun. **44**, 719.

Depmeier, W., and S.A. Mason, 1983, Solid State Commun. **46**, 409.

Depmeier, W., J. Felsche and G. Wildermut, 1977, J. Solid State Chem. **21**, 57.

Kind, R., 1980, Ferroelectrics **24**, 81.

Kind, R., S. Plesko and J. Roos, 1978, Phys. Stat. Sol. (a) **47**, 233.

Landau, L.D., and E.M. Lifshitz, 1958, Statistical Physics (Pergamon Press, London).

Levanyuk, A.P., and D.G. Sannikov, 1976a, Sov. Phys. Solid State **18**, 245.

Levanyuk, A.P., and D.G. Sannikov, 1976b, Sov. Phys. Solid State **18**, 1122.

Muralt, P., 1984, PhD Thesis, Diss. Nr. 7524 (Swiss Federal Institute of Technology, Zürich) to be published.

Muralt, P., and R. Kind, 1983, Helv. Phys. Acta **56**, 689.

Muralt, P., R. Kind, R. Blinc and B. Zeks, 1982, Phys. Rev. Lett. **49**, 1019.

Peterson, E.R., and R.D. Willet, 1972, J. Chem. Phys. **56**, 1879.

CHAPTER 21

Incommensurate Structures in Liquid Crystals

S.A. PIKIN

Institute of Crystallography
Academy of Sciences of the USSR
Moscow 117333
USSR

Incommensurate Phases in Dielectrics 2
Edited by
R. Blinc and A.P. Levanyuk

Contents

1. Liquid crystals and effects of incommensurability . . . 321
2. Incommensurate nonpolar structures . . . 329
   2.1. Phenomenological description . . . 329
   2.2. One-dimensional cholesteric phases . . . 331
   2.3. Blue phases . . . 333
3. Phase transition cholesteric–smectic A . . . 334
4. Incommensurate structures in cholesteric and smectic liquid crystals induced by external fields . . . 339
5. Incommensurate flexoelectric structure in nematic liquid crystals . . . 345
   5.1. The flexoelectric effect . . . 345
   5.2. The linear problem . . . 347
   5.3. The nonlinear problem . . . 348
6. Ferroelectric structures in liquid crystals . . . 352
   6.1. Symmetry properties . . . 352
   6.2. The helical ferroelectric . . . 355
   6.3. The behaviour of liquid–crystalline ferroelectrics in external fields . . . 357
   6.4. Results of experimental investigations of ferroelectric liquid crystals . . . 359
7. Incommensurate smectic structures as polymorphous modifications . . . 362
References . . . 365

1. *Liquid crystals and effects of incommensurability*

Evidently for the first time incommensurate structures were observed optically by Reinitzer in 1888. Reinitzer had discovered cholesteric and blue phases of liquid crystalline substances. However, the structures of these phases were deciphered and described much later when physicists understood what is a liquid crystalline state of substance in general and what properties are usual for it. At present one considers liquid crystalline phases or mesophases as thermodynamic states intermediate in properties between solid crystalline and isotropic liquid states. Mesophases, formed by anisotropic molecules, are classified by their kind of density function $\rho(\boldsymbol{r})$, characterizing the distribution of the centers of gravity of molecules, and by the kind of local orientation of specific molecular axes.

The most "simple" mesophase is the "nematic" phase of substance consisting of rod-like molecules. A nematic phase (or "nematic") is characterized by the homogeneous function $\rho(\boldsymbol{r}) = \text{const.}$, as in an ordinary isotropic liquid, and by a specific direction along which at any point $\boldsymbol{r}$ long molecular axes are oriented in average. Thus, a nematic is an anisotropic liquid. Since a nematic has a single specific direction which causes corresponding anisotropic optical properties, as in an anisotropic solid crystal, such a liquid crystalline state is uniaxial. One can imagine a more complex nematic state, for example biaxial, if an anisotropic molecular object has two specific axes. The biaxial nematic is characterized by two specific directions in space: here the different molecular axes are oriented in average along these directions.

The parameter of orientational order in an uniaxial nematic is a tensor $\hat{Q}$ with components

$$Q_{ik}(\boldsymbol{r}) = \langle e_i e_k \rangle - \tfrac{1}{3}\langle e_j e_j \rangle \delta_{ik} = Q(\boldsymbol{r})\left(n_i n_k - \tfrac{1}{3}\delta_{ik}\right),$$

which are the local averages of quadratic combinations formed by projections of the unit vector $\boldsymbol{e}$ parallel to the long molecular axis. Here $Q(\boldsymbol{r})$ determines the share of the molecules oriented along some common axis, labelled by a unit vector (or "director") $\boldsymbol{n}$, at the point $\boldsymbol{r}$. The states with directors $\boldsymbol{n}$ and $-\boldsymbol{n}$ are physically indistinguishable. The value of $Q(\boldsymbol{r})$ is 0 in the isotropic phase of such substance. In the general case, for example in biaxial nematics, the order parameter, describing the nematic–isotropic phase transition, is a symmetric tensor $\hat{Q}$ with zero trace ($\mathrm{Sp}\,\hat{Q} = 0$), i.e., $\hat{Q}$ has five independent components. Since the density of free energy F in nematics contains the invariant $\mathrm{Sp}\,\hat{Q}^3 \neq 0$, the isotropic–nematic phase transformation is a first-order transition.

Another type of mesophase—a "smectic" liquid crystal (or "smectic")—is characterized by the orientational order parameter $\hat{Q}$ and by the function $\rho(\boldsymbol{r})$ which is periodic along a specific axis z, but $\rho(\boldsymbol{r}) = \text{const.}$ along smectic planes xy. Smectics possess a fluidity along planes xy, but they are similar to solids in relation to the uniaxial load along the crystalline axis z. In the uniaxial smectic A the directions n and z are collinear, and the density function $\rho(z)$ takes the form

$$\rho(z) = \rho_0 + |\psi|\cos(kz + \omega),$$

where $|\psi|$ and ω are the amplitude and the phase of the density wave correspondingly. One can consider the two-component order parameter

$$\psi(z) = |\psi|\exp[\mathrm{i}(kz + \omega)]$$

as the parameter of the nematic–smectic A phase transition. The value of ψ is 0 in nematics. Since the value F in smectics contains only invariants, even in powers of ψ, one can have the nematic–smectic A phase transformation as a second–order transition.

At lower temperature the uniaxial smectic A can transform to the biaxial smectic C which is characterized by a finite angle θ between director $\boldsymbol{n}$ and crystalline axis z. The presence of the angle θ corresponds to the arising of a certain tilt of molecules in smectic planes. One can suppose that the parameter of orientational order $\hat{Q}$ and the parameter of positional order $|\psi|$ are the same in smectics A and C. So one can describe the A–C phase transition by the two-component order parameter

$$\xi = (\boldsymbol{nk})[\boldsymbol{nk}]/k^2.$$

Since invariants in F must contain even powers of the components $\xi_1 = n_z n_x$ and $\xi_2 = n_z n_y$, this phase transition can be of the second kind.

Uniaxial nematic and smectic A phases have a point group of symmetry $D_{\infty h}$. The biaxial smectic C has point symmetry C_{2h}. Such structures are formed by molecules which have only one plane of symmetry if any. If the molecules are chiral, i.e., they have no center and planes of symmetry, the point symmetry of these liquid crystals falls to D_∞ and C_2 correspondingly. This decrease of symmetry is the physical reason for the appearance of incommensurate structures in liquid crystals. In fact, the spatially homogeneous orientational state of nematics with point symmetry D_∞ is instable with respect to spatial modulations of the director $\boldsymbol{n}(\boldsymbol{r})$, since this symmetry admits the existence of Lifshitz' invariant

$$(\boldsymbol{n}\,\mathrm{rot}\,\boldsymbol{n}) = n_x \frac{\partial n_y}{\partial z} - n_y \frac{\partial n_x}{\partial z}$$

in the free energy expansion. Therefore a "cholesteric" liquid crystal (or "cholesteric") must have a spontaneous spiral (helical) modulation of molecular orientations in space (fig. 1).

In smectics A formed by chiral molecules such modulation $\boldsymbol{n}(\boldsymbol{r})$ does not arise, since here the director $\boldsymbol{n}$ is obviously parallel to the crystalline axis z and the modulation $\boldsymbol{n}(\boldsymbol{r})$ could induce strong deformations of smectic planes. In smectics C with point symmetry C_2 (chiral smectics $\tilde{C}$) the orientational modulation $\boldsymbol{n}(\boldsymbol{r})$ is possible without perturbations of the smectic positional order due to rotations of the director $\boldsymbol{n}$ around the crystalline axis z (fig. 2). Thus, in the $\tilde{C}$ phase projections of $\boldsymbol{n}$ on smectic planes form a spontaneous helical structure in space similar to the cholesteric spiral structure.

The large spatial periods h and large angles $\varphi(z)$ of rotation of $\boldsymbol{n}$ around the helix axis z are typical for the orientational structures shown: $h \gg l$ and $h \gg 2\pi k^{-1}$, where l is the length of a molecule, $2\pi k^{-1}$ is the thickness of a smectic monolayer, and $\varphi(z+h) - \varphi(z) = 2\pi$. The large values of h and φ in liquid crystals remind of the situation in solid helical ferromagnets

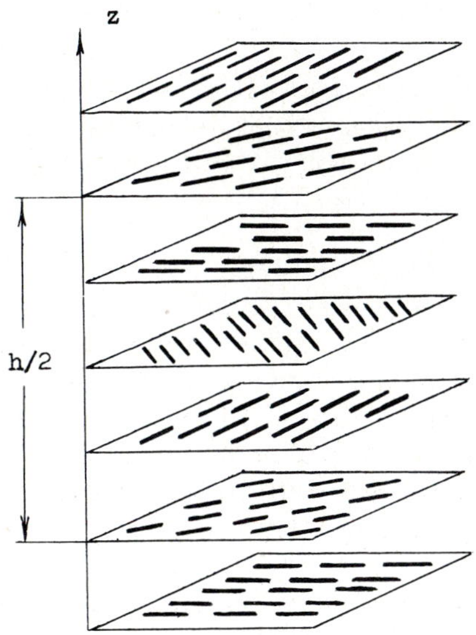

Fig. 1. Helical cholesteric structure.

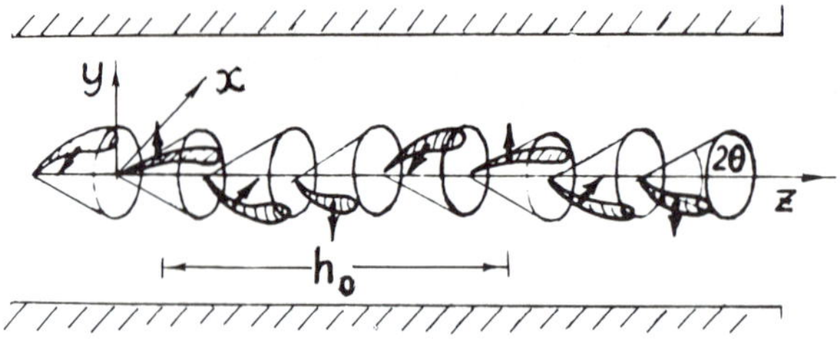

Fig. 2. Helical structure in the chiral smectic phase C.

(Dzyaloshinskii 1964), where rotations of the magnetic moment are caused by anisotropic relativistic interactions which are weak in comparison with the basic exchange interaction. Liquid crystals have no crystal anisotropy and possess chiral interaction which is weak in comparison with the basic interaction causing the orientational ordering of molecules. In fact, all the director orientations around the helix axis in cholesteric and chiral smectic $\tilde{C}$ phases are energetically equal at constant polar angle θ (figs. 1, 2). Usually the angle between directors of neighbour planes *xy* is in the order of $l/h \sim 10^{-2}$. Solid dielectrics have large crystal anisotropy and correspondingly small values of h and φ.

Upon consideration of the properties of cholesterics one can see some more possibilities for incommensurate structures attributed now to "blue" phases existing in a very narrow temperature interval (~ 1 K) between the isotropic and cholesteric phases (Hornreich and Shtrikman 1980, 1981, 1983, Meiboom et al. 1981, 1983, Grebel et al. 1983). Namely, the helical orientational structure can reform in such interval and new superstructures can arise. These superstructures possess a cubic spatial symmetry and cell parameters greater than the initial helix pitch (fig. 3). This possibility stems from the character of the isotropic–cholesteric phase transition which is as close to the second-order transition as the weak first-order isotropic–nematic phase transition. An additional scale ζ (besides l and h) makes sense for this transition: $\zeta \gg l$ is the correlation length of molecular orientations, it is the characteristic of short-range order in the system. Here one can consider two possibilities: (1) $\zeta \ll h$ if the phase transition is not close to the second-order transition, (2) $\zeta \gg h$ if the phase transition is very close to the second-order transition. There is uniaxial nematic ordering on the scale h in the first case. In the second case the incommensurate structure differs from the uniaxial nematic on the scale h, for example it can be biaxial and the biaxiality increases if the pitch h decreases (Brazovskii and Dmitriev 1975, Brazovskii and Filev 1978). In the second case there is also the possibility of spatial twisting of the director $\boldsymbol{n}$ in two independent directions in locally uniaxial environment (Meiboom et al. 1983, see fig. 3). On the basis of these considerations one can study non one-dimen-

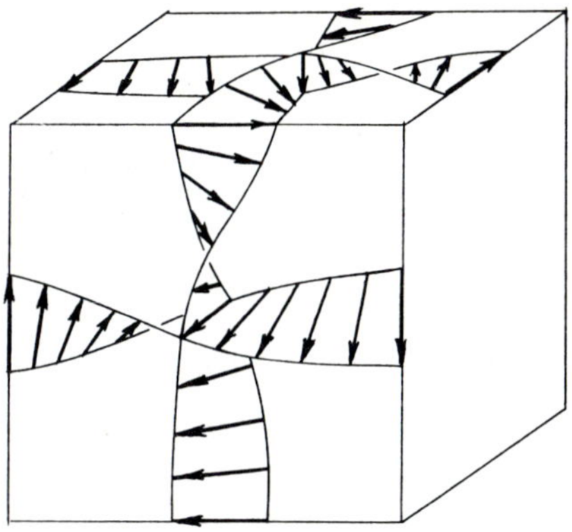

Fig. 3. Double orientational spiral in the blue phase with the group of spatial symmetry O^5. (Meiboom et al. 1983.)

sional incommensurate structures with flat hexagonal and cubic lattices.

The existence of incommensurate structures described above is caused by a weak chirality of molecules. In liquid crystalline systems of achiral molecules one can consider incommensurate smectic phases which are very similar to incommensurate structures in solid dielectrics (Prost 1979, Prost and Barois 1983). Usually such structures arise in smectics when a system has two (or more) order parameters, for example density waves with different wave numbers (mass density, charge density, electric potential). The interaction of these order parameters causes various transitions between commensurate and incommensurate phases observed recently. The periods of these structures are comparable with the period of the one-dimensional crystalline lattice $2\pi k^{-1}$ but these periods are not commensurate to $2\pi k^{-1}$.

Besides incommensurate structures, arising as a result of phase transitions at fixed temperatures, thermodynamically equilibrium incommensurate structures can appear in liquid crystals under the action of external fields. The formation

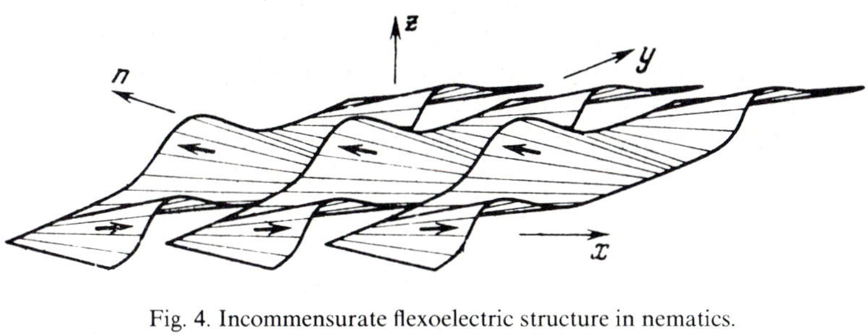

Fig. 4. Incommensurate flexoelectric structure in nematics.

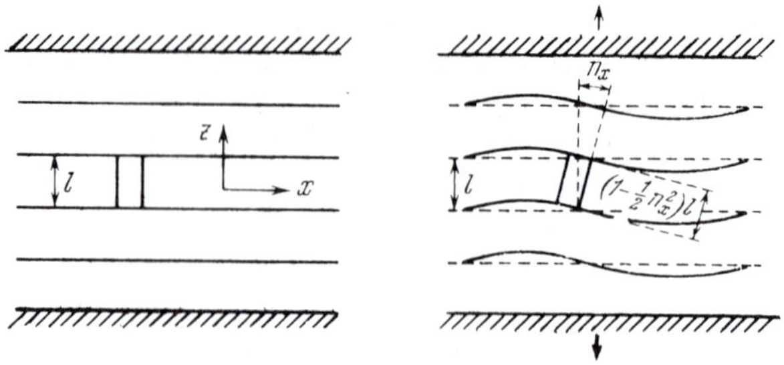

Fig. 5. Bend of smectic layers under mechanical stress.

of such structures is a peculiar phase transition: above some threshold value of the external field the amplitude and the period of the corresponding wave change in accordance with predictions of the phenomenological theory which takes into account essential invariants in F. An interesting example of such orientational structure is the periodic flexoelectric modulation of nematics (fig. 4). Here spatial orientational perturbations are related to the induced polarization $\boldsymbol{P}_{\mathrm{i}}$, arising in the external electric field $\boldsymbol{E}$ and containing the terms $\boldsymbol{n}\,\mathrm{div}\,\boldsymbol{n}$ and $\boldsymbol{n}\,\mathrm{rot}\,\boldsymbol{n}$ with some coefficients. The period of this incommensurate flexoelectric structure, $2\pi q_{\mathrm{c}}^{-1}$, is proportional to the sample thickness d at the threshold value $E = E_{\mathrm{c}}$, however the period does not depend on thickness and $q \sim E$ at $E \gg E_{\mathrm{c}}$.

Other examples of incommensurate structures arising in external fields are undulations of smectic planes in smectics A under the action of uniaxial stretching (fig. 5) and undulations of quasi-nematic layers in cholesterics under the action of electric fields (fig. 6). Here the threshold values of the period s_{c} obey the proportionalities $s_{\mathrm{c}} \sim (dl)^{1/2}$ for smectics and $s_{\mathrm{c}} \sim (dh)^{1/2}$ for cholesterics. The change of the periods of incommensurate structures in nematics, smectics A and cholesterics above corresponding thresholds is connected with

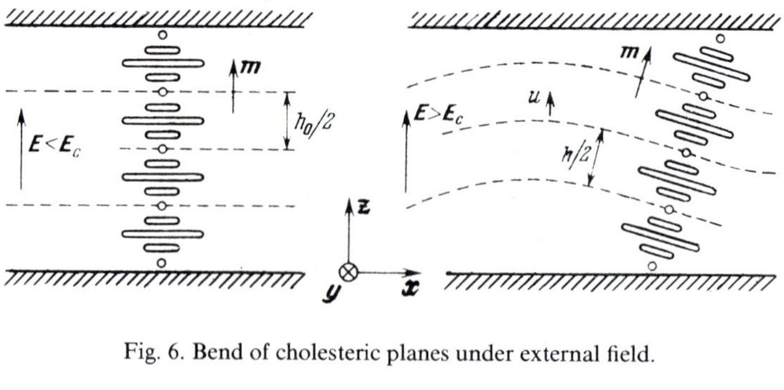

Fig. 6. Bend of cholesteric planes under external field.

the formation of various defects, dislocations for example. This is similar to the situation in solid dielectrics (Kawasaki 1983). The formation of defects in liquid crystals is easily observed by optical methods (fig. 7).

Orientational helixes in cholesterics and smectics C change their pitch under the action of external fields and temperature variations (de Gennes 1974, Pikin 1981). The helixes can be totally unwound at determinate critical values of external fields. The change of period of these incommensurate structures is also accompanied by the formation of various defects, for example some walls arise, reminding of Bloch walls in ferromagnets in the character of the director orientation change $\boldsymbol{n}(\boldsymbol{r})$. One can easily estimate the threshold values of electric (magnetic) field $E_c(H_c)$ in a nonpolarized (nonmagnetized) substance with dielectric anisotropy ε_a (diamagnetic anisotropy χ_a) from a balance of electric, $\int F_e \mathrm{d}V$ (magnetic, $\int F_m \mathrm{d}V$), and elastic, $\int F_0 \mathrm{d}V$, terms in the free energy:

$$F_e = -\frac{\varepsilon a}{8\pi}(\boldsymbol{nE})^2, \qquad F_m = -\frac{\chi a}{2}(\boldsymbol{nH})^2, \qquad F_0 = \tfrac{1}{2}K_2(q + \boldsymbol{n}\,\mathrm{rot}\,\boldsymbol{n})^2,$$

$$E_c^2 \sim \frac{K_2 q^2}{\varepsilon_a}, \qquad H_c^2 \sim \frac{K_2 q^2}{\chi_a},$$

where K_2 is the coefficient of orientational elasticity, and $q = 2\pi h^{-1}$ is the wave number of the incommensurate structure. If the spiral structure in a smectic $\tilde{\mathrm{C}}$ is ferroelectric and if the amplitude of the spontaneous polarization

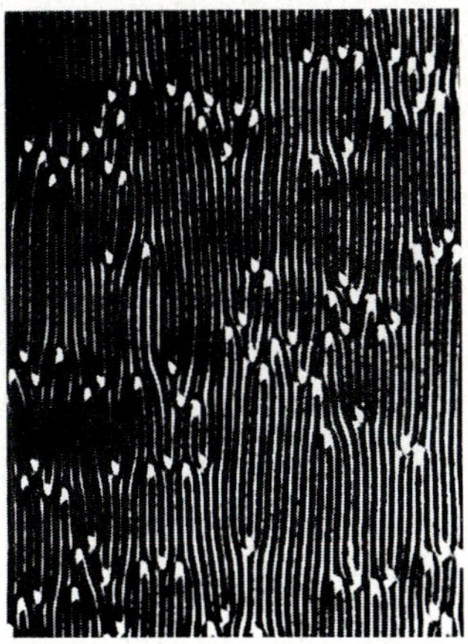

Fig. 7. Defects in liquid crystals. Flexoelectric structure in *p*-*n*-butyl-*p*-metoxyazoxybenzene. (Barnik et al. 1978.)

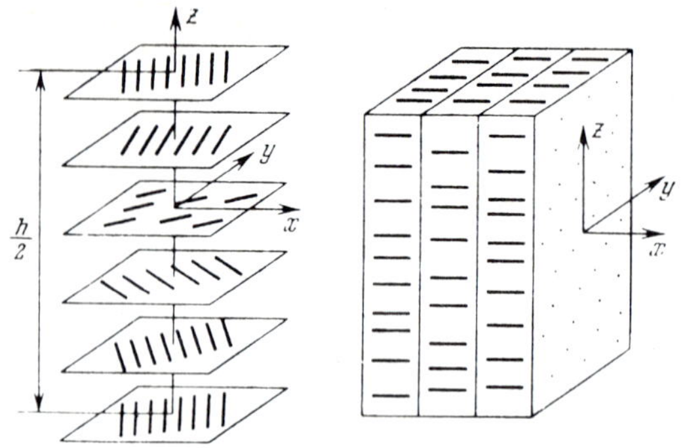

Fig. 8. Cholesteric–smectic A phase transition: director orientations in cholesteric and smectic planes.

is P_s, then total unwinding of the helix takes place at the threshold value $E = E_c \sim P_s^{-1} K q^2 \theta^2$, where θ ($\theta^2 \ll 1$) is the projection of $\boldsymbol{n}$ on the smectic plane (fig. 2), K is the elastic constant. It is also clear that the cholesteric pitch must change significantly when temperature decreases to the smectic–cholesteric phase transition point (figs. 8, 9).

The examples mentioned above show a variety of incommensurate structures in liquid crystals, broad possibilities of operation on the periods of such structures by external action, and also the essential influence of various defects on the formation of real incommensurate structures in liquid crystals. The

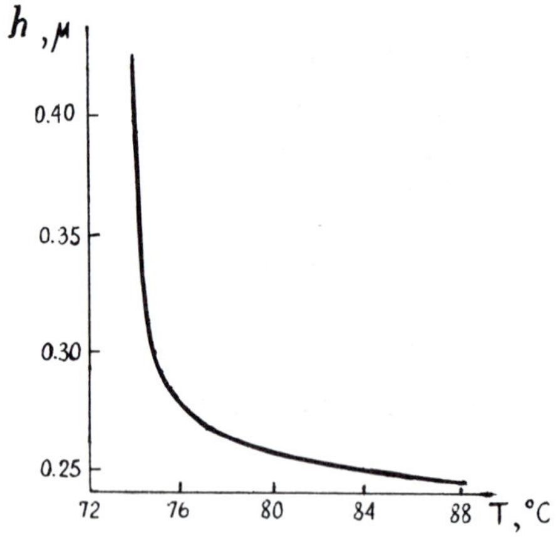

Fig. 9. Experimental temperature dependence of the pitch of cholesteric spiral $h = 2/q$ in cholesteryl nonanoate near the smectic A-cholesteric phase transition point (74°C). (Kassubek and Meier 1969).

present review is devoted to the qualitative and quantitative description of the liquid crystalline incommensurate phases.

2. Incommensurate nonpolar structures

2.1. Phenomenological description

The symmetry group of an isotropic liquid with chiral molecules is the complete group of rotations around a point (***K***). A decrease of symmetry as the result of phase transitions usually results in either cholesteric or smectic A or blue phases. A smectic A and a cholesteric with large pitch of the helix have the point group of symmetry D_∞ that is typical for uniaxial liquid crystals. In that case the orientational order parameter $\hat{Q}$ depends on Q and $\boldsymbol{n}$ only. $Q(T)$ is a weak function of temperature far from phase transition points. But orientations of $\boldsymbol{n}$ are strongly perturbed by thermal fluctuations in intermediate temperature regions and also by external fields.

In the general case the tensor $\hat{Q}(\boldsymbol{r})$ has five independent components transforming according to the five-dimensional representation ***G*** of the group ***K***. This representation describes the ordering both of long and of short molecular axes. The antisymmetric square of ***G*** contains a vector representation. Consequently the Lifshitz invariant exists and has the general form

$$e_{ijk}Q_{im}\frac{Q_{jm}}{x_k}, \tag{1}$$

where $\hat{e}$ is the totally antisymmetric unit tensor. The invariant (1) determines the character of incommensurate structures. In uniaxial cholesterics the invariant (1) results in the term $(\boldsymbol{n}\operatorname{rot}\boldsymbol{n})$ and corresponds to formation of a helical structure with director components

$$n_x = \cos(qz), \qquad n_y = \sin(qz), \qquad n_z = 0. \tag{2}$$

Since the symmetric cube of ***G*** ($[\boldsymbol{G}^3]$) contains the unit representation A_1, the phase transition between an isotropic liquid and a cholesteric must be of first order; that is the consequence of existence of the invariant

$$\operatorname{Sp}\hat{Q}^3 = \tfrac{2}{9}\hat{Q}^3.$$

In the general case of a multicomponent order parameter $\hat{Q}$ the existence of the invariant $\operatorname{Sp}\hat{Q}^3$ does not forbid a second-order incommensurate phase transition over certain components of $\hat{Q}$. Such leading components enter the corresponding invariants in even powers only and the other components (as a perturbation) form invariants of odd powers (Pikin 1981).

The incommensurate structures of this type can be described in detail by consideration of the free energy $\mathscr{F}$ as the functional (Brazovskii and Dmitriev 1975)

$$\mathscr{F}[\hat{Q}] = \int \left\{ a \operatorname{Sp} \hat{Q}^2 + b \operatorname{Sp} \hat{Q}^3 + c \operatorname{Sp}^2 \hat{Q}^2 + g_1 \left(\frac{\partial Q_{\gamma\delta}}{\partial x_\alpha} \right)^2 \right.$$

$$\left. + g_2 \frac{\partial Q_{\alpha\gamma}}{\partial x_\alpha} \frac{\partial Q_{\beta\gamma}}{\partial x_\beta} + 2 g_1 q Q_{\alpha\delta} \frac{\partial Q_{\beta\delta}}{\partial x_\gamma} e_{\alpha\beta\gamma} \right\} \mathrm{d}V, \tag{3}$$

which contains the invariants discussed above. Here a, b, c, g_1, g_2 and q are the phenomenological parameters of a system, $c > 0$, $g_1 > 0$, g_1 and g_2 characterize short-range orientational correlations (the scale ζ).

It is convenient to turn into the impulse representation

$$\hat{Q}_p = \frac{1}{V} \int \hat{Q}(\boldsymbol{r}) \mathrm{e}^{-\mathrm{i}\boldsymbol{p}\boldsymbol{r}} \mathrm{d}V$$

and to determine the correlation functions in the Gaussian approximation according to the elegant technique of Brazovskii and Dmitriev (1975):

$$G_{\alpha\beta}^{\gamma\delta}(\boldsymbol{p}) = \frac{\int Q_{\alpha\beta}(\boldsymbol{p}) Q_{\gamma\delta}(-\boldsymbol{p}) \exp\left[-\frac{V}{T} \sum_{p} t_\lambda^\mu(\boldsymbol{p})\, \delta_{\zeta\nu} Q_{\lambda\zeta}(\boldsymbol{p})\, Q_{\mu\nu}(-\boldsymbol{p}) \right] \mathrm{d}\hat{Q}}{\int \exp\left[-\frac{V}{T} \sum_{p} t_\lambda^\mu(\boldsymbol{p})\, \delta_{\zeta\nu} Q_{\lambda\zeta}(\boldsymbol{p})\, Q_{\mu\nu}(-\boldsymbol{p}) \right] \mathrm{d}\hat{Q}}, \tag{4}$$

where

$$t_\alpha^\gamma(\boldsymbol{p}) = (a + g_1 \boldsymbol{p}^2)\, \delta_{\alpha\gamma} + g_2 p_\alpha p_\gamma + \mathrm{i} 2 g_1 q e_{\alpha\gamma\nu} p_\nu. \tag{5}$$

Since the tensor $\hat{Q}$ has five independent components, one can find five eigenfunctions $\sigma^s(\boldsymbol{p})$ and five corresponding eigenvalues $\tau^s(\boldsymbol{p})$ of the operator $\hat{t}(\boldsymbol{p})$ ($s = 0, 1, 2, 3, 4$):

$$t_{\alpha\beta}^{\gamma\delta}(\boldsymbol{p}) \sigma_{\alpha\beta}^s(\boldsymbol{p}) = \tau^s(\boldsymbol{p}) \sigma_{\gamma\delta}^s(\boldsymbol{p}). \tag{6}$$

One can easily find the solution of this problem if one introduces the right triad of mutually orthogonal real unit vectors $\boldsymbol{p}p^{-1}$, $\boldsymbol{n}'(\boldsymbol{p})=\boldsymbol{n}'(-\boldsymbol{p})$ and $\boldsymbol{n}''(\boldsymbol{p})=-\boldsymbol{n}''(-\boldsymbol{p})$, and also the complex vector $\boldsymbol{m}(\boldsymbol{p})=2^{-1/2}(\boldsymbol{n}'+\mathrm{i}\boldsymbol{n}'')=\boldsymbol{m}(-\boldsymbol{p})$:

$$\sigma^0_{\alpha\beta}=\frac{1}{\sqrt{6}\,p^2}\left(3p_\alpha p_\beta-p^2\delta_{\alpha\beta}\right),\qquad \sigma^1_{\alpha\beta}=m_\alpha m_\beta,\qquad \sigma^2_{\alpha\beta}=m^*_\alpha m^*_\beta,$$

$$\sigma^3_{\alpha\beta}=\frac{\mathrm{i}}{\sqrt{2}\,p}\left(m_\alpha p_\beta+m_\beta p_\alpha\right),\qquad \sigma^4_{\alpha\beta}=\frac{\mathrm{i}}{\sqrt{2}\,p}\left(m^*_\alpha p_\beta+m^*_\beta p_\alpha\right). \tag{7}$$

Here the tensors $\hat{\sigma}^s(\boldsymbol{p})$ as the orthonormal basis of $\hat{t}$ obey the conditions

$$\sigma^{s_1}_{\alpha\beta}(\boldsymbol{p})\sigma^{s_2*}_{\alpha\beta}(\boldsymbol{p})=\delta_{s_1s_2},\qquad \sum_s\sigma^s_{\alpha\beta}(\boldsymbol{p})\,\sigma^{s*}_{\gamma\delta}(\boldsymbol{p})=I^{\gamma\delta}_{\alpha\beta},$$

$$\sigma^s_{\alpha\beta}(-\boldsymbol{p})=\sigma^{s*}_{\alpha\beta}(\boldsymbol{p}),\qquad I^{\gamma\delta}_{\alpha\beta}=\tfrac{1}{2}\delta_{\alpha\gamma}\,\delta_{\beta\delta}+\tfrac{1}{2}\delta_{\alpha\delta}\,\delta_{\beta\gamma}-\tfrac{1}{3}\delta_{\alpha\beta}\,\delta_{\gamma\delta}.$$

Equations (7) in fact correspond to a subdivision of the matrix $\hat{t}(\boldsymbol{p})$ into the unit matrix $\hat{I}$, the projection matrix onto the $\boldsymbol{p}$-direction and the rotation matrix in the plane perpendicular to $\boldsymbol{p}$. The fundamental values $\tau^s(\boldsymbol{p})$ are equal to

$$\begin{aligned}\tau^0(\boldsymbol{p})&=a+\left(g_1+\tfrac{2}{3}g_2\right)p^2,\\ \tau^{1,2}(\boldsymbol{p})&=a+g_1\left(p^2\mp 2qp\right),\\ \tau^{3,4}(\boldsymbol{p})&=a+\left(g_1+\tfrac{1}{2}g_2\right)p^2\mp g_1qp.\end{aligned} \tag{8}$$

Using eqs. (6)–(8), one can write the matrixes $\hat{Q}(\boldsymbol{p})$ as

$$Q_{\alpha\beta}(\boldsymbol{p})=\sum_s\varphi^s_p\sigma^s_{\alpha\beta}(\boldsymbol{p}), \tag{9}$$

where $\varphi^s_{-p}=\varphi^{s*}_p$ are real quantities.

2.2. *One-dimensional cholesteric phases*

Since a system must be stable with respect to small-scale perturbations (with large p), one obtains from eq. (8) that $g_1+\frac{1}{2}g_2>0$ and $g_1+\frac{2}{3}g_2>0$, i.e., the sign of g_2 is not fixed. If $g_2>0$ and $q>0$ the rotational orientational mode $\varphi^{1,2}_p$ has the minimal energy, but at $g_2<0$ the mode $\varphi^{3,4}_p$ is favourable. The mode $\varphi^{1,2}_p$ describes the simple spiral (2), since the tensor $\hat{\sigma}^1$ is formed by the components of vectors $\boldsymbol{n}'\perp\boldsymbol{p}$ and $\boldsymbol{n}''\perp\boldsymbol{p}$. The mode $\varphi^{3,4}_p$ describes a conical spiral, since the tensor $\hat{\sigma}^3$ is formed by the components of vectors $\boldsymbol{n}'(\boldsymbol{n}'')$ and $\boldsymbol{p}$. In both cases the spiral axis z is parallel to the wave vector $\boldsymbol{p}$, the components of vectors $\boldsymbol{n}'$ and $\boldsymbol{n}''$ depend on z as harmonic functions. Up until now conical spirals have not been observed experimentally, so one can suppose that $g_2 > 0$.

The equilibrium values $Q_{\alpha\beta}(\boldsymbol{r})$ are determined by minimization of the functional (3) with respect to the amplitudes $\varphi_{\boldsymbol{p}}^s$ after substitution of $\hat{Q}(\boldsymbol{p})$ into $F[\hat{Q}]$. At $g_2 > 0$ the functional F in fact depends on the amplitudes $\varphi_0^0 = \varphi_0$ and $\varphi_{\boldsymbol{q}}^1 = \varphi_1$, since the conditions $\boldsymbol{p} = 0$ and $\boldsymbol{p} = \boldsymbol{q}$ ensure the minimal values $\tau_0 = \tau^0(0)$ and $\tau_1 = \tau^1(\boldsymbol{q})$ of the energies $\tau^0(\boldsymbol{p})$ and $\tau^1(\boldsymbol{p})$. Therefore the quantity $F = \mathscr{F}V^{-1}$ can be written as (Brazovskii and Dmitriev 1975)

$$F = \tau_0\varphi_0^2 + 2\tau_1\varphi_1^2 - \sqrt{6}\, b\varphi_0\varphi_1^2 + (b/\sqrt{6})\varphi_0^3 + c(\varphi_0^2 + 2\varphi_1^2)^2. \tag{10}$$

Variation of eq. (10) with respect to φ_0 and φ_1 gives the values

$$\varphi_0 = \frac{9b}{8\sqrt{6}\, c}\frac{\alpha - \alpha^2 + \beta}{1 + 3\alpha}, \qquad \varphi_1^2 = \tfrac{3}{2}\varphi_0^2\left(1 + \frac{2\sqrt{2}}{3\sqrt{3}}\frac{g_1 q^2}{b\varphi_0}\right), \tag{11}$$

where $\alpha = 8cg_1q^2/3b^2$, $\beta = 16c\tau_1/3b^2$; the phase boundary is determined by the equation

$$9\beta^2 + 2\beta(9\alpha - 1) - 3\alpha(1 - \alpha^2) = 0;$$

the value φ_0 from eq. (11) is determined on this phase boundary.

At $\boldsymbol{q} = 0$ one obtains from eqs. (11) and (9) that $\varphi_1 = \pm(\frac{3}{2})^{1/2}\varphi_0$ and the tensor $\hat{Q}(\boldsymbol{r})$ describes a uniaxial liquid crystal. In fact this case corresponds to the formation of a nematic liquid crystal. At $\boldsymbol{q} \neq 0$, i.e., at $\alpha \neq 0$, one deals with a biaxial cholesteric liquid crystal and the biaxiality increases if the wave number $q = 2\pi h^{-1}$, i.e., the parameter α increases.

In the ordered phase the temperature dependence of φ_0 is described by

$$\varphi_0^2 - \frac{3b}{8\sqrt{6}\, c}(1 - \alpha)\varphi_0 + \frac{3\tau_1}{c} = 0, \tag{12}$$

which shows that one has: at $\alpha = 1$, the tricritical point $\tau_1 = 0$; at $\alpha < 1$ and $b > 0$, the first-order transition boundary; at $\alpha > 1$ and $b > 0$, the second-order transition boundary between isotropic and ordinary (one-dimensional) spiral phases. In the last case eq. (11) shows that there is a proportionality $\varphi_0 \sim (b/g_1q^2)\varphi_1^2$ i.e., here φ_0 is the perturbation and φ_1 is the basic harmonic entering the expansion (10) in even powers only. This corresponds to the mentioned situation of the large correlation length $\zeta \sim g_1^{1/2} \sim \alpha^{1/2}h \gg h$.

2.3. *Blue phases*

Brazovskii and Dmitriev (1975) have shown that non one-dimensional incommensurate structures are energetically favourable at $\tau_1 > 0$. For example, the flat hexagonal lattice with the planar wave vector $\pm \boldsymbol{p}_1$, $\pm \boldsymbol{p}_2$ and $\pm \boldsymbol{p}_3$ ($|\boldsymbol{p}_i| = q$, $\boldsymbol{p}_1 + \boldsymbol{p}_2 + \boldsymbol{p}_3 = 0$) is favourable at $\tau_1 = 0.044b^2/c$. This conclusion has prompted the beginning of investigations of incommensurate structures with double spatial twist of biaxial molecules. At present such structures are attributed to blue phases. Hornreich and Shtrikman (1980, 1981) have advanced cubic structures. They have shown that the cubic structures with the space groups of symmetry O^5, O^2 and O^8 are more favourable from an energetic point of view. There is only one set of $\hat{Q}(\boldsymbol{p})$ for a given symmetry group. The biaxial order parameter $\hat{Q}(\boldsymbol{r})$ is an analytic function of position. One can decompose this function into the eigenfunctions $\sigma^s(\boldsymbol{r})$ and obtain the eigenvalues τ^s at any point $\boldsymbol{r}$. The equilibrium values of $\hat{Q}(\boldsymbol{r})$ are determined by the minimization of F with respect to the amplitudes φ^s.

Calculations show that the basic contribution in F is given by so called "curling modes" (Hornreich and Shtrikman 1981) or "double twist" regions (Meiboom, Sammon and Brinkman 1983), where the cholesteric is practically uniaxial and the director orientation changes in space as is shown in fig. 10a (the distribution $\boldsymbol{n}(\boldsymbol{r})$ on coaxial cylindrical surfaces is shown). Such cylindrical regions form, for example, the simple cubic lattice of spatial symmetry O^2 (fig. 10b). Narrow biaxial regions are located along half of the body

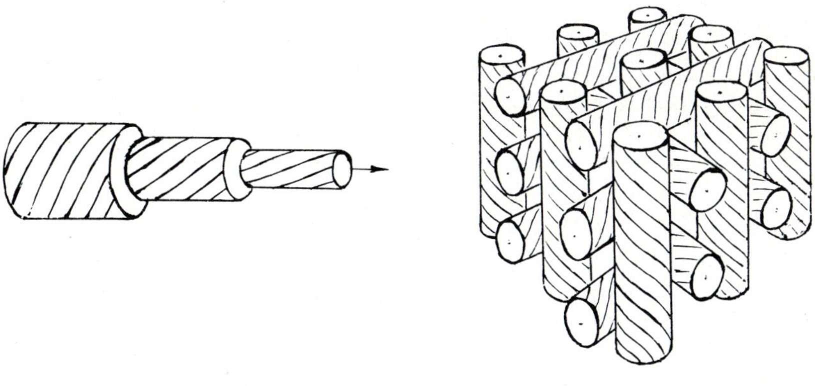

(a) (b)

Fig. 10. (a) The director orientation describing a curling mode configuration cylinder, and (b) the O^2 cubic structure, showing interplaced assemblies of the curling mode configuration cylinders. (Grebel et al. 1984).

diagonals of the unit cell of the O^2 structure. The biaxial regions occupy only 3% of the unit cell volume (Hornreich and Shtrikman 1981).

Meiboom, Sammon and Brinkman (1983) assumed that blue phases are uniaxial and that the observed narrow regions in the unit cell are linear defects of orientational structure (disclinations). They started from general topological considerations that one cannot continue rotating the director along two independent directions without eventually either undoing one rotation or introducing defects in the liquid crystal. The formation of disclinations, in the vicinity of which the order parameter $\hat{Q}(\boldsymbol{r})$ changes considerably, is most probable near the isotropic–cholesteric phase transition point where the correlation length ξ can be large. Probably therefore blue phases are observed in a narrow temperature interval of 1–2 K near this transition point.

It should be pointed out that both approaches give convergent results describing the experimental data on structure and optical properties of blue phases. The calculations show that the O^8 structure corresponds to the low-temperature blue phase BPI. The simple-cubic structure O^2 corresponds to the high-temperature blue phase BPII. The blue phases demonstrate the "red shift", i.e., an expansion (20–40%) of the unit cell. Thus, the structures of cholesteric, blue I and blue II phases are incommensurate. The phase transitions cholesteric–BPI–BPII are of the first kind.

Hornreich, Kugler and Shtrikman (1984) have shown recently that the two-dimensional hexagonal structure suggested by Brazovskii and Dmitriev (1975) should be thermodynamically stable in weak electric and magnetic fields due to dielectric and diamagnetic anisotropy. One can consider this structure as an assembly of parallel infinite "double twist" cylinders whose axes lie on the sites of a planar triangular lattice.

3. *Phase transition cholesteric (Ch)–smectic A*

The mechanism of the strong temperature dependence of the period of incommensurate structure in the vicinity of the Ch–A phase transition point is considered below. The Ch–A phase transition is a very weak transition of the first kind as experiments show. In that case a formal analogy between the smectic A and the superfluid state is valid. Starting from this analogy, Vigman and Filev (1975) have given an interpretation of the observed strong temperature dependence of the pitch h of the cholesteric spiral near the phase transition point T_c. This phenomenon has a simple physical cause related to the increase of smectic fluctuations in a cholesteric while T decreases to T_c. In fact, molecules must be oriented along the x axis in a smectic A (fig. 8). Thus, the average quadratic fluctuations of the positional order $\langle|\Psi|^2\rangle$ along the x axis of a one-dimensional crystal must strongly disturb a cholesteric spiral with axis z, since these fluctuations result in large molecular rotations in cholesteric planes around the axis z. However there is a sufficiently broad temperature region near the phase transition point, where fluctuations are small and therefore the phenomenological Landau theory is correct.

Small deviations of the director $\boldsymbol{n}$ from the axis x are equivalent to a phase change $\pm k((n_x - 1)x + n_y y + n_z z)$ in the basis functions $|\psi|\exp(\pm ikx)$. Since F must not be changed at rotations of the system as a whole, the expansion of F takes the form (Vigman and Filev 1975)

$$F = a|\psi|^2 + c|\psi|^4 + g_{\parallel}\left|\left[\frac{\partial}{\partial x} - ik(1-n_x)\right]\psi\right|^2 + g_{\perp}\left[\left|\left(\frac{\partial}{\partial y} - ikn_y\right)\psi\right|^2 + \left|\left(\frac{\partial}{\partial z} - ikn_z\right)\psi\right|^2\right] + \tfrac{1}{2}\left[K_1(\operatorname{div}\boldsymbol{n})^2 + K_2\left(\boldsymbol{n}\operatorname{rot}\boldsymbol{n} + \frac{2\pi}{h_0}\right)^2 + K_3[\boldsymbol{n}\operatorname{rot}\boldsymbol{n}]^2\right], \tag{13}$$

where K_1, K_2 and K_3 are the elastic constants, h_0 is the pitch of the helix in absence of smectic fluctuations, $2\pi k^{-1}$ is the thickness of a smectic monolayer, and $a = a'(T - T_0)$. In a smectic A one has $n_x = 1$, $n_y = n_z = 0$, $|\psi|^2 = -a/2c$ and

$$F = F_A = -\frac{a^2}{4c} + \tfrac{1}{2}K_2\left(\frac{2\pi}{h_0}\right)^2.$$

In a cholesteric one has $\psi = 0$ and $F = 0$ if one does not take account of smectic fluctuations. From the condition $F_A = 0$ at $T = T_c$ one can find T_c:

$$a_c = -\frac{2\pi}{h_0}(2cK_2)^{1/2}.$$

Thus, the formation of a smectic A with the homogeneous order parameter ψ as the result of the first-order phase transition is totally analogous to the phase transition between a normal metal and a homogeneous superconducting metal in the presence of an external magnetic field. However, taking fluctuations into account changes the transition picture, since in fact the pitch of the helix strongly changes near T_c.

One can obtain the condition of the free energy minimum by putting to zero the derivative of $\mathscr{F} = \int F \mathrm{d}V$ with respect to $q = 2\pi/h$:

$$K_2(q-q_0)V + \frac{\partial \mathscr{F}[\psi]}{\partial q} = 0, \tag{14}$$

where V is the volume of the system, $\mathscr{F}[\psi]$ is the free energy part related to fluctuations of the order parameter ψ. In the mean field approximation one has

$$\delta \boldsymbol{n}^2|\psi|^2 \approx \langle \delta \boldsymbol{n}^2\rangle|\psi|^2 + \delta \boldsymbol{n}^2\langle|\psi|^2\rangle - \langle \delta \boldsymbol{n}^2\rangle\langle|\psi|^2\rangle,$$

where $\langle \cdots \rangle$ is the thermodynamical average. So expression (13) can be rewritten as

$$F_{\mathrm{Ch}} = F_\psi + F_{\boldsymbol{n}},$$

$$F_\psi = \left[a + k^2\left(g_\parallel\langle\delta n_x^2\rangle + g_\perp\langle\delta n_y^2\rangle + g_\perp\langle\delta n_z^2\rangle\right)\right]|\psi|^2 + g_\perp\left|\frac{\partial\psi}{\partial z}\right|^2$$

$$+ g_\perp\left|\left(\frac{\partial}{\partial y} - ik\sin(qz)\right)\psi\right|^2 + g_\parallel\left|\left(\frac{\partial}{\partial x} - ik(1-\cos(qz))\right)\psi\right|^2,$$

$$F_{\boldsymbol{n}} = k^2\langle|\psi|^2\rangle\left(g_\parallel\delta n_x^2 + g_\perp\delta n_y^2 + g_\perp\delta n_z^2\right) + F_0, \tag{15}$$

where F_0 is the sum of three last terms of eq. (13), and the $|\psi|^4$ term is omitted, since in cholesterics it is small in comparison with the $|\psi|^2$ terms. The coefficient of $|\psi|^2$ in F_ψ corresponds to a renormalization of T_{c}, it is marked as a.

A cholesteric is characterized by some parameters with the dimension of length:

$$\lambda = \left(\frac{cK_2}{ag_\perp k^2}\right)^{1/2}, \qquad \xi = \left(\frac{g_\perp}{a + g_\perp kq}\right)^{1/2}, \qquad \eta = \left(\frac{1}{2qk}\right)^{1/2}.$$

The parameter λ is the penetration depth of the director field $\boldsymbol{n}(\boldsymbol{r})$ in a smectic with $|\psi| \sim ac^{-1}$, it characterizes the maximum scale on which both the term F_0 and terms proportional to $|\psi|^2\delta\boldsymbol{n}^2$ are essential. The correlation length ξ characterizes a scale in the plane xy perpendicular to the helix axis z; on the scale ξ smectic clusters arise and decay in consequence of thermodynamic fluctuations. The parameter η is a scale of heterogeneity of the function $\psi(\boldsymbol{r})$ along the axis z. Since in cholesterics the value q/k is small ($\sim 10^{-2}$), the quantity $qz \sim q\eta \sim qk^{-1}$ is also small, and therefore in (15) one can approximately write $\cos(qz) \simeq 1$ and $\sin(qz) \simeq qz$. As a result of this approximation the functional $\mathscr{F}|\psi|$ has the aspect of the free energy of a normal metal in the presence of a magnetic field near the temperature of the transition to a superconducting state:

$$\mathscr{F}[\psi]=\int F_{\psi}\mathrm{d}V$$
$$=\int\left\{a|\psi|^2+g_{\|}\left|\frac{\partial\psi}{\partial x}\right|^2+g_{\perp}\left|\left(\frac{\partial}{\partial y}-\mathrm{i}kqz\right)\psi\right|^2+g_{\perp}\left|\frac{\partial\psi}{\partial z}\right|^2\right\}\mathrm{d}V. \tag{16}$$

Here the quantity kq is analogous to a magnetic field parallel to the axis x.

It should be underlined that the quantity $M=-\partial\mathscr{F}[\psi]/\partial q$ has the meaning of magnetization in presence of an magnetic field. The well-known result of calculations of M (Schmid 1969) reads

$$M=-\frac{VT}{24\pi}\left(\frac{g_{\perp}}{g_{\|}}\right)^{1/2}qk^2f\left(\frac{a}{\tilde{\mathscr{E}}},\frac{g_{\perp}kq}{\tilde{\mathscr{E}}}\right),$$

$$f(x,y)=3\left(\frac{g_{\perp}}{\pi\tilde{\mathscr{E}}}\right)^{1/2}\sum_{m=1}^{\infty}\frac{\exp(-mx)}{\sqrt{m}\,\sinh(my)}\left[\coth(my)-\frac{1}{my}\right].$$

Here $\tilde{\mathscr{E}}$ is the cut-off parameter. In the limiting cases $q=0$ and $a=0$ one has correspondingly

$$f_{q=0}=\xi(q=0)=\left(\frac{g_{\perp}}{a}\right)^{1/2},\qquad f_{a=0}=1.55(kq)^{-1/2}.$$

In the case

$$\frac{\eta^2}{\xi^2}=\tfrac{1}{2}\left(1+\frac{a}{g_{\perp}kq}\right)\ll 1, \tag{17}$$

corresponding to "strong fields" q, one obtains

$$f\approx 6\left[kq\left(1+\frac{a}{g_{\perp}kq}\right)\right]^{-1/2}=6\xi(a,q). \tag{18}$$

At $\eta^2/\xi^2<\tfrac{1}{2}$ one gets the approximate expression

$$f=\xi(a,q)\left[1+\frac{1}{2}\left(\frac{\xi^2}{2\eta^2}\right)+\frac{1}{5}\left(\frac{\xi^2}{2\eta^2}\right)^2-\cdots\right]. \tag{19}$$

Thus, the temperature dependence of the pitch of the helix is determined, in accordance with eqs. (14) and (19), by

$$q_0 - q = \frac{1}{24\pi}\frac{T}{K_2}\left(\frac{g_\perp}{g_\parallel}\right)^{1/2} qk^2 f, \tag{20}$$

where the quantity f is approximately given by the expressions (17)–(19). It follows from eq. (20) that at $a < 0$ there is a point (a^*, q^*) on the curve $q(a)$ in which the function $q(a)$ has an infinite derivative. This point is the boundary of absolute instability of cholesterics, since at $a < a^*$ eq. (20) has no real solutions q (fig. 11). In fact the Ch–A first order phase transition takes place before, at the point (a_c, q_c), where $a_c > a^*, q_c > q^*$.

To find the point (a_c, q_c) it is necessary to calculate the free energy F_A of the smectic A and to solve the equation $F_A(a_c) = F_{Ch}(a_c, q_c)$. Here one can neglect the quantity q_0 and density fluctuations but it is necessary to take into account the director fluctuations $\delta \boldsymbol{n}$. At the condition $\partial F_A/\partial|\psi| = 0$ and $K_j = K_2$ the quantity F_A is given by the expression

$$F_A = -\frac{a^2}{4c}\left(1 + \frac{\Lambda}{3\pi\sqrt{2}}\left(\frac{\xi}{\lambda}\right)^3\right) + \tfrac{1}{2}K_2 q_0^2,$$

where $\Lambda = cT_c(|a|g_\perp^3)^{-1/2}$ is the Levanyuk–Ginzburg parameter. The condition $\Lambda\xi^3 \ll 10\lambda^3$, fulfilled in experiments, allows the mean field approximation. $F_{Ch}(a_c, q_c) = \frac{1}{2}K_2(q_c - q_0)^2$, since $\psi = 0$ in cholesterics. Therefore one has a second equation for determination of the point (a_c, q_c):

$$\tfrac{1}{2}K_2(q_c^2 - 2q_c q_0) = -a_c^2/4c. \tag{21}$$

These results show that the small fluctuational term $\mathscr{F}[\psi]$ in the free energy of a cholesteric induces the essential change of the pitch h. The pretransitional

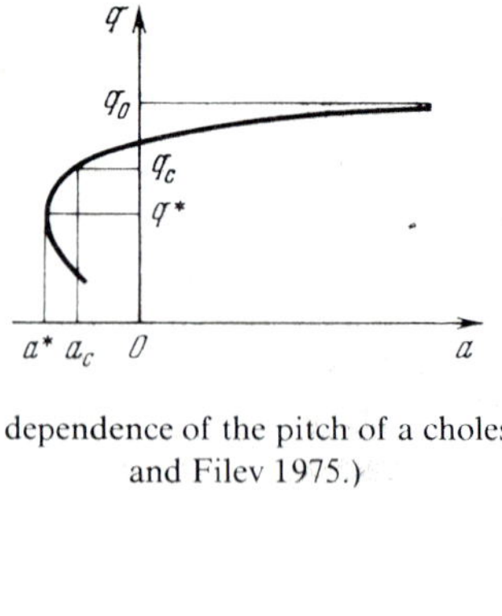

Fig. 11. Theoretical temperature dependence of the pitch of a cholesteric spiral $h = 2\pi/q$. (Vigman and Filev 1975.)

increase of the pitch of the cholesteric spiral is not described by a universal power function. It should be noted that eqs. (20) and (21) are correct until the correction $V^{-1}\mathscr{F} \ll K_2 q_c^2$, i.e., the region of validity of Landau theory depends on the parameters $\eta\xi^{-1}$ and $\lambda\eta^{-1}$. For example, this region strongly narrows at $\eta < \xi$ and it widens at $\lambda \ll \eta$. One can see from eqs. (20) and (21) that the "velocity" of the temperature change of the pitch $h = 2\pi q^{-1}$ is very large if the points a_c and a are close (fig. 11). Vigman and Filev (1975) have shown starting from equations (20) and (21) that experimental data are well described by this phenomenological theory at $T - T_c > 0.5\,\mathrm{K}$. In particular $(q_0 - q_c)/q_0 \sim 1$ if $T - T_c \simeq 5\,\mathrm{K}$, $K_2 \simeq 3\times10^{-7}\,\mathrm{dyn}$, $k \sim 10^7\,\mathrm{cm}^{-1}$, $g_\perp/g_\parallel \sim 0.1$ and $g_\perp \sim 10^{-14}\,\mathrm{cm}^2$. Since the parameters of cholesterics (q_c, c, etc) strongly depend on the impurity concentration, an increase of the concentration can sharply intensify the temperature dependence of pitch near the phase transition point if the values a_c and a^* approach each other due to impurities (Pindak et al. 1974).

4. *Incommensurate structures in cholesteric and smectic liquid crystals induced by external fields*

In cholesteric and smectic liquid crystals, as layered systems, at certain threshold conditions spatial periodic structures arise under the action of external fields. The spatial periods of these structures are incommensurate to the thicknesses of smectic monolayers and to the pitches of cholesteric spirals. Such structures are bends of smectic and cholesteric planes modulated along directions perpendicular to crystal and spiral axes correspondingly. These incommensurate structures are not related to the existence of Lifshitz invariants.

First of all let us consider the threshold deformation of a smectic A at homogeneous stretching along the crystal axis. It is useful to note that in the case under consideration the local symmetry of a smectic A is preserved, i.e., as always the director $\boldsymbol{n}$ is perpendicular to smectic planes (fig. 5). Assuming the possibility of displacements u_z of smectic planes one can easily obtain, at the conditions mentioned above, that

$$n_z \approx 1, \qquad n_x = \frac{\partial u_z}{\partial x}, \qquad n_y = \frac{\partial u_z}{\partial y}, \qquad \operatorname{div} \boldsymbol{n} = \frac{\partial^2 u_z}{\partial x^2} + \frac{\partial^2 u_z}{\partial y^2}. \tag{22}$$

The free energy density F as a function of the elastic deformations u_{ik} in a smectic A must contain five elastic constants $(C_{11}, C_{12}, C_{13}, C_{33}, C_{44})$ as in the case of a hexagonal crystal. However, here the constants C_{44} and $C_{11} - C_{12}$ are zero in accordance with the consideration of a one-dimensional crystal and with experimental data. As a result the contribution $F_{\boldsymbol{u}}$ of elastic deformations to the free energy of a smectic A can be written in the form (de Gennes 1974)

$$F_{\boldsymbol{u}} = \tfrac{1}{2}C_{33}u_{zz}^2 + \tfrac{1}{2}C_{11}(u_{xx} + u_{yy})^2 + C_{13}u_{zz}(u_{xx} + u_{yy}), \tag{23}$$

where the components u_{ik} are

$$u_{ik} = \frac{1}{2}\left(\frac{\partial u_i}{\partial x_k} + \frac{\partial u_k}{\partial x_i} + \frac{\partial u_j}{\partial x_i}\frac{\partial u_j}{\partial x_k}\right). \tag{24}$$

Taking into account the relations (22) and the invariance of F with respect to director sign changes, one should include in F the term

$$\tfrac{1}{2}K_1(\operatorname{div}\boldsymbol{n})^2 = \tfrac{1}{2}K_1\left(\frac{\partial^2 u_z}{\partial x^2} + \frac{\partial^2 u_z}{\partial y^2}\right)^2.$$

One can neglect in F the terms proportional to the fourth power of $\boldsymbol{u}$, as well as spatial derivatives, for example the quantities $(\boldsymbol{n}\operatorname{rot}\boldsymbol{n})^2$ and $[\boldsymbol{n}\operatorname{rot}\boldsymbol{n}]^2$. This is correct for small long-wave perturbations. Thus, one has the final expression for F:

$$F = F_{\boldsymbol{u}} + \tfrac{1}{2}K_1\left(\frac{\partial^2 u_z}{\partial x^2} + \frac{\partial^2 u_z}{\partial y^2}\right)^2. \tag{25}$$

One can see from fig. 5 that at the given type of perturbations a monolayer bend induces the local dilatation

$$-\tfrac{1}{2}\left(n_x^2 + n_y^2\right)$$

if the bend does not change the number and dimensions of monolayers along the axis z. This dilatation is not zero in absence of a spatial heterogeneity along the axis z, i.e., at $\partial u_z/\partial z = \partial u_x/\partial z = \partial u_y/\partial z = 0$. The local dilatation is the sum $u_{xx} + u_{yy}$. Note that the quantity

$$u_{xx} + u_{yy} = -\tfrac{1}{2}\left(n_x^2 + n_y^2\right) = -\frac{1}{2}\left[\left(\frac{\partial u_z}{\partial x}\right)^2 + \left(\frac{\partial u_z}{\partial y}\right)^2\right] \tag{26}$$

is proportional to the second power of spatial derivatives. Substitution of eqs.

(26) and (23) into eq. (25) gives

$$F=\tfrac{1}{2}C_{33}u_{zz}^2+\tfrac{1}{8}C_{11}\left[\left(\frac{\partial u_z}{\partial x}\right)^2+\left(\frac{\partial u_z}{\partial y}\right)^2\right]^2$$

$$+\tfrac{1}{2}C_{13}u_{zz}\left[\left(\frac{\partial u_z}{\partial x}\right)^2+\left(\frac{\partial u_z}{\partial y}\right)^2\right]+\tfrac{1}{2}K_1\left(\frac{\partial^2 u_z}{\partial x^2}+\frac{\partial^2 u_z}{\partial y^2}\right)^2. \tag{27}$$

It is useful to write the quantity u_{zz} as

$$u_{zz}=\lambda+\frac{\partial u_1(x,z)}{\partial z},$$

where $\lambda=\mathrm{d}u_0/\mathrm{d}z$ is the homogeneous deformation, and $u_1(x,z)$ is the modulated perturbation.

It is possible to show, starting from eq. (27), that a critical value λ_c of the homogeneous deformation λ exists, and that at $\lambda>\lambda_c$ the perturbations

$$u_1=a\sin(\pi z/d)\sin(qx),\qquad u_1(x,0)=u_1(x,d)=0. \tag{28}$$

increase inevitably (here d is the sample thickness). In fact, taking into account eqs. (27) and (28), one can write the free energy change per unit area as

$$\Delta\mathscr{F}=\int_0^d F\,\mathrm{d}z-\frac{d}{2}C_{33}\lambda^2=\frac{d}{4}\left[C_{33}\left(\frac{\pi}{d}\right)^2-C_{13}\lambda q^2+K_1q^4\right]a^2\sin^2(qx). \tag{29}$$

$\Delta\mathscr{F}\leqslant 0$, if

$$\lambda\geqslant\lambda_c=\frac{2\pi}{d}(K_1C_{33}/C_{13}^2)^{1/2}\sim\frac{2\pi l}{d},$$

$$q_c=\left(\frac{\pi}{d}\right)^{1/2}\left(\frac{C_{33}}{K_1}\right)^{1/4}\sim\left(\frac{1}{dl}\right)^{1/2}\ll\frac{1}{l}. \tag{30}$$

Thus, a small but finite homogeneous stretching of a smectic sample along the z axis induces instability of the macroscopically homogeneous state with respect to the modulated bend of smectic planes.

It has been shown (Delrieu 1974) that a square lattice of perturbations

$$u_1\sim\sin(qx/\sqrt{2})\sin(qy/\sqrt{2}).$$

is the most energetically favourable. In fact, the type of modulated structure is determined by free energy terms proportional to the fourth power of u_1. Since in accordance with (30) $q_c \gg \pi/d$, the most essential term in (27) is

$$\tfrac{1}{8}C_{11}\left[\left(\frac{\partial u_1}{\partial x}\right)^2+\left(\frac{\partial u_1}{\partial y}\right)^2\right]^2.$$

After averaging over a unit area this term is equal to the sum

$$\sum_{q\neq q',-q'}\sum |u_{1q}|^2|u_{1q'}|^2\left[\boldsymbol{q}^4+2(\boldsymbol{q}\boldsymbol{q}')^2\right],$$

where

$$u_1(x,y)=\sum_q u_{1q}\exp\left[\mathrm{i}(q_x x+q_y y)\right], \qquad |\boldsymbol{q}|=|\boldsymbol{q}'|.$$

Evidently, the sum is minimal if $\boldsymbol{q}\perp\boldsymbol{q}'$, i.e., for a square lattice.

Usually with instability phenomena the increase of perturbations above a threshold is limited because of nonlinear effects. In the case under consideration such limit is not achieved, since before the onset of saturation the amplitude of deformations increases sufficiently to break up the layer structure and to create dislocations which correspond to an increase of the number of smectic planes. Thus, the smectic structure becomes defect and the number of defects will increase until the deformation λ becomes smaller than λ_c (Liao et al. 1973, Clark and Pershan 1973, Ribotta et al. 1973, Delaye et al. 1973).

A similar instability of a smectic A can be induced in another way. If the sample thickness is kept constant, i.e., in first approximation one has $\partial u_z/\partial z = 0$ inside a layer, and uniform compression $\partial u_x/\partial x+\partial u_y/\partial y=\lambda<0$ takes place in smectic planes, an analogous calculation shows that $\Delta\mathscr{F}\leqslant 0$ for

$$u_{zz}=\frac{\partial u_z}{\partial z}, \qquad u_z=u_1=a\sin\left(\frac{\pi z}{d}\right)\sin(qx)\sin(qy),$$

$$u_{xx}+u_{yy}=\lambda+\frac{1}{2}\left[\left(\frac{\partial u_z}{\partial x}\right)^2+\left(\frac{\partial u_z}{\partial y}\right)^2\right]$$

at

$$|\lambda|>|\lambda_c|, \qquad \lambda_c=-\frac{2\pi}{d}\left(\frac{K_1C_{33}}{C_{11}^2}\right)^{1/2}, \qquad q_c=\left(\frac{\pi}{2d}\right)^{1/2}\left(\frac{C_{33}}{K_1}\right)^{1/4}. \tag{31}$$

Thus, uniform compression in smectic planes induces the incommensurate structure considered (Pikin 1981). In experiments this situation arises as a result of the compression of a local smectic area achieved by a sharp local temperature drop due to the change in the intensity of a laser beam. For the first time this phenomenon was observed by Kahn (1973).

The consideration of instability (in certain conditions) of a smectic sample can be extended to cholesterics (de Gennes 1974) which are also layered systems. The thickness of "monolayers" limited by cholesteric planes, is equal to half the pitch of the cholesteric spiral.

Large-scale perturbations of such layered system at $kh \ll 1$ (k^{-1} is the perturbation scale) are described (as in smectics) by the expansion of the free energy density F in powers of spatial derivatives of monolayer displacements u along the spiral axis z. In that case the pitch of the spiral is a function $h(z)$ which differs from the nonperturbed pitch value h_0. For dimensions which are much smaller than k^{-1} but much larger than h_0, the free energy term related to the heterogeneous director twisting is given by

$$\tfrac{1}{2}K_2(q-q_0)^2 = \tfrac{1}{2}K_2 q_0^2\left(\frac{h_0}{h}-1\right)^2 = \tfrac{1}{2}K_2 q_0^2\left(\frac{\partial u}{\partial z}\right)^2. \tag{32}$$

The quantity $K_2 q_0^2$ in eq. (32) plays the part of the elastic constant C_{33} which characterizes compression or stretching of the cholesteric spiral along its axis.

Introducing the unit vector $\boldsymbol{m}$ perpendicular to cholesteric planes, one can write by analogy with (22):

$$m_z \approx 1, \qquad m_x = \frac{\partial u}{\partial x}, \qquad m_y = \frac{\partial u}{\partial y}, \qquad \operatorname{div}\boldsymbol{m} = \frac{\partial^2 u}{\partial x^2} + \frac{\partial^2 u}{\partial y^2}. \tag{33}$$

Evidently, the quantity F must also contain the term

$$\tfrac{1}{2}\tilde{K}_1(\operatorname{div}\boldsymbol{m})^2 = \tfrac{1}{2}\tilde{K}_1\left(\frac{\partial^2 u}{\partial x^2} + \frac{\partial^2 u}{\partial y^2}\right)^2. \tag{34}$$

de Gennes (1974) has shown that the elastic constant $\tilde{K}_1$ is related to the modulus K_3 by the equation $\tilde{K}_1 = \frac{3}{8}K_3$. One can introduce for cholesterics, by analogy with (23), the elastic constants $\tilde{C}_{11}$ and $\tilde{C}_{13}$ which should be taken into account in presence of the dilatation $u_{xx} + u_{yy}$. Making the substitutions $C_{11} \to \tilde{C}_{11}$, $C_{33} \to \tilde{C}_{33}$, $C_{13} \to \tilde{C}_{13}$ and $K_1 \to \tilde{K}_1$ in eqs. (30) and (31), one can find the conditions of mechanical instability of cholesterics as of some layered systems. It is seen from expressions (30) and (31) that the orders of magnitude of the critical deformations are equal:

$\lambda_c \sim l/d \sim 10^{-4}$ for smectics,

$\lambda_c \sim h/d \sim 10^{-1}$ for cholesterics,

if $l \sim 10^{-7}$ cm, $h \sim 10^{-4}$ cm, $d \sim 10^{-3}$ cm, $\tilde{C}_{11} \sim \tilde{C}_{13} \sim \tilde{C}_{33} \sim K_2 h^{-2}$. Thus, the layered structures of cholesterics are more stable than those of smectics with respect to the incommensurate modulated perturbations considered.

There are experimental observations of another type of instability in cholesterics under the action of external electric and magnetic fields along the axis of the cholesteric spiral if the dielectric and diamagnetic anisotropies (ε_a and χ_a) are positive (de Gennes 1974, Gerritsma and Van Zanten 1971, Rondelez et al. 1972, Rondelez and Hulin 1972, Kahn 1970). In that case the incommensurate structure arising is similar to the smectic plane modulation induced by mechanical deformations. Here one can also introduce the deviations (m_x, m_y) of the axis of the spiral from the normal to unperturbed cholesteric planes which coincides with the axes z, $\boldsymbol{E}$ and $\boldsymbol{H}$ (fig. 6). One has the relations (33) and the relation

$$\tfrac{1}{2}\left(m_x^2 + m_y^2\right) = \overline{n_z^2} = \frac{1}{h_0}\int_0^{h_0} n_z^2 \, \mathrm{d}z$$

between the deviations $\boldsymbol{m}$ and the pitch perturbation $\partial u/\partial z = (q_0 - q)/q_0 = (h - h_0)/h_0$. Therefore one can easily obtain the expression for $\mathscr{F}$ in the perturbed cholesteric using eqs. (32)–(34) and, for example, the quantity F_e:

$$\mathscr{F} = \int \left\{ \tfrac{3}{16} K_3 \left(\frac{\partial^2 u}{\partial x^2} + \frac{\partial^2 u}{\partial y^2} \right)^2 + \tfrac{1}{2} K_2 q_0^2 \left(\frac{\partial u}{\partial z} \right)^2 - \frac{\varepsilon_a}{16\pi} E^2 \left[\left(\frac{\partial u}{\partial x} \right)^2 + \left(\frac{\partial u}{\partial y} \right)^2 \right] \right\} \mathrm{d}V. \tag{35}$$

Expressions (35) and (25) are similar.

Assuming hard boundary conditions for molecular orientations on solid surfaces at $z = 0$ and $z = d$, one can write the function $u(x, y, z)$ as

$$u = u_0 \sin(\pi z/d) \sin(\pi x/s) \sin(\pi y/s), \tag{36}$$

where $2s$ is the period of the incommensurate structure. Substitution of eq. (36) into eq. (35) gives an expression similar to eq. (29), from which after redetermination of parameters,

$$K_1 \to \tfrac{3}{8} K_3, \qquad C_{33} \to K_2 q_0^2, \qquad C_{13}\lambda \to \frac{\varepsilon_a}{8\pi} E^2,$$

one obtains the threshold values of electric field E_c and spatial semiperiod s_c

(see eq. 30) at $\varepsilon_a \ll \varepsilon_\perp$:

$$E_c^2 = \frac{(2\pi)^3(6K_2K_3)^{1/2}}{\varepsilon_a h_0 d}, \qquad s_c^2 = \frac{1}{2}\left(\frac{3K_3}{2K_2}\right)^{1/2} h_0 d. \tag{37}$$

In the case considered, at $E > E_c$ a square lattice of perturbations of cholesteric planes arises (similar to the perturbations of stretched smectics). These perturbations are local bends at a constant number of planes in the fixed thickness d. At $E - E_c \ll E_c$ the observed perturbations are small. At $E - E_c > E_c$ the linear theory used is not correct. In the last case a heterogeneous cholesteric texture is observed (Rondelez and Hulin 1972, Kahn 1970). This texture is characterized by local deviations of the axis of the spiral up to angles of the order of $\pi/2$, which is accompanied by the formation of defects of the incommensurate structure.

5. *Incommensurate flexoelectric structure in nematic liquid crystals*

5.1. *The flexoelectric effect*

In the absence of electric fields a nematic has the point group of symmetry $D_{\infty h}$ and the homogeneous director distribution $\boldsymbol{n}(\boldsymbol{r}) = \boldsymbol{n}_0$. Application of an electric field $\boldsymbol{E}$ can induce the formation of a modulated orientational structure $\boldsymbol{n}(\boldsymbol{r}) \neq \boldsymbol{n}_0$ at constant density of the nematic. This is a special phenomenon attributed to liquid crystals with a center of symmetry and related to the formation of induced polarization in a system of special molecules (Meyer 1969). Such molecules have an anisotropic form and electric dipole (or multipole) moments. Here it is assumed that the system has no spontaneous ordering of constant electric dipole moments. Correspondingly the polarizational term in the free energy density has no temperature peculiarity in absence of some orientational transformations. This term can be written as

$$\tfrac{1}{2}(\hat{\chi}_0^{-1})_{ij}P_iP_j - P_jE_j, \tag{38}$$

where $\boldsymbol{P}$ is the electric polarization, $\hat{\chi}_0 = (\hat{\varepsilon}_0 - \hat{1})/4\pi$ is the dielectric susceptibility tensor, $\hat{\chi}_0^{-1}$ is the inverse tensor ($\hat{\chi}_0^{-1}\hat{\chi}_0 = \hat{1}$), and the main values of $\hat{\chi}_0$ and $\hat{\chi}_0^{-1}$ are positive at any temperature. In some cases one can neglect the anisotropy of $\hat{\chi}_0$ and write $\chi_0^{-1}\delta_{ij}$ instead of $(\hat{\chi}_0^{-1})_{ij}$ in expression (38).

Meyer (1969) has shown that in nematics the polarization $\boldsymbol{P}$ can arise as a result of bend and twist deviations of $\boldsymbol{n}$ in a nematic from the nonperturbed state $\boldsymbol{n}_0$. In fact, it is always possible to build the true vectors $\boldsymbol{n}\,\mathrm{div}\,\boldsymbol{n}$ and $[\boldsymbol{n}\,\mathrm{rot}\,\boldsymbol{n}]$ which are transformed as the polarization vector $\boldsymbol{P}$ under actions of symmetry elements from the point group $D_{\infty h}$. Therefore, if molecules possess constant electric dipole moments, one can write for a nematic:

$$\boldsymbol{P} = f_1 \boldsymbol{n} \operatorname{div} \boldsymbol{n} + f_2 [\operatorname{rot} \boldsymbol{n} \cdot \boldsymbol{n}], \tag{39}$$

where f_1 and f_2 are the flexoelectric coefficients. Equation (39) is allowed simultaneously with the invariant

$$-\mu_1 (\boldsymbol{Pn}) \operatorname{div} \boldsymbol{n} - \mu_2 (\boldsymbol{P}[\operatorname{rot} \boldsymbol{n} \cdot \boldsymbol{n}]) \tag{40}$$

in the free energy density F. The coefficients (f_1, f_2) and (μ_1, μ_2) have the order of magnitude $p_0 k^2$, where p_0 is the molecular dipole moment, k^{-1} is the molecular length. If $p_0 \sim 10^{-18}$ (cgs units) and $k^{-1} \sim 10^{-7}$ cm, one obtains the estimation $\mu_{1,2} \sim 10^{-4}$ (cgs units).

In a uniaxial nematic the tensor $\hat{\chi}_0$ has the components

$$(\hat{\chi}_0)_{ij} = \chi_{0\perp} \delta_{ij} + (\chi_{0\parallel} - \chi_{0\perp}) n_i n_j,$$

where

$$\chi_{0\perp} = \frac{\varepsilon_{0\perp} - 1}{4\pi}, \qquad \chi_{0\parallel} = \frac{\varepsilon_{0\parallel} - 1}{4\pi}.$$

Therefore expression (38) can be written as

$$\frac{1}{2}\left(\frac{1}{\chi_{0\perp}} \boldsymbol{P}_\perp^2 + \frac{1}{\chi_{0\parallel}} \boldsymbol{P}_\parallel^2 \right) - \boldsymbol{PE}, \tag{41}$$

where $\boldsymbol{P}_\perp$ and $\boldsymbol{P}_\parallel$ are correspondingly perpendicular and parallel to $\boldsymbol{n}$. Besides, $\boldsymbol{P}_\perp \| [\operatorname{rot} \boldsymbol{n} \cdot \boldsymbol{n}]$ and $\boldsymbol{P}_\parallel \| \boldsymbol{n} \operatorname{div} \boldsymbol{n}$.

After the minimization of the sum of expressions (40) and (41) with respect to the variable $\boldsymbol{P}$ one obtains the relations

$$f_1 = \chi_{0\parallel} \mu_1, \qquad f_2 = \chi_{0\perp} \mu_2.$$

and the free energy term δF depending on $\boldsymbol{n}$ and $\boldsymbol{E}$:

$$\delta F = -\frac{\varepsilon_a}{8\pi} (\boldsymbol{En})^2 - f_1 (\boldsymbol{En}) \operatorname{div} \boldsymbol{n} - f_2 (\boldsymbol{E}[\operatorname{rot} \boldsymbol{n} \cdot \boldsymbol{n}])$$
$$- \tfrac{1}{2} \chi_{0\parallel} \mu_1^2 (\operatorname{div} \boldsymbol{n})^2 - \tfrac{1}{2} \chi_{0\perp} \mu_2^2 [\operatorname{rot} \boldsymbol{n} \cdot \boldsymbol{n}]^2, \tag{42}$$

where $\varepsilon_a = \varepsilon_{0\parallel} - \varepsilon_{0\perp}$. The last two terms in (42) are the corrections to corresponding terms in the expression

$$F_0 = \tfrac{1}{2}K_1(\operatorname{div} \boldsymbol{n})^2 + \tfrac{1}{2}K_2(\boldsymbol{n} \operatorname{rot} \boldsymbol{n})^2 + \tfrac{1}{2}K_3[\boldsymbol{n} \operatorname{rot} \boldsymbol{n}]^2$$

(see eq. 13). In fact, the values

$$K_1 = K_1^0 - \chi_{0\parallel}\mu_1^2, \qquad K_3 = K_3^0 - \chi_{0\perp}\mu_2^2$$

are determined experimentally, here K_1^0 and K_3^0 are the elastic constants in absence of the flexoelectric effect. At $K_{1,3} \sim 10^{-6}$dyn and $\mu_{1,2} \sim 10^{-4}$ (cgs units) these corrections are equal to $10^{-2}K_{1,3}$.

5.2. *The linear problem*

Expression (42) shows that commensurate structure $\boldsymbol{n}(\boldsymbol{r})$ is induced by a homogeneous electric field in nematics. This is the linear flexoelectric effect. Bobylev and Pikin (1977) have shown that the instability arising in a nematic sample with a finite thickness at the threshold field E_c is characterized by small director deviations from the x axis in two planes: by an angle θ in the xz plane and by an angle φ in the xy plane. In that case a periodic (along the y axis) system of domains parallel to the x axis arises (fig. 4). Note that this effect occurs at hard boundary conditions, i.e., at invariable planar molecular orientations on sample boundaries.

Supposing $n_x \simeq 1$, $n_y = \varphi(y, z)$, $n_z = \theta(y, z)$, $E_x = E_y = 0$, $E_z = E$, $K_1 = K_2 = \overline{K}$ and minimizing the quantity $\mathscr{F} = \int(F_0 + F)\mathrm{d}V$ with respect to the variables θ and φ, one obtains the following equations describing the incommensurate structure in nematics:

$$\overline{K}\left(\frac{\partial^2\varphi}{\partial y^2} + \frac{\partial^2\varphi}{\partial z^2}\right) - \tilde{f}E\frac{\partial\theta}{\partial y} = 0, \qquad \overline{K}\left(\frac{\partial^2\theta}{\partial y^2} + \frac{\partial^2\theta}{\partial z^2}\right) + \tilde{f}E\frac{\partial\varphi}{\partial y} + \frac{\varepsilon_a}{4\pi}E^2\theta = 0, \tag{43}$$

where $\tilde{f} = f_1 - f_2$. Taking account of the boundary conditions $\varphi = \theta = 0$ at $z = \pm d/2$, where d is the sample thickness, one finds the solutions of equations (43) as

$$\theta = \theta_0\cos(qy)\cos\frac{\pi z}{d}, \qquad \varphi = \varphi_0\sin(qy)\cos\frac{\pi z}{d}.$$

Demanding nonzero solutions θ_0 and φ_0, one obtains the dispersion equation

$$E^2 = \left(\frac{\bar{K}}{\tilde{f}}\right)^2 \frac{\left[q^2 + (\pi/d)^2\right]^2}{q^2 + \nu\left[q^2 + (\pi/d)^2\right]}, \tag{44}$$

where $\nu = \varepsilon_a \bar{K}/4\pi\tilde{f}^2$, and the relation

$$\varphi_0 = \frac{\tilde{f}E}{\bar{K}} \frac{q}{q^2 + (\pi/d)^2} \theta_0. \tag{45}$$

The threshold values E_c and q_c are obtained by minimization of the function $E(q)$:

$$E_c = \frac{2\pi\bar{K}}{|\tilde{f}|(1+\nu)d}, \qquad q_c = \frac{\pi}{d}\left(\frac{1-\nu}{1+\nu}\right)^{1/2}, \tag{46}$$

It is seen from eq. (46) that the instability can arise only at the condition $|\nu| < 1$, i.e., at

$$|\varepsilon_a| < 4\pi\tilde{f}^2/K. \tag{47}$$

Figure 12 shows the calculated dependences of the threshold characteristics of the incommensurate structure on the dielectric anisotropy at the real values of elastic constants $K_1 = 0.78 \times 10^{-6}$dyn and $K_2 = 0.55 \times 10^{-6}$dyn, which demanded numerical calculations (Bobylev et al. 1979). Experimental data from Barnik et al. (1978) are indicated by crosses. These results show the fast increase of $U_c = E_c d$ and q_c at $\nu \to -1$ and the asymptotic approach of U_c and q_c to $U_c = \text{const.}$ and $q_c = 0$ at $\nu \to +1$, in accordance with eq. (46).

5.3. *The nonlinear problem*

It is interesting to analyze nematic behaviour under the action of electric fields above the threshold at $E > E_c$. One can think that in that case an incommensurate structure with large wave numbers q is energetically favourable. It follows from (44) and (45) that $q \sim E$ and $\varphi_0 \sim (\tilde{f}E/\bar{K}q)\theta_0$ at $E \gg E_c$ in linear approximation. Terent'ev and Pikin (1982) have shown that these conclusions are qualitatively correct for the nonlinear problem as well. Some results of this consideration will be given below.

Since at $E \gg E_c$ the period $2\pi/q$ of the incommensurate structure is much smaller than the layer thickness d, one can neglect in the first-order approximation the derivatives with respect to the coordinate z in the expression for $\mathscr{F}$.

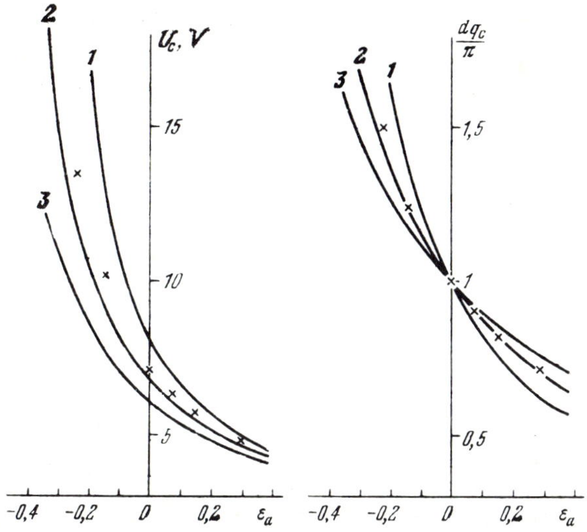

Fig. 12. Dependence of threshold characteristics of the flexoelectric effect on the dielectric anisotropy. Calculations (Bobylev et al. 1979) are carried out for the following f values (cgs units): (curve 1) 1.5×10^{-4}; (2) 1.8×10^{-4}; (3) 2.1×10^{-4}. Experimental data (Barnik et al. 1978) are designated by crosses.

Using the relations

$$n_x = \cos\theta\cos\varphi, \qquad n_y = \cos\theta\sin\varphi, \qquad n_z = \sin\theta,$$

with allowance made for nonlinear corrections up to terms of sixth order in the angles θ and φ, one can write down the functional $\mathscr{F}$. The dependences of the angles θ and φ on the coordinate z are substituted in the form of the expansions

$$\theta(y, z) = \theta(y)\cos\frac{\pi z}{d} + v(y)\cos\frac{3\pi z}{d} + \cdots,$$

$$\varphi(y, z) = \varphi(y)\cos\frac{\pi z}{d} + w(y)\cos\frac{3\pi z}{d} + \cdots.$$

The problem is solved first by confining ourselves to the first harmonics in this expansion. Next the corrections are calculated for making allowance for the higher harmonics and it is verified that these are small. Variation of the functional $\mathscr{F}$ with respect to the variables θ and φ leads to a system of equations for the functions $\theta(y)$ and $\varphi(y)$. The solution is sought in the form

$$\theta(y)=\theta_1\cos(qy)+\theta_3\cos(3qy)+\cdots,$$
$$\varphi(y)=\varphi_1\sin(qy)+\varphi_3\sin(3qy)+\cdots. \tag{48}$$

The solution of the considered nonlinear problem at $K_j=K$, $\varepsilon_a=0$ and $f_1=-f_2=f$ shows that $\varphi_1\sim\theta_1$ and $\varphi_3\sim\theta_3\ll\theta_1$. After substituting expansions (48) in the corresponding equations one then arrives at the algebraic relations

$$\varphi\simeq\tfrac{3}{128}\theta_1^3,\qquad \theta_3\simeq\tfrac{1}{128}\theta_1^3,\qquad \varphi_1\simeq\theta_1+\tfrac{3}{32}\theta_1^3,\qquad \frac{q(E)}{q}=1+\tfrac{9}{32}\theta_1^2, \tag{49}$$

where $q(E)=2fE/K$. The density of the free energy is then

$$F\simeq\tfrac{1}{4}Kq^2(E)\left(-\tfrac{9}{64}\theta_1^4+0.055\theta_1^6\right). \tag{50}$$

From this one obtains a value $\theta_1=\theta^*$ that satisfies the condition $\mathrm{d}F/\mathrm{d}\theta_1=0$, for the minimum of the free energy, as well as the corresponding value $q=q^*(E)$: $|\theta^*|\simeq1.3$, $q^*(E)\simeq0.6q(E)$, $F(q^*)\simeq-0.04Kq^2(E)$. Assuming that $v\sim w\ll\theta$, we confine ourselves in this functional only to the terms quadratic in v and w. The variational equations are then linear in these variables and one readily obtains

$$v=v_1\cos(qy)+\cdots,\qquad w=w_1\sin(qy)+\cdots,$$

with

$$v_1=-\tfrac{3}{26}\theta_1-0.007\theta_1^3,\qquad w_1=-\tfrac{3}{26}\theta_1+0.021\theta_1^3.$$

It can be seen that allowing for the next higher harmonics in the coordinate z cannot qualitatively change the results. The approximate solution describes the dependence of the spatial period of the incommensurate structure, Y, on the electric field E above the threshold at $E\gg E_c$:

$$Y=\frac{2\pi}{q^*(E)}=\frac{\pi K}{0.6fE}. \tag{51}$$

These qualitative results do not change if $\varepsilon_a\neq0$ and $f_1+f_2\neq0$, provided the condition $|\varepsilon_aK/4\pi(f_1-f_2)^2|<1$ is satisfied.

The director-field distortions given by relations (48) are directly observed in polarized light as a system of opaque stripes parallel to the x axis (fig. 7, above). Defects of the stripe system similar to edge dislocations in crystals appear already at $E > E_c$. In stronger fields the number of defects increases greatly. It is possible to show (Terent'ev and Pikin 1982) that the mechanism whereby the period of the considered flexoelectric incommensurate structure is decreased with increasing field is the formation of dislocations in such effectively one-dimensional lattices. It is easy to determine the distortion energy of this lattice by introducing the domain-distortion field $u(x, y)$, reckoned along the y axis. The components of the stress tensor of this effective crystal are

$$\sigma_{12} = \sigma_{21} = \sigma_{xy} = 2\chi_1 \partial u/\partial x, \qquad \sigma_{22} = \sigma_{yy} = 2\chi_2 \partial u/\partial y,$$

where $\chi_1 \sim \chi_2 \sim 4^{-1}Kq^2\theta_1^2$. Defects in the lattice are similar to edge dislocations in a solid with zero Poisson coefficient.

Note that in an ideal (defect-free) system of domains even a small change of the period changes the direction of the director by angles of the order of $2\theta^* \sim \pi$ in practically the entire nematic volume. This process requires a macroscopically high energy. If, however, an increase of the electric field from E_c to E leaves the period equal to $Y_c = 2\pi q_c^{-1}$, where $q_c = q^*(E_c)$, this means the appearance of a distortion $u = y(q^*(E) - q_c)/q_c$ and of stresses

$$\sigma_{yy} \sim 0.8 f^2 K^{-1} \theta^{*2} E_c (E - E_c).$$

These distortions can be offset completely by creation of $N = N(E)$ dislocations, which leads to the appearance of additional domains and to a change of

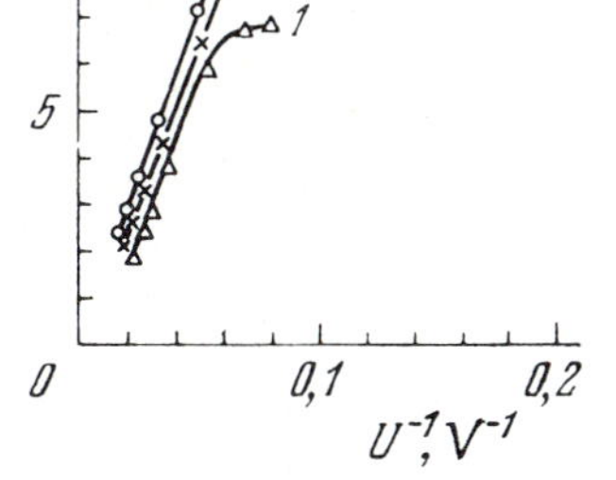

Fig. 13. Dependence of the semiperiod of the flexoelectric structure shown in fig. 7 on voltage, at: (curve 1) $\varepsilon_a = -0.25$; (2) $\varepsilon_a = 0$; (3) $\varepsilon_a = 0.15$. (Barnik et al. 1978.)

the flexostructure period to $Y = 2\pi/q^*(E)$. Dislocations are produced by fluctuations at any inhomogeneity of the director distribution on the boundary. They "break away" from the pinned ends at some critical conditions and move into the interior under the influence of the force acting on them. Each such source produces a group of dislocations along the x axis and each group of dislocations produces the reverse (blocking) action on the source. The equilibrium number of dislocations is determined from the balance between the critical stress necessary to tear a dislocation away from the source and the stress acting on the source. In that case it is possible to estimate the critical value E_1 of the electric field at which dislocation production is possible at all. This is the field corresponding to the end point of the plateau on the plot of $\frac{1}{2}Y(1/E)$ (fig. 13). One can also estimate the number of dislocations $N(E) \sim 0.6fL(E - E_c)/\pi K$, where L is the lattice length. Thus, the production of dislocations, grouped in definite clusters, leads to a dependence, dictated by thermodynamic considerations, of the period of the incommensurate flexoelectric structure on the field (see eq. 51).

6. *Ferroelectric structures in liquid crystals*

6.1. *Symmetry properties*

The possibility of ferroelectricity in liquid crystals was predicted by Meyer (Meyer et al. 1975) for the smectic phase C with chiral molecules. The ferroelectric state arises as the result of the phase transition A → C and of the smectic symmetry lowering $D_\infty \to C_2$ (fig. 14). Note that measurements of the spontaneous polarization show a small effective dipole moment p_0 per molecule: $p_0 \sim 10^{-20}$ (cgs units). This means that the dipole–dipole interaction ($p_0^2(k/2\pi)^3 \sim 10^{-16}$erg) cannot determine the observable phase transition

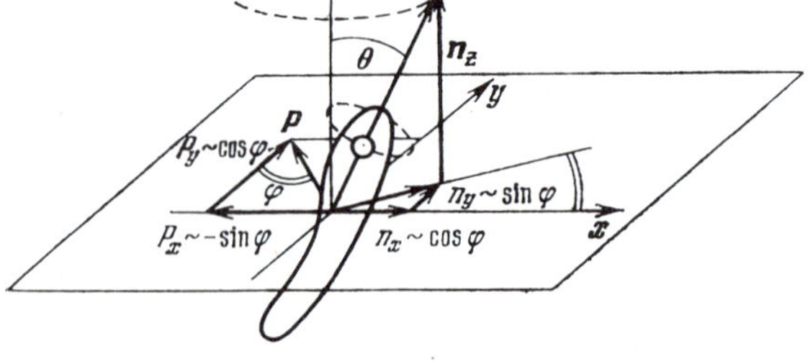

Fig. 14. Two-component order parameter in the chiral smectic phase C.

temperature $T_c \sim 350\,\mathrm{K} \sim 5\times10^{-14}$erg. In that case the order parameter is rather a physical quantity describing the spatial director orientation. The macroscopic distribution of polarization arises, as a secondary phenomenon, due to the linear relation between such order parameter and the polarization.

Both the components of the spontaneous polarization (P_x, P_y) and the components of the orientational order parameter (ξ_1, ξ_2), where

$$\xi_1 = n_z n_x, \qquad \xi_2 = n_z n_y, \tag{52}$$

transform according to the irreducible representation E_1 of the transformation group D_∞ (Pikin and Indenbom 1978a, Pikin 1981). The order parameter (52) describes the tilt of molecules on the angle θ between director $\boldsymbol{n}$ and the z axis. In that case one has

$$n_x = \sin\theta\cos\varphi, \qquad n_y = \sin\theta\sin\varphi, \qquad n_z = \cos\theta, \tag{53}$$

where φ is the azimuth angle in the plane xy (fig. 14). One can see from fig. 14 that there are the following relations between (P_x, P_y) and (ξ_1, ξ_2):

$$(P_x, P_y) \sim (\xi_2, -\xi_1) \sim \sin 2\theta\,(\sin\varphi, -\cos\varphi). \tag{54}$$

The relations (54) show that P_x and P_y transform correspondingly as ξ_2 and $-\xi_1$ under the action of the elements $\mathrm{C}(\varphi)$ and U_2 of the group D_∞. This relation between the two parameters exists in chiral smectics only and is called the piezoelectric effect.

A proportionality similar to (54) must occur at the substitution of $\boldsymbol{E}$ for $\boldsymbol{P}$. In that case relation (54) describes the effect of the average molecular tilt in the smectic A under the action of the external field $\boldsymbol{E}$. This effect, named the electrocline effect, was observed for the first time by Garoff and Meyer (1977).

Simultaneously with (54) the density of the free energy F of chiral smectics contains the invariant (Pikin and Indenbom 1978a)

$$-\mu_\mathrm{p}(P_x\xi_2 - P_y\xi_1) = -\tfrac{1}{2}\mu_\mathrm{p} P \sin 2\theta, \tag{55}$$

where P is the projection of $\boldsymbol{P}$ on the polar axis. Expressions (55) and (38) show that spontaneous polarization arises in the chiral smectic C phase at any weak piezoelectric coefficient μ_p.

It was noted above that there is spatial twisting of the director $\boldsymbol{n}$ in chiral smectics C due to the existence of the Lifshitz invariant

$$\lambda\left(\xi_1\frac{\partial\xi_2}{\partial z} - \xi_2\frac{\partial\xi_1}{\partial z}\right) = \lambda\sin^2\theta\frac{\partial\varphi}{\partial z} \tag{56}$$

and of the analogous invariant

$$P_x \frac{\partial P_y}{\partial z} - P_y \frac{\partial P_x}{\partial z}. \tag{57}$$

These invariants are in F because of the relation $\{E_1^2\} = A_2$; this relation means that the antisymmetric square of the two-dimensional vector representation E_1 of the group D_∞ contains the vector representation A_2. It is well known that such invariants induce the helical twisting of molecular axes and of dipole moments shown in fig. 2; the latter is called the cholesteric effect.

The spontaneous heterogeneous distribution $\boldsymbol{n}(\boldsymbol{r})$ must induce a spontaneous flexoelectric effect due to the relation (Pikin and Indenbom 1978b, Pikin 1981)

$$(P_x, P_y) \sim \left(\frac{\partial \xi_1}{\partial z}, \frac{\partial \xi_2}{\partial z} \right) \tag{58}$$

that is allowed simultaneously with the invariant

$$-\mu_f \left(P_x \frac{\partial \xi_1}{\partial z} + P_y \frac{\partial \xi_2}{\partial z} \right) = \tfrac{1}{2} \mu_f P \sin 2\theta \frac{\partial \varphi}{\partial z} \tag{59}$$

in the expression for F in accordance with symmetry properties described by the group D_∞. The expressions (58) and (59) are the consequences of the relation between the irreducible representations E_1 and A_2: $A_2 \times E_1 = E_1$. Thus, there is a flexoelectric term (58) in the spontaneous polarization $\boldsymbol{P}$. The flexoeffect in the phase C can be induced also by orientational perturbations which are described by the representation A_2. The corresponding invariant is

$$P_z \left(\frac{\partial \xi_1}{\partial x} + \frac{\partial \xi_2}{\partial y} \right). \tag{60}$$

At lowering of the temperature T, i.e., at increase of the angle θ the invariants

$$(\xi_1^2 + \xi_2^2) \left(P_x \frac{\partial \xi_1}{\partial z} + P_y \frac{\partial \xi_2}{\partial z} \right), \qquad \left(\xi_2 \frac{\partial \xi_1}{\partial z} - \xi_1 \frac{\partial \xi_2}{\partial z} \right) (P_x \xi_2 - P_y \xi_1). \tag{61}$$

become essential; they have flexoelectric origin and are proportional to a higher than second power of θ. It can be easily seen that in the presence of heterogeneous perturbation of the angle $\theta(z)$ the invariant (59) and the first of the invariants (61) give the terms

$$\tfrac{1}{2} \mu_f P \sin(2\theta) \frac{\partial \varphi}{\partial z} \quad \text{and} \quad -\mu_f P' \cos(2\theta) \frac{\partial \theta}{\partial z},$$

where P' is the polarization projection on the line of intersection of the smectic plane xy with the plane perpendicular to the axis C_2 (fig. 14). The second of the invariants (61) is proportional to $P\sin^3\theta\,\partial\varphi/\partial z$. Thus, at appreciable angles θ there are different flexoelectric coefficients μ_f and μ'_f in the flexoelectric terms, which correspond to different polarization components, proportional to $\theta\,\partial\varphi/\partial z$ and to $\partial\theta/\partial z$. At small angles θ the difference $(\mu_f-\mu'_f)\sim\theta^2$, i.e., it is small near the phase transition point.

6.2. *The helical ferroelectric*

Since the symmetric cube of the representation E_1 does not contain the unit representation ($[E_1^3]=E_3+3E_1$) the A–C phase transition can be second order. One can write the density of the free energy of smectics A, considering constant orientational order parameter Q, small polar angle θ and taking into account the invariants (55), (56) and (59)–(61), in the form (Pikin 1981):

$$
\begin{aligned}
F={}& a_0\theta^2+b\theta^4+\tfrac{1}{2}g_{\parallel}^0\left(\frac{\partial\theta}{\partial z}\right)^2+\tfrac{1}{2}g_{\perp}^0\left[\left(\frac{\partial\theta}{\partial x}\right)^2+\left(\frac{\partial\theta}{\partial y}\right)^2\right]\\
&+\tfrac{1}{2}K^0\theta^2\left(\frac{\partial\varphi}{\partial z}\right)^2+\tfrac{1}{2}K'\left(\frac{\partial\xi_1}{\partial x}+\frac{\partial\xi_2}{\partial y}\right)^2+\tfrac{1}{2}K''\left(\frac{\partial\xi_1}{\partial y}-\frac{\partial\xi_2}{\partial x}\right)^2\\
&+K'''\left(\xi_1\frac{\partial\xi_2}{\partial z}-\xi_2\frac{\partial\xi_1}{\partial z}\right)\left(\frac{\partial\xi_1}{\partial y}-\frac{\partial\xi_2}{\partial x}\right)+\lambda\theta^2\frac{\partial\varphi}{\partial z}\\
&+\frac{1}{2\chi_{\perp}^0}(P^2+P'^2)+\frac{1}{2\chi_{\parallel}^0}P_z^2-\mu_pP\theta+\mu_fP\theta\frac{\partial\varphi}{\partial z}\\
&-\mu'_fP'\frac{\partial\theta}{\partial z}-\mu''_fP_z\left(\frac{\partial\xi_1}{\partial x}+\frac{\partial\xi_2}{\partial y}\right),
\end{aligned}
\tag{62}
$$

where $a_0=a'_0(T-T_0)$, $a'_0>0, b>0$; $g_{\parallel}^0, g_{\perp}^0, K^0, K', K'', K'''$ are the elastic constants. The first eight terms in (62) correspond to ordinary invariants proportional to even powers of the components (ξ_1,ξ_2) and of their derivatives. Here invariants proportional to higher powers of polarization components and of their derivatives (for example, expression 57) are not taken into account. These effects and the invariants proportional to higher powers of θ^2 do not change results of the consideration and give some terms in effective values of coefficients. Particularly one can consider that the corrections of the order of value θ^2 are effectively taken into account in the coefficients g, K, λ, μ and χ.

Note that at temperatures $T<T_c$, when $\theta\sim1$, the terms proportional to the constants K', K'', K''' are essential to the same extent as the K^0 term. Near

the A–C phase transition point one can neglect the K', K'' and K''' terms, since they are small in comparison with the other terms in eq. (62). In fact, they contain additional powers of θ, $\nabla\theta$ and $\nabla\varphi$.

In the absence of external fields the components of polarization P_z and P', corresponding to a minimum of F, are equal to

$$P' = \chi_{\perp}^{0}\mu_{\mathrm{f}}'\frac{\partial\theta}{\partial z}, \qquad P_z = \chi_{\parallel}^{0}\mu_{\mathrm{f}}''\left(\frac{\partial\xi_1}{\partial x} + \frac{\partial\xi_2}{\partial y}\right). \tag{63}$$

The substitution of eqs. (63) into eq. (62) in fact gives the renormalization of the coefficients $g_{\parallel}^{0}$ and K'; for example:

$$g_{\parallel}^{0} \rightarrow g_{\parallel} = g_{\parallel}^{0} - \chi_{\perp}^{0}\mu_{\mathrm{f}}'^{2}.$$

Thus, a minimum of the free energy is achieved at homogeneous distribution of θ in the smectic volume and at a constant value of φ in the smectic plane xy.

Carrying out the minimization of F with respect to P, one gets at $\theta \ll 1$:

$$P = \chi_{\perp}^{0}\left(\mu_{\mathrm{p}} - \mu_{\mathrm{f}}\frac{\partial\varphi}{\partial z}\right)\theta,$$

$$F = a\theta^2 + b\theta^4 + \tfrac{1}{2}K\theta^2\left(\frac{\partial\varphi}{\partial z} - q_0\right)^2, \tag{64}$$

$$q_0 = -\frac{\lambda + \chi_{\perp}^{0}\mu_{\mathrm{p}}\mu_{\mathrm{f}}}{K},$$

where $a = a_0 - 2^{-1}\chi_{\perp}^{0}\mu_{\mathrm{p}}^{2} - 2^{-1}Kq_{\mathrm{c}}^{2}$, $K = K^{0} - \chi_{\perp}^{0}\mu_{\mathrm{f}}^{2}$, $q_{\mathrm{c}} = q_0(T_{\mathrm{c}})$. It is seen from (64) that the minimum of F is achieved at $\varphi = \varphi_0(z) = q_0 z$. In that case a helical polarization distribution arises, with the amplitude

$$P_{\mathrm{s}} = \chi_{\perp}^{0}|\mu_{\mathrm{p}} - \mu_{\mathrm{f}}q_0|\theta. \tag{65}$$

The temperature T_{c} is determined by the expression

$$T_{\mathrm{c}} = T_0 + \frac{1}{2a_0'}\left(\chi_{\perp}^{0}\mu_{\mathrm{p}}^{2} + Kq_{\mathrm{c}}^{2}\right). \tag{66}$$

At $T < T_{\mathrm{c}}$ the finite tilt angle of molecules is equal to

$$\theta_0 = (-a/2b)^{1/2}.$$

The helical polarization distribution is related to the orientational spiral $\boldsymbol{n}(z)$: $\boldsymbol{P}$ is perpendicular to the axis z and rotates around this axis with the pitch $h_0 = 2\pi/q_0$.

Note that, in general, the incommensurate structure arises due to two Lifshitz invariants in (62) which correspond to different physical effects. If the piezoeffects (54) and (58) are the most important, one can neglect the invariant (56) and suppose $\lambda = 0$: $q_0 \sim \mu_p \mu_f / K$. On the contrary, if the flexoelectric effect (58) is small, one can suppose $\mu_f = 0$. Then the helical twisting is determined by the cholesteric effect: $q_0 \sim \lambda/K$. At the condition

$$\lambda + \chi_\perp^0 \mu_p \mu_f = 0 \tag{67}$$

two Lifshitz invariants compensate each other. In that case the existence of a macroscopically homogeneous ferroelectric state $P_x = 0$ and $P_y = -\chi_\perp^0 \mu_p \theta$ is possible. This situation really exists, since the parameters λ and $\mu_p \mu_f$ are independent. This effect is realized in experiments by mixing of liquid crystals.

It should be underlined that the expansion (62) makes sense for small values of q_0 only, i.e., for small values λ/K and of $\chi_\perp^0 \mu_p \mu_f / K$. If the quantity q_0^{-1} would be comparable with the smectic monolayer thickness $2\pi k^{-1}$, the second-order phase transition under consideration would be forbidden, since gradient and nongradient terms would be of the same order at all powers of θ^2.

6.3. *The behaviour of liquid–crystalline ferroelectrics in external fields*

It is necessary to include the term $-\boldsymbol{PE}$ in the expression (62) to consider the behaviour of chiral smectics under the action of the electric field $\boldsymbol{E}$. The variation of $F - \boldsymbol{PE}$ with respect to $\boldsymbol{P}$, θ and φ gives the necessary equations for the functions $\theta(\boldsymbol{r})$ and $\varphi(\boldsymbol{r})$. The action of $\boldsymbol{E}$ disturbs the incommensurate structure of the phase C. The perturbations $\theta(\boldsymbol{r}) - \theta_0$ and $\varphi(\boldsymbol{r}) - \varphi_0(z)$ are small if the field $\boldsymbol{E}$ is weak. The perturbations considered induce untwisting of the spiral of dipole moments and increase of the tilt angle in strong fields. At $E \ll E_c$, where E_c is the critical value at which total untwisting takes place, the orientational perturbations can be subdivided into the azimuth perturbation $\varphi - \varphi_0$ (at the homogeneous distribution $\theta = \theta_0$) and the polar distribution $\theta - \theta_0$ (at the nonperturbed pitch $h_0 = 2\pi/q_0$).

Supposing $\boldsymbol{E}$ is oriented in the smectic plane xy ($E_z = 0$), one can neglect the effects related to the polarization component P_z and to the orientational disturbances in the xy plane. In that case θ and φ are functions of the coordinate z. Here linear response of the system is considered. The correspondent correction in the dielectric susceptibility $\delta\chi_\perp = \chi_\perp - \chi_\perp^0$ is related to the orientational perturbations; both azimuth and polar deformations give contributions to $\delta\chi_\perp$ due to the piezoeffect.

If the field $\boldsymbol{E}$ is oriented along the y axis, the variation of $F-\boldsymbol{PE}$ with respect to P_x, P_y, $\theta(z)$ and $\varphi(z)$ results in the expressions

$$P_x=\chi_\perp^0\left(\mu_p-\mu_f\frac{d\varphi}{dz}\right)\theta\sin\varphi+\chi_\perp^0\mu_f'\frac{d\theta}{dz}\cos\varphi,$$

$$P_y=\chi_\perp^0 E-\chi_\perp^0\left(\mu_p-\mu_f\frac{d\varphi}{dz}\right)\theta\cos\varphi+\chi_\perp^0\mu_f'\frac{d\theta}{dz}\sin\varphi, \qquad (68)$$

where the functions $\theta(z)=\theta_0+\theta_1$ and $\varphi(z)$ at $E\ll E_c$ and $\theta_1\ll\theta_0$ are to be found from the equations

$$K\theta_0\frac{d^2\varphi}{dz^2}+\chi_\perp^0 E\mu_p\sin\varphi=0,$$

$$g_\parallel\frac{d^2\theta_1}{dz^2}+4a\theta_1-\chi_\perp^0 E\left[\mu_p-(\mu_f-\mu_f')\frac{d\varphi}{dz}\right]\cos\varphi=0. \qquad (69)$$

The first of equations (69) describes the disturbance of the helix in the electric field, $\varphi(z)=\varphi_0(z)$ at $E=0$. This problem is similar to that of deformations of a cholesteric spiral (de Gennes 1968). Using the existing solution, one can find

$$E_c=\frac{\pi^2}{16}\frac{Kq_0^2\theta_0}{\chi_\perp^0\mu_p}. \qquad (70)$$

At $E\ll E_c$ the periodic solution $\theta_1(z)$ takes the form

$$\theta_1(z)=\frac{\chi_\perp^0 E(\mu_p-(\mu_f-\mu_f')q_0)\cos(q_0z)}{4a-g_\parallel q_0^2} \qquad (71)$$

and the function $\varphi(z)$ is given by the expression

$$\varphi(z)=q_0z+\frac{\chi_\perp^0 E\mu_p}{Kq_0^2\theta_0}\sin(q_0z). \qquad (72)$$

With the help of eqs. (68), (71) and (72) one obtains the macroscopic average value of polarization $\mathscr{P}$ along the y axis in the chiral phase C (Pikin 1981):

$$\mathscr{P}=(\chi_\perp^0+\delta\chi_\perp)E,$$

$$\delta\chi_\perp|_C=\tfrac{1}{2}\chi_\perp^{02}\left[\frac{\mu_p^2}{Kq_0^2}+\frac{(\mu_p-(\mu_f-\mu_f')q_0)^2}{g_\parallel q_0^2-4a}\right]. \qquad (73)$$

At $T > T_c$ the smectic structure is homogeneous in absence of the electric field: $q = d\varphi/dz = 0$. Therefore, in linear approximation with respect to E, one obtains by means of eq. (64) (Pikin and Indenbom 1978b)

$$\varphi = \begin{cases} 0, & \mu_p < 0, \\ \pi, & \mu_p > 0, \end{cases}$$

$$\theta_1 = \theta = \left| \frac{\chi^0_\perp \mu_p}{2a + Kq_c^2} \right| E, \qquad \delta\chi_\perp|_A = \frac{\chi^{02}_\perp \mu_p^2}{2a + Kq_c^2}. \tag{74}$$

Thus, in the phase A the quantity $\delta\chi_\perp$ is determined by the piezoelectric coefficient μ_p only, since here the macroscopic orientational twist and, consequently, the flexoelectric effect are absent. It should be underlined that the dielectric susceptibility is finite at $T = T_c$, i.e., at $a = 0$, since the low-temperature phase C is modulated and, consequently, only the response of the system to the correspondent modulated external field can be infinite. At $a \gg Kq_c^2$ expression (74) reminds of the Curie law, since in that case the correlation length $r_c \sim (K/|a|)^{1/2}$ is much smaller than the pitch of the helix $h_c = 2\pi/q_c$, i.e., the medium is orientationally homogeneous in the fluctuation area with the radius r_c.

Note that expression (73) is obtained at $E \ll E_c$. According to eq. (70) the critical value E_c is zero at $T = T_c$. Therefore at $T = T_c$ the inequality $E \ll E_c$ is incorrect and expression (73) is not applicable; instead of (73) one should use formula (74) which is correct at $T - T_c > -1\,\mathrm{K}$ as estimations show.

6.4. *Results of experimental investigations of ferroelectric liquid crystals*

The character of investigated A–C phase transitions shows that in many cases these transitions are second order or nearly second order (Meyer et al. 1975, Garoff and Meyer 1977, Ostrovskii et al. 1978a, b). Ostrovskii et al. (1978a) have given an estimation of the critical index β which describes the temperature dependence $\theta_0(T)$ and $P_s(T)$. At $T - T_c < -1\,\mathrm{K}$ the values of β for $\theta_0(T)$ and $P_s(T)$ are different: 0.31 ± 0.05 and 0.37 ± 0.01 correspondingly. Garoff and Meyer (1977), applying the measurements of the linear electro-optic response, estimated the index γ_A for the temperature dependence of the relaxation time $\tau \sim (T - T_c)^{-\gamma_A}$ in the phase A at $T - T_c > 1\,\mathrm{K}$. γ_A is near the phenomenological value: $\gamma_A = 1.11 \pm 0.06$. Beresnev et al. (1978), applying the measurements of pyroelectrical effects in the phase $\tilde{C}$, estimated the index γ_C for the dependence $\tau \sim (T_c - T)^{-\gamma_C}$: $\gamma_C = 0.83 \pm 0.16$. This behaviour of helical liquid crystalline ferroelectrics is similar to the behaviour of solid ferroelectrics near the Curie point. However, it should be pointed out that up until now there are no sufficient experimental data on critical indexes to draw a definitive conclusion about the character of the A–$\tilde{C}$ transition as similar to the phase transition in liquid helium, for example. So it is reasonable to carry out qualitative and quantitative descriptions of such phenomena in the framework of the phenomenological theory.

The fast increase of the relaxation time τ near T_c agrees with the observed increase of the dielectric constant in the vicinity of the phase transition point. Note that, according to (74), the quantities τ and $\varepsilon_\perp$ must be finite at $T=T_c$ as is observed in experiments. Figure 15 shows the temperature dependence $\varepsilon_\perp(T)$ at low frequences (Ostrovskii et al. 1978a). This dependence is essentially nonmonotonous, and it is similar to the temperature dependence of the pitch $h_0(T)$ (fig. 16). It is clear from eqs. (64), (73) and (74) that the maxima of the temperature dependences $\varepsilon_\perp(T)$ and $h_0(T)$ do not coincide with T_c and correspond to different temperatures in general. Levstik et al. (1984) have pointed out this experimental fact for some liquid crystalline ferroelectrics. It should be noted that in a narrow temperature region near T_c the simple phenomenological theory, including the expansion of h_0 in ascending powers of θ^2, is not reasonable (Osipov and Pikin 1982). In fact, this last expansion needs the assumption that coefficients in terms powers higher than unity in (θ^2) are anomalously large (10^2 times larger than $h_0(T_c)$).

Anomalies in the temperature dependences $\varepsilon_\perp(T)$ and $h_0(T)$ are not related to a critical behaviour or to thermal fluctuations, but they can be caused by the strong influence of steric molecular interactions on the temperature dependences of the flexoelectric coefficients μ_f and μ'_f. According to eqs. (64) and (73) the quantities h_0 and $\varepsilon_\perp$ effectively depend on μ_f and μ'_f. Osipov and Pikin (1982) have shown that the narrow temperature region of the

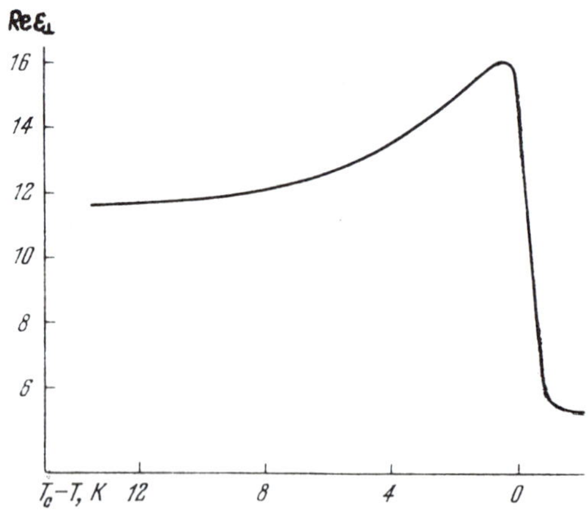

Fig. 15. Temperature dependence of the dielectric constant in decyloxybenzylideneamino-cinnamate (DOBAMBC). (Ostrovskii et al. 1978a.)

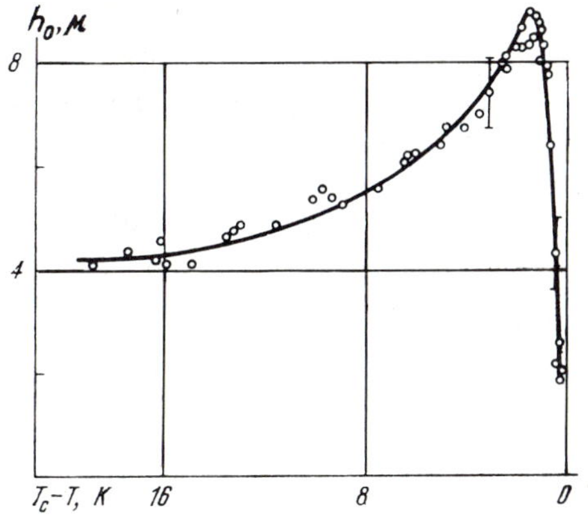

Fig. 16. Temperature dependence of the pitch of the helix in the C phase of DOBAMBC. (Ostrovskii et al. 1978a.)

anomalous dependences $\delta\mu_f(T)$ and $\delta\mu'_f(T)$ is determined by shape anisotropy and by the dimensions of molecules. Particularly, $\delta\mu_f \sim \nu[1+\nu^{-2}(T_c - T)]^{1/2}$, where $\nu \sim 10^{-2}$ is the degree of bend of the molecular shape. At $T_c - T \leqslant 10^{-2}T_c$ we have $|d\mu_f/dT| \sim \nu^{-1} \sim 10^2$, which explains the observed phenomenon. Another possible reason of the anomalous temperature dependence of the pitch near T_c is the unwinding of the helix via defects of the incommensurate structure (Glogarova et al. 1984).

It should be underlined that the temperature behaviour of $\varepsilon'_\perp(T)$ outside this narrow temperature region in the $\tilde{C}$ phase is determined by absolute values and by signs of μ_p and $(\mu_f - \mu'_f)q_0$ according to eq. (73). Since here $(\mu_f - \mu'_f) \sim \theta^2$, one can observe both descending and ascending functions $\varepsilon_\perp(T)$ at low temperatures. Levstik et al. (1984) observed a slightly ascending temperature dependence $\varepsilon_\perp(T)$ with decreasing temperature in a liquid crystalline mixture.

Muševič et al. (1982) obtained interesting experimental data on the temperature dependence of the critical magnetic field $H_c(T)$ corresponding to the total unwinding of the helix in the $\tilde{C}$ phase. The dependence $H_c(T)$ is similar to the anomaly in the temperature dependence of the helical pitch. One can easily understand this result from the qualitative relation $\chi_a H_c^2 \sim K q_0^2$, where χ_a is the diamagnetic anisotropy, i.e., $H_c(T) \sim q_0(T) \sim h_0^{-1}(T)$.

Some estimations of fundamental material parameters of ferroelectric liquid crystals were made by Ostrovskii et al. (1979a). These estimations show that

$$\chi_\perp^0 \sim 0.3, \qquad |\mu_p| \sim 3\times10^{-8}\,\mathrm{C\,cm^{-2}}, \qquad |\mu_p| \sim |\mu_f q_0|,$$

$$|\lambda| \lesssim |\chi_\perp^0 \mu_p \mu_f|.$$

The value of the spontaneous polarization essentially depends on the molecular structure. It is maximum in L-4-hexyloxybenzylidene-4′-amino-2-chlorinepropylcinnamate ($P_s = 1.5 \times 10^{-8}\,\mathrm{C\,cm^{-2}}$, Beresnev and Blinov 1981) and in some mixtures of smectics ($P_s = 2.2 \times 10^{-2}\,\mathrm{C\,cm^{-2}}$, Beresnev et al. 1982). The spatial period h_0 of the incommensurate structure of liquid crystalline ferroelectrics can be changed by changing the composition of the liquid crystalline mixtures. In the latter case, a macroscopic polarization should be obtained for certain compositions; this has been confirmed experimentally (Beresnev et al. 1981).

7. *Incommensurate smectic structures as polymorphous modifications*

Smectic phases provide various possibilities for the formation of incommensurate structures if these phases are characterized by several density waves with different wave numbers. This situation differs from the ordinary smectic A (A_1), where only the substance density wave $\Psi = \Psi_0 \cos(k_1 z)$ exists. In the general case, modulation of the electric charge density and modulation of the electric potential $\Phi(z) = \Phi_0 \cos(k_2 z)$ can also take place if the molecules of the liquid crystals have large electric dipole moments (Prost 1979, Prost and Barois 1983). The spatial periods of structures $2\pi/k_1$ (for the density $\Psi(z)$) and $2\pi/k_2$ (for the potential $\Phi(z)$) are not obviously commensurate if interaction between these order parameters is absent. Particularly $2\pi/k_1$ is comparable with the molecular length l_1 while the period $2\pi/k_2$ is determined by the dimensions of some molecular clusters, for example by the antiparallel arrangement of two molecules with total length l_2. The length l_2 has an intermediate value: $l_1 < l_2 < 2l_1$ (fig. 17). Switching on the interaction between Ψ and Φ gives the possibility of the formation of new commensurate and incommensurate structures as a result of phase transitions.

Prost and Barois (1983) have proposed a simple model describing such phase transitions. This model is based on the phenomenological form of the free energy density F as the functional $F[\Psi, \Phi]$:

$$
\begin{aligned}
F = {} & \tfrac{1}{2}a_1\Psi^2 + \tfrac{1}{4}b_1\Psi^4 + \tfrac{1}{2}c_1\left(\left(\nabla^2 + k_1^2\right)\Psi\right)^2 \\
& + \tfrac{1}{2}a_2\Phi^2 + \tfrac{1}{4}b_2^2\Phi^4 + \tfrac{1}{2}c_2\left(\left(\nabla^2 + k_2^2\right)\Phi\right)^2 + d_1\Phi^2\Psi \\
& + d_2\Phi^2\Psi^2 + g_1 Q_{ik}\nabla_i\Psi \cdot \nabla_k\Psi + g_2 Q_{ik}\nabla_i\Phi \cdot \nabla_k\Phi,
\end{aligned}
$$

where $a_1 \sim (T - T_1)$, $a_2 \sim (T - T_2)$, $b_1 > 0, b_2 > 0, c_1 > 0, c_2 > 0, d_2 > 0$, the constants g_1 and g_2 characterize the interactions of the density waves with the orientational order parameter Q_{ik}; here $2k_2 \neq k_1$ but $|2k_2 - k_1| \ll k_1$, the sums of wave vectors are zero in each term of the expansion.

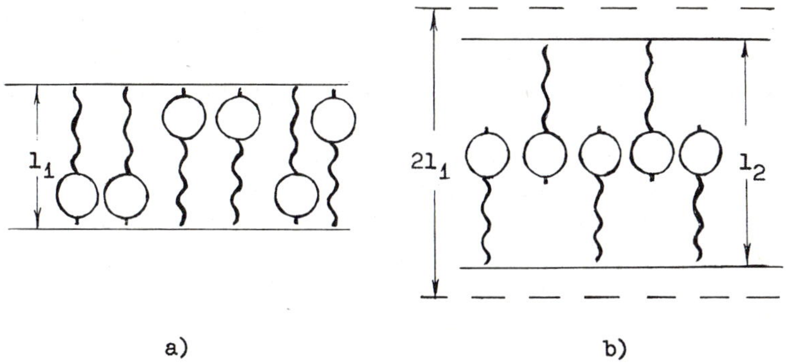

Fig. 17. Formation of the uniaxial incommensurate structure in smectics A: (a) ordinary phase A_1; (b) incommensurate phase A_2.

Minimizing F with respect to Ψ and Φ one can obtain, for example, the following possible types of phase transitions and structures;

phase A_1 (fig. 17a) with $\Psi = \Psi_0 \exp(ik_1 z)$, $\Phi = 0$;
phase A_2 (fig. 17b) with $\Psi = \Psi_0 \exp(2ikz)$, $\Phi = \Phi_0 \exp(ikz + \alpha)$;
phase A_d with $\Psi = 0$ and $\Phi = \Phi_0 \exp(ik_2 z)$;
phase A_a (fig. 18a) with $\Psi = \Psi_0 \exp(2ikz + \alpha)$,

$$\Phi_1 = \Phi_0 \exp(i(kz + qx)), \qquad \Phi_2 = \Phi_0 \exp(i(kz - qx));$$

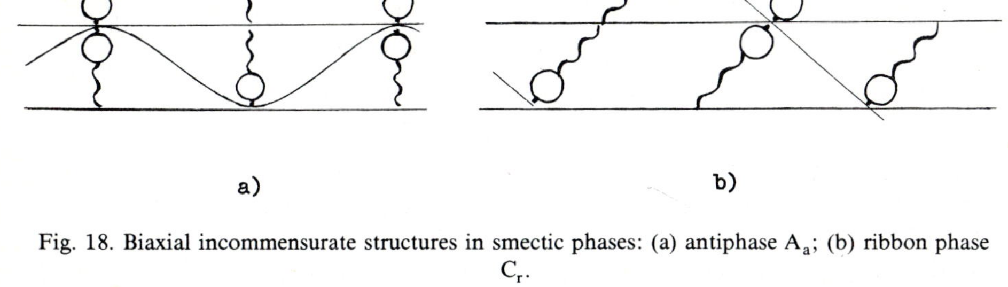

Fig. 18. Biaxial incommensurate structures in smectic phases: (a) antiphase A_a; (b) ribbon phase C_r.

phase C_r (fig. 18b) with $\Psi = \Psi_0 \exp(2ikz + \alpha)$,

$$\Phi_1 = \Phi_0^+ \exp(i(k+\delta k)z + iqx), \qquad \Phi_2 = \Phi_0^- \exp(i(k-\delta k)z - iqx).$$

The smectic phase A_2 has an intermediate value of the wave number k which is incommensurate to the wave numbers k_2 and $k_1/2$. The smectics A_1 and A_d have the same symmetry (the point group $D_{\infty h}$). The so called smectic "antiphase" A_a (with the point group of symmetry D_{2h}) and the "ribbon" phase C_r (with the point group of symmetry C_{2h}) are characterized by two-dimensional ordering.

There is a periodic modulation of the potential Φ in smectic planes of the phase A_a, the corresponding wave number q is incommensurate to k. In the phase C_r long molecular axes makes some angle with the crystal axis z; here the planes of equal phases and of equal amplitudes of the potential Φ are described by the equation

$$\delta kz + qx = \text{const.}$$

The analysis of phase diagrams (Prost and Barois 1983) shows that in most cases the phase transitions between smectic modifications are of the first kind, however, one cannot exclude the possibility of the second-order phase transitions A_1–A_2, A_a–C_r and A_2–A_a. Experimentally the first-order phase transitions A_1–A_d (Hardouin et al. 1983, Levelut et al. 1981), A_d–C_r (Sigaud et al. 1981), A_1–C_r (Pershan and Prost 1979), A_1–A_2 (Hardouin et al. 1983, Levelut et al. 1981), A_2–A_a (Sigaud et al. 1981) and A_1–A_a (Levelut et al. 1981, Sigaud et al. 1981, Pershan and Prost 1979) were observed.

Prost and Barois (1983) predicted also the existence of the incommensurate phases A_{j1} and A_{j2}, correspondingly at $k_1 \approx k_2$ and $k_1 \approx 2k_2$ with the order parameters

$$\{\Psi = \Psi_0 \exp[i(kz + \beta(z))], \quad \Phi = \Phi_0 \exp[i(kz + \alpha(z))]\}$$

and

$$\{\Psi = \Psi_0 \exp[i(2kz + \beta(z))], \quad \Phi = \Phi_0 \exp[i(kz + \alpha(z))]\}.$$

These phases arise as a result of the phase transitions A_1–A_{j1} and A_2–A_{j2}. One can see that in the framework of a phenomenological description of these structural transitions the terms $d_1\Phi\Psi \sim d_1\Phi_0\Psi_0 \sin^2(\alpha/2 - \beta/2)$ and $d_1\Phi^2\Psi \sim d_1\Phi_0^2\Psi_0 \sin^2(\alpha - \beta/2)$, which are similar to the term $F_e = -(\varepsilon_a/8\pi)(\boldsymbol{En})^2$, play an essential role in corresponding expansions of F. There is a total "unwinding" of the phases $(\alpha-\beta)/2$ and $(\alpha - \beta/2)$ at some critical values of the amplitudes $\Phi_0\Psi_0$ and $\Phi_0^2\Psi_0$ in smectics A_{j1} and A_{j2} like the total unwinding of the cholesteric helix taking place at the threshold value $E = E_c$. The phases $(\alpha - \beta)/2$ and $(\alpha - \beta/2)$ are constant in smectics A_{j1} and A_{j2} correspondingly. Near the A_1–A_{j1} and A_2–A_{j2} phase transition points there are relatively large regions with the A_1 and A_2 smectic structures in the smectics A_{j1} and A_{j2} correspondingly. These large regions are periodically devided by narrow walls in which the phases $(\alpha - \beta)/2$ and $(\alpha - \beta/2)$ change practically by jump on π. The corresponding periods are incommensurate to the periods of the one-dimensional crystallines lattices of the smectics A_1 and A_2. The periods of the incommensurate structures A_{j1} and A_{j2} depend on temperature. Gray et al. (1984) have recently discovered a smectic phase of this kind.

References

Barnik, M.I., L.M. Blinov, A.N. Trufanov and B.A. Umanski, 1978, J. Physique **39**, 417.

Beresnev, L.A., and L.M. Blinov, 1981, Ferroelectrics **33**, 129.

Beresnev, L.A., L.M. Blinov and E.B. Sokolova, 1978, Zh. Eksp. Teor. Fiz. Pis'ma **28**, 340.

Beresnev. L.A., V.A. Baikalov, L.M. Blinov, E.P. Pozhydaev and G.V. Purvanezskas, 1981, Zh. Eksp. Teor. Fiz. Pis'ma **33**, 553.

Beresnev, L.A., E.P. Pozhydaev, L.M. Blinov, A.M. Pavlyuchenko and N.B. Etingen, 1982, Zh. Eksp. Teor. Fiz. Pis'ma **35**, 430.

Bobylev, Y.P., and S.A. Pikin, 1977, Zh. Eksp. Teor. Fiz. **72**, 369.

Bobylev, Y.P., V.G. Chigrinov and S.A. Pikin, 1979, J. Physique Colloque **40**, C3-331.

Brazovskii, S.A., and S.G. Dmitriev, 1975, Zh. Eksp. Teor. Fiz. **69**, 979.

Brazovskii, S.A., and V.M. Filev, 1978, Zh. Eksp. Teor. Fiz. **75**, 1140.

Clark, N.A., and P.S. Pershan, 1973, Phys. Rev. Lett. **30**, 3.

de Gennes, P.G., 1968, Solid State Commun. **6**, 163.

de Gennes, P.G., 1974, The Physics of Liquid Crystals (Clarendon Press, Oxford).

Delaye, M., R. Ribotta and G. Durand, 1973, Phys. Lett. **44**, 139.

Delrieu, J.M., 1974, J. Chem. Phys. **60**, 1081.

Dzyaloshinskii, I.E., 1964, Zh. Eksp. Teor. Fiz. **46**, 1420.

Garoff, S., and R.B. Meyer, 1977, Phys. Rev. Lett. **38**, 848.

Gerritsma, C.J., and P. Van Zanten, 1971, Phys. Lett. **A37**, 47.

Glogarova, M., J. Fousek, L. Lejček and J. Pavel, 1984, Ferroelectrics **58**, 161.

Gray, G.W., A.J. Leadbetter, A.R. Tajbakhsh and P.A. Tucker, 1984, Abstracts 10th Int. Liquid Crystal Conf., York, UK, E7.

Grebel, H., R.M. Hornreich and S. Shtrikman, 1983, Phys. Rev. **A28**, 1114.

Hardouin, F., A.M. Levelut, M.F. Achard and G. Sigaud, 1983, J. Chim. Phys. **80**, 47.

Hornreich, R.M., and S. Shtrikman, 1980, J. Physique **41**, 335.

Hornreich, R.M., and S. Shtrikman, 1981, Phys. Rev. **A24**, 635.

Hornreich, R.M., and S. Shtrikman, 1983, Phys. Rev. **A28**, 1791.

Hornreich, R.M., M. Kugler and S. Shtrikman, 1984, Abstracts 10th Int. Liquid Crystal Conf., York, UK, C66.

Kahn, F., 1973, Appl. Phys. Lett. **22**, 111.

Kahn, F.J., 1970, Phys. Rev. Lett. **24**, 209.

Kassubek, P., and G. Meier, 1969, Mol. Cryst. Liq. Cryst. **8**, 305.

Levelut, A.M., R.J. Tarento, F. Hardouin, M.F. Achard and G. Sigaud, 1981, Phys. Rev. **A24**, 2180.

Levstik, A., B. Žekš, C. Filipič, R. Blinc and I. Levstik, 1984, Ferroelectrics **58**, 33.
Liao, Y., N. Clark and P. Pershan, 1973, Phys. Rev. Lett. **30**, 639.
Meiboom, S., J.P. Sethna, P.W. Anderson and W.F. Brinkman, 1981, Phys. Rev. Lett. **46**, 1216.
Meiboom, S., M. Sammon and W.F. Brinkman, 1983, Phys. Rev. **A27**, 438.
Meyer, R.B., 1969, Phys. Rev. Lett. **22**, 918.
Meyer, R.B., L. Liebert, L. Strzelecki and P. Keller, 1975, J. Physique Lett. **36**, L-69.
Muševič, I., B. Žekš, R. Blinc, T. Rasing and P. Wyder, 1982, Phys. Rev. Lett. **48**, 192.
Osipov, M.A., and S.A. Pikin, 1982, Zh. Eksp. Teor. Fiz. **82**, 774.
Ostrovskii, B.I., A.Z. Rabinovich, A.S. Sonin and B.A. Strukov, 1978a, Zh. Eksp. Teor. Fiz. **74**, 1748.
Ostrovskii, B.I., A.Z. Rabinovich, A.S. Sonin, B.A. Strukov and S.A. Taraskin, 1978b, Ferroelectrics **20**, 189.
Ostrovskii, B.I., S.A. Pikin and V.G. Chigrinov, 1979a, Zh. Eksp. Teor. Fiz. **77**, 1631.
Ostrovskii, B.I., A.Z. Rabinovich, A.S. Sonin and B.A. Strukov, 1979b, Fiz. Tverd. Tela **21**, 917.
Pershan, P.S., and J. Prost, 1979, J. Phys. Lett. **40**, L-27.
Pikin, S.A., 1981, Strukturnyie Prevrascheniya v Zhydkih Kristallah (Nauka, Moscow).
Pikin, S.A., and V.L. Indenbom, 1978a, Usp. Fiz. Nauk **125**, 251.
Pikin, S.A., and V.L. Indenbom, 1978b, Ferroelectrics **20**, 151.
Pindak, R.S., C.C. Huang and J.T. Ho, 1974, Phys. Lett. **A47**, 263.
Prost, J., 1979, J. Physique **40**, 581.
Prost, J., 1980, in: Proc. Conf. on Liquid Crystals of One and Two Dimensional Order, Garmisch Partenkirchen (Springer Berlin) p. 125.
Prost, J., and P. Barois, 1983, J. Chim. Phys. **80**, 65.
Reinitzer, F., 1888, Montash Chem. **9**, 421.
Ribotta, R., G. Durand and J.D. Litster, 1973, Solid State Commun. **12**, 27.
Rondelez, F., and J. Hulin, 1972, Solid State Commun. **10**, 1009.
Rondelez, F., H. Arnould and C.J. Gerritsma, 1972, Phys. Rev. Lett. **28**, 735.
Schmid, A., 1969, Phys. Rev. **180**, 527.
Sigaud, G., F. Hardouin, M.F. Achard and A.M. Levelut, 1981, J. Phys. **42**, 107.
Terent'ev, E.M., and S.A. Pikin, 1982, Zh. Eksp. Teor. Fiz. **83**, 1038.
Vigman, P.B., and V.M. Filev, 1975, Zh. Eksp. Teor. Fiz. **69**, 1466.

APPENDIX

List of Incommensurate Crystals

A.U. SHELEG and V.V. ZARETSKII

Institute for Semiconductor Physics
Academy of the Belorussian Soviet Socialist Republic
Minsk
USSR

Incommensurate Phases in Dielectrics 2
Edited by
R. Blinc and A.P. Levanyuk

Incommensurate crystals.

Crystal	References[a]
$(N(CH_3)_4)_2MnCl_4$	[GeI80b], [MaT80], [Ii81], [Ge82]
$(N(CH_3)_4)_2FeCl_4$	[GeI80b], [MaT80], [Ii81], [Ge82]
$(N(CH_3)_4)_2CoCl_4$	[GeI80b], [MaT80], [Ii81], [Ge82]
$(N(CH_3)_4)_2CuCl_4$	[GeI80b], [MaT80], [Ii81], [Ge82]
$(N(CH_3)_4)_2ZnCl_4$	[GeI80b], [MaT80], [Ii81], [Ge82]
$(N(CD_3)_4)_2ZnCl_4$	[MaAL81], [IiG83b]
$(N(CD_3)_4)_2CoCl_4$	[HaMT84]
$(N(CH_3)_4)_2CuBr_4$	[Ge82], [Ha82], [JaS83]
Rb_2ZnCl_4	[MaTH81], [Be83], [MeA83], [ShVS83], [YaH83], [PrB84], [BlLM84]
Rb_2ZnBr_4	[RaWJ82], [UeIT82], [BeJB83], [DaJB83], [IiG83a], [NoAC83], [YaH83]
K_2ZnCl_4	[Ku82], [ShGP83], [ShVS83], [VtS83], [YaH83], [Ge84]
$(NH_4)_2BeF_4$	[LeS76], [UeOH81], [HiTI82], [Ku82], [StAU82], [EsL83], [Hu83], [StGS83] [LePZ84]
$(ND_4)_2BeF_4$	[IiG77], [Ku82]
$(NH_4)_2ZnCl_4$	[Ma82], [SaEO82], [WaK82], [MeIZ84]
$(NH_4)_2CoCl_4$	[Br84]
$(NH_4)_2ZnBr_4$	[OsKM82], [SaOM83]
$RbH_3(SeO_3)_2$	[GlKM77], [SaL77], [GeI80a], [EsLM81], [EsL83]
$RbD_3(SeO_3)_2$	[SaKT79], [ViR79], [BlBL82]
$(C_3H_7NH_3)_2MnCl_4$	[DeM82], [MuKB82], [De83], [Ki83], [MuK83]
$(C_2H_5NH_3)_2FeCl_4$	[KlSL84]
$NaNO_2$	[HoS61], [Ta61], [CaH64], [LeS76], [YaHM81], [FiM83], [Ha84]
$SC(NH_2)_2$	[LeS77], [GoL81], [PaM82], [ReV82]
$SC(ND_2)_2$	[MoDL79], [JaL84], [PaM84]
$Hg_{3-\delta}AsF_6$	[BrCD74], [JaJ79], [AxB82], [BuWM83], [FiR83]
$Cu_{1.8}Se$	[To81], [IwIT83], [TeIM83]
$AlPO_4$	[DuLC83]
Biphenyl	[CaMZ81], [Cu81],
Biphenyl (*cont'd*)	[PoP81], [WaNB81], [EcSP83]
K_2SeO_4	[TeTS75], [SaL78], [Ku79], [ChSH81], [Ku82], [EsL83], [Hu83], [HeHB83], [MaU83], [NoAC83], [ShVS83], [Un83], [LuHR84], [MaEG84], [SaG84],[YaI84]
$ThBr_4$	[Cu81], [BeCD83]
Quartz	[GoLK83], [BeVM84]
$RbHSeO_4$	[RoMS83]
NH_4HSeO_4	[RoMS83], [AlMR84]
$Sr_2Nb_2O_7$	[Ya82], [AkKO83], [La84]
$BaMnF_4$	[BaAS83], [CoSN83], [Go83], [Lo83], [HiSS84], [PiKM84]
$KFeF_4$	[Ma83], [SaP83]
Cu_3BiS_3	
Ag_3AsS_3	[BeGM84], [TaEH84]
p-$C_6F_4Br_2$	[OvPM80]
Barium Sodium Niobate	[ScTA81]
Polyacetylene	[Ba83], [BaM83], [HoS83]
K_2ZnCl_6 (?)	[KuKP83]
$TlInS_2$	[VaZK84]
α-ZnP_2	[ShZ84,85]
$NaNH_4C_4H_4O_6\cdot4H_2O$	[SaL77]
$RbLiSO_4$	[FrJP83]
CrMn	[GeGB82]
NiNSal	[StA83]
α-CuNSal	[AdBJ82]
Na_2CO_3	[vaAdeHP76], [dePH79], [MeHJ84]
2H–$NbSe_2$	[MuHP84]
2H–$TaSe_2$	[BaMV79], [SuMU81], [SiJD82], [DoW83], [JaPW83]
1T–$TiSe_2$	[SuYN84]
1T–VSe_2	[Mu84]
1T–$TaSe_2$	[WaW83]
2H–TaS_2	[Mu84]
1T–TaS_2	[WaM83], [Ya83], [TaST84]
NbS_3	[SaIT79]
TaS_3	[SaIT79], [HiG83], [MoRR83]
$NbSe_3$	[HiG83], [MoRR83], [JoM84]
$(TaSe_2)_4I$	[MoRR83]
V_2O_3	[SiC79]
$K_{0.30}MoO_3$	[ChSF84]

[a] Given are the first two (and further, if necessary) letters of the first author's last name, followed by the first letters of the second and third authors' names. The number refers to the year of publication.

References

Adlhart, W., H. Blauk and H. Jagodzinski, 1982, Acta Cryst. **A38**, 505.
Akishige, Y., T. Kubota and K. Ohi, 1983, J. Phys. Soc. Japan **52**, 1290.
Aleksandrova, I.P., Yu.N. Moskvich, O.V. Rozanov, A.F. Sadreev and A.A. Sukhovskii, 1984, Zh. Eksp. Teor. Fiz. Pis'ma **40**, 129.
Axe, J.D., and P. Bak, 1982, Phys. Rev. **26**, 4963.
Baeriswyl, D., 1983, Helv. Phys. Acta **56**, 639.
Baeriswyl, D., and K. Maki, 1983, Phys. Rev. **28**, 2068.
Bak, P., D. Mukamel, J. Villain and K. Wentowska, 1979, Phys. Rev. **19**, 1610.
Barthes-Regis, M., R. Almairac, P. St.-Gregoire, C. Filippini, U. Steigenberger, J. Nouet and Y. Gesland, 1983, J. Physique **44**, L829.
Beest, B.W., A. Janner and R. Blinc, 1983, J. Phys. **C16**, 5409.
Beliaev, A.D., Yu.P. Gololobov, V.F. Machulin, I.M. Miselyuk and I.M. Nekrasova, 1984, Fiz. Tverd. Tela **26**, 1349.
Berge, B., M. Vallade and G. Martinez, 1984, Ferroelectrics Lett. **1**, 135.
Bernard, L., R. Currat, P. Delamoye, C.M.E. Zeyen, S. Hubert and R. de Kouchkovsky, 1983, J. Phys. **C16**, 433.
Blinc, R., M.I. Burgar, B. Lozar and B. Topic, 1982, Phys. Rev. **26**, 6159.
Blinc, R., B. Lozar, F. Milia and R. Kind, 1984, J. Phys. **C17**, 241.
Broda, H., 1984, Phys. Stat. Sol. (a) **82**, K117.
Brown, I.O., B.D. Cutforth, C.G. Davies, R.J. Gillespie, P.R. Freland and J.E. Verkis, 1974, Can. J. Chem. **52**, 791.
Bulting, J.M., M. Weger and F.M. Mueller, 1983, Solid State Commun. **46**, 857.
Cailleau, H., F. Moussa, C.M.E. Zeyen and J. Bouillot, 1981, J. Physique **42**, 704.
Canut, M., and R. Hosemann, 1964, Acta Cryst. **17**, 973.
Chen, C.-E., Y. Schlesinger and A.J. Heeger, 1981, Phys. Rev. **24**, 5139.
Chen, C.H., L.F. Schneemeyer and R.M. Fleming, 1984, Phys. Rev. **29**, 3765.
Cox, D.E., S.M. Shapiro, R.J. Nelmes, T.W. Ryan, H.J. Bleif, R.A. Cowley, M. Eibschutz and H.J. Guggenheim, 1983, Phys. Rev. **B28**, 1640.
Currat, R., 1981, J. Physique **42**, 693.
Dam, B., A. Janner, P. Bennema, W.H. Linden and Th. Rasing, 1983, Phys. Rev. Lett. **50**, 849.
de Pater, C.J., and R.B. Helmholdt, 1979, Phys. Rev. **B19**, 5735.
Depmeier, W., 1983, Solid State Commun. **45**, 1089.
Depmeier, W., and S.A. Mason, 1982, Solid State Commun. **44**, 719.
Doran, N.J., and A.M. Wooley, 1983, J. Phys. **C16**, L675.
Durand, J., M. Lopez, L. Cot, O. Retout and P. St.-Gregoire, 1983, J. Phys. **C16**, L311.
Ecolivet, C., M. Sanguer, J. Pellegrin and J. DeWitte, 1983, J. Chem. Phys. **78**, 6317.
Esayan, S.H., and V.V. Lemanov, 1983, Izv. Akad. Nauk SSSR Ser. Fiz. **47**, 591.
Esayan, S.H., V.V. Lemanov, P. Mamatkulov and L.A. Shuvalov, 1981, Kristallografiya **26**, 1086.
Finger, W., and T.M. Rice, 1983, Phys. Rev. **B28**, 340.

Fivez, J., and K.H. Michel, 1983, Z. Phys. **B51**, 127.
Frey, F., H. Jagodzinski, A. Pietraszko and C. Zeyen, 1983, Solid State Commun. **46**, 557.
Geerken, B.M., R. Griessen, G. Benediktsson, H.U. Astrom and C. Dijk, 1982, J. Phys. **F12**, 1603.
Gesi, K., 1982, J. Phys. Soc. Japan **51**, 203.
Gesi, K., 1984, J. Phys. Soc. Japan **53**, 62.
Gesi, K., and M. Iizumi, 1980a, J. Phys. Soc. Japan **48**, 697.
Gesi, K., and M. Iizumi, 1980b, J. Phys. Soc. Japan **48**, 1775.
Gladkii, V.V., V.A. Kirikov, U.K. Magataev and L.A. Shuvalov, 1977, Fiz. Tverd. Tela **19**, 291.
Golovko, V.A., 1983, Fiz. Tverd. Tela **25**, 557.
Golovko, V.A., and A.P. Levanyuk, 1981, Fiz. Tverd. Tela **23**, 3170.
Gouhara, K., Y.H. Li, N. Kato, 1983, J. Phys. Soc. Japan **52**, 3697.
Hasebe, K., H. Mashiyama, S. Tanisaki and K. Gesi, 1982, J. Phys. Soc. Japan **51**, 1045.
Hasebe, K., H. Mashiyama, S. Tanisaki and K. Gesi, 1984, J. Phys. Soc. Japan **53**, 1863.
Hatta, I., 1984, J. Phys. Soc. Japan **53**, 635.
Henning, I., E. Hegenbarth and B. Brezina, 1983, Phys. Stat. Sol. (a) **78**, K167.
Hidaka, M., J.F. Scott and J.S. Storey, 1984, Physica **123B**, 291.
Higgs, A.W., and J.C. Gill, 1983, Solid State Commun. **47**, 737.
Hikita, T., T. Tsukahara and T. Ikeda, 1982, J. Phys. Soc. Japan **51**, 2900.
Horovitz, B., and B. Schaub, 1983, Phys. Rev. Lett. **50**, 1942.
Hoshino, S., and I. Shibuya, 1961, J. Phys. Soc. Japan **16**, 1254.
Hudak, O., 1983, J. Phys. **C16**, 2641, 2659.
Iizumi, M., 1981, Phase Transitions in Incommensurate Insulators. JAERI-M9843.
Iizumi, M., and K. Gesi, 1977, Solid State Commun. **22**, 37.
Iizumi, M., and K. Gesi, 1983a, J. Phys. Soc. Japan **52**, 2526.
Iizumi, M., and K. Gesi, 1983b, Physica **120B**, 291.
Iwazumi, T., S. Iida and H. Terauchi, 1983, J. Phys. Soc. Japan **52**, 1884.
Jacobs, A.E., P.D. Poll and M.B. Walker, 1983, Phys. Rev. **B28**, 1621.
Jakubas, R., and L. Sobczyk, 1983, Phys. Stat. Sol. **A78**, K147.
Jamet, J.P., and P. Lederer, 1984, Ferroelectrics Lett. **1**, 139.
Janner, A., and T. Janssen, 1979, Lect. Notes Phys. **4**, 82.
Joos, B., and D. Murray, 1984, Phys. Rev. **B29**, 1094.
Kind, R., 1983, Ber. Bunsenges. Phys. Chem. **87**, 248.
Kleeman, W., F.J. Schafer, T. Levola and R. Laiho, 1984, Ferroelectrics **55**, 139.
Kucharczyk, D., W. Paciorek and J. Kalieinska-Karut, 1982, Phase Transitions **2**, 277.
Kudo, S., 1982, J. Phys. Soc. Japan **21**, 255.
Kugler, W., K. Kuorr and W. Prandl, 1983, Solid State Commun. **47**, 163.
Kumar, J.A., 1979, Solid State Commun. **31**, 237.
Lavrencic, B.B., 1984, Ferroelectrics **53**, 265.
Levanyuk, A.P., and D.G. Sannikov, 1976, Fiz. Tverd. Tela (USSR) **18**, 423.
Levanyuk, A.P., and D.G. Sannikov, 1977, Fiz. Tverd. Tela (USSR) **18**, 1927.
Levstik, A., P. Prelovšek, B. Žekš, C. Filipić and A. Kanduser, 1984, Ferroelectrics **53**, 269.
Lorenc, J., 1983, J. Phys. **C16**, 1807.
Lupsin, Y., G. Hauret, A.M. Robinet and J.P. Benoit, 1984, J. Phys. **C17**, 2203.
Makovicky, E., 1983, J. Solid State Chem. **49**, 85.
Marion, G., R. Almairac, J. Lefebvre and M. Ribet, 1981, J. Phys. **C14**, 3177.
Mashiyama, H., and S. Tanisaki, 1980, Phys. Lett. **76A**, 347.
Mashiyama, H., and H.-G. Unruh, 1983, J. Phys. **C16**, 5009.
Mashiyama, H., S. Tanisaki and K. Hamano, 1981, J. Phys. Soc. Japan **50**, 2139.
Massa, N., P. Echegut and F. Gervais, 1984, Ferroelectrics **53**, 281.
Matsunaga, H., 1982a, J. Phys. Soc. Japan **51**, 864.
Matsunaga, H., 1982b, J. Phys. Soc. Japan **51**, 873.
Meekes, H., K. Hanssen, A. Janner, T. Janssen, P. Wyder and Th. Rasing, 1984, Ferroelectrics **53**, 285.

Melnikova, S.V., and A.T. Anistratov, 1983, Fiz. Tverd. Tela **25**, 848.
Melnikova, S.V., I.M. Iskornev, L.I. Zherebtsova, L.A. Kot and A.T. Anistratov, 1984, Fiz. Tverd. Tela **26**, 776.
Moncean, R., M. Renard, J. Richard, M.C. Saint-Lager, H. Salva and Z.Z. Wang, 1983, Phys. Rev. **B28**, 1646.
Moudden, A.H., F. Dénoyer, M. Lambert and W. Fitzgerald, 1979, Solid State Commun. **32**, 933.
Muralt, P., and R. Kind, 1983, Helv. Phys. Acta **56**, 689.
Muralt, P., R. Kind, R. Blinc and B. Žekš, 1982, Phys. Rev. Lett. **49**, 1019.
Mutka, H., N. Houssean, J. Pelissier, R. Ayroles and C. Roncan, 1984, Solid State Commun. **50**, 161.
Nomoto, K., T. Atake, B.K. Chaudhuri and H. Chihara, 1983, J. Phys. Soc. Japan **52**, 3475.
Osaka, T., M. Komukae and Y. Makita, 1982, J. Phys. Soc. Japan **51**, 3409.
Overal, J.S.W., G.S. Pewley, G.A. Mackenzie, W.J.J. Fitzgerald and D. Kirin, 1980, Chem. Phys. Lett. **74**, 373.
Parlinski, K., and K.H. Michel, 1982, Rap Inst. Fiz. Jadr. Krakov N 1164øS.
Parlinski, K., and K.H. Michel, 1984a, Phys. Rev. **B29**, 396.
Parlinski, K., and K.H. Michel, 1984b, Phys. Rev. **B29**, 410.
Pisarev, R.V., B.B. Kaichevizov, P.A. Markovin, O.Yu. Korshunov and J.F. Scott, 1984, Ferroelectrics **56**, 99.
Poulet, H., and R.M. Pick, 1984, J. Physique **42**, 701.
Prelovšek, P., and R. Blinc, 1984, J. Phys. **C17**, 577.
Rasing, Th., P. Wyder, A. Janner and T. Janssen, 1982, Phys. Rev. **B25**, 7504.
Rehwald, W., and A. Vonlanthen, 1982, J. Phys. **C15**, 5361.
Rozanov, O.V., Yu.N. Moskvich, A.A. Sukhovskii and I.P. Aleksandrova, 1983, Izv. Akad. Nauk SSSR Ser. Fiz. **47**, 719.
Sakai, A., M. Kasahara and I. Tatsuzaki, 1979, J. Phys. Soc. Japan **47**, 161.
Sambongi, T., M. Ido, K. Tsutsumo, M. Yamamoto, T. Takoshima and A. Abe, 1979, Lect. Notes Phys. **95**, 349.
Sannikov, D.G., and V.A. Golovko, 1984, Kristallografiya **26**, 1117.
Sannikov, D.G., and A.P. Levanyuk, 1977, Fiz. Tverd. Tela **19**, 118.
Sannikov, D.G., and A.P. Levanyuk, 1978, Fiz. Tverd. Tela **20**, 1005.
Sato, T., M. Endo, T. Osaka and Y. Makita, 1982, J. Phys. Soc. Japan **51**, 3411.
Sato, T., T. Osaka and Y. Makita, 1983, J. Phys. Soc. Japan **52**, 3297.
Schneck, J., J.C. Tolédano, J. Aubrée, B. Joukoff and C. Joffrin, 1981, J. Physique **42**, 734.
Sheleg, A.U., and V.V. Zaretskii, 1984a, Zh. Eksp. Teor. Fiz. Pis'ma **39**, 166.
Sheleg, A.U., and V.V. Zaretskii, 1984b, Ferroelectrics, in press.
Shkuryaev, P.G., A.N. Vtyurin and V.F. Shabanov, 1983, Kristallografiya **28**, 1140.
Shuvalov, L.A., S.A. Gridnev, B.N. Prasolov and V.G. Sannikov, 1983, Ferroelectrics **1**, 85.
Simpson, A.M., M.H. Jericho and F.J. DiSalvo, 1982, Solid State Commun. **44**, 1543.
Sinclair, F., and R. Colella, 1979, Solid State Commun. **31**, 359.
St.-Gregoire, P., and A. Ferez, 1983, Z. Kristallogr. **163**, 135.
Steurer, W., and W. Adlhart, 1983, Acta Crystallogr. **B39**, 349.
Straube, U., D. Grutzmann and G. Sorge, 1983, Acta Phys. Slovaca **33**, 315.
Strukhov, B.A., V.M. Arutiunova and U. Uesu, 1982, Fiz. Tverd. Tela **24**, 3061.
Sugai, S., K. Murase, S. Uchida and S. Tanake, 1981, J. Physique **42**, 728.
Suzuki, N., A. Yamamoto and K. Notizuki, 1984, Solid State Commun. **49**, 1039.
Tanda, S., T. Sambongi, T. Tani and S. Tanaka, 1984, J. Phys. Soc. Japan **53**, 476.
Tanisaki, S., 1961, J. Phys. Soc. Japan **16**, 579.
Taylor, W., P.J.S. Ewen and T. Han, 1984, Ferroelectrics **55**, 83.
Terauchi, H., H. Takenaka and K. Shimaoka, 1975, J. Phys. Soc. Japan **39**, 435.
Terauchi, H., S. Iida, S. Motomi, H. Maeda and T. Ishii, 1983, J. Phys. Soc. Japan **52**, 1882.
Tonejc, A., 1981, Recent Dev. Condens. Matter Phys. **4**, 337.
Ueda, T., S. Iida and H. Terauchi, 1982, J. Phys. Soc. Japan **51**, 3953.

Uesu, Y., J. Ogawa, N. Hanami, B.A. Strukov and J. Kobayashi, 1981, Phys. Stat. Sol. (a) **67**, 693.
Unruh, H.-G., 1983, J. Phys. **C16**, 3245.
Vakhrushev, S.B., V.V. Zhdanova, B.E. Kvyatkovskii, N.M. Okuneva, K.R. Allakhverdiev, R.A. Aliev and R.M. Sardarly, 1984, Zh. Eksp. Teor. Fiz. Pis'ma **39**, 245.
van Aalst, W., J. den Hollander, W.J.A.M. Peterse and P.M. de Wolff, 1976, Acta Cryst. **B32**, 47.
Vinogradova, I.S., and A.I. Roslyakov, 1979, Kristallografiya **24**, 284.
Vtyurin, A.N., and P.G. Shkuryaev, 1983, Izv. Akad. Nauk SSSR. Ser. Fiz **47**, 727.
Walker, M.B., and R.L. Withers, 1983, Phys. Rev. **B28**, 2766.
Warczewskii, J., and D. Kucharczyk, 1982, Phase Transitions **2**, 255.
Wasiutynski, T., I. Natkaniec and A.I. Belushkin, 1981, J. Physique **42**, 599.
Yagi, T., Y. Hidaka and K. Miura, 1981, J. Physique **42**, 731.
Yamada, N., and T. Ikeda, 1984, Ferroelectrics **53**, 323.
Yamada, Y., and N. Hamaya, 1983, J. Phys. Soc. Japan **52**, 3466.
Yamamoto, A., 1983, Phys. Rev. **B27**, 7823.
Yamamoto, N., 1982, Acta Crystallogr. **A38**, 780.

AUTHOR INDEX

Abe, A., *see* Sambongi, T. 369
Abe, N., *see* Shimizu, H. 26, 27, 29, 66, 67
Abe, R., *see* Horioka, M. 65
Abrahams, S.C., *see* Jamieson, P.B. 235, 240
Abrahams, S.C., *see* Keve, E.T. 285
Achard, M.F., *see* Hardouin, F. 364
Achard, M.F., *see* Levelut, A.M. 364
Achard, M.F., *see* Sigaud, G. 364
Adlhart, W. 369
Adlhart, W., *see* Steurer, W. 369
Aiki, K. 6, 60, 61, 62, 67
Ailion, D.C., *see* Blinc, R. 189, 193
Ainger, W., *see* Schneck, J. 250
Aizu, K. 208, 277
Akishige, Y. 369
Albuquerque, E.L. 295
Aleksandrova, I.P. 107, 369
Aleksandrova, I.P., *see* Rozanov, O.V. 369
Aliev, R.A., *see* Vakhrushev, S.B. 369
Allakhverdiev, K.R., *see* Vakhrushev, S.B. 369
Allen, G., *see* Takeuchi, H. 76, 82
Almairac, R., *see* Barthes-Regis, M. 286, 289, 292, 293, 294, 369
Almairac, R., *see* Marion, G. 31, 43, 369
Almlöf, J. 81
Als-Nielsen, J., *see* Moudden, A.H. 107, 137, 141
Al'Shin, B.I. 297
Ambrosetti, R. 107
Amelinckx, S., *see* Dolino, G. 215, 216, 217
Amelinckx, S., *see* Van Tendeloo, G. 209, 216, 217
Anderson, P.W., *see* Meiboom, S. 324
André, G. 132, 154, 173
Andrews, S.R. 142
Angelone, R., *see* Ambrosetti, R. 107
Anistratov, A.T., *see* Melnikova, S.V. 369
Arnold, H. 212, 219, 227
Arnould, H., *see* Rondelez, F. 344
Arutiunova, V.M., *see* Strukhov, B.A. 369
Aslanyan, T.A. 140, 207, 209, 215, 219, 220, 223, 225
Astrom, H.U., *see* Geerken, B.M. 369
Astrov, D.N., *see* Al'Shin, B.I. 297
Atake, T. 56, 85, 88, 89
Atake, T., *see* Nomoto, K. 56, 57, 369
Atake, T., *see* Saito, K. 97
Aubrée, J., *see* Errandonéa, G. 244, 247, 248, 249
Aubrée, J., *see* Schneck, J. 236, 237, 238, 239, 240, 250, 369
Aubrey, S., *see* Axel, F. 45
Aubry, S. 136, 146
Aubry, S., *see* Axel, F. 291
Auderset, H., *see* Steigmeier, E.F. 173
Auld, B.A. 294
Axe, J.D. 38, 46, 116, 177, 194, 198, 369
Axe, J.D., *see* de Pater, C.J. 13, 14, 22, 45, 136, 291
Axe, J.D., *see* Harada, J. 18
Axe, J.D., *see* Iizumi, M. 3, 6, 9, 12, 13, 16, 17, 18, 19, 21, 22, 42, 45, 51, 52, 63, 107, 143, 291
Axe, J.D., *see* Majkrzak, C.F. 34, 36
Axe, J.D., *see* Moncton, D.E. 3, 42, 143
Axe, J.D., *see* Moudden, A.H. 153
Axe, J.D., *see* Press, W. 23, 67
Axe, J.P. 219, 225, 227
Axel, F. 45, 291
Ayroles, R., *see* Mutka, H. 369

Bachheimer, J.P. 208, 209, 210, 211, 212, 219, 227
Bachheimer, J.P., *see* Dolino, G. 207, 208, 209, 210, 211, 212, 213, 215, 216, 217, 218, 225, 226
Bachheimer, J.P., *see* Zeyen, C.M.E. 209
Baeriswyl, D. 369
Bagley, B.G., *see* Disalvo, F.J. 237
Baikalov, V.A., *see* Beresnev, L.A. 362
Bailey, D.M., *see* Mason, J.T. 166
Bak, P. 42, 43, 68, 215, 226, 291, 369
Bak, P., *see* Axe, J.D. 369
Bak, P., *see* Hoegh-Jensen, M. 142
Ballantyne, J.M., *see* Barnoski, M.K. 111
Barnik, M.I. 327, 348, 349, 351
Barnoski, M.K. 111
Barois, P., *see* Prost, J. 325, 362, 364
Barreto, M. 142, 153, 154
Barthes-Regis, M. 286, 289, 292, 293, 294, 369
Bastiansen, O. 81
Bastie, P. 215
Bates, J.B., *see* Brooker, M.H. 275
Baudour, J.L. 76, 82, 84
Baudour, J.L., *see* Cailleau, H. 73, 82, 83, 84, 86
Bauer, K.H.W. 227
Becht, H.Y. 256, 257, 262
Beest, B.W. 369
Beliaev, A.D. 369
Bellamy, A., *see* Dénoyer, F. 143, 146, 147, 149, 152
Belushkin, A.I., *see* Wasiutynski, T. 369
Benediktsson, G., *see* Geerken, B.M. 369
Bennema, P., *see* Dam, B. 369
Benoit, J.P. 134, 149
Benoit, J.P., *see* Cao, A.X. 151
Benoit, J.P., *see* Lupsin, Y. 369
Benoit, J.P., *see* Moudden, A.H. 156
Beresnev, L.A. 359, 362
Berge, B. 212, 221, 223, 227, 228, 369
Berge, B., *see* Bachheimer, J.P. 219
Berge, B., *see* Dolino, G. 212, 213, 215, 216, 217, 225
Berger, J., *see* Benoit, J.P. 149
Bernard, L. 165, 166, 169, 171, 177, 178, 179, 180, 196, 197, 369
Bernard, L., *see* Delamoye, P. 170, 183
Bernard, L., *see* Durand, D. 103, 106, 115, 119, 120, 121, 123, 124, 125, 141, 226
Bernstein, J.L., *see* Jamieson, P.B. 235, 240
Bernstein, J.L., *see* Keve, E.T. 285
Bethke, J., *see* Arnold, H. 219
Bielushkin, A.V., *see* Natkaniec, I. 76, 82
Bielushkin, A.V., *see* Plakida, N.M. 82, 83
Bielushkin, A.V., *see* Wasiutynski, T. 76, 82
Bijk, P. 258, 262
Bilz, H., *see* Bussmann-Holder, A. 16
Blaschke, H., *see* Meissner, G. 173
Blauk, H., *see* Adlhart, W. 369
Bleif, H., *see* Cox, D.E. 286, 287, 289
Bleif, H.J., *see* Cox, D.E. 369
Blinc, R. 107, 189, 193, 275, 309, 369
Blinc, R., *see* Aleksandrova, I.P. 107
Blinc, R., *see* Beest, B.W. 369
Blinc, R., *see* Levstik, A. 287, 289, 360, 361
Blinc, R., *see* Muralt, P. 306, 307, 310, 312, 314, 369
Blinc, R., *see* Muševič, I. 361
Blinc, R., *see* Prelovšek, P. 369
Blinov, L.M., *see* Barnik, M.I. 327, 348, 349, 351
Blinov, L.M., *see* Beresnev, L.A. 359, 362
Bobylev, Y.P. 347, 348, 349
Böhm, H. 108
Boissier, M., *see* Berge, B. 212, 221, 227
Bonadeo, H. 82
Bonadeo, H., *see* Burgos, E. 82
Boudou, A. 236, 239
Bouillot, J., *see* Cailleau, H. 76, 79, 83, 85, 86, 89, 90, 91, 92, 196, 369
Boysen, H. 227
Bradley, C.J. 306
Brazovskii, S.A. 324, 330, 332, 333, 334
Bree, A. 73, 95
Bréhat, F. 134, 156
Brezina, B., *see* Henning, I. 369
Briat, B. 189
Bridgman, P.W. 145
Brillard, L., *see* Hussonnois, M. 166
Brinkman, W.F., *see* Meiboom, S. 324, 325, 333, 334
Brock, C.P. 81
Brockway, L.O., *see* Karle, I.L. 81
Broda, H. 369
Brooker, M.H. 275

Brouns, E. 255, 257
Brown, D. 166, 167
Brown, I.O. 369
Bruce, A.D. 13, 34, 42, 46, 114, 115, 141, 142, 195, 226
Bruce, A.D., *see* Cowley, R.A. 106, 137
Bruce, A.D., *see* Majkrzak, C.F. 34, 36
Brunskill, I.H. 315
Buchheit, W. 107, 122
Buchheit, W., *see* Ishibashi, Y. 107, 122, 141
Bugaut, F., *see* Cailleau, H. 99
Bulou, A. 181
Bulting, J.M. 369
Burgar, M.I., *see* Blinc, R. 369
Burgeat, J. 236
Burgos, E. 82
Burgos, E., *see* Bonadeo, H. 82
Busch, M., *see* Tolédano, J.C. 236, 244
Busing, W.R. 81, 82, 83
Bussmann-Holder, A. 16
Büttner, H., *see* Bussmann-Holder, A. 16

Cailleau, H. 73, 76, 79, 82, 83, 84, 85, 86, 89, 90, 91, 92, 99, 196, 369
Cailleau, H., *see* Baudour, J.L. 82
Cailleau, H., *see* Dworkin, A. 89
Cailleau, H., *see* Toudic, B. 82
Calvarin, G., *see* Schneck, J. 158, 237, 244, 245, 246, 248
Calvo, C. 133
Candille, M., *see* Regis, M. 288, 289
Cantor, S. 256
Canut, M. 369
Cao, A.X. 151
Caravatti, P. 307
Carlier, R., *see* Hussonnois, M. 166
Casalone, G. 81
Cerisier, P. 257, 279
Chang, S.S. 136
Chanvin, M., *see* Lederer, P. 125
Chapelle, J.P., *see* Benoit, J.P. 134
Chapelle, J.P., *see* Cao, A.X. 151
Chaplot, S.L., *see* Kirin, D. 84
Chapuis, G. 303
Charbonneau, G.P. 81, 82
Chaudhuri, B.K., *see* Atake, T. 56
Chaudhuri, B.K., *see* Nomoto, K. 56, 57, 369
Chaves, C.M., *see* Lederer, P. 118, 139, 151
Chen, C.-E. 369
Chen, C.H. 217, 369
Chernevich, J.G., *see* Shustin, O.A. 218
Chiang, S., *see* Fleury, P.A. 39
Chiba, A., *see* Futama, H. 131
Chigrinov, V.G., *see* Bobylev, Y.P. 348, 349
Chigrinov, V.G., *see* Ostrovskii, B.I. 361
Chihara, H., *see* Atake, T. 56, 85, 88, 89
Chihara, H., *see* Nomoto, K. 56, 57, 369
Chihara, H., *see* Saito, K. 97
Chiotti, P., *see* Mason, J.T. 165, 166
Christmann, M. 255, 257
Cizikov, S., *see* Levstik, A. 287, 289
Clark, N., *see* Liao, Y. 342
Clark, N.A. 342
Claudel, J., *see* Bréhat, F. 134, 156
Cohen, L.H., *see* Klement Jr, W. 256
Colella, R., *see* Sinclair, F. 369
Colligiani, A., *see* Ambrosetti, R. 107
Comès, R. 136
Conway, J.G., *see* Delamoye, P. 182, 183, 188
Conway, J.G., *see* Genet, M. 181
Cot, L., *see* Durand, J. 369
Courtens, E. 288
Coutinho-Filho, M.D. 151
Coutinho-Filho, M.D., *see* Yokoi, C.S.O. 151
Cowley, J.A., *see* Sakurai, J. 111
Cowley, R., *see* Bruce, A.D. 195
Cowley, R.A. 79, 106, 137, 164
Cowley, R.A., *see* Bruce, A.D. 13, 34, 42, 46, 114, 115, 142, 226
Cowley, R.A., *see* Cox, D.E. 286, 289, 294, 295, 369
Cox, D.E. 286, 287, 289, 294, 295, 369
Cracknell, A.P., *see* Bradley, C.J. 306
Crooijmans, P.C.B. 274, 279
Cruickshank, D.W.J., *see* Kalman, A. 6, 7
Cullick, A.S. 73
Cummins, H.Z., *see* Qiu, S.L. 119
Cummins, H.Z., *see* Shapiro, S.M. 208, 209, 227
Currat, R. 139, 369
Currat, R., *see* André, G. 154, 173
Currat, R., *see* Bernard, L. 165, 166, 169, 171, 177, 178, 179, 180, 196, 197, 369
Currat, R., *see* de Pater, C.J. 13, 14, 22, 45,

136, 291
Currat, R., *see* Delamoye, P. 170, 182, 183, 184, 186, 188
Currat, R., *see* Dénoyer, F. 143, 146, 147, 149, 152
Currat, R., *see* Durand, D. 103, 106, 115, 119, 120, 121, 123, 124, 125, 141, 153, 154, 155, 215, 226
Currat, R., *see* Quilichini, M. 41, 196
Currat, R., *see* Simonson, T. 142, 158
Cutforth, B.D., *see* Brown, I.O. 369

D'Alessio, E., *see* Burgos, E. 82
Dam, B. 369
Davies, C.G., *see* Brown, I.O. 369
De Gennes, P.G. 327, 340, 343, 344, 358
De Jong, F. 257, 262, 264, 270
De Kouchkovsky, R. 169
De Kouchkovsky, R., *see* Bernard, L. 165, 166, 169, 171, 177, 178, 179, 180, 196, 369
De Moura, M.A., *see* Coutinho-Filho, M.D. 151
De Pater, C.J. 13, 14, 21, 22, 45, 136, 258, 263, 264, 271, 272, 273, 274, 276, 279, 291, 369
De Witte, J., *see* Ecolivet, C. 77, 79, 80, 90, 97
De Wolff, P.M. 269
De Wolff, P.M., *see* Brouns, E. 255, 257
De Wolff, P.M., *see* Dubbeldam, G.C. 262, 268
De Wolff, P.M., *see* Hogervorst, A.C.R. 43, 170
De Wolff, P.M., *see* Hogervorst, A.J. 268
De Wolff, P.M., *see* Janner, A. 79
De Wolff, P.M., *see* van Aalst, W. 257, 265, 266, 276, 369
Delahaigue, A. 157
Delahaigue, A., *see* Khelifa, B. 156
Delamoye, P. 170, 182, 183, 184, 185, 186, 188
Delamoye, P., *see* Bernard, L. 165, 166, 169, 171, 177, 178, 179, 180, 196, 197, 369
Delamoye, P., *see* Briat, B. 189
Delamoye, P., *see* de Kouchkovsky, R. 169
Delamoye, P., *see* Genet, M. 181
Delamoye, P., *see* Hubert, S. 165, 168, 169, 170, 182, 198
Delaye, M. 342
Delrieu, J.M. 341
Delugeard, Y., *see* Charbonneau, G.P. 81, 82
Den Hollander, J., *see* van Aalst, W. 257, 265, 266, 276, 369
Deniau, M., *see* Benoit, J.P. 134
Dénoyer, F. 135, 137, 143, 146, 147, 149, 152
Dénoyer, F., *see* André, G. 132, 154, 173
Dénoyer, F., *see* Durand, D. 103, 106, 115, 119, 120, 121, 123, 124, 125, 136, 141, 153, 154, 155, 215, 226
Dénoyer, F., *see* Moudden, A.H. 135, 139, 146, 156, 369
Dénoyer, F., *see* Schneck, J. 136, 235, 236, 237, 238, 239, 250
Dénoyer, F., *see* Simonson, T. 142, 158
Depmeier, W. 303, 304, 306, 309, 314, 315, 316, 369
Depmeier, W., *see* Brunskill, I.H. 315
Dewar, M.J.S. 81
DeWitte, J., *see* Ecolivet, C. 369
Dianoux, A.J., *see* Takeuchi, H. 76, 82
Dijk, C., *see* Geerken, B.M. 369
Disalvo, F.J. 237
DiSalvo, F.J., *see* Moncton, D.E. 3, 42, 143
DiSalvo, F.J., *see* Simpson, A.M. 369
Dmitriev, S.G., *see* Brazovskii, S.A. 324, 330, 332, 333, 334
Dolino, G. 140, 207, 208, 209, 210, 211, 212, 213, 215, 216, 217, 218, 225, 226
Dolino, G., *see* Bachheimer, J.P. 208, 209, 219
Dolino, G., *see* Bastie, P. 215
Dolino, G., *see* Berge, B. 212, 221, 227
Dolino, G., *see* Zeyen, C.M.E. 209
Dolling, G., *see* Sakurai, J. 111
Doran, N.J. 369
Dorner, B. 76
Dorner, B., *see* Bauer, K.H.W. 227
Dorner, B., *see* Boysen, H. 227
Dorner, B., *see* Grimm, H. 208, 225
Draegert, D.A., *see* Singh, S. 235
Dryden, J.S., *see* MacKenzie, D.R. 131
Dubbeldam, G.C. 262, 268
Durand, D. 103, 106, 115, 119, 120, 121, 123, 124, 125, 136, 141, 153, 154, 155, 215, 226
Durand, D., *see* André, G. 132, 154, 173
Durand, G., *see* Delaye, M. 342

Durand, G., *see* Ribotta, R. 342
Durand, J. 369
Duxbury, P.M., *see* Selke, W. 124
Dvořàk, V. 79, 89
Dvořák, V. 41, 45, 61, 63, 223, 294, 295
Dvořák, V., *see* Ishibashi, Y. 61, 77
Dvoryankin, V.F. 133
Dworkin, A. 89
Dworkin, A., *see* Cailleau, H. 73, 82, 83
Dzyaloshinskii, I.E. 142, 324

Echegut, P., *see* Massa, N. 369
Eckold, G., *see* Arnold, H. 219
Ecolivet, C. 77, 79, 80, 90, 97, 218, 219, 369
Edelson, M., *see* Bree, A. 73, 95
Edelstein, N., *see* Delamoye, P. 182, 183, 186, 188
Edelstein, N., *see* Genet, M. 181
Ehrhardt, H. 262
Ehrhardt, K.D. 118
Eibschutz, M., *see* Cox, D.E. 286, 287, 289, 294, 295, 369
Eibschutz, M., *see* Glass, A.M. 295
Eibschutz, M., *see* Holmes, L.M. 295
Elcombe, M.M. 132, 133, 134
Eller, W., *see* Unruh, H.-G. 65
Ema, K. 111
Ema, K., *see* Hamano, K. 111, 293
Emery, J. 189, 191, 192
Emery, J., *see* Hubert, S. 189, 191, 193, 194
Emery, V.J., *see* Bak, P. 42
Endo, M., *see* Sato, T. 369
Endoh, Y., *see* Itoh, H. 14, 26, 32
Englman, R. 297
Errandonéa, G. 244, 247, 248, 249
Errandonéa, G., *see* Hebbache, M. 243
Errandonéa, G., *see* Tolédano, J.C. 188
Esayan, S.H. 369
Esayan, S.Kh. 114, 293
Esayan, S.Kh., *see* Dvořák, V. 294
Etingen, N.B., *see* Beresnev, L.A. 362
Evesque, P., *see* Briat, B. 189
Ewen, P.J.S., *see* Taylor, W. 369
Eysel, W. 255, 257, 262

Fayet, J.C., *see* Emery, J. 189, 191, 192
Fayet, J.C., *see* Hubert, S. 189, 191, 193, 194
Feigel'man, M.V. 142
Felsche, J., *see* Depmeier, W. 315
Ferez, A., *see* St.-Gregoire, P. 369
Figuière, P. 145
Filev, V.M., *see* Brazovskii, S.A. 324
Filev, V.M., *see* Vigman, P.B. 334, 335, 338, 339
Filipic, C., *see* Levstik, A. 45, 62, 287, 360, 361, 369
Filippini, C., *see* Barthes-Regis, M. 286, 289, 292, 293, 294, 369
Finger, W. 369
Finger, W., *see* Zeyher, R. 40, 89, 195
Fisher, D., *see* Fisher, M.E. 142
Fisher, M.E. 142
Fitzgerald, W., *see* Moudden, A.H. 135, 156, 369
Fitzgerald, W.J.J., *see* Overal, J.S.W. 369
Fivez, J. 119, 369
Fleming, R.M., *see* Chen, C.H. 217, 369
Fleury, P.A. 39, 239
Fortuin, C.M. 267
Fousek, J. 288, 289
Fousek, J., *see* Glogarova, M. 361
Fox, D.L. 295, 297, 298
Franchini, P. 81
Frazer, B.C., *see* Kay, M.I. 104
Freland, P.R., *see* Brown, I.O. 369
Frey, F. 369
Frey, F., *see* Boysen, H. 227
Friedman, P.S. 73
Fritz, I.J. 293
Fujii, Y., *see* Yamada, Y. 112
Fujimoto, S., *see* Shimizu, H. 26, 27, 29, 66, 67
Fujimoto, T., *see* Hamano, K. 293
Fung, K.K. 217
Furuichi, J., *see* Futama, H. 131
Furuichi, J., *see* Mitsui, T. 131
Futama, H. 131, 133, 135, 150

Gabelotaud, A., *see* Schneck, J. 236, 237, 238, 239, 240
Gallier, J., *see* Toudic, B. 82
Garito, A.F., *see* Comès, R. 136
Garland, C.W. 226
Garoff, S. 353, 359
Gatebois, L., *see* Moudden, A.H. 146
Gatehouse, B.M. 262

Geerken, B.M. 369
Genet, M. 181
Genet, M., *see* Delamoye, P. 182, 186
Genet, M., *see* Hussonnois, M. 166
Gerkin, R.E., *see* Cullick, A.S. 73
Gerritsma, C.J. 344
Gerritsma, C.J., *see* Rondelez, F. 344
Gervais, F., *see* Dolino, G. 208, 209
Gervais, F., *see* Massa, N. 369
Gesi, K. 13, 14, 21, 23, 29, 31, 51, 119, 145, 149, 152, 291, 369
Gesi, K., *see* Hasebe, K. 369
Gesi, K., *see* Iizumi, M. 13, 14, 22, 23, 31, 43, 52, 170, 369
Gesi, K., *see* Itoh, H. 14, 26, 32
Gesi, K., *see* Mashiyama, H. 31
Gesi, K., *see* Wada, M. 21
Gesland, J.-Y., *see* Barthes-Regis, M. 286, 289, 292, 293, 294
Gesland, Y., *see* Barthes-Regis, M. 369
Geusic, J.E., *see* Singh, S. 235
Gibson, J.M., *see* Chen, C.H. 217
Gill, J.C., *see* Higgs, A.W. 369
Gillespie, R.J., *see* Brown, I.O. 369
Ginzburg, V.L. 209
Girard, A. 95
Girard, A., *see* Cailleau, H. 82, 89
Gladkii, V.V. 369
Glass, A.M. 295
Glogarova, M. 361
Goldsmith, G.J. 131, 133
Gololobov, Yu.P., *see* Beliaev, A.D. 369
Golovko, V.A. 40, 41, 61, 62, 89, 195, 369
Golovko, V.A., *see* Sannikov, D.G. 369
Gordon, M.B., *see* Villain, J. 291
Gorelik, V.S. 111
Gottlieb, K., *see* Takahashi, H.B. 134
Gouhara, K. 212, 215, 218, 223, 225, 369
Gray, G.W. 365
Grebel, H. 324, 333
Gridnev, S.A., *see* Shuvalov, L.A. 369
Griessen, R., *see* Geerken, B.M. 369
Grimm, H. 208, 225
Grimm, H., *see* Bauer, K.H.W. 227
Grimm, H., *see* Boysen, H. 227
Grutzmann, D., *see* Straube, U. 369
Guelfenstein, M., *see* Figuière, P. 145
Guggenheim, H.J., *see* Cox, D.E. 286, 287, 289, 294, 295, 369
Guggenheim, H.J., *see* Fox, D.L. 295, 297, 298
Guggenheim, H.J., *see* Glass, A.M. 295
Guggenheim, H.J., *see* Holmes, L.M. 295
Guggenheim, H.J., *see* Lyons, K.B. 289
Guggenheim, H.J., *see* Spencer, E.G. 286
Guibé, L., *see* Khan Malek, C. 189, 190, 191
Guilluy, M., *see* Toledano, P. 79

Habbal, F., *see* Scott, J.F. 289, 290
Hadni, A., *see* Bréhat, F. 134, 156
Hahn, Th., *see* Arnold, H. 219
Hall, T.L., *see* Brown, D. 167
Ham, N.S., *see* MacKenzie, D.R. 134
Hamano, K. 111, 119, 124, 158, 188, 217, 293
Hamano, K., *see* Ema, K. 111
Hamano, K., *see* Hatta, I. 114
Hamano, K., *see* Mashiyama, H. 173, 188, 369
Hamano, K., *see* Mashiyama, M. 244
Hamaya, N., *see* Yamada, Y. 43, 286, 291, 292, 293, 369
Han, T., *see* Taylor, W. 369
Hanami, M., *see* Hatta, I. 114
Hanami, N., *see* Uesu, Y. 369
Hanssen, K., *see* Meekes, H. 369
Hao Li, Y., *see* Gouhara, K. 212, 215, 218
Haque, M.S. 16, 17, 20, 67
Harada, J. 18
Harada, J., *see* Axe, J.P. 219
Harada, M. 108
Harada, M., *see* Axe, J.D. 116
Hardouin, F. 364
Hardouin, F., *see* Levelut, A.M. 364
Hardouin, F., *see* Sigaud, G. 364
Hardy, J.R., *see* Haque, M.S. 16, 17, 20, 67
Hardy, J.R., *see* Press, W. 23, 67
Hargreaves, A. 75, 81
Hartwig, C.M. 111
Hasebe, K. 25, 369
Hasebe, K., *see* Mashiyama, H. 24, 25
Hashimoto, T., *see* Midorikawa, M. 274, 276
Hatta, I. 110, 111, 112, 113, 114, 369
Hatta, I., *see* Ema, K. 111
Hatta, I., *see* Hoshizaki, H. 61, 62
Hauret, G., *see* Cao, A.X. 151
Hauret, G., *see* Lupsin, Y. 369

Heatherly, D.E., *see* Cantor, S. 256
Hebbache, M. 243
Hebbache, M., *see* Errandonéa, G. 244
Heeger, A.J., *see* Chen, C.-E. 369
Heeger, A.J., *see* Comès, R. 136
Hegenbarth, E., *see* Henning, I. 369
Heide, K., *see* Jäger, C. 278, 279
Heine, V. 103, 108, 116, 117, 119, 138, 174
Heine, V., *see* McConnell, J.D.C. 174
Helmholdt, R.B., *see* de Pater, C.J. 271, 273, 274, 276, 279, 369
Henning, I. 369
Herget, A.J., *see* Dewar, M.J.S. 81
Herth, G., *see* Buchheit, W. 107
Hertogs, P.H.S. 256, 257, 258
Hewat, A.W., *see* Bulou, A. 181
Hidaka, M. 290, 369
Hidaka, M., *see* Scott, J.F. 289, 290
Hidaka, Y., *see* Yagi, T. 114, 369
Higgs, A.W. 369
Hikita, T. 369
Hill, A.E., *see* Waldeck, W.F. 277
Hirotsu, S., *see* Hamano, K. 293
Ho, J.T., *see* Pindak, R.S. 339
Höchli, U.T. 208
Hochstrasser, R.M. 75
Hoegh-Jensen, M. 142
Hogervorst, A.C.R. 43, 170, 275
Hogervorst, A.J. 268
Holah, G.D. 111
Holmes, L.M. 295
Horioka, M. 65
Hornreich, R.M. 117, 324, 333, 334
Hornreich, R.M., *see* Grebel, H. 324, 333
Horovitz, B. 369
Hosemann, R., *see* Canut, M. 369
Hoshino, S. 105, 369
Hoshino, S., *see* Yamada, Y. 105, 107
Hoshino, T., *see* Iio, K. 297
Hoshizaki, H. 61, 62
Hosoya, M., *see* Mitsui, T. 131
Hosoya, M., *see* Shiozaki, Y. 135
Houssean, N., *see* Mutka, H. 369
Hsu, F.S.L., *see* Glass, A.M. 295
Huang, C.C., *see* Pindak, R.S. 339
Hubert, S. 165, 168, 169, 170, 182, 189, 191, 193, 194, 198
Hubert, S., *see* Bernard, L. 165, 166, 169, 171, 180, 196, 369
Hubert, S., *see* Briat, B. 189
Hubert, S., *see* Emery, J. 189, 191, 192
Hudak, O. 68, 369
Hudak, O., *see* Dvořák, V. 61
Hukuda, K., *see* Aiki, K. 60, 61, 62, 67
Hukudu, K., *see* Aiki, K. 6
Hulin, J., *see* Rondelez, F. 344, 345
Huse, D.A. 293
Hussonnois, M. 166
Hussonnois, M., *see* Hubert, S. 165, 168, 169, 170, 182, 198

Ichikawa, M., *see* Mitsui, T. 131
Ida, S., *see* Ueda, T. 22
Ido, M., *see* Sambongi, T. 369
Iida, A., *see* Inoue, N. 218
Iida, S., *see* Iwazumi, T. 369
Iida, S., *see* Terauchi, H. 369
Iida, S., *see* Ueda, T. 291, 369
Iio, K. 297
Iizumi, M. 3, 6, 9, 12, 13, 14, 16, 17, 18, 19, 21, 22, 23, 31, 32, 42, 43, 45, 51, 52, 63, 107, 143, 170, 291, 369
Iizumi, M., *see* Gesi, K. 13, 14, 21, 23, 31, 51, 152, 291, 369
Iizumi, M., *see* Itoh, H. 14, 26, 32
Ikeda, T., *see* Hikita, T. 369
Ikeda, T., *see* Kudo, S. 67
Ikeda, T., *see* Yamada, N. 19, 20, 369
Ikeda, Y., *see* Hamano, K. 293
Ikushima, A., *see* Hatta, I. 111
Indenbom, V.L., *see* Pikin, S.A. 353, 354, 359
Inoue, K., *see* Hidaka, M. 290
Inoue, N. 218
Ishibashi, Y. 54, 55, 56, 60, 61, 77, 79, 103, 107, 118, 122, 139, 141, 151, 157, 238, 240, 241
Ishibashi, Y., *see* Dvořàk, V. 79, 89
Ishibashi, Y., *see* Dvořák, V. 223
Ishibashi, Y., *see* Hoshizaki, H. 61, 62
Ishibashi, Y., *see* Midorikawa, M. 274, 276, 277, 278, 279
Ishibashi, Y., *see* Petzelt, J. 65
Ishibashi, Y., *see* Sawada, A. 23, 45, 51, 62, 63
Ishibashi, Y., *see* Shimizu, H. 114
Ishibashi, Y., *see* Shiozaki, S. 7, 36

Ishibashi, Y., *see* Sugiyama, J. 60, 291
Ishibashi, Y., *see* Wada, M. 21, 38, 39, 63, 64, 65, 67, 95, 96
Ishibashi, Y., *see* Wada, M.A. 157
Ishii, T., *see* Terauchi, H. 369
Iskornev, I.M., *see* Melnikova, S.V. 369
Itoh, H. 14, 26, 32
Ivanov, S.A., *see* Shustin, O.A. 218
Iwasaki, H., *see* Yamada, T. 236
Iwata, Y., *see* Harada, M. 108
Iwazumi, T. 369

Jacobs, A.E. 369
Jacoby, Y. 173
Jäger, C. 278, 279
Jagodzinski, H., *see* Adlhart, W. 369
Jagodzinski, H., *see* Bauer, K.H.W. 227
Jagodzinski, H., *see* Frey, F. 369
Jakubas, R. 369
Jamet, J.P. 136, 150, 152, 153, 158, 246, 249, 369
Jamet, J.P., *see* Barreto, M. 142, 153, 154
Jamet, J.P., *see* Lederer, P. 125
Jamet, J.P., *see* Regis, M. 37
Jamieson, P.B. 235, 240
Janner, A. 79, 369
Janner, A., *see* Beest, B.W. 369
Janner, A., *see* Dam, B. 369
Janner, A., *see* Meekes, H. 369
Janner, A., *see* Rasing, Th. 369
Janovec, V. 217
Janovec, V., *see* Dvořàk, V. 79
Janovec, V., *see* Dvořák, V. 223
Janssen, T., *see* Janner, A. 79, 369
Janssen, T., *see* Meekes, H. 369
Janssen, T., *see* Rasing, Th. 369
Jericho, M.H., *see* Simpson, A.M. 369
Jha, M.C., *see* Mason, J.T. 165, 166
Joffrin, C., *see* Schneck, J. 236, 237, 238, 239, 240, 250, 369
Joos, B. 369
Joukoff, B., *see* Schneck, J. 235, 236, 237, 238, 239, 240, 249, 250, 369
Jouve, P., *see* Delahaigue, A. 157
Jouve, P., *see* Khelifa, B. 156

Kabalkina, S.S. 146
Kadaba, P., *see* Levstik, A. 287, 289
Kahn, F. 343
Kahn, F.J. 344, 345
Kaichevizov, B.B., *see* Pisarev, R.V. 369
Kalieinska-Karut, J., *see* Kucharczyk, D. 369
Kalman, A. 6, 7
Kanduser, A., *see* Levstik, A. 369
Karataeva, I.M. 256, 258
Karle, I.L. 81
Kasahara, M., *see* Sakai, A. 369
Kassubek, P. 328
Kato, N., *see* Gouhara, K. 212, 215, 218, 223, 225, 369
Kawasaki, X. 327
Kay, M.I. 104
Keller, P., *see* Meyer, R.B. 352, 359
Keve, E.T. 285
Khan Malek, C. 189, 190, 191
Khelifa, B. 156
Khelifa, B., *see* Delahaigue, A. 157
Kiat, J.M., *see* Schneck, J. 158, 237, 244, 245, 246, 248
Kim, Q., *see* Haque, M.S. 67
Kind, R. 303, 307, 369
Kind, R., *see* Blinc, R. 369
Kind, R., *see* Muralt, P. 306, 307, 310, 312, 314, 369
Kirf, G., *see* Unruh, H.-G. 65
Kirikov, V.A., *see* Gladkii, V.V. 369
Kirin, D. 84
Kirin, D., *see* Overal, J.S.W. 369
Kittel, C. 16
Kizhaev, A. 295, 298
Kleeman, W. 288, 369
Kleeman, W., *see* Schafer, F.J. 286, 288, 289, 296, 297, 298
Klein, M.L. 119
Klein, M.L., *see* Lynden-Bell, R. 103, 119
Klement Jr, W. 256
Klimowski, J. 145, 149
Kobayashi, J., *see* Uesu, Y. 369
Kobayashi, T., *see* Aiki, K. 6, 60, 61, 62, 67
Koga, H., *see* Aiki, K. 6, 60, 61, 62, 67
Kohra, K., *see* Inoue, N. 218
Kokubo, N., *see* Shimizu, H. 26, 29, 66, 67
Kolkman, G.H., *see* Crooijmans, P.C.B. 274, 279
Kominiak, G.J., *see* Spencer, E.G. 286
Komukae, M., *see* Osaka, T. 369

Kopelman, R., *see* Friedman, P.S. 73
Korshunov, O.Yu., *see* Pisarev, R.V. 286, 288, 289, 294, 369
Kot, L.A., *see* Melnikova, S.V. 369
Kovalev, O.V. 9, 135
Koyano, N., *see* Harada, M. 108
Kozlov, G.V. 295, 296
Kozlov, G.V., *see* Petzelt, J. 65
Krichene, S., *see* Cao, A.X. 151
Krichevtzov, B.B., *see* Pisarev, R.V. 286, 288, 289, 294
Krüger, J.K., *see* Rehwald, W. 62
Kruhler, W.W., *see* Jacoby, Y. 173
Krupa, J.C., *see* Delamoye, P. 170, 182, 183, 188
Krupa, J.C., *see* Hussonnois, M. 166
Kubo, R., *see* Suzuki, M. 103, 112, 114
Kubota, T., *see* Akishige, Y. 369
Kucharczyk, D. 103, 105, 108, 117, 124, 369
Kucharczyk, D., *see* Warczewskii, J. 369
Kudo, S. 21, 23, 67, 369
Kugler, M., *see* Hornreich, R.M. 334
Kugler, W. 369
Kulczycki, A., *see* Natkaniec, I. 82
Kumar, J.A. 369
Kunchur, N.R. 133
Kuorr, K., *see* Kugler, W. 369
Kvyatkovskii, B.E., *see* Vakhrushev, S.B. 369

Laiho, R., *see* Kleeman, W. 369
Lajzérowicz, J., *see* Aslanyan, T.A. 209, 215, 220, 223
Lambert, M., *see* Dénoyer, F. 135, 137, 143, 146, 147, 149, 152
Lambert, M., *see* Durand, D. 106, 115, 119, 125, 215
Lambert, M., *see* Moudden, A.H. 135, 139, 146, 369
Lamborelle, C., *see* Tolédano, J.C. 240, 241
Landau, L.D. 310
Lavrencic, B.B. 287, 369
Lax, M., *see* Nelson, D.F. 294
Le Chatelier, H. 208
Le Cloarec, M.F., *see* de Kouchkovsky, R. 169
Le Fur, D., *see* Durand, D. 226
Leadbetter, A.J., *see* Gray, G.W. 365
Lederer, P. 118, 125, 139, 151
Lederer, P., *see* Barreto, M. 142, 153, 154
Lederer, P., *see* Jamet, J.P. 152, 153, 158, 246, 249, 369
Lefebvre, J., *see* Marion, G. 31, 43, 369
Lefrant, S., *see* Hubert, S. 165, 168, 169, 170, 182, 198
Lefur, D., *see* Durand, D. 103, 119, 120, 121, 123, 124, 141
Lehmann, M.J., *see* Wright, A.F. 208
Lejček, L., *see* Glogarova, M. 361
Lemanov, V.V., *see* Esayan, S.H. 369
Lemanov, V.V., *see* Esayan, S.Kh. 114, 293
Leonidova, G.G. 145
Lepeshkov, I.L., *see* Karataeva, I.M. 256, 258
Lepostollec, M., *see* Hubert, S. 165, 168, 169, 170, 182, 198
Levanyuk, A.P. 55, 61, 77, 103, 108, 115, 116, 119, 140, 142, 156, 207, 219, 238, 240, 244, 287, 310, 311, 369
Levanyuk, A.P., *see* Aslanyan, T.A. 140, 207, 209, 215, 219, 220, 223, 225
Levanyuk, A.P., *see* Ginzburg, V.L. 209
Levanyuk, A.P., *see* Golovko, V.A. 40, 41, 89, 195, 369
Levanyuk, A.P., *see* Sannikov, D.G. 369
Levelut, A.M. 364
Levelut, A.M., *see* Hardouin, F. 364
Levelut, A.M., *see* Sigaud, G. 364
Levin, E.M., *see* Schneider, S.J. 256
Levola, T., *see* Kleeman, W. 369
Levstik, A. 45, 62, 287, 289, 360, 361, 369
Levstik, I., *see* Levstik, A. 287, 360, 361
Li, Y.H., *see* Gouhara, K. 369
Liao, Y. 342
Liebert, L., *see* Meyer, R.B. 352, 359
Lifshitz, I.M., *see* Landau, L.D. 310
Linden, W.H., *see* Dam, B. 369
Lindeyer, G. 278
Lines, M.E., *see* Glass, A.M. 295
Litster, J.D., *see* Ribotta, R. 342
Lloyd, D.J., *see* Gatehouse, B.M. 262
Lopez, M., *see* Durand, J. 369
Lorenc, J. 369
Lozar, B., *see* Blinc, R. 369
Luban, M., *see* Hornreich, R.M. 117
Lukaszewicz, K., *see* Kucharczyk, D. 103, 105, 108, 117, 124

Lupsin, Y. 369
Lynden-Bell, R. 103, 119
Lynden-Bell, R.M., *see* Heine, V. 103, 117, 119
Lynn, G., *see* Waldeck, W.F. 277
Lyons, K., *see* Fleury, P.A. 239
Lyons, K.B. 289
Lyons, K.B., *see* Fleury, P.A. 39
Lyuksyutov, I.F., *see* Feigel'man, M.V. 142

Machulin, V.F., *see* Beliaev, A.D. 369
Maciel, A. 271, 275, 276, 280
MacKenzie, D.R. 131, 134, 135, 139
Mackenzie, G.A., *see* Kirin, D. 84
Mackenzie, G.A., *see* Overal, J.S.W. 369
Maeda, H., *see* Terauchi, H. 369
Magataev, U.K., *see* Gladkii, V.V. 369
Majkrzak, C.F. 34, 36
Majkrzak, C.F., *see* Press, W. 23, 67
Maki, K., *see* Baeriswyl, D. 369
Makita, Y., *see* Osaka, T. 369
Makita, Y., *see* Sato, T. 369
Makovicky, E. 369
Malov, Y.U. 209, 216
Mamatkulov, N., *see* Esayan, S.Kh. 114, 293
Mamatkulov, P., *see* Esayan, S.H. 369
Manolikas, C. 236, 237
Marakov, S.Z. 256
Mariani, C., *see* Casalone, G. 81
Marion, G. 31, 43, 369
Markovin, P.A., *see* Pisarev, R.V. 286, 288, 289, 294, 369
Martinez, G., *see* Berge, B. 227, 228, 369
Mashiyama, H. 21, 23, 24, 25, 26, 31, 36, 43, 45, 107, 173, 188, 369
Mashiyama, H., *see* Andrews, S.R. 142
Mashiyama, H., *see* Hasebe, K. 25, 369
Mashiyama, H., *see* Tanisaki, S. 24, 52
Mashiyama, M. 244
Mason, J.T. 165, 166
Mason, S.A., *see* Depmeier, W. 303, 306, 315, 316, 369
Massa, N. 369
Massa, N.E., *see* Press, W. 23, 67
Massa, N.E., *see* Samara, G.A. 83
Matsuda, T., *see* Hoshizaki, H. 61, 62
Matsumura, O., *see* Aiki, K. 6
Matsunaga, H. 14, 43, 369
Matsuo, M., *see* Sawada, S. 23
Mayer, J., *see* Natkaniec, I. 82
McAlpine, R.D., *see* Hochstrasser, R.M. 75
McConnell, J.D.C. 108, 174
McConnell, J.D.C., *see* Heine, V. 103, 108, 116, 117, 119, 138, 174
McDonald, I.R., *see* Heine, V. 103, 117, 119
McDonald, I.R., *see* Klein, M.L. 119
McDonald, I.R., *see* Lynden-Bell, R. 103, 119
McKernan, S., *see* Fung, K.K. 217
McMillan, W.L. 42, 141, 142, 238
Meekes, H. 369
Megaw, H. 208
Meiboom, S. 324, 325, 333, 334
Meier, G., *see* Kassubek, P. 328
Meier, W., *see* Takahashi, H.B. 134
Meinnel, J., *see* Cailleau, H. 73, 82, 83
Meissner, G. 173
Mellet, R., *see* Schneck, J. 235, 249, 250
Melnikova, S.V. 369
Menyhárd, N., *see* Meissner, G. 173
Messager, J.C. 81
Messager, J.C., *see* Cailleau, H. 99
Meyer, R.B. 345, 352, 359
Meyer, R.B., *see* Garoff, S. 353, 359
Mezei, F., *see* Durand, D. 115
Michel, K.H. 109, 118
Michel, K.H., *see* Ehrhardt, K.D. 118
Michel, K.H., *see* Fivez, J. 119, 369
Michel, K.H., *see* Parlinski, K. 140, 144, 369
Michelson, A. 139
Midorikawa, M. 58, 59, 67, 274, 276, 277, 278, 279
Mikheeva, L.F., *see* Yakolev, I.A. 209, 216, 218
Milia, F., *see* Blinc, R. 369
Min Hua, J., *see* Arnold, H. 219
Miselyuk, I.M., *see* Beliaev, A.D. 369
Mitani, S., *see* Harada, M. 108
Mitsui, T. 131
Mitsui, T., *see* Futama, H. 131
Miura, K., *see* Yagi, T. 114, 369
Moncean, R. 369
Moncton, D.E. 3, 42, 143
Moncton, D.E., *see* Moudden, A.H. 153
Mons, J., *see* Cailleau, H. 73, 79, 85
Montambaux, G., *see* Lederer, P. 125

Morgenthaler, F.R., *see* Venturini, E.L. 295
Moseley, P.T., *see* Brown, D. 167
Moskvich, Yu.N., *see* Aleksandrova, I.P. 369
Moskvich, Yu.N., *see* Rozanov, O.V. 369
Motegi, H., *see* Hoshino, S. 105
Motegi, H., *see* Mitsui, T. 131
Motomi, S., *see* Terauchi, H. 369
Moudden, A.H. 107, 135, 137, 139, 141, 146, 152, 153, 156, 157, 369
Moudden, A.H., *see* Dénoyer, F. 135, 137, 143, 146, 147, 149, 152
Moudden, A.H., *see* Jamet, J.P. 136, 152, 153
Moussa, F., *see* André, G. 154, 173
Moussa, F., *see* Cailleau, H. 73, 76, 79, 82, 83, 85, 86, 89, 90, 91, 92, 99, 196, 369
Moussa, F., *see* Durand, D. 136, 141
Mueller, F.M., *see* Bulting, J.M. 369
Mugnoli, A., *see* Casalone, G. 81
Mukamel, D., *see* Bak, P. 215, 226, 369
Muralt, P. 305, 306, 307, 308, 309, 310, 312, 314, 315, 316, 317, 318, 369
Murase, K., *see* Sugai, S. 369
Murray, A.F., *see* Bruce, A.D. 42, 115, 142, 226
Murray, D., *see* Joos, B. 369
Muševič, I. 361
Mutka, H. 369
Muto, S., *see* Ueno, Y. 109

Nagata, K., *see* Iio, K. 297
Nakamura, E., *see* Matsunaga, H. 43
Nakamura, E., *see* Mitsui, T. 131
Nakamura, N., *see* Tanisaki, S. 135
Nakamura, T., *see* Takashige, M. 63
Nartowski, M., *see* Natkaniec, I. 82
Nassau, K., *see* Pouget, J.P. 219
Natkaniec, I. 76, 82
Natkaniec, I., *see* Plakida, N.M. 82, 83
Natkaniec, I., *see* Wasiutynski, T. 76, 82, 369
Nedungadi, T.M.K., *see* Raman, C.V. 226
Negran, T.J., *see* Lyons, K.B. 289
Nekrasova, I.M., *see* Beliaev, A.D. 369
Nelmes, R.J. 286
Nelmes, R.J., *see* Cox, D.E. 286, 287, 289, 369
Nelson, D.F. 294
Niizeki, R., *see* Yamada, T. 236
Noda, Y. 106
Nomoto, K. 56, 57, 369
Nomoto, K., *see* Atake, T. 56
Notizuki, K., *see* Suzuki, N. 369
Nouet, J., *see* Barthes-Regis, M. 286, 289, 292, 293, 294, 369
Nouet, J., *see* Bulou, A. 181
Nouet, J., *see* Kleeman, W. 288

Ogawa, J., *see* Uesu, Y. 369
Oguchi, T., *see* Ueno, Y. 109
Oguri, A., *see* Shimizu, H. 26
Ohi, K., *see* Akishige, Y. 369
Okuneva, N.M., *see* Vakhrushev, S.B. 369
Osaka, T. 369
Osaka, T., *see* Sato, T. 369
O'Shea, D.C., *see* Shapiro, S.M. 208, 227
Osipov, M.A. 360
Osipov, V.V., *see* Levanyuk, A.P. 287
Ostlund, S. 293
Ostrovskii, B.I. 359, 360, 361
Overal, J.S.W. 369
Overhauser, A.W. 177
Ozaki, Y., *see* Klein, M.L. 119
Ozawa, K., *see* Gesi, K. 119
Ozgo, D., *see* Klimowski, J. 145, 149

Paciorek, W., *see* Kucharczyk, D. 369
Pannetier, J., *see* Bernard, L. 177, 178, 179
Papin, G. 256, 257
Papin, G., *see* Christmann, M. 255, 257
Parlinski, K. 140, 144, 369
Pateau, L., *see* Tolédano, J.C. 237
Pavel, J., *see* Glogarova, M. 361
Pavlyuchenko, A.M., *see* Beresnev, L.A. 362
Pawley, G.S., *see* Kirin, D. 84
Pelissier, J., *see* Mutka, H. 369
Pellegrin, J., *see* Ecolivet, C. 77, 79, 80, 90, 97, 369
Péneau, A., *see* Khan Malek, C. 189, 190, 191
Pershan, P.S. 364
Pershan, P.S., *see* Clark, N.A. 342
Pershan, P.S., *see* Liao, Y. 342
Peterse, W.J.A.M., *see* van Aalst, W. 257, 265, 266, 276, 369
Peterson, E.R. 306
Petersson, J., *see* Buchheit, W. 107, 122
Petersson, J., *see* Ishibashi, Y. 107, 122, 141

Petitgrand, D., *see* Delamoye, P. 170, 183
Petzelt, J. 65, 156
Petzelt, J., *see* Dvořák, V. 41, 45, 61, 63
Petzelt, J., *see* Fousek, J. 288, 289
Pewley, G.S., *see* Overal, J.S.W. 369
Pick, R., *see* Poulet, H. 41
Pick, R.M., *see* Poulet, H. 369
Pietrasko, A., *see* Kucharczyk, D. 103, 105, 108, 117, 124
Pietraszko, A., *see* Frey, F. 369
Pikin, S.A. 327, 330, 342, 353, 354, 355, 358, 359
Pikin, S.A., *see* Bobylev, Y.P. 347, 348, 349
Pikin, S.A., *see* Osipov, M.A. 360
Pikin, S.A., *see* Ostrovskii, B.I. 361
Pikin, S.A., *see* Terent'ev, E.M. 348, 351
Pindak, R.S. 339
Pisarev, R.V. 286, 288, 289, 294, 369
Plakida, N.M. 82, 83
Plesko, S., *see* Kind, R. 307
Poignant, H., *see* Ecolivet, C. 218, 219
Pokrovsky, V.L. 142
Poll, P.D., *see* Jacobs, A.E. 369
Porto, S.P.S., *see* Hartwig, C.M. 111
Pouget, J.P. 219
Poulet, H. 41, 369
Pozhydaev, E.P., *see* Beresnev, L.A. 362
Prandl, W., *see* Kugler, W. 369
Prasad, P.N., *see* Friedman, P.S. 73
Prasolov, B.N., *see* Shuvalov, L.A. 369
Prelovšek, P. 61, 62, 369
Prelovšek, P., *see* Blinc, R. 189, 193
Prelovšek, P., *see* Levstik, A. 45, 62, 369
Press, W. 23, 67
Primot, J., *see* Schneck, J. 235, 250
Prost, J. 325, 362, 364
Prost, J., *see* Pershan, P.S. 364
Purvanezskas, G.V., *see* Beresnev, L.A. 362

Qiu, S.L. 119
Quilichini, M. 41, 196
Quittet, A.M., *see* Jamet, J.P. 136

Rabinovich, A.Z., *see* Ostrovskii, B.I. 359, 360, 361
Rajnak, K., *see* Delamoye, P. 182, 186
Raman, C.V. 226
Randa, J. 293
Rapoport, E. 119
Rasing, T., *see* Muševič, I. 361
Rasing, Th. 369
Rasing, Th., *see* Dam, B. 369
Rasing, Th., *see* Meekes, H. 369
Ravez, J., *see* Schneck, J. 235, 250
Regis, M. 37, 288, 289
Rehwald, W. 62, 369
Reinitzer, F. 321
Reisman, A. 255, 256, 258
Renard, M., *see* Moncean, R. 369
Retout, O., *see* Durand, J. 369
Ribet, J.L., *see* Regis, M. 37
Ribet, M., *see* Marion, G. 31, 43, 369
Ribotta, R. 342
Ribotta, R., *see* Delaye, M. 342
Rice, T.M., *see* Finger, W. 369
Richard, J., *see* Moncean, R. 369
Richards, P.M., *see* Samara, G.A. 287, 295, 298
Ridou, C., *see* Bulou, A. 181
Rigamonti, A., *see* Aleksandrova, I.P. 107
Rigamonti, A., *see* Ambrosetti, R. 107
Rivet, P., *see* Toudic, B. 82
Rivoal, J.C., *see* Briat, B. 189
Rizvi, S.H., *see* Hargreaves, A. 75, 81
Robertson, G.B. 75, 81
Robinet, A.M., *see* Lupsin, Y. 369
Roncan, C., *see* Mutka, H. 369
Rondelez, F. 344, 345
Roos, J., *see* Kind, R. 307
Roslyakov, A.I., *see* Vinogradova, I.S. 369
Rousseau, J.J., *see* Hubert, S. 189, 191, 193, 194
Rousseau, M., *see* Bulou, A. 181
Roux, F., *see* Cerisier, P. 257, 279
Rozanov, O.V. 369
Rozanov, O.V., *see* Aleksandrova, I.P. 369
Rutar, V., *see* Blinc, R. 189, 193
Ryan, J.F. 286, 287
Ryan, J.F., *see* Maciel, A. 271, 275, 276, 280
Ryan, T.W., *see* Cox, D.E. 286, 287, 289, 369

Sadreev, A.F., *see* Aleksandrova, I.P. 369
Saint-Lager, M.C., *see* Moncean, R. 369
Saito, K. 97
Saito, K., *see* Atake, T. 85, 88, 89
Sakai, A. 369

Sakudo, T., *see* Wada, M. 39, 64, 65
Sakurai, J. 111
Salinas, S.R., *see* Yokoi, C.S.O. 151
Salva, H., *see* Moncean, R. 369
Samara, G.A. 83, 287, 295, 298
Sambongi, T. 369
Sambongi, T., *see* Tanda, S. 369
Sammon, M., *see* Meiboom, S. 324, 325, 333, 334
Sanguer, M., *see* Ecolivet, C. 369
Sannikov, D.G. 61, 62, 103, 118, 369
Sannikov, D.G., *see* Levanyuk, A.P. 55, 61, 77, 103, 108, 115, 116, 119, 140, 142, 156, 207, 219, 310, 311, 369
Sannikov, G.D., *see* Levanyuk, A.P. 238, 240, 244
Sannikov, V.G., *see* Shuvalov, L.A. 369
Sanquer, M., *see* Baudour, J.L. 76, 84
Sanquer, M., *see* Ecolivet, C. 77, 79, 80, 90, 97
Sapriel, J., *see* Boudou, A. 236, 239
Sardarly, R.M., *see* Vakhrushev, S.B. 369
Sato, T. 369
Savary, H. 244
Savary, H., *see* Errandonéa, G. 244, 247, 248, 249
Sawada, A. 23, 45, 51, 62, 63
Sawada, A., *see* Horioka, M. 65
Sawada, A., *see* Hoshizaki, H. 61, 62
Sawada, A., *see* Ishibashi, Y. 55, 60
Sawada, A., *see* Shiozaki, S. 7, 36
Sawada, A., *see* Sugiyama, J. 60, 291
Sawada, A., *see* Wada, M. 21, 38, 39, 63, 64, 65, 67, 95, 96
Sawada, A., *see* Wada, M.A. 157
Sawada, S. 23
Sawada, S., *see* Shimizu, H. 26, 27, 29, 66, 67
Sawada, S., *see* Takashige, M. 63
Sbrana, G., *see* Franchini, P. 81
Schader, B., *see* Takahashi, H.B. 134
Schafer, F.J. 286, 288, 289, 296, 297, 298
Schafer, F.J., *see* Kleeman, W. 288, 369
Schaub, B., *see* Horovitz, B. 369
Schlesinger, Y., *see* Chen, C.-E. 369
Schmid, A. 337
Schneck, J. 136, 158, 235, 236, 237, 238, 239, 240, 241, 242, 244, 245, 246, 248, 249, 250, 369
Schneck, J., *see* Errandonéa, G. 244, 247, 248, 249
Schneck, J., *see* Tolédano, J.C. 188, 236, 238, 240, 241, 243, 244
Schneemeyer, L.F., *see* Chen, C.H. 369
Schneider, S.J. 256
Schweer, H., *see* Ehrhardt, H. 262
Scott, J.F. 208, 227, 244, 287, 289, 290, 293, 295, 298
Scott, J.F., *see* Fox, D.L. 295, 297, 298
Scott, J.F., *see* Hidaka, M. 290, 369
Scott, J.F., *see* Höchli, U.T. 208
Scott, J.F., *see* Kozlov, G.V. 295, 296
Scott, J.F., *see* Lavrencic, B.B. 287
Scott, J.F., *see* Pisarev, R.V. 286, 288, 289, 294, 369
Scott, J.F., *see* Randa, J. 293
Scott, J.F., *see* Ryan, J.F. 286, 287
Scott, J.F., *see* Samara, G.A. 295
Scott, J.F., *see* Tilley, D.R. 295, 298
Seehra, M., *see* Scott, J.F. 295
Seidel, H., *see* Ehrhardt, H. 262
Sekido, T., *see* Mitsui, T. 131
Selke, W. 124
Sethna, J.P., *see* Meiboom, S. 324
Shabanov, V.F., *see* Shkuryaev, P.G. 369
Shapiro, S.M. 208, 209, 227
Shapiro, S.M., *see* Comès, R. 136
Shapiro, S.M., *see* Cox, D.E. 286, 287, 289, 294, 295, 369
Shapiro, S.M., *see* Pouget, J.P. 219
Shapiro, S.M., *see* Qiu, S.L. 119
Sheldrick, G.M. 177
Sheleg, A.U. 369
Shiba, H., *see* Ishibashi, Y. 77, 103, 118, 139, 141, 151
Shibukawa, K., *see* Mitsui, T. 131
Shibuya, I., *see* Harada, M. 108
Shibuya, I., *see* Hoshino, S. 369
Shibuya, I., *see* Yamada, Y. 105, 107
Shichi, H., *see* Wada, M. 67
Shimaoka, K., *see* Iizumi, M. 3, 6, 9, 12, 13, 16, 17, 18, 19, 21, 22, 42, 45, 51, 52, 63, 107, 143, 291
Shimaoka, K., *see* Terauchi, H. 13, 369
Shimizu, H. 26, 27, 29, 66, 67, 114
Shiozaki, S. 7, 36
Shiozaki, Y. 135, 139

Shiozaki, Y., *see* Futama, H. 131
Shiozaki, Y., *see* Mitsui, T. 131
Shirane, G., *see* Axe, J.D. 116
Shirane, G., *see* Axe, J.P. 219, 225, 227
Shirane, G., *see* Comès, R. 136
Shirane, G., *see* Harada, J. 18
Shirane, G., *see* Iizumi, M. 3, 6, 9, 12, 13, 16, 17, 18, 19, 21, 22, 42, 45, 51, 52, 63, 107, 143, 291
Shirane, G., *see* Moudden, A.H. 152
Shiroichi, Y., *see* Shimizu, H. 26, 27
Shiroishi, Y., *see* Sawada, S. 23
Shkuryaev, P.G. 369
Shkuryaev, P.G., *see* Vtyurin, A.N. 369
Shtrikman, S., *see* Grebel, H. 324, 333
Shtrikman, S., *see* Hornreich, R.M. 117, 324, 333, 334
Shulgina, M.P., *see* Marakov, S.Z. 256
Shustin, O.A. 218
Shuvalov, L.A. 369
Shuvalov, L.A., *see* Esayan, S.H. 369
Shuvalov, L.A., *see* Esayan, S.Kh. 293
Shuvalov, L.A., *see* Gladkii, V.V. 369
Siapkas, D.I. 156
Sigaud, G. 364
Sigaud, G., *see* Hardouin, F. 364
Sigaud, G., *see* Levelut, A.M. 364
Sigov, A.S., *see* Levanyuk, A.P. 287
Simonetta, M., *see* Casalone, G. 81
Simonson, T. 142, 158
Simpson, A.M. 369
Sinclair, F. 369
Singh, S. 235
Sobczyk, L., *see* Jakubas, R. 369
Sobyanin, A.A., *see* Levanyuk, A.P. 287
Sokolova, E.B., *see* Beresnev, L.A. 359
Solomon, L. 131
Sonin, A.S., *see* Ostrovskii, B.I. 359, 360, 361
Sonyushkin, V.E., *see* Malov, Y.U. 209, 216
Sorge, G., *see* Straube, U. 369
Sosman, R.B. 208
Spencer, E.G. 286
Srishchinskii, M.M., *see* Gorelik, V.S. 111
St.-Gregoire, P. 369
St.-Grégoire, P., *see* Bachheimer, J.P. 219
St.-Gregoire, P., *see* Barthes-Regis, M. 286, 289, 292, 293, 294, 369
St.-Grégoire, P., *see* Dolino, G. 213, 225
St.-Gregoire, P., *see* Durand, J. 369
St.-Gregoire, P., *see* Regis, M. 288, 289
Steeds, J.W., *see* Fung, K.K. 217
Steigenberger, U., *see* Barthes-Regis, M. 286, 289, 292, 293, 294, 369
Steigmeier, E.F. 173
Steiner, J.C., *see* Qiu, S.L. 119
Steinwehr, E.H. 212
Stephens, J.S., *see* Kalman, A. 6, 7
Steurer, W. 369
Storey, J.S., *see* Hidaka, M. 369
Straube, U. 369
Strimer, P., *see* Bréhat, F. 134, 156
Struikmans, R., *see* Becht, H.Y. 256, 257, 262
Strukhov, B.A. 369
Strukov, B.A., *see* Ostrovskii, B.I. 359, 360, 361
Strukov, B.A., *see* Uesu, Y. 369
Strzelecki, L., *see* Meyer, R.B. 352, 359
Sudnik-Hrynkiewicz, M., *see* Natkaniec, I. 82
Sugai, S. 369
Sugiyama, J. 60, 291
Sugiyama, J., *see* Ishibashi, Y. 55, 60
Sugiyama, J., *see* Sawada, A. 23, 45, 51, 62, 63
Sukhovskii, A.A., *see* Aleksandrova, I.P. 369
Sukhovskii, A.A., *see* Rozanov, O.V. 369
Suzuki, H. 81
Suzuki, M. 103, 112, 114
Suzuki, M., *see* Wada, M. 21
Suzuki, N. 369
Suzuki, S., *see* Takeuchi, H. 76, 82
Svensson, E.C., *see* Moudden, A.H. 152
Sźepfalusy, P., *see* Meissner, G. 173
Szwarc, H., *see* Figuière, P. 145

Tajbakhsh, A.R., *see* Gray, G.W. 365
Takagi, Y., *see* Gesi, K. 119
Takagi, Y., *see* Ishibashi, Y. 157
Takagi, Y., *see* Midorikawa, M. 274, 276
Takagi, Y., *see* Shiozaki, S. 7, 36
Takagi, Y., *see* Wada, M. 38, 39, 63, 64, 65
Takagi, Y., *see* Wada, M.A. 157
Takahashi, H.B. 134
Takama, T., *see* Mitsui, T. 131
Takashige, M. 63

Takashige, M., *see* Sawada, S. 23
Takashige, M., *see* Shimizu, H. 26
Takenaka, H., *see* Terauchi, H. 13, 369
Takeuchi, H. 76, 82
Takoshima, T., *see* Sambongi, T. 369
Talapov, A.L., *see* Pokrovsky, V.L. 142
Tanaka, E., *see* Futama, H. 131
Tanaka, S., *see* Tanda, S. 369
Tanake, S., *see* Sugai, S. 369
Tanda, S. 369
Tani, T., *see* Tanda, S. 369
Tanisaki, S. 24, 52, 103, 105, 135, 369
Tanisaki, S., *see* Hasebe, K. 25, 369
Tanisaki, S., *see* Mashiyama, H. 23, 24, 25, 26, 31, 43, 173, 188, 369
Tanizaki, S., *see* Mashiyama, M. 244
Taraskin, S.A., *see* Ostrovskii, B.I. 359
Tarento, R.J., *see* Levelut, A.M. 364
Tatsuzaki, I., *see* Sakai, A. 369
Taylor, J.C., *see* Elcombe, M.M. 132, 133, 134
Taylor, W. 369
Terauchi, H. 13, 369
Terauchi, H., *see* Iwazumi, T. 369
Terauchi, H., *see* Ueda, T. 22, 291, 369
Terent'ev, E.M. 348, 351
Thomas, F., *see* Benoit, J.P. 149
Tilley, D.R. 295, 298
Tilley, D.R., *see* Albuquerque, E.L. 295
Tilley, D.R., *see* Fox, D.L. 295, 297, 298
Tilley, D.R., *see* Scott, J.F. 295
Tolédano, J.C. 188, 236, 237, 238, 240, 241, 243, 244
Tolédano, J.C., *see* Burgeat, J. 236
Tolédano, J.C., *see* Errandonéa, G. 244, 247, 248, 249
Tolédano, J.C., *see* Savary, H. 244
Tolédano, J.C., *see* Schneck, J. 236, 237, 238, 239, 240, 250, 369
Toledano, P. 79
Tonejc, A. 369
Topic, B., *see* Aleksandrova, I.P. 107
Topic, B., *see* Blinc, R. 369
Toudic, B. 82
Traetteberg, M., *see* Bastiansen, O. 81
Trotter, J. 75, 81
Trufanov, A.N., *see* Barnik, M.I. 327, 348, 349, 351
Truter, M.R. 133
Truter, M.R., *see* Kunchur, N.R. 133
Tsuboi, T., *see* Schafer, F.J. 286, 288, 289, 296, 297, 298
Tsukahara, T., *see* Hikita, T. 369
Tsukamoto, M., *see* Shimizu, H. 114
Tsunoda, K., *see* Midorikawa, M. 276, 277, 278, 279
Tsutsumo, K., *see* Sambongi, T. 369
Tucker, P.A., *see* Gray, G.W. 365
Tuinstra, F., *see* Crooijmans, P.C.B. 274, 279
Tuinstra, F., *see* de Jong, F. 262, 264
Tuinstra, F., *see* de Wolff, P.M. 269

Uchida, S., *see* Sugai, S. 369
Ueda, R., *see* Kay, M.I. 104
Ueda, T. 22, 291, 369
Ueno, Y. 109
Uesu, U., *see* Strukhov, B.A. 369
Uesu, Y. 369
Ullman, F.G., *see* Haque, M.S. 67
Ullman, F.G., *see* Press, W. 23, 67
Ullman, F.G., *see* Samara, G.A. 83
Umanski, B.A., *see* Barnik, M.I. 327, 348, 349, 351
Ume, H., *see* Wada, M. 64, 65
Umeno, M., *see* Shimizu, H. 114
Unruh, H.-G. 65, 158, 246, 369
Unruh, H.-G., *see* Mashiyama, H. 369
Unruh, H.-G., *see* Rehwald, W. 62
Uwe, H., *see* Wada, M. 39

Vacher, R., *see* Berge, B. 212, 221, 227
Vainshtein, B.K., *see* Dvoryankin, V.F. 133
Vakhrushev, S.B. 369
Vallade, M. 223
Vallade, M., *see* Aslanyan, T.A. 209, 215, 220, 223
Vallade, M., *see* Berge, B. 212, 221, 227, 228, 369
Van Aalst, W. 257, 265, 266, 276, 369
Van den Berg, A.J. 276
Van Dijk, C., *see* de Pater, C.J. 13, 14, 21, 22
Van Landuyt, J., *see* Dolino, G. 215, 216, 217
Van Landuyt, J., *see* Van Tendeloo, G. 209, 216, 217
Van Tendeloo, G. 209, 216, 217

Van Tendeloo, G., *see* Dolino, G. 215, 216, 217
Van Zanten, P., *see* Gerritsma, C.J. 344
Velichkina, T.S., *see* Yakolev, I.A. 209, 216, 218
Venturini, E.L. 295
Verkis, J.E., *see* Brown, I.O. 369
Vettier, C., *see* Cailleau, H. 99
Vettier, C., *see* Dénoyer, F. 143, 146, 147, 149, 152
Vettier, C., *see* Durand, D. 153, 154, 155
Vettier, C., *see* Simonson, T. 142, 158
Vigman, P.B. 334, 335, 338, 339
Villain, J. 291
Villain, J., *see* Bak, P. 215, 226, 369
Vinogradova, I.S. 369
Visser, J.W. 257, 262
Visser, J.W., *see* Bijk, P. 258, 262
Visser, J.W., *see* Brouns, E. 255, 257
Volkov, A.A., *see* Kozlov, G.V. 295, 296
Volkov, A.A., *see* Petzelt, J. 65
Von Boehm, J., *see* Bak, P. 291
Von Der Muhl, R., *see* Schneck, J. 235, 250
Von Gattow, G. 6
Vonlanthen, A., *see* Rehwald, W. 62, 369
Vtyurin, A.N. 369
Vtyurin, A.N., *see* Shkuryaev, P.G. 369

Wada, M. 21, 38, 39, 63, 64, 65, 67, 95, 96
Wada, M., *see* Sawada, A. 23, 45, 51, 62, 63
Wada, M., *see* Sugiyama, J. 60, 291
Wada, M.A. 157
Wadhawan, V.K. 258
Wakaki, N., *see* Mitsui, T. 131
Waldeck, W.F. 277
Walker, M.B. 223, 369
Walker, M.B., *see* Jacobs, A.E. 369
Walker, P.J., *see* Maciel, A. 271
Wallerius, R., *see* Rehwald, W. 62
Wanarski, W., *see* Klimowski, J. 145, 149
Wang, Z.Z., *see* Moncean, R. 369
Warczewskii, J. 369
Wasiutynski, T. 76, 82, 369
Wasiutynski, T., *see* Natkaniec, I. 76, 82
Wasiutynski, T., *see* Plakida, N.M. 82, 83
Waszczak, J.V., *see* Disalvo, F.J. 237
Weger, M., *see* Bulting, J.M. 369
Wentowska, K., *see* Bak, P. 215, 226, 369
Westrum, E.F., *see* Chang, S.S. 136
Whatmore, R., *see* Schneck, J. 250
White, J.G., *see* Goldsmith, G.J. 131, 133
Whiteman, J.D., *see* Hochstrasser, R.M. 75
Whitfield, H.J., *see* MacKenzie, D.R. 134
Wiener-Avnear, E., *see* Hartwig, C.M. 111
Wildermut, G., *see* Depmeier, W. 315
Willet, R.D., *see* Peterson, E.R. 306
Wilson, J.A., *see* Disalvo, F.J. 237
Wilson, J.A., *see* Fung, K.K. 217
Withers, R.L., *see* Walker, M.B. 369
Wooley, A.M., *see* Doran, N.J. 369
Wright, A.F. 208
Wright, A.F., *see* Dolino, G. 208, 209
Wyder, P., *see* Meekes, H. 369
Wyder, P., *see* Muševič, I. 361
Wyder, P., *see* Rasing, Th. 369

Yagi, T. 114, 369
Yakolev, I.A. 209, 216, 218
Yakolev, I.A., *see* Shustin, O.A. 218
Yamada, N. 19, 20, 369
Yamada, T. 236
Yamada, T., *see* Yamada, Y. 108, 109
Yamada, Y. 43, 105, 107, 108, 109, 112, 119, 286, 291, 292, 293, 369
Yamada, Y., *see* Noda, Y. 106
Yamaguchi, T., *see* Shimizu, H. 26, 29, 66, 67
Yamamoto, A. 369
Yamamoto, A., *see* Sawada, S. 23
Yamamoto, A., *see* Suzuki, N. 369
Yamamoto, M., *see* Sambongi, T. 369
Yamamoto, N. 369
Yamashita, S., *see* Hidaka, M. 290
Yasuda, N., *see* Shimizu, H. 26, 27, 29, 66, 67
Yatom, H., *see* Englman, R. 297
Yelon, W.B., *see* Baudour, J.L. 82
Yokoi, C.S.O. 151
Young, R.A. 209
Yust, S., *see* Jacoby, Y. 173

Zaretskii, V.V., *see* Sheleg, A.U. 369
Zarka, A. 218
Žekš, B., *see* Levstik, A. 45, 62, 360, 361, 369
Žekš, B., *see* Muralt, P. 306, 307, 310, 312, 314, 369
Žekš, B., *see* Muševič, I. 361
Zeyen, C., *see* Frey, F. 369

Zeyen, C.M.E. 209
Zeyen, C.M.E., *see* Bachheimer, J.P. 219
Zeyen, C.M.E., *see* Bernard, L. 165, 166, 169, 171, 177, 178, 179, 180, 196, 369
Zeyen, C.M.E., *see* Cailleau, H. 73, 76, 79, 82, 83, 84, 85, 86, 89, 90, 91, 92, 99, 196, 369
Zeyen, C.M.E., *see* Dolino, G. 207, 209, 211, 212, 213, 215, 216, 217, 225
Zeyher, R. 40, 89, 195
Zhdanova, V.V., *see* Vakhrushev, S.B. 369
Zheludev, I.S., *see* Gorelik, V.S. 111
Zherebtsova, L.I., *see* Melnikova, S.V. 369
Zorin, R.V., *see* Al'Shin, B.I. 297
Zubov, V.G. 216
Zumer, S., *see* Aleksandrova, I.P. 107

SUBJECT INDEX

Amplitude
complex, of the incommensurate wave 52
local 34
mode 64, 196
dispersion relation 194, 200
of modulation
in β-$ThBr_4$ 179
in $NaNO_2$ 106
in thiourea 137
Amplitudon 38
Atomic displacement analysis in K_2SeO_4 18

$BaMnF_4$ 285–300
Barium sodium niobate (BSN) 233–251
Biphenyl 73–127
Birefringence
$BaMnF_4$ 288
in BSN 247
in Na_2CO_3 277
Blue phases in liquid crystals 333
Brillouin scattering in quartz 227

Commensurate
domains 42
field-induced phase 151
phase
in AFB 23
in thiourea 142
wavevector 52
Constant amplitude approximation 56
Crystal structure
β-$ThBr_4$ 165
BSN 241
Na_2CO_3 258, 262
$NaNO_2$ 104
Curie–Weiss dielectric susceptibility 110

Debye
susceptibility 112

Debye (*cont'd*)
–Waller factor 176
Defect-stabilized structure in incommensurate $BaMnF_4$ 293
Devil's staircase 43
Dielectric susceptibility
Curie–Weiss 110
in $NaNO_2$ 110
temperature variation in the N–INC–C transition sequence 60
Diffuse scattering in Na_2CO_3 273
Discommensuration 55
Displacements
atomic, analysis in K_2SeO_4 18
static 37, 137
Dynamical
matrix 17
fluctuations in K_2SeO_4 37
Dynamics near the soliton limit 45

Edge singularities of NMR and NQR spectrum 107, 190
Elastic
constant
temperature variation in the N–INC–C transition sequence 60
strain tensor 116
structure factor 176
Electric conductivity in Na_2CO_3 279
Electron microscopy in quartz 216
EPR measurements in $ThBr_4$ 191
Euler equation 314
Excitations in incommensurate phases 89
Extended zone scheme in K_2SeO_4 10

Ferroelasticity in Na_2CO_3 227
Ferroelectric phase
in thiourea 157
in liquid crystals 352

Field-induced commensurate phase 151
Flexoelectric
  modulation of nematics 326
  effect 345
Floating phase 192
Fluctuations in the normal phase 12
Fluorescence spectrum of β-$ThBr_4$ 187
Free energy density 81, 116, 121, 220, 222, 241, 330
Frequency
  gap 93
  distribution of NMR and NQR spectrum 107

Gap of a phonon dispersion relation 38
  in biphenyl 93
Global hysteresis 136
Goldstone mode 40
Group theoretical analysis of K_2SeO_4 7

Heat capacity
  in biphenyl 85
Hydrostatic pressure
  effect on the modulated phases 145

Incommensurate
  amplitude 34
  crystals, list of 369
  excitations of biphenyl 89
  fluctuations 33
  instability 15
  liquid crystals 321–365
  modulation 20
  parameter 22
  phase 34
    in BSN 236
    in Na_2CO_3 255–281
    in quartz 209, 223
  state, 3
  wave vector 5, 306
Inelastic structure factor 195
Infrared spectroscopy in Na_2CO_3 275
Interlayer force constants 16
Irreducible representations
  in thiourea 137
  of normal modes in K_2SeO_4 9

K_2SeO_4 1–48

Lagrange–Euler equation 54
Landau theory for the incommensurate transition 51
Lifshitz
  gradient term 115
  invariant 53, 209, 310, 323
  point 117
Liquid crystals and effects of incommensurability 321
Local
  amplitude 34
  coordinate 35
  phase 34
Lock-in phase 5, 42
  macroscopic properties 45
  pressure-induced 146
  transition
    in Na_2CO_3 279
    in quartz 216
    in thiourea 142

Magneto-electric phenomena in $BaMnF_4$ 294
Misfit parameter 22
Modulated phase
  in thiourea 157
  static aspects in β-$ThBr_4$ 173, 181, 194
Modulation
  wave in thiourea 135
  wave number 55
  wave vector
    in BSN 236
    in thiourea 141

Na_2CO_3 255–281
$NaNO_2$ 101–127
Neutron scattering
  function 114
  in β-$ThBr_4$ 196
  in biphenyl 85
  in perovskites 316
  in thiourea 135, 146
  spin echo in $NaNO_2$ 114
Normal modes in β-$ThBr_4$ 168
NQR study of β-$ThBr_4$ and β-$ThCl_4$ 190

Ornstein–Zernicke susceptibility 111

Perovskite 303–318
Phase diagrams
  electric field–temperature 67, 119, 150
  of thiourea 135
  pressure–temperature 28, 66
Phase-mode dispersion modulation 194
  dispersion relation 200
  mode in $ThBr_4$ 196

Phason 38, 90
  gap 198
Phonons
  dispersion analysis 16
  in the normal phase 12
Pinning of modulation
  in β-$ThBr_4$ 188
  in BSN 248
Polarization
  temperature variation in the N–INC–C transition sequence 58
Pressure-induced lock-in transition 146
Pseudo-phase mode 38

Quartz 207–230
  dynamical properties 226–229
Quasi-commensurate phase 237

Raman spectra
  of β-$ThBr_4$ 169
  of biphenyl 96
  of Na_2CO_3 275
  of quartz 227
Relaxation
  effects in BSN 244
Relaxation time
  amplitudon 112
  of order parameter 114
  phason 112
Rotation-like wave 175

Satellite reflections
  in β-$ThBr_4$ 169–170
  in biphenyl 85
Sine-Gordon equation 241
Soliton
  dynamics near the soliton limit 45
  limit 42
Soft phonon mode
  behaviour in the α and β phases of quartz 226
  dispersion relation in BSN 238
Soft phonon mode (*cont'd*)
  in β-$ThBr_4$ 170
  in K_2SeO_4 12
  in thiourea 156
  phase transition in biphenyl 74
  temperature variation in the N–INC–C transition sequence 64
Space group symmetry of β-$ThBr_4$ 167
Specific heat
  temperature variation in the N–INC–C transition sequence 56
Strain
  temperature variation in the N–INC–C transition sequence 58
Structure factor 195
Susceptibility
  Curie–Weiss 110
  Debye 112
  Ornstein–Zernicke type 111

Temperature dependence of modulation wave in thiourea 135
Thermal hysteresis in BSN 244
Thermodynamic potential 52, 139
  average 54
Thiourea 131–160
  Twist-like wave 175

Ultrasonic dispersion in $NaNO_2$ 114
Umklapp lock-in terms in thiourea 143

Waller–Debye factor 176
Wave
  modulation wave in thiourea 135
  rotation-like 175
  twist-like 175
Wave vector
  commensurate 52
  incommensurate 5

X-ray diffraction in thiourea 153

Zernicke–Ornstein susceptibility 111

CUMULATIVE INDEX, VOLUMES 1–14

Monographs

Agranovich, V.M., and M.D. Galanin, Electronic excitation energy transfer in condensed matter **3**

Galanin, M.D., *see* Agranovich, V.M. **3**

Mashkova, E.S., and V.A. Molchanov, Medium-energy ion reflection from solids **11**
Molchanov, V.A., *see* Mashkova, E.S. **11**

Contributed Volumes

Abelés, F., and T. Lopez-Rios, Surface polaritons at metal surfaces and interfaces **1**, 239
Agranovich, V.M., Effects of the transition layer and spatial dispersion in the spectra of surface polaritons **1**, 187
Agranovich, V.M., V.E. Kravtsov and T.A. Leskova, Scattering of surface polaritons by order parameter fluctuations near phase transition points **1**, 511
Agranovich, V.M., Biphonons and Fermi resonance in vibrational spectra of crystals **4**, 83
Agranovich, V.M., Surface excitons **9**, 513
Agranovich, V.M., and V.V. Kirsanov, Production of radiation defects by collision cascades in metals **13**, 117
Altshuler, B.L., and A.G. Aronov, Electron–electron interaction in disordered conductors **10**, 1
Aoyagi, K., *see* Tanabe, Y. **2**, 603
Aronov, A.G., and A.S. Ioselevich, Exciton electrooptics **2**, 267
Aronov, A.G., *see* Altshuler, B.L. **10**, 1
Aronov, A.G., Yu.M. Gal'perin, V.L. Gurevich and V.I. Kozub, Nonequilibrium properties of superconductors (transport equation approach) **12**, 325
Aslamazov, L.G., and A.F. Volkov, Nonequilibrium phenomena in superconducting weak links **12**, 65
Axe, J.D., M. Iizumi and G. Shirane, Phase transformations in K_2SeO_4 and structurally related insulators **14.2**, 1

Bagaev, V.S., T.I. Galkina and N.N. Sibeldin, Interaction of EHD with deformation field, ultrasound and non-equilibrium phonons **6**, 267
Beasley, M.R., *see* De Lozanne, A.L. **12**, 111
Belousov, M.V., Vibrational Frenkel excitons **2**, 771

Belyakov, V.A., and E.I. Kats, Light scattering and phase transitions in liquid crystals **5**, 227
Benoit à la Guillaume, C., *see* Planel, R. **8**, 353
Bereznyak, P.A., *see* Slyozov, V.V. **13**, 575
Berkovits, V.L., *see* Paget, D. **8**, 381
Bernard, L., *see* Currat, R. **14.2**, 161
Bersuker, I.B., and V.Z. Polinger, Theoretical background of the Jahn–Teller effect **7**, 21
Beyer, J., C.J. Pethick, J. Rammer and H. Smith, Pair breaking in nonequilibrium **12**, 129
Bhatt, R.N., *see* Milligan, R.F. **10**, 231
Bill, H., Observation of the Jahn–Teller effect with electron paramagnetic resonance **7**, 709
Birman, J.L., Electrodynamic and non-local optical effects mediated by exciton polaritons **2**, 27
Borstel, G., *see* Wöhlecke, M. **8**, 423
Bullough, R., and M.H. Wood, Theory of microstructural evolution **13**, 189
Burns, M.J., W.K. Liu and A.H. Zewail, Nonlinear laser spectroscopy and dephasing of molecules: An experimental and theoretical overview **4**, 301
Burstein, E., *see* Chen, Y.J. **1**, 587

Cailleau, H., Incommensurate phases in an aromatic molecular crystal: biphenyl **14.2**, 71
Celotta, R.J., *see* Pierce, D.T. **8**, 259
Challis, L.J., and A.M. de Goër, Phonon spectroscopy of Jahn–Teller ions **7**, 533
Chang, Jhy-Jiun, Kinetic equations in superconducting thin films **12**, 453
Chen, Y.J., and E. Burstein, Three wave nonlinear interactions involving surface polaritons; Raman scattering, light diffraction and parametric mixing **1**, 587
Clarke, J., Experiments on charge imbalance in superconductors **12**, 1
Clinard Jr, F.W., and L.W. Hobbs, Radiation effects in non-metals **13**, 387
Cottam, M.G., and A.A. Maradudin, Surface linear response functions **9**, 1
Cummins, H.Z., Brillouin scattering studies of phase transitions in crystals **5**, 359
Currat, R., *see* Durand, D. **14.2**, 101
Currat, R., *see* Dénoyer, F. **14.2**, 129
Currat, R., L. Bernard and P. Delamoye, Incommensurate phase in β-$ThBr_4$ **14.2**, 161

de Goër, A.M., *see* Challis, L.J. **7**, 533
De Lozanne, A.L., and M.R. Beasley, Time-dependent superconductivity in SNS bridges, and examples of TDGL theory **12**, 111
de Wolff, P.M., and F. Tuinstra, The incommensurate phase of Na_2CO_3 **14.2**, 253
Delamoye, P., *see* Currat, R. **14.2**, 161
deMartini, F., *see* Shen, Y.R. **1**, 629
Dénoyer, F., *see* Durand, D. **14.2**, 101
Dénoyer, F., and R. Currat, Modulated phases in thiourea **14.2**, 129
Dmitriev, V.M., V.N. Gubankov and F. Ya. Nad', Experimental study of enhanced superconductivity **12**, 163
Dolgopolov, V.T., and I.L. Landau, Resistive states in type-I superconductors **12**, 641
Dolino, G., The incommensurate phase of quartz **14.2**, 205
Durand, D., F. Dénoyer, R. Currat and M. Lambert, Incommensurate phase in $NaNO_2$ **14.2**, 101
Dyakonov, M.I., and V.I. Perel', Theory of optical spin orientation of electrons and

nuclei in semiconductors **8**, 11

Efros, A.L., and B.I. Shklovskii, Coulomb interaction in disordered systems with localized electronic states **10**, 483
Ehrhart, P., K.H. Robrock and H.R. Schober, Basic defects in metals **13**, 3
Eliashberg, G.M., and B.I. Ivlev, Theory of superconductivity stimulation **12**, 211
Errandonéa, G., *see* Tolédano, J.C. **14.2**, 233

Fayer, M.D., Exciton coherence **4**, 185
Fischer, B., *see* Lagois, J. **1**, 69
Fleisher, V.G., and I.A. Merkulov, Optical orientation of the coupled electron–nuclear spin system of a semiconductor **8**, 173
Fleury, P.A., and K.B. Lyons, Central peaks near structural phase transitions **5**, 449
Flores, F., and F. Garcia-Moliner, Electronic surface excitations **9**, 441
Fukuyama, H., Interaction effects in the weakly localized regime of two- and three-dimensional disordered systems **10**, 155

Galaĭko, V.P., and N.B. Kopnin, Theory of the resistive state in narrow superconducting channels **12**, 543
Galitskii, V.M., and Yu.V. Kopaev, The kinetic theory of superconductors with excess quasiparticles **12**, 377
Galkina, T.I., *see* Bagaev, V.S. **6**, 267
Gal'perin, Yu.M., *see* Aronov, A.G. **12**, 325
Garcia-Moliner, F., *see* Flores, F. **9**, 441
Gershenzon, E.M., A.P. Mel'nikov and R.I. Rabinovich, H^--like impurity centers, molecular complexes and electron delocalization in semiconductors **10**, 483
Ginzburg, V.L., A.A. Sobyanin and A.P. Levanyuk, General theory of light scattering near phase transitions in ideal crystals **5**, 3
Glinchuck, M.D., Paraelectric resonance of off-center ions **7**, 819
Goldburg, W.I., Light scattering investigations of the critical region in fluids **5**, 531
Goldman, A.M., *see* Kadin, A.M. **12**, 253
Golovko, V.A., and A.P. Levanyuk, Light scattering from incommensurate phases **5**, 169
Gor'kov, L.P., Disorder and interactions in the system of quasi one-dimensional electrons **10**, 619
Grun, J.B., B. Hönerlage and R. Lévy, Biexcitons in CuCl and related systems **2**, 459
Gubankov, V.N., *see* Dmitriev, V.M. **12**, 163
Gurevich, V.L., *see* Aronov, A.G. **12**, 325
Guseva, M.I., and Yu.V. Martynenko, Blistering **13**, 621

Haarer, D., and M.R. Philpott, Excitons and polarons in organic weak charge transfer crystals **4**, 27
Hanson, D.M., J.S. Patel, I.C. Winkler and A. Morrobel-Sosa, Effects of electric fields on the spectroscopic properties of molecular solids **4**, 621
Hermann, C., and C. Weisbuch, Optical detection of conduction electron spin resonance in semiconductors and its application to $\boldsymbol{k}\cdot\boldsymbol{p}$ perturbation theory **8**, 463
Hesselink, W.H., and D.A. Wiersma, Theory and experimental aspects of photon echoes in molecular solids **4**, 249
Hizhnyakov, V.V., and N.N. Kristoffel, Jahn–Teller mercury-like impurities in

ionic crystals **7**, 383
Hobbs, L.W., *see* Clinard Jr, F.W. **13**, 387
Hönerlage, B., *see* Grun, J.B. **2**, 459

Iizumi, M., *see* Axe, J.D. **14.2**, 1
Ioselevich, A.S., *see* Aronov, A.G. **2**, 267
Ishibashi, Y., Phenomenology of incommensurate phases in the A_2BX_4 family **14.2**, 49
Ivchenko, E.L., Spatial dispersion effects in the exciton resonance region **2**, 141
Ivchenko, E.L., *see* Pikus, G.E. **2**, 205
Ivlev, B.I., *see* Eliashberg, G.M. **12**, 211

Jeffries, C.D., *see* Wolfe, J.P. **6**, 431
Judd, B.R., Group theoretical approaches **7**, 87

Kadin, A.M., and A.M. Goldman, Dunamical effects in nonequilibrium superconductors, some experimental perspectives **12**, 253
Kalia, R.K., *see* Vashishta, P. **6**, 1
Kamimura, H., Electron–electron interactions in the Anderson-localised regime near the metal–insulator transition **10**, 555
Kats, E.I., *see* Belyakov, V.A. **5**, 227
Kholodar, G.A., *see* Vinetskii, V.L. **13**, 283
Kind, R., and P. Muralt, Unique incommensurate–commensurate phase transitions in a layer-structure perovskite **14.2**, 301
Kirsanov, V.V., *see* Agranovich, V.M. **13**, 117
Klein, M.V., Light scattering studies of incommensurate transitions **5**, 503
Kolb, D.M., The study of solid–liquid interfaces by surface plasmon polariton excitation **1**, 299
Kopaev, Yu.V., *see* Galitskii, V.M. **12**, 377
Kopelman, R., Energy transport in mixed molecular crystals **4**, 139
Kopnin, N.B., *see* Galaĭko, V.P. **12**, 543
Koteles, E.S., Investigation of exciton-polariton dispersion using laser techniques **2**, 83
Kozub, V.I., *see* Aronov, A.G. **12**, 325
Kravtsov, V.E., *see* Agranovich, V.M. **1**, 511
Kristoffel, N.N., *see* Hizhnyakov, V.V. **7**, 383
Kulakovskii, V.D., and V.B. Timofeev, Thermodynamics of electron–hole liquid in semiconductors **6**, 95

Lagois, J., and B. Fischer, Surface exciton polaritons from an experimental viewpoint **1**, 69
Lambert, M., *see* Durand, D. **14.2**, 101
Landau, I.L., *see* Dolgopolov, V.T. **12**, 641
Larkin, A.I., and Yu.N. Ovchinnikov, Vortex motion in superconductors **12**, 493
Lemaistre, J., *see* Zewail, A.H. **2**, 665
Leskova, T.A., *see* Agranovich, V.M. **1**, 511
Levanyuk, A.P., *see* Ginzburg, V.L. **5**, 3
Levanyuk, A.P., A.S. Sigov and A.A. Sobyanin, Light scattering anomalies due to defects **5**, 129
Levanyuk, A.P., *see* Golovko, V.A. **5**, 169

Lévy, R., *see* Grun, J.B. **2**, 459
Litster, J.D., Scattering spectroscopy of liquid crystals **5**, 583
Liu, W.K., *see* Burns, M.J. **4**, 301
Lopez-Rios, T., *see* Abelés, F. **1**, 239
Loudon, R., *see* Ushioda, S. **1**, 535
Loudon, R., Ripples on liquid interfaces **9**, 589
Luhman, T., *see* Snead Jr, C.L. **13**, 345
Lushchik, Ch.B., Free and self-trapped excitons in alkali halides: spectra and dynamics **2**, 505
Lushchik, Ch.B., Creation of Frenkel defect pairs by excitons in alkali halides **13**, 473
Lyons, K.B., *see* Fleury, P.A. **5**, 449

Maradudin, A.A., Interaction of surface polaritons and plasmons with surface roughness **1**, 405
Maradudin, A.A., *see* Cottam, M.G. **9**, 1
Markiewicz, R.S., and T. Timusk, Interaction of electromagnetic radiation with electron–hole droplets **6**, 543
Martynenko, Yu.V., *see* Guseva, M.I. **13**, 621
Meier, F., and D. Pescia, Spin-polarized photoemission by optical orientation **8**, 295
Mel'nikov, A.P., *see* Gershenzon, E.M. **10**, 483
Merkulov, I.A., *see* Fleisher, V.G. **8**, 173
Milligan, R.F., T.F. Rosenbaum, R.N. Bhatt and G.A. Thomas, A review of the metal–insulator transition in doped semiconductors **10**, 231
Mills, D.L., Surface spin waves on magnetic crystals **9**, 379
Mirlin, D.N., Surface phonon polaritons in dielectrics and semiconductors **1**, 3
Mirlin, D.N., Optical alignment of electron momenta in GaAS-type semiconductors **8**, 133
Morrobel-Sosa, A., *see* Hanson, D.M. **4**, 621
Moskalova, M.A., *see* Zhizhin, G.N. **1**, 93
Muralt, P., *see* Kind, R. **14.2**, 301

Nad', F.Ya., *see* Dmitriev, V.M. **12**, 163
Natadze, A.L., A.I. Ryskin and B.G. Vekhter, Jahn–Teller effects in optical spectra of II–IV and III–V impurity crystals **7**, 347
Nelson, R.J., Excitons in semiconductor alloys **2**, 319
Nizzoli, F., *see* Stegeman, G.I. **9**, 195

Ortuño, M., *see* Pollak, M. **10**, 287
Osad'ko, I.S., Theory of light absorption and emission by organic impurity centers **4**, 437
Ovchinnikov, Yu.N., *see* Larkin, A.I. **12**, 493

Paget, D., and V.L. Berkovits, Optical investigation of hyperfine coupling between electronic and nuclear spins **8**, 381
Patel, J.S., *see* Hanson, D.M. **4**, 621
Pearlstein, R.M., Excitons in photosynthetic and other biological systems **2**, 735
Perel', V.I., and B.P. Zakharchenya, Major physical phenomena in the optical orientation and alignment in semiconductors **8**, 1
Perel', V.I., *see* Dyakonov, M.I. **8**, 11

Perlin, Yu.E., and M. Wagner, Introduction **7**, 1
Perlin, Yu.E., and B.S. Tsukerblat, Optical bands and polarization dichroism of Jahn–Teller centers **7**, 251
Permogorov, S., Optical emission due to exciton scattering by LO phonons in semiconductors **2**, 177
Personov, R.I., Site selection spectroscopy of complex molecules in solutions and its applications **4**, 555
Pescia, D., *see* Meier, F. **8**, 295
Pethick, C.J., *see* Beyer, J. **12**, 129
Philpott, M.R., *see* Haarer, D. **4**, 27
Pierce, D.T., and R.J. Celotta, Applications of polarized electron sources utilizing optical orientation in solids **8**, 259
Pikin, S.A., Incommensurate structures in liquid crystals **14.2**, 319
Pikus, G.E., and E.L. Ivchenko, Optical orientation and polarized luminescence of excitons in semiconductors **2**, 205
Pikus, G.E., and A.N. Titkov, Spin relaxation under optical orientation in semiconductors **8**, 73
Planel, R., and C. Benoit à la Guillaume, Optical orientation of excitons **8**, 353
Pokrovskii, Ya.E., Transport phenomena in electron–hole liquid **6**, 509
Polinger, V.Z., *see* Bersuker, I.B. **7**, 21
Pollak, M., and M. Ortuño, The effect of Coulomb interactions on electronic states and transport in disordered insulators **10**, 287
Pooler, D.R., Numerical diagonalization techniques in the Jahn–Teller effect **7**, 199

Rabinovich, R.I., *see* Gershenzon, E.M. **10**, 483
Raether, H., Surface plasmons and roughness **1**, 331
Rammer, J., *see* Beyer, J. **12**, 129
Rashba, E.I., Self-trapping of excitons **2**, 543
Reik, H.G., Non-adiabatic systems: analytical approach and exact results **7**, 117
Reznichenko, E.A., *see* Zelensky, V.F. **13**, 527
Robrock, K.H., *see* Ehrhart, P. **13**, 3
Rosenbaum, T.F., *see* Milligan, R.F. **10**, 231
Ryskin, A.I., *see* Natadze, A.L. **7**, 347

Schneck, J., *see* Tolédano, J.C. **14.2**, 233
Schober, H.R., *see* Ehrhart, P. **13**, 3
Schön, G., Collective modes in superconductors **12**, 589
Scott, J.F., Raman spectroscopy of structural phase transitions **5**, 291
Scott, J.F., Statics and dynamics of incommensurate $BaMnF_4$ **14.2**, 283
Sheleg, A.U., and V.V. Zaretskii, List of incommensurate crystals **14.2**, 367
Shen, Y.R., and F. deMartini, Nonlinear wave interaction involving surface polaritons **1**, 629
Shirane, G., *see* Axe, J.D. **14.2**, 1
Shklovskii, B.I., *see* Efros, A.L. **10**, 483
Shomina, E.V., *see* Zhizhin, G.N. **1**, 93
Shustin, O.A., *see* Yakovlev, I.A. **5**, 605
Sibeldin, N.N., *see* Bagaev, V.S. **6**, 267
Sigmund, E., Phonon scattering at Jahn–Teller defects in semiconductors **7**, 495

Sigov, A.S., *see* Levanyuk, A.P. **5**, 129
Silbey, R., Theories of energy transport **4**, 1
Silin, A.P., Electron–hole liquid in a magnetic field **6**, 619
Singwi, K.S., *see* Vashishta, P. **6**, 1
Sipe, J.E., and G.I. Stegeman, Nonlinear optical response of metal surfaces **1**, 661
Slyozov, V.V., and P.A. Bereznyak, Irradiation creep in metals **13**, 575
Small, G.J., Persistent nonphotochemical hole burning and the dephasing of impurity electronic transitions in organic glasses **4**, 515
Smith, D.D., *see* Zewail, A.H. **2**, 665
Smith, H., *see* Beyer, J. **12**, 129
Snead Jr, C.L., and T. Luhman, Radiation damage and stress effects in superconductors: Materials for high-field applications **13**, 345
Sobyanin, A.A., *see* Ginzburg, V.L. **5**, 3
Sobyanin, A.A., *see* Levanyuk, A.P. **5**, 129
Stegeman, G.I., *see* Sipe, J.E. **1**, 661
Stegeman, G.I., and F. Nizzoli, Surface vibrations **9**, 195
Sturge, M.D., Introduction **2**, 1
Sugakov, V.I., Excitons in strained molecular crystals **2**, 709

Tanabe, Y., and K. Aoyagi, Excitons in magnetic insulators **2**, 603
Thewalt, M.L.W., Bound multiexciton complexes **2**, 393
Thomas, G.A., *see* Milligan, R.F. **10**, 231
Timofeev, V.B., Free many particle electron–hole complexes in an indirect gap semiconductor **2**, 349
Timofeev, V.B., *see* Kulakovskii, V.D. **6**, 95
Timusk, T., *see* Markiewicz, R.S. **6**, 543
Titkov, A.N., *see* Pikus, G.E. **8**, 73
Tolédano, J.C., J. Schneck and G. Errandonéa, Incommensurate phase of barium sodium niobate **14.2**, 233
Tsukerblat, B.S., *see* Perlin, Yu.E. **7**, 251
Tuinstra, F., *see* de Wolff, P.M. **14.2**, 253

Ulrici, W., Manifestations of the Jahn–Teller effect in the optical spectra of transition metal impurities in crystals **7**, 439
Ushioda, S., and R. Loudon, Raman scattering by surface polaritons **1**, 535

Vashishta, P., R.K. Kalia and K.S. Singwi, Electron–hole liquid: theory **6**, 1
Vekhter, B.G., *see* Natadze, A.L. **7**, 347
Vinetskii, V.L., and G.A. Kholodar, Quasichemical reactions involving point defects in irradiated semiconductors **13**, 283
Vinogradov, E.A., G.N. Zhizhin and V.I. Yudson, Thermally stimulated emission of surface polaritons **1**, 145
Volkov, A.F., *see* Aslamazov, L.G. **12**, 65

Wagner, M., *see* Perlin, Yu.E. **7**, 1
Wagner, M., Unitary transformation methods in vibronic problems **7**, 155
Weisbuch, C., *see* Hermann, C. **8**, 463
Westevelt, R.M., Kinetics of electron–hole drop formation and decay **6**, 187

Wiedersich, H., Phase stability and solute segregation during irradiation **13**, 225
Wiersma, D.A., *see* Hesselink, W.H. **4**, 249
Winkler, I.C., *see* Hanson, D.M. **4**, 621
Wöhlecke, M., and G. Borstel, Spin-polarized photoelectrons and crystal symmetry **8**, 423
Wolfe, J.P., and C.D. Jeffries, Strain-confined excitons and electron–hole liquid **6**, 431
Wood, M.H., *see* Bullough, R. **13**, 189

Yakovlev, I.A., and O.A. Shustin, Light scattering in quartz and ammonium chloride and its peculiarities in the vicinity of phase transition of crystals. A retrospective view and recent results **5**, 605
Yakovlev, V.A., *see* Zhizhin, G.N. **1**, 93
Yakovlev, V.A., *see* Zhizhin, G.N. **1**, 275
Yudson, V.I., *see* Vinogradov, E.A. **1**, 145

Zakharchenya, B.P., *see* Perel', V.I. **8**, 1
Zaretskii, V.V., *see* Sheleg, A.U. **14.2**, 367
Zelensky, V.F., and E.A. Reznichenko, Irradiation growth of metals and alloys **13**, 527
Zewail, A.H., D.D. Smith and J. Lemaistre, Dynamics of molecular excitons: disorder, coherence and dephasing **2**, 665
Zewail, A.H., *see* Burns, M.J. **4**, 301
Zhizhin, G.N., M.A. Moskalova, E.V. Shomina and V.A. Yakovlev, Surface electromagnetic wave propagation on metal surfaces **1**, 93
Zhizhin, G.N., *see* Vinogradov, E.A. **1**, 145
Zhizhin, G.N., and V.A. Yakovlev, Resonance of transition layer excitations with surface polaritons **1**, 275